T. C. R. White

The Inadequate Environment

Nitrogen and the Abundance of Animals

With 41 Figures

Springer-Verlag
Berlin Heidelberg New York
London Paris Tokyo
Hong Kong Barcelona
Budapest

Thomas C.R. White
Dept. of Crop Protection
Waite Agricultural Research Institute
Glen Osmond, South Australia
5064 Australia

ISBN-13:978-3-642-78301-2 e-ISBN-13:978-3-642-78299-2
DOI: 10.1007/978-3-642-78299-2

Softcover reprint of the hardcover 1st edition 1993

Production Editor: Renate Münzenmayer
Typesetting: K+V Fotosatz GmbH, 64743 Beerfelden
31/3145-5 4 3 2 1 0 – Printed on acid-free paper

To the memory
of my two mentors
H. G. (Andy) Andrewartha and
G. B (Joe) Rawlings

Acknowledgements

My story is, inevitably, built upon the ideas and research of others – from Darwin to Dawkins – as reflected in my references. However, there are some whose influence has been general rather than specific, others of whose direct influence I am no longer conscious; still others whose relevant work I have never encountered. To all these I apologize for failing to acknowledge them.

The basic outline of my thesis finally crystallized while I was teaching ecology at the University of the South Pacific in Fiji, and it was Hermann Remmert's enthusiastic badgering which started me putting it into book form. But I was only able to properly address the task after retirement, and with the support of an honorary research fellowship at the Waite Agricultural Research Institute.

For many years I worked where ready access to the literature was limited. ISI's *Current Contents*, and the hundreds of reprints sent to me by colleagues throughout the world enabled me to keep up.

Too many colleagues to list lent photographs to illustrate the book – mostly colour slides. Jennie Groom did a grand job producing black and white prints from the latter. Many of these same colleagues, and several others, also helped with comments on appropriate sections. Tom Browning read much of an early draft, alerting me to grammatical slips and lapses into teleology. But I continue to deliberately split an occasional infinitive, and resort to teleological analogy if I think this makes for clearer exposition!

However, it has been Piet den Boer, with his detailed and wise criticisms via our lengthy discussions by post, who has done more than anybody else to improve both what I have to say, and the way in which I say it. I am especially grateful to him.

To all, my very sincere thanks, coupled with the usual exoneration from any responsibility for what I have said. And in this age of word processors I can blame nobody but myself for typing and spelling errors!

Last, my very special thanks to my wife, Janice, for grammatical advice, painstaking assistance with final much valuable proof-reading, and, above all, for continuing to believe in the worth of the project, and for keeping me at it on the numerous occasions when I lost my confidence.

Preface

This is not a text book. Nor is it an objective review. It is a personal – and thus biased – view of some facets of ecology; a view I vigorously advocate. More specifically, it is an attempt to report patterns that I perceive threaded through the fabric of the living world, the common cause of those patterns, and some of the implications for our understanding of ecology that flow from this perception. As a result I come to conclusions that sometimes differ from much that is generally accepted, and with which many will not agree. But I have not felt obliged to put the case for alternative hypotheses or interpretations. I leave those who support them to expound their own viewpoints.

In advocating my viewpoint I have taken examples from studies that cover a wide geographic and taxonomic range. However, I have not attempted to be comprehensive. Such would have been a futile task, even had I the energy, or time left, to try to do so. Every story I follow up uncovers several new ones. Yet there remains a majority of the literature that I have never seen – and never will.

Apart from this, however, there is simply not room for all the examples that I *have* found. If I reported these, the book would be double its size, so I have had to knowingly put much good evidence on one side. For example, I have left out major groups like the mites and the nematodes, and the sawflies and leafbeetles among the insects. Some, like the bacteria, get inadequate coverage; others, like the bark beetles, get but passing mention. And within the groups from which I have taken examples, there are many more cases that I could equally have reported.

Furthermore, I have not tried to be even-handed in selecting which out of so many examples to use. Rather, I have tried to select studies which tell a reasonably complete story, or which make a particular point unambiguously. For all these reasons what I have to say is incomplete and patchy, but I hope that what I have chosen illustrates the generalities sufficiently, and that others will fill in the gaps.

In attempting to write clearly and simply I have eliminated acronyms, all but common abbreviations, and as much jargon, technical language, and quantitative detail as possible. I have balanced the latter, however, with a very full bibliography wherein the reader who so wishes can find these details.

One of the temptations in modern ecological research is to retreat within the protective jungle of complexity. For some, the natural world seems *so* diverse, and interactions within it *so* varied, that they believe we cannot expect to find general patterns there. Unlike physics, they say, there can be no hard laws of ecology that will encompass the ever-growing diversity of life; every case will be different. Many delight in this diversity, and present evidence for it in detailed descriptions of the structure of, and interactions within, ecosystems, communities, food webs, and populations.

However, I believe that the task of ecologists is to do more than describe – no matter how precisely and quantitatively. Like other scientists we must search for rules that are more general than others, and which *explain* the complexity and variability we observe. This book is my attempt to contribute to that process.

My colleague Piet den Boer warns me, however, that we cannot make science by piling up selected examples. I agree. But by presenting a broad cross-section of examples I can best demonstrate the existence of general patterns in nature. A knowledge of these patterns allows the derivation of new, testable hypotheses about their cause. These, in turn, will lead to the making of more science.

This is especially so if they are primary patterns. Feeding on flush tissues, coprophagy, or the early death of most young, for example, are some of the primary ecological patterns that I perceive to be reflecting the common scarcity of nitrogenous food. They are analogous to primary evolutionary patterns, like the skeletal structure of vertebrates, which reflect a common ancestry. Understanding them may also help interpret secondary ecological patterns, like those found in the interrelations within communities and food webs, or in guilds of animals associated with vegetational succession.

Others will protest that there are many exceptions which I have not presented. But, as Peter Price (1991a) recently pointed out, exceptions cannot destroy the existence of a pattern – they cannot *disprove* it. This can be done only when the preponderance of evidence shows that another pattern – or no pattern – exists. Furthermore, many exceptions, when more carefully investigated, prove not to be exceptions at all. They are either reconcilable with the rule they seemed to contradict, or they lie outside of it.

There are two basic concepts that underpin what I have to say: the oneness of life; and the importance of the individual. And I do not use either expression in any emotional or anthropomorphic sense; quite the reverse.

The enormous diversity of form, function, and interaction of life that we observe today is all part of one pullulating, mindless continuum; the whole evolved from the same simple, early replicators, and programmed and driven by their descendent genes to produce even more genes in the future. [Dawkins (1986, 1989) says all this much better than I]. Natural selection operates upon the consequent "oversupply" of individual phenotypes, eliminating all that cannot cope. The death of so many is not "wasteful", nor cause for sentimental concern. Nature has no "thought" for the "welfare" of future generations. Natural selection, while determining the future, is yet a process for the moment.

Because this is the way nature works, the individual phenotype should be paramount in ecological studies. The environment of an organism is *everything* that impinges upon it, including those of its own kind (Andrewartha 1970). As Einstein is purported to have said, "The environment is everything that is not *me*". All of nature comprises individual phenotypes each struggling to survive in its own indifferently harsh environment. It is essential that we *think* about ecological interactions from the point of view of that individual's struggle.

All that said, this book is mostly about herbivores. Not because they are more "important" than carnivores, but because it is not widely recognized or accepted, as it is for carnivores, that herbivores are limited by their food – even less that it is nitrogen, not energy, which is in critically short supply for them.

This is the heart of my thesis. A lack of access to nitrogen in a form that can be used for the production and growth of young is the major restriction on the abundance of *all* animals. This shortage is reflected in a host of common structural, physiological, and behavioural adaptations, which all serve the same function of helping alleviate this "universal nitrogen hunger" (Keeble 1910).

Many years ago my friend and colleague Alan Newsome, who has worked all his life with mammals, told me that, for him, insects might just as well be Men from Mars! Mindful that there are many ecologists like him, I have arranged and subdivided the text so that those who wish to find out what I have to say can do so without reading about very much other than "their" animals. But I would plead with them not to stop there. I think they will find that animals which they have never encountered, or have had little to do with since their undergraduate days have, ecologically, much in common with the animals that they work with every day. Then, if I have been half-way successful in gathering and presenting the evidence, they will begin to see the generality of my thesis.

Finally, I did not write this book solely for my fellow ecologists. I hope that many others will find what I have to say both useful and enjoyable to read. And students, too, both postgraduate and undergraduate – those who are not yet set in their thinking like we older ones. To them I say: read this, think about it, and then go and look at the world from this point of view. The evidence is there.

Adelaide, November 1992 T. C. R. White

Contents

Part I: The Inadequate Environment

Introduction 3

Chapter 1. The Environment of All Organisms Is Inadequate 5

1.1 Natural Selection Is a Negative Process 6
1.2 Populations Press Against Limits of a Minimum Resource 8
1.3 What Essential Resource Is Most Likely to Be Limiting? 11
1.3.1 Nitrogen the Most Limiting Chemical 11
1.3.2 Nitrogen Limiting Plants 12
1.3.3 Nitrogen Limiting Animals 13
1.3.4 Energy Not Limiting 14
1.4 Competition a Consequence Not a Cause 14
1.4.1 Intra-specific Competition 14
1.4.2 Inter-specific Competition and "Competitive Exclusion" 16
1.5 Self-regulation Does Not Exist 20

Chapter 2. Plants as Food for Herbivores 22

2.1 Why Is the World Green? 22
2.2 How Might Plants Be an Inadequate Source of Food? .. 23
2.3 How and When Might Nutrients in Plants Be Too Dilute? 23
2.4 When Is a Minimum Supply of Nitrogen Critically Important? 24
2.5 How Might Herbivores Counter the Plants' Evolved Strategies? 24
2.6 How General Is Dilution of Nitrogen in Herbivore Food, and What Adaptations Have Evolved to Counter it? ... 26

Part II: Herbivores in an Inadequate Environment

Chapter 3. Insects 31

3.1 Flush and Senesence Feeders 32
3.1.1 Two Australian Psyllids on *Eucalyptus* 34
3.1.2 Two African Scale Insects on Californian Ice Plants 37
3.1.3 Two Aphids on Scots Pine 38
3.1.4 The Green Spruce Aphid in Scotland 39
3.1.5 Aphids on Sycamore, Apple, Wheat, and Alfalfa 41
3.1.6 Scale Insects on *Euphorbia* and *Euonymus* 42
3.1.7 A Leafhopper with Alternate Generations on Brambles and Oak 43
3.1.8 Two Species of Caterpillars Eating Oak Leaves .. 44
3.1.9 Two Species of Sawflies Mining in Birch Leaves . 45
3.1.10 A Chewer and a Sucker on Poplar Leaves 46
3.2 Leaf-miners 48
3.2.1 The Switch from Flush to Senescence Feeding ... 48
3.2.2 Leaf-miners Which Induce "Green Islands" in Leaves 49
3.2.3 A New Zealand Weevil Mining in Fallen Beech Leaves 49
3.2.4 The American Holly Leaf-miner 50
3.3 Gall Makers 52
3.3.1 Physiological Galls 53
3.3.2 Nutritional Benefits of Galling 54
3.3.3 Adaptive Nature of Galls Debated Anew 55
3.3.4 Double-dipping: Prolonged Growth plus Hastened Senescence 57
3.3.5 Selection for High Nitrogen and Survival of the Young 58
3.3.6 Selection of Growing Tissues for Proliferation of Galls 61
3.4 Chewing Insects 62
3.4.1 Creaming-off as a Tactic to Increase Access to Nitrogen: White Butterflies on Crucifers 63
3.4.2 Early Instars Need More Nitrogen: Gypsy Moth on Artificial Diet 65
3.4.3 Illustrations from the Life Cycles of Economically Unimportant Butterflies 65
3.4.4 Further Examples from Forest Defoliators 70
3.4.5 Pests of Crops also Reveal the Need for Nitrogen 75
3.4.6 Examples from Biological Control of Weeds 77
3.4.7 Locusts and Grasshoppers 80
3.5 Sap-Sucking Insects 85

3.5.1 Aphids 85
3.5.2 Psyllids 89
3.5.3 Scale Insects 92
3.5.4 Planthoppers, Leafhoppers, and a Mirid 93
3.5.5 Xylem-feeders 95
3.6 Fruit Flies 97
3.7 Wood-eating Insects 98
3.7.1 The Key Role of Fungi: Increasing Nitrogen in Wood 99
3.7.2 Termites and Woodroaches: Gut Fauna, Coprophagy, and Recycled Nitrogen 100
3.7.3 Furniture and Longhorn Beetles 104
3.7.4 Woodwasps 106
3.7.5 Borers Which Do Not Ingest Wood 106

Chapter 4. Crustaceans 108

4.1 Microcrustaceans 108
4.1.1 Distribution and Abundance of Food Limited by Nitrogen 108
4.1.2 Nitrogen Content of Food also Important 110
4.1.3 Microcrustaceans Feed Selectively for Nitrogen 112
4.1.4 Coprorhexy 115
4.2 Macrocrustaceans: Land Crabs, Lobsters, and Shrimps 116
4.3 Terrestrial Crustaceans – The Isopods 117
4.3.1 The Case of a Common Woodlouse 117
4.3.2 The Role of Microorganisms and Coprophagy 119

Chapter 5. Molluscs 123

5.1 Some Examples of Freshwater Snails 123
5.2 Marine Limpets and the Flow of Nitrogen Through the Food Chain 125
5.3 Death of the Young, Selective Feeding, and Animal Protein in the Diet 127
5.4 Detritus Feeders Feed Selectively, and Depend upon Microorganisms and Coprophagy 129
5.5 Terrestrial Snails Live with the Same Constraints 130
5.6 Cannibalism by Young Snails Illustrates Shortage of Nitrogen 131
5.7 Teredo Shipworms Depend upon Microorganisms Which Fix Atmospheric Nitrogen 133

Chapter 6. Mammals 135

6.1 Large Mammals 135
6.1.1 Feral Donkeys in Australia 136

6.1.2 Red Deer in Scotland ... 137
6.1.3 Antelope, Giraffe, and Greater Kudu in Africa ... 138
6.1.4 Deer in North America ... 141
6.1.5 The Giant Panda in China ... 142
6.1.6 Domestic Stock ... 144
6.2 Rodents ... 144
6.2.1 Squirrels ... 144
6.2.1.1 True Squirrels ... 145
6.2.1.2 Chipmunks and Ground Squirrels ... 148
6.2.2 Rats and Mice ... 150
6.2.2.1 The House Mouse ... 151
6.2.2.2 The Australian Smokey Mouse ... 154
6.2.2.3 The American White-Footed Mouse ... 155
6.2.2.4 American Woodrats ... 156
6.2.3 Voles ... 156
6.2.4 Supplemental Feeding of Small Rodents ... 162
6.2.5 Rabbits and Hares ... 164
6.2.5.1 The European Mountain Hare ... 165
6.2.5.2 The European Rabbit ... 166
6.2.5.3 The North American Snowshoe Hare ... 167
6.3 Primates ... 172
6.3.1 Colobine Monkeys ... 172
6.3.2 Cercopithecid Monkeys ... 174
6.3.3 Howler Monkeys ... 174
6.3.4 The Gorilla ... 175
6.4 Fruit and Flower Bats ... 178
6.5 Marsupials ... 181
6.5.1 The Koala ... 181
6.5.2 Possums and Gliders ... 184
6.5.3 The Habitat of Possums and Gliders ... 187
6.5.4 Kangaroos and Wallabies ... 189

Chapter 7. Birds ... 193

7.1 Birds Eating Green Leaves ... 193
7.1.1 Geese in Europe and North America ... 193
7.1.2 European Grouse, Ptarmigan, and Capercaillie ... 200
7.1.3 North American Grouse ... 207
7.1.4 Partridges and Pheasants ... 210
7.1.5 Galliforms as Hindgut Fermenters ... 211
7.1.6 Changes in Abundance of *Lagopus* Species ... 212
7.1.7 The Takahe ... 215
7.1.8 The Hoatzin ... 217
7.2 Birds Eating Nectar and Fruit ... 217
7.3 Birds Eating Seeds ... 221
7.3.1 Columbids ... 221

7.3.2 African Queleas, European Finches, and the Great Tit 222
7.3.3 Darwin's Galapagos Finches 226
7.3.4 The Australian Galah 230

Chapter 8. Reptiles 233

8.1 The Giant Tortoises of Aldabra Atoll 233
8.2 The Green Turtle of the Bahamas Islands 236
8.3 The Marine and Terrestrial Iguanids of the Galapagos Islands 237
8.4 The Desert Iguanid of California 239
8.5 The Green Iguanid of Panama 240

Chapter 9. Fish 242

9.1 The Carnivorous Young of Fish 242
9.2 Fish Which Eat Detritus 244
9.3 Fish Which Eat Algae 245
9.4 Gut Microbes and Coprophagy in Fish 247

Part III: Survival in an Inadequate Environment

Chapter 10. Strategies to Counter Shortage of Nitrogen 253

10.1 Strategy A: Synchronize the Life Cycle with Availability of Good Food 253
10.2 Strategy B: Concentrate or Prolong Availability of Nitrogen in Food 254
10.3 Strategy C: Eat More Food More Quickly, and Digest More Efficiently 255
10.4 Strategy D: Enlist the Help of Microorganisms 256
10.5 Strategy E: Supplement Plant Food with Animal Protein 258
10.6 Strategy F: Apportion and Concentrate the Limited Food to a Select Few 260

Chapter 11. Territorial and Social Behaviours 261

11.1 Territorial Behaviour in Carnivores 264
11.1.1 Birds 264
11.1.2 Lizards 265
11.1.3 Insects 266
11.1.4 Spiders 267
11.2 Territorial Behaviour in Herbivores 268
11.2.1 Mammals 269

11.2.2 Birds 271
11.2.3 Insects 274
11.2.4 Fish 278
11.2.5 Molluscs 280
11.3 Surplus Young, Dispersal, and Philopatry 281
11.4 Social Structures and Dominance Hierarchies 287

Chapter 12. Cannibalism 291

12.1 Cannibalism by Females Producing Young 292
12.2 Cannibalism by Growing Young 295
12.3 Cannibalism, Warfare, and Protein 297

Part IV: Predators in an Inadequate Environment

Chapter 13. Vertebrates 304

13.1 Lions, Lynx, and Feral Cats 304
13.2 Coyotes, Wolves, and Foxes 306
13.3 Stoats, Mice, and Seed Mast 307
13.4 Pelicans, Puffins, and Other Sea Birds 308

Chapter 14. Invertebrates 314

14.1 Triclad Worms 314
14.2 Spiders and Scorpions 315
14.3 Ground Beetles, Tiger Beetles, and Ant-lions 317
14.4 Praying Mantids 319
14.5 Parasitoids, Parasites, and Diseases 323

Part V: The Alleviation of an Inadequate Environment: Outbreaks

Chapter 15. What is an Outbreak 333

15.1 Some Examples 333
15.2 What Causes Outbreaks? 336
15.3 The Paradox of Enrichment and "r" and "K" Strategists 338

Chapter 16. The Interaction of Food, Prey, and Predators in Outbreaks 341

16.1 Bacteria and Protozoa 341
16.2 Rabbits, Foxes, Cats, and Dingoes 342
16.3 The Varying Response of Predators to Changes in Prey 343

16.4 A Natural Experiment: Guano-Algae-Limpets-Oystercatchers 345
16.5 A Thought Experiment: Hot Spots in a Box of Wadding 347

Chapter 17. Cyclic Outbreaks 350

Chapter 18. The Influence of Weather on the Generation of Outbreaks 356

18.1 Hot Spots Again: Outbreak Centres and Boundaries .. 357
18.2 Spruce Budworm Outbreaks Revisited 359
18.3 Patchy Environments and Metapopulations 361
18.4 The Role of Viral Diseases 362
18.5 The Link to Climatic Oscillations 363
18.6 Major Outbreaks Which Are Independent of the Weather 365

References 367

Subject Index 413

Part I: The Inadequate Environment

Introduction

Over 80 years ago the Professor of Botany at the then University College of Reading in England, Frederick Keeble, published a small book (Keeble 1910). I discovered this book in 1985! In it he described his simple but elegant and imaginative experiments with the green and the yellow-brown marine platyhelminth worms *Convoluta roscoffensis* and *C. paradoxa*. He demonstrated, inter alia, that the relationship between these worms and their symbiotic photosynthetic algae is driven by the lack of available nitrogen in the environment.

The algae are mostly free-living photosynthetic flagellates. They exist precariously in an environment notoriously deficient in nitrogen, but some become embedded in the mucilagenous egg capsules of the worms. There, with access to a much improved supply of nitrogen, they rapidly multiply. Some are ingested by the newly hatched worms, and divide repeatedly to form a dense mass of photosynthetic cells throughout the worms' bodies, using the metabolic waste of the worms as their source of nitrogen (the worms have no excretory system and, in the experimental absence of algae, accumulate vacuoles full of crystals of nitrogenous waste).

The worms are colourless when they hatch and feed by ingesting single-celled plants and animals in the water. Once their contained algae have multiplied and started to photosynthesize they reduce, or in the case of *C. roscoffensis*, cease their feeding. They continue to grow, and eventually produce eggs, using the products of algal photosynthesis for their nutrition. (When artificially prevented from acquiring the algae they whither and die.) After the production of eggs, however, in this closed system of exchange of nitrogen back and forth between worm and alga, nitrogen again becomes limiting. The worms then digest their algae before themselves dying.

Keeble pointed out that to label the association of worm and alga as symbiosis is to miss the significance of the association. For the worm it is obligate dependence upon the alga – parasitism. Without the alga it can neither rid itself of nitrogenous wastes nor gain further nutrition. For the alga, on the other hand, that some few of its individuals enter this association with a worm is "...an episode without significance...". Most of them remain free as green photosynthesizers or colourless saprophytes "...which batten on the offal of the sea...". But for an ingested cell the association with the animal solves the problem of acquiring enough nitrogen. As Keeble so colourfully expressed it, "It sacrifices its independence for a life of plenty. This universal nitrogen-hunger is a misery which makes strange bedfellows." .

Indeed it does. It is a misery which drives the ecology of all organisms, and Keeble was at pains to emphasize this fact. "The nitrogen problem" as he called it, "...stems from the shortage of nitrogen which is *available* for the formation of nitrogenous food for plants", and as such is, "...the problem which besets all living organisms". This nitrogen hunger, he said, is "...no small matter of mere academic importance". It "...presents a problem which every living organism must solve. *The supply of available nitrogen is a limiting factor of life*" (my emphases).

An awareness of the fundamental role that this limitation of nitrogen plays in the ecology of all organisms should be as much a part of each ecologist's intellectual equipment as is the awareness of the fact of evolution by means of natural selection. What follows is *my* attempt, (having painfully reinvented the wheel!), to get this message across to future ecologists. Please do not let it languish for another 80 years!.

Chapter 1 The Environment of All Organisms Is Inadequate

This chapter title seems a fairly obvious statement to make about a world in which only a small fraction of the offspring of all organisms survive to pass on their genes to the next generation. The world must be a harsh and inhospitable place for this to happen. Yet there still seems to be plenty of space, and organisms rarely use more than a fraction of the resources in their habitats.

Much of conventional ecology says that this is because there are many and varied environmental factors, and numerous competitive and social interactions, which regulate numbers of organisms below those which their resources would allow; which contain the innate tendency of all organisms to increase progressively. The environment is therefore not limiting or inadequate. Exhaustion of a resource, probably food, would ultimately limit further increase of any organism, but this does not happen – or happens only rarely when the regulators temporarily cease to operate. For most of the time there are processes operating which keep populations in balance around equilibrium levels well below those which would use up all the resources and destroy the environment.

This view has it that the struggle for existence in nature follows remorselessly from the capacity of organisms to increase their numbers exponentially.

But rather the reverse is true. The capacity of all organisms to increase their numbers exponentially follows remorselessly from the struggle for existence. Surviving on this earth is, and always has been, especially for the very young, a struggle, a chancy business. The huge "biotic potential" of all organisms is the universal illustration of that fact. The capacity to increase exponentially did not evolve to provide a struggle for existence as a vehicle for evolution. Nor need it be espoused as the reason why populations must be regulated. It evolved because, only in populations in which females produced many offspring did sufficient individuals gain access to sufficient resources, and survive in each generation, to enable the population to persist.

How did this come about? Because this is a finite world, and in a finite world there was one sure consequence of the evolution of the first self-replicating entities. Sooner or later a resource, essential to their growth and replication, ran out. There was no longer enough for all seeking to use it. No longer could all replicate. No longer were all potentially immortal; differential survival had arrived. From the moment a resource became limiting the environment became less adequate, and natural selection began to operate. The struggle for existence – to gain enough of that resource to replicate – had begun. The "lucky" few that happened to get enough replicated; those that failed perished. Then, as now, *anything* which lessened the chance of an individual being one of those few ensured the eventual elimination of that individual's genes from the gene pool. *All else in the living world has followed from this.*

In such a world balance is an unnecessary idea. The usual and variable shortage of a resource provides an alternative explanation. The tacit assumption in the balance of nature is that all species of organisms tend to produce an abundance of progeny that would survive to reproduce and lead to ever increasing numbers unless controlled by negative feedback mechanisms. As a consequence the mean observable density over a number of generations is seen as an optimum or equilibrium density. The further the numbers depart from this mean the greater the density-dependent pressures in the system pressing them back toward this mean.

However, there is no such thing as a mean population in nature. The mean is merely a statistic – an arbitrary, albeit frequently useful, abstraction – derived from a series of samples of the numbers in a population in which numbers are continually changing. These numbers are not deviations from a predetermined mean; they define it. They result from the members of the population struggling, generation after generation, to exploit the limited opportunities afforded by the inadequate environment. Nor is there an "optimum" or "equilibrium" density of a population in nature – only the maximum number that can survive each generation in a population that is pressing hard against the variable but limited supply of resources in its environment.

To use the original engineering analogy of steam generated by heat being released through a negative feedback valve, ecologists should be looking to see what generates the steam, not how much steam passes through the regulating valve. In nature rarely is enough steam generated to make the valve operate. I (White 1978) and others before (Den Boer 1968) and since (Dempster and Pollard 1981; Den Boer 1987) have made these points, but the arguments go on (Wolda 1991)!

1.1 Natural Selection Is a Negative Process

The role of natural selection in deciding which individuals survive does not involve *positive, active selection* of those lucky few. It is a *negative, passive* process which eliminates all the rest; in Den Boer's (1985) words "the non-survival of the non-fit". The environment is inadequate to support all those seeking to live in it, so most will fail to do so. They will be selected *out* or selected *against.* Wallace (1866) made this very point in a letter to Darwin. He was urging him to adopt Spencer's term "the survival of the fittest" in place of "natural selection", and expressed concern at the misunderstanding of the latter, and the constant "...comparing it in its effects to man's selection...". "Nature", he said, "does not so much select variations as exterminate the most unfavourable ones". Some years ago I made the same point to a colleague. His response was "So what? The glass is half empty or half full!" But I think it is important that we *do* make the distinction, for two reasons.

First, those few left after all others which could not cope had died, may not necessarily have survived because they had what *we* can see would be the optimum strategy for survival in that environment. They needed only to have some attribute which was *good enough*; what Den Boer (1986a) calls an "effective strategy". That is sufficient. In fact, there may well be more than one attribute which is good enough – Gould's (1989) "contingency"; one or more fortuitous outcomes of many possible outcomes. (The malarial mosquito which rested under the table rather than on the wall painted with DDT, survived to reproduce just as well as the one which rested on the wall but had the physiological capacity to detoxify the DDT. Each gave rise to a resistant strain of mosquito: one behavioural, one physiological.)

Second, those individuals which died need not have done so because they were actively killed (by predators or weather) or because they were out-competed (by others of their own or another species). Most die just because they fail, right at the start, ever to gain a foothold – to gain access to enough of the limiting resource to survive and grow. For most the "struggle for existence" is not a tooth and claw business. It is a lonely struggle to live in an inadequate world. They die young, and their passing is passive, solitary, and unnoticed.

I found a good example of each of these points in the life history of the psyllid *Cardiaspina densitexta* (see Sects. 3.1.1 and 3.5.1). Its females will lay their eggs, and its first instar nymphs will feed, only on mature leaves of one species of *Eucalyptus*. The nymphs can hasten the rate at which these leaves senesce and release soluble nutrients into the phloem from which they feed. Still, most die from starvation, and the few survivors take 3 to 6 months to grow to adults. Yet only a short and easily achievable walk up the stem there are new flushing leaves with much greater amounts of nutrients in their phloem, and upon which other species of psyllids complete their development in less than 3 weeks. It seems *C. densitexta* would do much better on this much richer source of food. But even when confined on young leaves females will not lay, and nymphs will not feed. Why not? Because they are "locked in" to a way of life which happened to prove successful for their ancestors when they were confronted with a choice of feeding on mature leaves or dying. Furthermore, this way of life will continue to be successful so long as it is a *sufficient* strategy for some psyllids to survive and reproduce in that environment. The new growth which we perceive as part of their environment is not. This potentially much better source of nutrition is not available to them. It might just as well be on another planet.

Most of the first-instar nymphs of *C. densitexta* die a day or two after they have settled and started to feed. No predators, parasites, or diseases attacked them, no competitors vied with them for any resource, nor was the weather too hot or too cold, too wet or too dry. And their food was abundant and palatable, but lethally inadequate, so they starved while eating their fill from the phloem. However, a few always chanced to settle where the concentration of nutrients in the phloem was greater. Those few survived. It was the *passive inadequacy* of the environment which killed most of those young animals, not the action of something, living or non-living, in their environment. Each individual nymph, no matter if it were the only animal of any species on the whole tree, was doomed from the moment it settled to feed where the food was too dilute to support it. I was alerted to all this because each nymph builds a long-lasting cover over itself. It is firmly attached to the leaf surface and can easily be seen many weeks after the dead nymph's body has vanished as a speck of dust. However, the young of most other animals disappear without leaving any such trace soon after their birth. The ecologist is left to *assume* that predators ate them, or rain washed them away, or they died of cold, or...

So, natural selection operates by eliminating all those that cannot cope, leaving only those that can. *How* they cope is of no consequence, only that they do. This is the reverse of our active selection of only the best plants and animals from which to breed the sorts of crops and stock that we want. I suspect that, all too often, ecologists think – albeit subconsciously – in these active, positive, human terms, and thus overlook the passive and negative realities of nature.

1.2 Populations Press Against Limits of a Minimum Resource

To return to the consequences of a resource running out for some primaeval replicating form of life. It follows that all organisms descended from these original ancestors were those which, for whatever reason, were best at gaining enough of the limited resource. And today it is no different. Evolution of any self-replicating system by means of natural selection can have no other outcome. All forms of life will continue to produce more of their own kind, and increase in abundance, until stopped by a decrease in the amount of whichever essential resource first runs out. *Only* then will their numbers stop increasing. The population is, from then on, pressing hard against the limits set by the shortage of that resource.

It is necessary to appreciate this concept of "pressing hard" for what it is – a metaphor, an analogy – and that it refers to the *population*, not the individual organisms. Individual organisms do not "press" against anything. They simply function in the way that they have evolved, each maximizing its chance of gaining access to essential resources.

On the other hand, it must be thought about from the point of view of the individual animals that are trying to survive in the habitat, not from that of a human observer. There may be – there *will* be – resources in the habitat which the animals could use. However, for various behavioural, physiological, topographic, and statistical reasons, all of a particular resource is usually not accessible to, or attainable by, any one individual. Only that part of the resource that it can find or process in time is available to it. The remainder which is not found, captured, assimilated, etc, is unavailable. So, while this means that a resource is rarely, if ever, exhausted, the production and dissemination of many individuals into the habitat ensures that, for the population as a whole, all of the resource which can be effectively used by individuals at the time is used.

This production of a "surplus" of individuals also means that any increase in the availability of the resource (for whatever reason) is quickly taken up by the population as a whole. It is in this sense that the population is pressing hard against the limits set by the changing availability of resources in the environment. That this is happening is, if you like, an inevitable consequence of the inexorable burgeoning of life on Earth.

Also, we can see that a habitat is fully occupied, but generally not saturated, in the sense that no more individuals of the species could survive and reproduce in it. Habitats are never totally saturated; there is always *some* resource left, and plenty of space. If only they could get at the resources we can see waiting and apparently available to them. But they cannot. So, as far as the animals are concerned, in terms of the essential resources that are accessible to them – of the number of animals that the habitat can support *at the moment* – most habitats *are* saturated for most of the time. As many as are able to survive in that habitat at that time do so. There may be lots of empty territories, lots of uneaten plants, lots of potential prey. It all looks to us as if the habitat is hardly being used at all, let alone being saturated, but for the animals trying to get enough good food to breed and raise young, there is no surplus capacity there. Within the restrictions imposed by the distribution and availability of their essential resources, and of their physiological and behavioural capability to get at and use those resources, the habitat *is* saturated (see also Sects. 10.2 and 11.3).

For any population of individuals the resource which runs out may change from habitat to habitat, or from time to time; and the amount of any particular resource that is available to them will vary from time to time. But it will always be an *essential* resource, not one for which there is an alternative, no matter how much less preferred or less advantageous the substitute might be.

Thus, rabbits in Australia mostly breed in burrows deep in major warrens situated in hard stony soil where the young are safe from being dug up by foxes. However, when there is a great deal of good food, and their numbers expand enormously, females driven from the warrens will dig shallow breeding burrows in sandy soils which confer little protection from foxes. Many species of Australian ducks which normally nest in holes in trees respond in a similar way when there is an abundance of food available. Some will even nest on the ground where their eggs and hatchlings are vulnerable to predation by feral foxes (J.M. White pers. comm.)

The response of the large cactus finch, *Geospiza conirostris*, to the enormous increase in its food supply in the Galapagos Islands during the 1982–1983 El Niño is another good example of what I mean (Sect. 7.3.3). The birds were thought to be limited by cactus bushes. They almost always nest in them and always established territories which included one or more of them. But when they bred so profusely in response to this increase in food, new breeders established territories in previously unoccupied areas with little or no cactus, and only 40% of all nests were built in cactus bushes. The remainder were in much more exposed sites in trees of other species.

If, however, it is not obvious that an essential resource is in short supply, then it is unlikely to be obvious to an ecologist that the number of individuals in the population is being limited by the incapacity of the habitat to support any more of them. It is much more likely to seem as if some directly acting factor in the environment is doing this. And this happens very commonly because the limit is being set by a *relative* shortage; there is plenty of the resource in the environment but most of it is not available for the individuals attempting to use it.

As I try to bring out in the many examples in this book, the availablity of essential food to individuals in a population of animals is usually constrained by this relative rather than absolute shortage. Most of it is not available because it is too "dilute" in either time or space, so most individuals cannot get enough of it, soon enough, and die long before it is all used up.

This concept of a relative shortage is fully treated by Andrewartha and Browning (1961), Andrewartha (1970), and Andrewartha and Birch (1984), but a few examples here may help.

First, a thought experiment with a larval mosquito. These aquatic insects feed by filtering the water in which they live through their mouthparts, extracting and eating particles of organic matter suspended in the water. I place a larva in a small container of water in which there are just enough particles of organic matter to enable the larva, filtering as fast as it is able for as long as it can, to obtain enough food to survive and grow to maturity. However, if I provided it with exactly the same quantity of food particles, but in a much larger container of water, then it will starve to death in a very short time. There is still enough food in the larger container to enable it to grow to maturity, but this food is too thinly spread in the water and the larva cannot get the particles fast enough or frequently enough to survive.

Secondly, the first instar nymphs of *C. densitexta*, which I discussed earlier in this chapter, provide a similar but real-life example. There are enough nutrients in the phloem sap of a tree, even of one leaf, to provide for many nymphs to grow to maturity, but these nutrients are too thinly spread – too dilute – for most individuals to ingest enough quickly enough to survive, let alone grow; so most quickly starve.

Other examples of this same dilution can be found in the ecologies of predators. When ticks and tsetse flies (Sect. 14.5) manage to find a host they have access to enough blood to feed hundreds of their kind. Because of their evolved life histories, physiologies and behaviours, however, they use only a small fraction of it. Many never find a host in the first place. Those that do, get one good meal, but may never find a second; similarly with larval carabid beetles (Sect. 14.3) and lionesses (Sect. 13.1). There is much more prey in their immediate habitat than they ever use, but, for various unalterable reasons, they are unable to catch most of it in time – so the surplus is not available.

1.3 What Essential Resource Is Most Likely to Be Limiting?

We can define a resource as any substance or thing, living or dead, complex or simple, that is essential for survival, growth, and reproduction; but the ultimate resource for all forms of life must be the supply of chemicals which are necessary for the biochemistry of life. In a word, food. From the time a mechanism of growth and replication chanced to work with a particular combination of particular chemicals, all descendants of the resulting life forms were locked in to using only those chemicals, and in that way. That there may have been other chemicals which were more abundant or available, or more efficient ways of combining those that were used, is irrelevant.

The presumed earliest organisms were heterotrophs using compounds like carbohydrates and amino acids already present in the "primaeval soup". Once the supply of any one of these compounds decreased only those organisms getting enough of it to grow and reproduce survived. The rest died, returning the components of their bodies to the soup to be recycled by those newly replicated individuals lucky enough to acquire them. There was differential survival, natural selection removing all but those able (by whatever means) to gain enough chemicals. Evolution's one-way arrow was launched.

And what has it produced? Organized life based upon hydrogen, oxygen, carbon, and nitrogen, and fuelled by energy from the sun. The supply of this energy, necessary to incorporate simple chemicals in copies of more complex and more ordered ones, is, to all intents and purposes, infinite. And, anyhow, only a very small fraction of it is incorporated in the biological cycle; most of it is re-radiated back to space as heat.

1.3.1 Nitrogen the Most Limiting Chemical

The supply of chemicals is not, however, unlimited, and some, although abundant, are less available than others. Carbon, oxygen, and hydrogen are all in great abundance and readily available. Nitrogen is equally abundant, but nearly all of it is unavailable (Sprent 1987). Of the total nitrogen in the biosphere, in excess of 99.95% is extremely inert nitrogen gas forming the major component of the Earth's atmosphere. It is unavailable for organic synthesis except for a tiny fraction added each year by nitrogen-fixing microorganisms (and latterly an even smaller fraction by industrial fixation for artificial fertilizers). Of the less than 0.5% of the nitrogen which is combined with other chemicals, and potentially available to living organisms, nearly half is in inorganic form. Furthermore, 95% of the half that is in organic form is present as dead material; in litter, soil, or (mostly) particulate and dissolved matter in the oceans. So, *relative* to the other major components of living organisms, nitrogen is in *very* short supply. Of all the elements essential to organic life on Earth it is the one which is least available and therefore most limiting (Delwiche 1970; Rosswall 1983; Stewart et al. 1983). What little is available tends to be thinly spread in the environment; there is a relative, rather than an absolute, shortage.

As a nutrient, nitrogen is required in quantities second only to carbon. It is a key constituent of all living cells. Without nitrogen proteins cannot be built. Proteins are the basic physico-chemical structures of all living things, and are made from amino acids. Nitrogen is the key component of these amino acids which all organisms must have. No organism – plant, animal, or protist – can survive, let alone grow, without an adequate supply of nitrogen for the synthesis of proteins. The productivity of all life on Earth, in both terrestrial and aquatic environments, is limited by biologically available nitrogen.

By comparison, the supply of the other major nutrient, carbon, is not limiting, and it is greatly in excess of nitrogen in all living tissues (Martin and Kukor 1984; Sprent 1987). Even in the basic compounds like amino acids and purines the ratio of carbon to nitrogen is greater than one (typically 1.1 – 1.7 : 1.0) and in the more complex proteins and nucleic acids it is of the order of 3.0 : 1.0. In whole organisms this ratio is lowest in the microorganisms – bacteria, algae, fungi – where, in some bacteria, it can be as low as that of proteins. At the other extreme secondary woody tissues of plants can have C : N ratios of 1000 : 1.

1.3.2 Nitrogen Limiting Plants

The very high ratio of carbon to nitrogen in wood reflects a major evolutionary development in plants – the use of the great surplus of available carbon to form nitrogen-free compounds for structural purposes, husbanding their scarce nitrogen resources to make protoplasm. Plants must obtain all their nitrogen in solution from the soil. The whole of agricultural and pastoral practice (including the large-scale use of artificial fertilizers obtained by industrial fixation of atmospheric nitrogen) attests to the acute shortage of nitrogen in the soil – that it is the major limitation to the growth and productivity of plants. The evolution of insectivorous and ant-fed plants in habitats where the soil is especially deficient in nitrogen provides an extreme illustration of this shortage (Thompson 1981). By digesting animal tissues such plants are able to survive and reproduce in habitats in which the amount of available nitrogen is at impossibly low levels. But becoming carnivorous has still not released these plants from being limited by a relative shortage of available nitrogen. When Thum (1988) fed *Drosera* plants with supplementary supplies of insects they grew bigger and produced more flowers and seeds than matched plants left to get by on whatever prey they could catch for themselves. That they are not limited by energy has been demonstrated by the fact that these plants do not use their prey as a source of carbon even when they are grown in artificially low light intensities.

As the primary producers, plants are, of course, responsible for fixing all the energy from the Sun which can be used by animals to fuel their metabolism. But it is not available energy which limits plants. Estimates of the efficiency with which plants can use carbon, light (= energy), and nitrogen, suggest that nitrogen is the most limiting, allowing rates of production an order of magnitude less than either energy or carbon (Agren 1985).

1.3.3 Nitrogen Limiting Animals

From the foregoing it should not be difficult to deduce that a lack of nitrogen is likely to loom large in the ecology of animals. The herbivores which eat the plants, the carnivores which eat the herbivores and other carnivores, and the detritivores and decomposers which eat both all depend upon the plants. Only plants can incorporate inorganic nitrogen in organic forms which animals can use. Plants absorb their nitrogen as ammonium or nitrate. Animals cannot do this. They must get their nitrogen as ready-made amino acids from plants and/or from other animals or microorganisms. And, unlike the plants, they use structural materials based upon proteins, not carbohydrates (Sprent 1987).

A lack of nitrogen looms largest of all, however, for the herbivores. As the first link in the chain from the plants, they are confronted with food which is readily accessible, and attainable with minimum effort, but is composed largely of carbon; and much of that is in the form of structural polymers, like cellulose and lignin, which are little, or not at all, attacked by enzymes. So, much of the often quite abundant nitrogen in their food is not available to them because it is too thinly spread in either or both time and space; or is locked away in a form that they cannot digest. At best they can eat reproductive tissues such as pollen and seeds, and get a food which has approximately 7% of its dry weight as nitrogen; at worst a diet of xylem sap could yield as little as 0.1%. Those that eat growing tissues, newly expanding leaves or cambium, will get about 5% nitrogen (Martin and Kukor 1984).

However, that is not the end of it. Not only is their food loaded with excess carbon, much of the limited nitrogen that it does contain is also present, for much of the time, in a structural rather than a metabolic form. As such, it too is much less readily broken down and assimilated. Small wonder that 90% of all plant material ends up with the decomposers rather than the herbivores.

As we will see in the forthcoming discussions, this has meant that herbivores must use every possible opportunity to find and eat those parts of the plants in which the nitrogen is present in its simpler, most readily assimilable forms. Furthermore, they must get to these tissues during the often very limited periods when they contain nitrogen in this form.

But they have a third problem. Not only is nitrogen scarce in their environment, and much of it still requiring the expenditure of considerable time and energy before it can be used, they cannot use all that they do manage to absorb. Their inherited metabolic biochemistry is such that in the process of incorporating nitrogen into their body tissues, some nitrogen must be discarded – excreted as metabolic waste. So they have had to evolve ways of saving and using some of this precious waste. Mostly they have done this with the "help" of microorganisms which can use animals' metabolic waste nitrogen as their nitrogen source when digesting structural polysaccharides of plants.

As we shall also see, the carnivores are confronted with the same problem of a lack of enough nitrogen for breeding and growth. Unlike the herbivores, their food is a concentrated source of readily assimilable nitrogen, but much of it is not available to be used because it is inaccessible or unattainable.

1.3.4 Energy Not Limiting

In spite of these well known and basic facts, one finds that nearly all studies which measure the food of animals, and its influence upon their growth and survival, look only at energy. The mental concepts of "ecosystem", "community", "trophic level", "food web", and ideas about the flow of solar energy through them, hold sway. But these *are* only mental concepts – abstractions – no different from a mathematical model, in spite of some who would have us believe the opposite, that these are the realities, and individual organisms the abstractions (Pimm et al. 1991)!

The supply of solar energy is essentially constant and limitless. Even so, less than 10% of it gets incorporated in plant tissue by photosynthesis (Radmer and Kok 1977), and it seems that only about one thousandth of that is converted to herbivores (McNaughton et al. 1989). And the loss continues unabated as herbivores are converted to carnivores, and so on until only the original chemicals are left.

If energy were the first to be limiting would so much go unused? And would the little of it which is trapped be so wantonly wasted? For example, the evolution of warm-blooded animals would have been a very improbable event had the energy needed for thermoregulation been limiting; similarly for long-distance migration. The "wastage" of the large investment in energy involved was apparently not selected against.

However, not all are wedded to the belief that the flow of energy is pre-eminent in influencing the success of animal populations, nor that communities, food webs, and their like are anything other than the arbitrary, but fallible, mental pictures of we human observers of the real world. Some are quite blunt in their rejection of such ideas (Boyd and Goodyear 1971; Maynard-Smith 1985; Gilbert and Owen 1990). Others are gentler, but nevertheless emphatic (Kokkinn and Davis 1986). Still others specifically identify nitrogen rather than energy as the limiting factor (e.g. Tenore 1977). I discuss the work of many of these in succeeding sections.

1.4 Competition a Consequence Not a Cause

1.4.1 Intra-specific Competition

One of the factors most commonly invoked as being responsible for keeping numbers in a population below the limits set by the environment is competition between members of that population. There is no doubt that competition is a reality. It is everywhere about us in multiple forms and at all levels of complexity of life. But it occurs only when one individual, by attempting to use something which is freely available to each individual, decreases the possibility of other individuals using it. It decides *which few* of the many obtain enough of whatever it is they are competing for to survive and pass on their genes to the next generation. Whenever this happens competition is a major influence upon the direction of the evolution of the form and function of species. But it does not determine *how many* survive. *Competition does not decide the distribution and abundance of organisms.* That is decided by the inadequacy of the environment.

Put this way it seems obvious, and I (White 1978) and others (e.g. Murray 1982) have stressed this point in the past. Perhaps a simplified hypothetical example of two populations of the same species of animal living in adjacent but different habitats will illustrate the point here. One has an acute shortage of nesting sites but abundant food; the other abundant nesting sites but little food. In both populations individuals will compete for the limited resource once numbers seeking to use it exceed the supply, and they can readily find each part of the resource, once there is an absolute shortage. From then only a fraction of the total population can survive and reproduce. All but the successful ones die without reproducing. But the *cause* of the level of abundance achieved by each population – how many survive – is different. For one it is food, for the other nesting sites. Both the competition and the numbers in each population are symptoms of the inadequacy of the environment. In both populations if there were no such thing as competition the end result – the number surviving in each population – would be the same.

Competition can only ever be a *consequence* of individuals trying to use more of a resource than is available for all those attempting to use it. In an environment in which self-replicating organisms are evolving, competition, like natural selection, can happen *only* when the environment becomes inadequate through the unavailability or inaccessibility of some resource. Before that, chance alone can influence events. There will be no pressure forcing an individual to expend time and energy pre-empting the use of a resource by another. Such behaviour would, anyhow, select *against* an individual to lessen its chance of survival and reproduction. The genes for such behaviour would quickly be eliminated from the gene pool, leaving only descendants of individuals which used all their time and energy to make maximum use of the freely available resource. *No* other outcome is possible for organisms which are evolving by Darwinian differential survival.

While there can be no competition without shortage, there can be shortage without competition. A resource can be in such critically short supply – so thinly spread through the environment – that most individuals will die because they do not get enough of it *whether or not* other individuals are attempting to use that resource.

A prime example is the one I have just discussed of the nymphs of *C. densitexta* feeding from phloem sap which is so dilute in essential nutrients that most starve to death. There is no sensible way that those nymphs can be considered to be competing with each other. Usually there are no others on the same leaf. Even if there were, however, the small amount of sap withdrawn by any one nymph from the volume of sap available could have no measurable, let alone meaningful, influence upon the amount available to any other nymph feeding from the same source. There is an enormous struggle for survival going on, but it is a struggle by each individual against the inadequacy of its environment, and the struggle of each one is independent of the struggle of every other one.

Furthermore, this sort of situation is just as common for carnivores as it is for herbivores. A good example is that of very young orb-web spiderlings (Wise 1983a, b; Sect. 14.2). It is at this stage that most of these spiders die. As in field studies of so many animals, most just disappear, and the observer has no knowledge of what happened to them. They die, however, because their webs do not catch enough prey fast enough. Their food, too, is so thinly spread – so dilute – that they starve. But their webs are sufficiently far apart that they have no influence on the rate at which nearby webs catch prey. There is no competition between individuals.

In all situations of this sort differential survival will occur. Some individuals will chance upon a greater concentration of resources than others; some will happen to be better than others at gathering what resources are attainable. In all, however, competition will have played no part in the resulting differential survival. Therefore, as well as playing no part in deciding how many will survive, competition cannot be a *necessary* condition for deciding which ones survive – for directing the evolutionary processes within a species (Den Boer 1985).

1.4.2 Inter-specific Competition and "Competitive Exclusion"

The other source of great debate about the role of competition in the ecology of organisms is that which occurs between members of different species attempting to use the same resource. There is growing evidence suggesting that this is a relatively uncommon phenomenon and, unlike competition between members of the same species, of debatable significance to the evolutionary process (Schoener 1982; Den Boer 1985, 1986c; Jermy 1985; Gilbert and Owen 1990).

Individuals of different species largely go about their independent business of surviving and reproducing, adapted to their different ecologies (different climatic tolerances, different habitats, different foods, different enemies, different behaviours) simply because they *are* different species. Often these differences are very slight. Very often they are so subtle that human observers are unaware of them. But they are real. So the activities of members of one species rarely impinge in any significant way upon the chances of survival and reproduction of members of other species with which, to our perception, they "coexist". Species are not excluding one another by competing, they live in separate but often spacially overlapping worlds.

A simple and quickly obvious example of this is that of the two sorts of psyllids: one living only on new leaves, the other only on mature ones (Sect. 1.1 and see Sect. 3.1.1). Also Chew's (1981) meticulously researched story of the presumed competitive exclusion of the indigenous white butterfly, *Pieris oleracea*, from southern New England, USA, by the introduced *P. rapae* nicely illustrates the point. From reports of naturalists and casual observers before and soon after the arrival of *P. rapae* in North America in 1860, it seemed the migrant rapidly ousted its indigenous relative from many habitats, and was well on the way to exterminating it altogether.

More than a 100 years later, however, Chew found both species to be abundant in Vermont, and with broadly overlapping distributions. His close studies of their respective ecologies revealed, however, that each occupies a habitat and eats plants not exploited by the other. Even when their adults are flying and actively mating close together in the same field, they completely ignore each other. The great changes in relative abundance of the two species, and the local extinctions of *P. oleracea* reported in the past, would have been caused by changes in the distribution and abundance of the cruciferous flora resulting from the major shifts in land use that were taking place at that time. What appeared from casual observation to be competitive exclusion proved on careful inspection to be change wrought by human alteration of habitats.

Janzen (1988) showed the same thing, but on a much grander scale, for the many species of caterpillars that live on the many species of plants in a Costa Rican forest. For an individual caterpillar most of the plants in the forest are just not part of its environment. The adult moths are programmed to lay their eggs, and their caterpillars to feed upon, only one or a few of the many species of plants in the forest. Consequently the caterpillars of most other species are not part of its environment either. Nor are most of the carnivores which eat caterpillars, for they too are mostly specialized to prey upon the caterpillars of only one or a few species. Similarly, for any individual carnivore the caterpillars of most species are not potential prey. What seems to human eyes a ready supply of food (for herbivore and carnivore) is not; what seems a large array of competitors is not; what seems a great many predators is not.

This sort of understanding of the separateness of the environment of different organisms exposes the concept of "competitive exclusion" as the non sequitur it is (and see Den Boer 1986c). There are many studies which reveal that there is no competition between apparently coexisting species (e.g. Becker (1990) for larval Trichoptera). A number of them are mentioned later in this book (e.g. Reynoldson 1983 for triclads in Sect. 14.1; Wise 1983a, b for spiders; Polis and McCormick 1987 for scorpions in Sect. 14.2; Milton 1981 for monkeys in Sect. 6.3.3), but further mention here of three of them will suffice to make the point.

Cockburn (1981a) showed that the Australian smokey mouse, *Pseudomys fumeus*, and three other species of small mammal only seem to be sharing the same heath and woodland habitat. They all select the particular microhabitat in which they live, sometimes overlapping with one or more of the other species, sometimes not. They do this on the basis of the quality of that microhabitat in terms of the availaility of their different foods, not because of any competitive exclusion by one of the other (Sect. 6.2.2.2).

Schluter (1982) found the same story when he investigated the distribution of four species of ground finches along an altitudinal gradient in the Galapagos Islands (Sect. 7.3.3). Their distribution overlapped considerably, and was not determined by either the physical or vegetational structure of the habitat, nor by territorial or other competition between species. Each species had a distinctly different diet, and both the distribution and the abundance of each were determined by the distribution and abundance of their respective foods. When all food was abundant there was some overlap of their diets, but this was minimal or absent when food was scarce.

Kearsley and Whitham's (1989) study of a gall-forming aphid and a leaf-eating beetle feeding upon the leaves of the same species of poplar illustrate a more subtle way in which apparently coexisting species are living in different worlds. The aphid is most often found, and survives better on leaves of mature trees; the beetle on those of young trees. Both insects can be found living on the same tree, but the abundance of each on that one tree will depend upon the tree's age, not upon the activity of the other species of insect (Sect. 3.1.10).

In each of these examples the members of each species are struggling in their own world, and are limited by their access to the resources which make up that world, not by the activities of the members of another species. Even when they live in only slightly different worlds there is no reason why they should be competitors.

There are, of course, many examples where individuals of two or more species do use the same resource. But usually its use by each species is separated in space or time – there is no competition between individuals. For an ecologist, the most fruitful way to think about such a situation is, again, the extent to which the resource is limiting. That is what is important to the individuals attempting to use it. What factors might be influencing the abundance of that resource may or may not be important to the ecologist. But it is of *no* significance to the animals. All that matters is how much of it is attainable.

The study by Crawley and Pattrasudhi (1988) of the relationship between the cinnabar moth, *Tyria jacobaeae*, and the ragwort seedfly, *Pegohylemyia seneciella*, on ragwort, *Senecio jacobaea*, in England is a good example. The cinnabar caterpillars eat the flowers, the fly larvae the seeds. When the caterpillars are sufficiently abundant in a patch of ragwort they eat most of the flowers, severely reducing the amount of food available for the flies' larvae. The feeding of the fly, on the other hand, has no measurable influence upon the amount of food for the caterpillars in either the present or future generations. Crawley and Pattrasudhi elegantly demonstrated this relationship by simply removing all cinnabar eggs and migrating caterpillars from one patch of ragwort, leaving them in another, and counting the number of flies produced in each patch. But they called what they found "asymmetric interspecific competition" between moth and fly.

I think this description deflects attention from what is most important in the ecology of each species. Their members remain totally "unaware" of each other. From a moth's point of view the flies do not exist. It is not competing with them and they have no measurable influence upon its chance to survive and reproduce. Nor is a fly competing with the moths. If a caterpillar should happen to eat a flowerhead containing seeds with developing fly grubs in them, it will kill those grubs. Browning (1963) likened this to an event such as you or me unknowingly stepping on a small fecund insect. The insect has no influence upon my chances of survival and reproduction – it is not part of my environment. Nor can I be thought of as competing with the insect, although I have totally destroyed its chance of reproducing by killing it. I am a hazard (a malentity if you will, vide Andrewartha 1970) in its environment, as are the cinnabar moths in that of the fly when they eat flowers which contain its larvae.

But the moths also have an indirect effect on the fly by reducing the amount of its food before it can lay its eggs. Yet had it been a year when there were few or no flowers, or had something else, such as a sheep, a cow, or a man with a mower, removed the flowers, the outcome would have been the same. Would the fly then have been competing with each of those, or they with the fly? And what was the authors' part in the ecology of moth and fly for the duration of their experiment? Make no mistake, in the plot where they removed all the offspring of the moth, they had a real influence upon both populations. A competitor or a predator of the moth? A symbiont to the fly? Clearly this starts to become absurd. The important questions are: How much food was available for individuals of each species? and what factors influence the supply of that food? Both can be answered more simply and less ambiguously in terms of the amount of resource that is available for each species rather than by invoking "competition" between the two species.

A more subtle yet very real division of a resource between two more closely related herbivores is that found by Straw (1989) in his elegant unravelling of the life histories of two tephritid flies, *Tephritis bardanae* and *Cerajocera tussilaginis*, in England. The larvae of both species live within the composite flowerheads of lesser burdock, *Arctium minus*. The females of both species lay their eggs in the green unopened flowerheads and their larvae complete their development within those heads. *T. bardanae* females lay their eggs earlier in the summer and tend to select smaller flowerheads than do female *C. tussilaginis*, but there is considerable overlap in both their time of oviposition and the size of heads selected; so both species appear to be sharing the same resource for much of the time.

However, before flowering, when the external dimensions of these composite heads remain constant, the individual florets within them are growing exponentially. Straw found that each species of fly can detect and lay its eggs in these expanding florets when they are at a quite different but precise stage of their development – one before, the other after fertilization. There is, in fact, no overlap of the place where these two species lay their eggs. This is born out by the subsequent feeding behaviour of their respective larvae. A larva of *T. bardanae* starts to feed in a floret as it is elongating most rapidly before and during flowering. It thus exploits the nutrients flowing to these growing structures. A *C. tussilaginis* larva, on the other hand, hatches when the seeds have started to develop after fertilization. It immediately eats out the embryo and endosperm of one almost full-sized seed, but without killing it. As a second instar it then occupies the position of the consumed embryo, and proceeds to divert to itself the nutrients which would have normally been used to form the cotyledon food reserves of the seed.

The larvae of these two flies therefore depend for their early growth upon the maximum inflow of concentrated soluble nutients to the flowerheads. However, each exploits one of the two different periods of physiological activity when this happens: flowering and maturation of the seed. Furthermore, the behaviour of female and larva of both species is finely tuned to ensure that each gets the resource it is evolved to use. So, what would seem from superficial inspection to be a close sharing of a common resource is not. Both species are using the same nutrients flowing to the same flowerheads, but at different times. Their worlds are quite separate.

There are, then, two important points to remember about competition. First, although there is no question that it is a real and potent cause of the direction in which many organisms evolve, it is not *necessary*. The inadequate environment will relentlessly select out all which cannot cope, or are just plain unlucky, without them having to interact, directly or indirectly, with others from their own or another species. Secondly, if we think about numbers and survival in terms of competition this deflects our attention from what actually decides the numbers – the capacity of the habitat to support them. Competition comes about *because* of the inadequacy of the environment, and it is this inadequacy, not the competition, which decides how many organisms can exist in any particular habitat at any particular time.

1.5 Self-regulation Does Not Exist

Other processes said to be operating to limit the abundance of many animals in nature are the so called self-regulating mechanisms (Chitty 1960; Wynne-Edwards 1962). Instead of individuals in a population using up all the available resource in the habitat, and in the process competing for a share of it, it is said that self-regulating mechanisms must function to conserve some of that resource for the next generation. To be self-regulating individuals must therefore possess some genetic or learned behaviour or other attribute which restrains them from using what is otherwise freely available. For replicating life, evolving in a finite world, this is clearly impossible. It is inconceivable that any mechanism could survive if it *reduced* the contribution of an individual's genes to the next generation below the level permitted by full use of the resources available to that individual at the time. To achieve such an outcome individuals would have to reduce their rate of reproduction – to practise birth control. Genes for any such trait would inexorably be eliminated from the gene pool, leaving only the descendants of those which used *all* the resource they could get – including that saved by the "altruists". *Homo sapiens* is the only species which can foresee the need to limit its rate of reproduction to within the bounds of the resources in its environment, and so far it has not shown any signs of doing so. So why do we imagine that other species, lacking our prescience, should be able to do so?

Another sort of self-regulation is that said to be generated by changes in the proportion of "inferior" and "superior" genotypes in a population. Similarly, this does not stand up when approached from the basic thesis of organisms evolving under the pressure of natural selection in a finite and variably inadequate environment. Numbers in a population will increase if the environment should become more favourable. The lessened harshness would mean that selection pressures were less. So, not only would more individuals survive, but also a wider range of genotypes. When the environment again became harsher, fewer of a narrower range of genotypes would survive – but not necessarily the original range. It is therefore misleading to talk of superior and inferior genotypes, and to further assume that a change in their proportion will change the numbers in a population. Any particular genotype will be more or less well adapted at different times as the environment changes. But a change in the proportion of genotypes does not influence the numbers in a population. Changes in both the number of individuals and the range of genotypes are decided by changes in the quality of the environment – by the extent of its inadequacy.

Ideas about self-regulating mechanisms have gained acceptance because ecologists have found evidence that predators and parasites are not able to limit the abundance of their prey, and they rarely *see* any evidence of a shortage of food or other resource for the prey, or of them starving. Nor have they found any other factor in the environment limiting their increase. Many think in terms of positive, not negative selection. It *must* be an advantage to leave some food for your children (unless your neighbour eats what you have saved, and he *will* do so!).

While most now reject ideas involving concepts which run contrary to the process of evolution, many still see territorial and social behaviours functioning to restrict access to resources at a level which prevents the resources being exhausted. But these behaviours, too, have nothing to do with conserving resources for the next generation, nor of restricting the current generation's use of them. They are *driven* by shortages of resources, not preventing them. Their importance lies in their influence upon the way in which the resources that *are* available are used – by deciding which of the many gain enough of them. I discuss these behaviours in more detail in Chapter 11.

Chapter 2 Plants as Food for Herbivores

2.1 Why is the World Green?

Why are the often densely growing plants on Earth not all destroyed by herbivorous animals? We know that at least some herbivores have the capacity to do so, for occasionally we see this happen in nature. And we know only too well that many can and would destroy our cultivated plants if we did not kill them first. Hairston et al. (1960) are frequently quoted as providing the explanation for the world being green – for why the great bulk of plants die and decay without being eaten by herbivores. These authors concluded that decomposers are limited by their food (otherwise fossil fuel would accumulate rapidly), plants are limited by their resources (as they are not all eaten up by the herbivores), herbivores are *not* limited by their food (because they are only seen to eat all their food plants when, rarely, they are "protected" by "man" or "natural events"). This left the authors with only "the vagaries of the weather" or "predators" to limit the number of herbivores. The hypothesis of direct control of herbivores by weather led them to "false or untenable implications", so they were left with only predators to limit the abundance of herbivores. From this they said it followed that because predators are controlling their prey they must, (by their own actions) be limited by their food resource. So, in their analysis terrestrial decomposers, plants, and carnivores are all limited by their food, and expand towards the limits it sets. Only herbivores are limited below the level of their available food by predators. That is why we rarely see other than relatively minor and trivial damage to plants by herbivores.

Murdoch (1966) and then Ehrlich and Birch (1967) extensively criticize Hairston et al., showing their thesis to be built on faulty premises and logic. But 20 years on, Hairston (1981) remained convinced of the validity of these conclusions. Today many ecologists, although not all (e.g. Crawley 1989), still accept them as axiomatic when seeking to understand what it is that limits the abundance of herbivores. They assume that there is always a surplus of unused food in the environment of a herbivore. Some (Fretwell 1987) have extended their deductive logic to build more elaborate explanatory models of ecological interactions.

If, however, the reason herbivores do not eat all of their food plants is that these plants can support but few herbivores because they are nutritionally inadequate; if, even in the absence of predators, few neonates that feed on plants survive, then all the deductive reasoning of Hairston and his colleagues falls down. We would be left with *all* living things, at *all* trophic levels, limited by their supply of food. This I believe to be the case, and I hope that the evidence I have gathered together in this book will convince others that it is so.

2.2 How Might Plants Be an Inadequate Source of Food?

The possible strategies which plants might adopt in order to avoid being eaten by herbivores can be summarized as follows:

1. Plants are nutritious, acceptable, non-toxic food for herbivores, but *are scattered, cryptic and rare* in the environment.
2. Plants are abundant, easy to find, nutritious food, but *distasteful or poisonous* to herbivores.
3. Plants are abundant, easy to find, acceptable, and non-toxic, but *inadequate food* for herbivores. There are two ways in which plants may achieve this. Either (a) because their tissues *lack* one or more *essential nutrients*, or (b) because one or more of these essential *nutrients* in their tissues are *too dilute.*

2.3 How and When Might Nutrients in Plants Be Too Dilute?

This dilution could take two forms:

1. In *volume.* The ratio of one or more nutrients to non-nutrients is too low in the tissues which the herbivore must eat. As I outlined in the previous chapter, of those elements which are essential to the evolved biochemistry of life on Earth, nitrogen, although abundant, is mostly present in forms which are not available to living things. It is the most likely candidate for the role of an essential resource which will first become limiting to the survival, growth and reproduction of present-day organisms, including plants. One adaptation of plants to the shortage of nitrogen in their environment is that their tissues are composed largely of structural carbohydrates. The relative amount of nitrogen in plants is therefore low, often very low, diluted by these carbohydrates. Much of this already limited supply is further locked away from herbivores in the form of unpalatable, indigestible, or poisonous substances. Dilution, then, is such that herbivores, ingesting and processing the maximum volume of plant material in the minimum possible time, still do not get enough nitrogen for growth and reproduction – indeed often not even enough for daily maintenance.
2. In *time.* The tissues are generally nutritious enough, but either are not available for long enough or cannot be broken down quickly enough in relation to when, and at what rate, the herbivore needs their contained nutrients for maintenance and growth.

However, plants too have their Achilles heel. They too must grow and reproduce. To do so they must concentrate a flow of nutrients into new, growing tissues, especially into reproductive structures. At such times there is a high concentration of soluble, easily digestible nitrogen readily available to herbivores chewing or sucking the newly formed cells, or the conducting tissues leading to them. However, plants mostly grow in short bursts, so they are adequate food for herbivores for only a limited time. Once growth stops, and seeds are formed, nutrients are again locked away in a form which cannot be absorbed and digested by the herbivores without a considerable expenditure of time and energy to again break them down.

Furthermore, plants conserve their limited supply of nutrients by translocating them from senescing tissues either to new growth or to storage. So, a similar situation develops. Nitrogen in a form which both sucking and chewing herbivores can readily absorb again becomes available in the tissues which are senescing, and in the conducting tissues leading from them; but, again, for only a limited time and usually in less concentrated form than the flow to new growth.

There is one other way in which more of the small amount of nitrogen in plants may become available to herbivores. That is when an internal physiological strain (a deviation from normal plant function) is induced in the tissues of plants, or parts of plants, in response to unpredictable stresses in the environment, such as a shortage of water in the soil (Welden and Slauson 1986; Lecowicz 1987). The basic effect of such strains is to hasten the senescence of the plant tissues. The potential benefit of this to herbivores has long been recognized (Fennah 1953; Kennedy 1958). I have discussed the phenomenon before (White 1969, 1984) and will deal with it again in Chapter 18.

2.4 When Is a Minimum Supply of Nitrogen Critically Important?

When it is needed by the herbivore to produce the next generation. The breeding females and the fast-growing neonates of *all* animals make the greatest demands for food which is rich in nitrogen for the synthesis of new protein. Nowhere is this more apparent than in herbivorous animals. For most of the time their food contains barely enough nitrogen in a form which they can process fast enough for the maintenance of existing individuals, let alone for the production and growth of new individuals.

2.5 How Might Herbivores Counter the Plants' Evolved Strategies?

Herbivores have evolved a seemingly endless variety of life cycles, behaviours, and physiologies in the face of the relentless selection pressure imposed by a relative shortage of nitrogenous food. All these adaptations are geared to help individual herbivores compensate wherever possible for the normal inadequacy of nitrogen in their food, and to take maximum advantage of any increase in its concentration or availability, no matter how ephemeral that might be. All are adaptations which help the individual maximize the *effective* use of this limited resource towards producing as many viable young as it can – towards contributing as much genetic material as possible to the next generation.

I said I could see three main ways in which plants might have evolved to lessen the prospects of being eaten. They might become harder to find, become distasteful or poisonous, or become an inadequate source of nutrition.

First, the food plants of an animal are widely scattered and/or concealed in the habitat. To human eyes individual plants are both rare and difficult to find; they are "unapparent" (Feeny 1976). However, natural selection will have eliminated all but those individual animals which produce the greatest number of most efficiently dispersing offspring with the best searching behaviour. Only the survivors, those that find their food plants, will reproduce. For most individuals of such species the consequences are grim; they never find a food plant and soon die. But for the population it is good. Most of the time most suitable food plants in the habitat are found by at least one individual. So, considered from the point of view of both the herbivores and their food plants, there is no such thing as an unapparent plant. As Crawley (1983) says, specialist herbivores of rare species of plants are *bound* to be good at finding their hosts.

There are many good illustrations in the literature of just *how* good they are. Andrewartha's oft-told anecdote of each year finding carrot seedlings hidden beneath a rank growth of weeds in his neglected garden, yet all heavily infested with the carrot aphid (*Cavariella aegopodii*), provides a vivid picture of this efficiency. So did my own experience of discovering how psyllids, by means of passive dispersal of their gravid females, can locate and colonize small and inconspicuous wisps of *Eucalyptus* foliage many kilometres from the nearest source of infestation (White 1970a, 1973).

Each individual herbivore that does find a plant has the capacity to quickly multiply and produce enough progeny to eat all the plant – to efficiently use all the resource in the habitat. Sometimes they do just that. To survive, the plant species must therefore be able to disperse more widely and faster than the herbivore. Usually, however, while numbers of the herbivore might flourish for a while, very little of the tissue of the plants that are found is destroyed. Why is this so?.

Secondly, plants may contain chemicals which repel or poison herbivores, or which reduce the efficiency with which they can absorb nitrogen from their food. Alternatively they may produce physical barriers against attack: thorns, thick cuticle, hard seed coats. In many cases direct attack by a herbivore will induce a concentration of such defences (Schultz 1988; Karban and Myers 1989).

The development of strategies to circumvent the effects of these defences has been a powerful driving force in the evolution of large numbers of species, especially among the insects. In the process many species have evolved the capacity to detoxify, sequester, or avoid particular chemicals, or overcome physical barriers produced by the plants. As a consequence those which have succeeded are usually specialists confined to feeding upon plants of one or only a few closely related species.

In other words, the plants have been successful in limiting the number of *species* that are able to feed upon them, but not how many *individuals* of an adapted species can survive by feeding upon such "protected" tissues. At best they may slow their development or reduce their fecundity. Mostly, however, adapted individuals thrive upon even those tissues with the highest levels of secondary compounds, provided that they are also high in nutrients. Then, having neutralized the deterrent, adapted herbivores have the potential to eat out their food plants, but they mostly do not. Why not?.

Thirdly, a plant may be an inadequate source of nutrition. There is, of course, *no* counter to the absence of any chemical which is essential to the evolved biochemistry of the growth and maintenance of an animal's body. Any plant lacking such a chemical would be immune from attack by all herbivores. In the light of the common origin of all life, however, it is not surprising that herbivores have metabolisms which, by and large, need only those chemicals which are also essential to the metabolism of the plants they eat.

Nor is there any counter to an essential chemical being below the minimum concentration (in volume or time) required for maintenance, growth, or reproduction of individual herbivores. It would seem that most plants, for most of the time, are likely to have a level of available nitrogen in their tissues which is below this minimum concentration. Such a situation is potentially insurmountable. How could a counter evolve? Only a quite different biochemistry could cope. Yet, while they rarely destroy all or even most of their food plants, herbivores persist and flourish in the world; so they must have found at least partial counters to this dilution of their food.

2.6 How General Is Dilution of Nitrogen in Herbivore Food, and What Adaptations Have Evolved to Counter it?

The best way to answer these questions is to review some of the many studies which reveal how universally the lives of herbivores are dominated by the need to gain access to enough nitrogen. A good part of this book is devoted to doing this. First, however, I want to very briefly outline the answer to the second question. To do so will, I think, enhance comprehension of the common threads which run through the accounts which follow.

The ways in which herbivores have evolved to try to cope with this universal nitrogen hunger may be classified under six broad strategies: .

A. The stages of the herbivores' life cycle when food high in readily digestible nitrogen is most vital to them (breeding by females and growth of the very young) are synchronized with a time when there is a high probability that the plants will provide such food. Such synchronization is, however, often imperfect.

B. The herbivores concentrate or prolong, in various ways, the availability of the supply of soluble nitrogen in their food plants. However, a critical minimum concentration must be exceeded at the appropriate time, and for long enough, if an individual is to breed or survive.

C. The herbivores eat more food more quickly, and become more efficient at extracting, absorbing and digesting the limited nitrogen in the food they do ingest. But for every individual there is a limit to the maximum bulk of food that it can process in the time available, and, again, a critical minimum of nitrogen that must be extracted from that food.

D. The herbivores enlist the help of microorganisms to break down otherwise indigestible components of their food and/or synthesize essential chemicals that they are unable to synthesize for themselves. And they use microorganisms directly as readily digestible protein. These microorganisms may operate upon the plant material either externally, or as endo- or ecto-cellular symbionts within the gut of a herbivore.

E. The herbivores supplement the limiting nitrogen in their plant food by eating other animals or their microbial "helpers". Individuals of many species of herbivore depend throughout their life on such supplementation, but virtually all neonate herbivores, in addition to adopting one or more of the above strategies, either eat animal protein or have it supplied to them by their mothers.

F. The herbivores apportion and concentrate the limited nitrogen that is available in the environment to a selected few individuals at the expense of the many. They do this by various forms of social behaviour.

I have simply listed these six strategies here, without examples or explanation, because I give both for the first five of them constantly and repeatedly throughout most of the rest of the book. Strategy F – territorial and social behaviours – is more economically considered in detail as a discrete topic. I do this in Chapter 11 after briefly recapitulating and reviewing all six in Chapter 10.

Many of the tactics incorporated in these six strategies are adopted alike by vertebrate and invertebrate, young and old, male and female. To simply survive in the short term – to maintain their metabolism in positive protein balance – they must do so, if not constantly, then at least at times when nitrogenous food is especially scarce. What is so revealing, however, is not just that these various adaptations increase nitrogen in the diet, but nitrogen in a quickly digestible form, and specifically in the diets of females when they are breeding and nurturing young, and in those of the neonates.

Yet, in spite of all these adaptations, the chance of any individual herbivore getting enough nitrogenous food in its first few hours or days of life is *still* very small. The typical type III survival curve of Deevey (1947), wherein most of the young of all species die at this time, bears witness to this, as does the universally large "biotic" or "reproductive potential" of all organisms. Each female that survives to breed produces a great "surplus" of young in her lifetime – surplus in the sense of the number of new individuals needed to replace her and her mate. But of course it is no surplus at all. It is the minimum necessary to keep the population extant in an inadequate environment, to keep it pressing hard against the limit of available food in the environment. That populations of herbivores *are* so pressing because individual herbivores suffer a chronic shortage of nitrogen in their diet is well illustrated by occasions when this lack of good food for the young is eased. Then their numbers quickly and explosively increase. How this may happen is the subject of the last part of this book.

Part II: Herbivores in an Inadequate Environment

Chapter 3 Insects

There are very many more species of insect which are herbivores than there are of any other taxonomic group of animals (there are very many more *insects* than there are all other sorts of animals). What is more there are species of insects specialized to feed on every conceivable species of plant; and on all different parts of plants. Any attempt to systematically review this entire vast array would be at best unwieldy, and ultimately tedious. However, there is no need for such extremes. The evidence for the generality of a relative shortage of nitrogen in the food of herbivorous insects is everywhere apparent. I can best illustrate this generality, and the adaptations that have evolved to alleviate the shortage, by taking examples from broad groupings of species according to what they eat and how they eat it.

All herbivorous insects are, to a greater or lesser degree, selective in their feeding. Many will eat the tissues of only one species of plant, some may range across a genus, or perhaps even one or more families of plants, but few have a truly catholic diet. Within these taxonomic constraints on what they will eat there are further specializations. Phytophagous insects may eat only seeds, developing reproductive tissues, leaves, wood, or detritus other than dead wood. Some chew and swallow pieces of a plant; others avoid the need to process a plant's structural tissues by sucking out the contents of its cells. Some chewers have become specialized to mine inside a plant's tissues where many – at least in their early instars – can function as "suckers" feeding entirely on cell sap. Others have induced plants to form galls and thus prolonged the availability of nutritious tissues for them to eat. Most will eat only the actively growing new tissues of a plant (or the liquid nutrients being translocated to that new growth). A smaller number of species have evolved to feed only on senescing tissues (or the nutrients being translocated from those tissues). Some manage to feed at different times in their life cycle on both growing and senescing tissues.

There seems little point in dwelling on specialization to feed on specific taxonomic groups of plants (although otherwise generalist feeders often show quite strong preferences for a group of plants, like the legumes, which are especially rich in nitrogen). But it will be worth taking a separate look at some of those that eat only certain tissues, e.g. wood or cell sap. First, however, I would like to emphasize the difference between those insects (both sap suckers and leaf chewers) which, on the one hand, have become specialized to eat newly growing tissues and, on the other, senescing tissues. Then I will look at other sorts of specializations like the leaf-miners, the gall makers, those that feed by either sucking sap or chewing the tissues of living plants, and finish with those which manage to subsist on wood.

3.1 Flush and Senescence Feeders

The difference between a flush and a senescence feeder is not simply that a flush feeder cannot eat the older growth on the same plant because of allelochemicals, or toughness, or because it does not contain enough nutrients; nor that a senescence feeder would feed on nearby new growth containing higher levels of available nitrogen were it not for allelochemicals preventing it from doing so (vide Feeny 1970; Ikeda et al. 1977). Rather, each is adapted to a different way of life, one dictated by whatever happened to allow more of its ancestors to survive best when eating that one sort of tissue. Today the tissues eaten by either one are not, in any significant way, part of the environment of the other (Sect. 1.4).

So, it does not matter to either sort of animal what happens in those other tissues, or how abundant or rare they are. And there is no point in demonstrating that, for example, a flush feeder will not thrive on old leaves. What is important to today's animals is how much nitrogen there is in the food they are "programmed" to eat, and for how long. The flush feeder is adapted to survive on a great concentration of readily assimilated nitrogen available for only a short time; the senescence feeder to a lesser amount available over a much longer period.

As a result there are fundamental, and often far-reaching differences between the ecologies of the two sorts of animals, especially in the way each will respond to changes in its food plant. Recognizing the distinction between the two may help in unravelling our understanding and experimental testing of ideas about basic relationships between (at least insect) herbivores and their food plants (Price 1991a, b, c). It may also aid our attempts to devise ways of combating those which are pests of our crops.

For example, I have previously discussed the confused and often contradictory responses of herbivorous insects to the application of fertilizer to their food plants (White 1984). Knowing whether an insect is a flush or a senescence feeder may lead to a better understanding of such responses.

The two sorts of animals are likely to respond (in terms of survival and growth) in opposite ways when nitrogen fertilizer is applied to a food plant that is deficient in nitrogen. The flush feeder would respond positively to the increased level of nitrogen being transported to the growing tissues, while the senescence feeder would respond negatively to the decreased export of nitrogen from senescing tissues (Sects. 3.5.1, 3.5.4 and 3.4.7).

If, on the other hand, the level of nitrogen in the plant was not deficient and the amount available to the flush feeder was adequate, then the flush feeder would respond little or not at all to increased nitrogen, unless this reached deterrent or toxic levels, when it would respond negatively. The senescence feeder would also not be likely to respond, unless the aging process was slowed, or the export of nitrogen was further decreased. Then it too would respond negatively.

Then there is the matter of how insect herbivores respond to food plants of different ages (e.g. Sect. 3.1.10) and to physiological strains induced in plants by external stresses such as poor site, drought, poor drainage, etc. (Chap. 18).

A flush feeder may thrive on a vigorous, growing young plant, but do less well on an aging one. A senescence feeder's response would likely be the reverse, faring poorly on a young plant but thriving on an old one.

A flush feeder may not respond at all to its food plant being stressed, unless this became so severe that growth was restricted or stopped. But a senescence feeder is likely to respond positively to even quite mild stress hastening the aging process and increasing the total amount of nitrogen exported. However, severe stress would again be detrimental.

Furthermore, there are likely to be many and varied responses by both sorts of insect to interactions of age and stress over time, and to variables associated with the frequency and extent of flowering of the plants (White 1984).

Most herbivorous animals, not just insects, are flush feeders seeking out new growth at every opportunity. They experience considerable variation in the success of their breeding, and in the survival of their young, as the amount, timing, and quality of the new growth of their food plants varies. Most of the animals I discuss in this book fall into this category.

Senescence feeders, on the other hand, seem to be less common. Most seem to be insects (see especially Sect. 3.2). Some are mites and a few may be found in other taxonomic groups (e.g. Sect. 6.2.2.4). I suppose detritivores could be thought of as extreme senescence feeders, but the plant is dead before they start to feed. Different adaptations are needed: on the one hand to extract nutrients from a living plant and on the other to digest dead tissues. The former requires adapting to (and often modifying) the plant's processes of transporting nutrients away from dying tissues. The latter requires access to microorganisms which break down the dead tissues left after the soluble nutrients have been removed. Browsing animals like the equids (horses and zebras), which feed on dead herbage, need to adopt the tactics of the detritivores rather than those of the senescence feeders.

The examples which follow are intended to illustrate how the two quite different adaptations of flush and senescence feeding occur across otherwise widely differing life histories. Some are sap suckers, others leaf chewers. There are others, which I discuss separately, that have evolved the capacity to benefit from feeding on both flushing and senescing tissue, either opportunistically (Sects. 3.4.7, 3.5.1 and 3.5.4) or at different stages of their life cycle (Sects. 3.2.1, 3.3.4 and 3.4.3). There are still others that cannot easily be classified. How, for example, do we classify bark beetles (Sect. 3.7.5)? Do they benefit from the flow of nutrients in the phloem both to and from storage in the roots, or does the fact that they increase to outbreak levels on old and/or drought-stressed trees indicate that they are solely senescence feeders?

Fig. 1. A A flush feeder. Mature *Glycaspis* on *Eucalyptus*. Nymphs feed on *inflowing* nutrients in main veins of *young* leaves. None of the leaf is killed. *Dark areas* are liquid honeydew. **B** A senescence feeder. Mature *Cardiaspina* on *Eucalyptus*. Nymphs feed *between* main veins on *outflowing* nutrient from *mature* leaves, enriching this outflow by hastening the senescence of surrounding tissues. *White areas* are dead; *dark areas* about to die. Remainder of the leaf stays green and healthy. Photos T.C.R. White

3.1.1 Two Australian Psyllids on Eucalyptus

I can perhaps best illustrate the differences between the two sorts of animals – and their indifference to each other's worlds – from some of my own experiences. A species of psyllid (Homoptera), *Cardiaspina densitexta*, lives on its specific host *Eucalyptus fasciculosa* in South Australia. From time to time this insect erupts into enormous numbers, defoliating trees over large areas of land. My job was to find out why (White 1969, 1970a, b). At the beginning of my study I needed to establish experimental populations and naively presumed the animals would do best on vigorous new growth, so I grew special plants for

Fig. 1B

them and caged females on the luxuriant new foliage. Nothing would induce them to accept these leaves. Repeatedly, all died without laying a single egg. I then tried releasing newly hatched nymphs on new leaves. After a few half-hearted probes they wandered aimlessly until they died. But either adults or first instars would immediately settle and feed on fully expanded mature leaves on any tree. The "best" most nutritious leaves, on the other hand, were just not a resource which they recognized in their world. This (I now know) is true for other species of the genus. There are many of them. Most are specific to one species of *Eucalyptus*, but will feed only on its mature leaves. They are senescence feeders.

Most *Eucalyptus* species also have one or more host-specific members of another genus of psyllid (*Glycaspis*) living on them, but these will lay their eggs and feed only on soft expanding leaves. They are flush feeders.

As members of the same subfamily living on the same genus of plant, these two psyllid genera have many features of their biology and behaviour in common. However, they also show very distinct differences associated with their very different ways of life. To illustrate these differences I have used as examples two species, *C. albitextura* and *G. brimblecombei*, which both live on the common and widespread red gum *E. camaldulensis* (White 1971). The newly hatched nymphs of both quickly settle and begin to feed. They insert their stylets through a stoma and into the phloem. They both build a cover over themselves, called a lerp, which is made from their own solidified faeces. Normally they will not move from the site where they settle, completing five instars before transforming into a winged adult. Usually great numbers of them die as first instars soon after they have built their tiny lerp. They remain with their mouthparts still inserted. They show no signs of physical damage or disease. Clearly they have fed, but nevertheless starved while eating an abundant but lethally deficient food (White 1970b). Both species at times do become enormously abundant, but mostly are quite scarce. There the similarities end.

The females of the senescence-feeding *Cardiaspina* will lay their eggs only on mature leaves, and the nymphs will settle only on those same leaves. *Glycaspis* adults and nymphs, on the other hand, will accept only soft new growth (there is one exception, see below). *Cardiaspina* nymphs settle on the leaf between the veins and feed from the fine ultimate phloem vessels. *Glycaspis* nymphs settle along the mid-rib and main veins and feed from the major phloem vessels in those veins. The faeces of *Cardiaspina* are sparse and consist of nothing but starch which solidifies on contact with the air to form the lerp, a tough, flat covering permanently fixed to the leaf's surface (White 1972). The nymph adds to this as it grows. The faeces of *Glycaspis* are voluminous, mostly liquid honeydew, and the lerp is soft, hygroscopic, easily and often detached, and quickly renewed by the nymph. Around the site where a *Cardiaspina* nymph feeds the leaf gradually turns yellow, then red, and soon after the adult has emerged and flown away it dies, leaving a brown patch of dead tissue in an otherwise green leaf. Attacked leaves may live for many more months and will often receive eggs from the newly emerged adults, and support the next generation of nymphs to maturity. If young nymphs are removed soon after they start to feed there is no discolouration, but after the third instar deterioration is rapid and irreversible. If two or more nymphs settle near each other the dead areas will coalesce. When there are very many nymphs on one leaf the whole leaf is killed, often before any nymphs reach adulthood. Then all the insects die too. Other than at such extreme densities, the hastened senescence of the leaf is clearly of benefit to the feeding nymphs. I reared nymphs at different densities on discs cut from freshly picked leaves floated on distilled water in the laboratory. The more nymphs there were feeding near it the more quickly a nymph developed, and its chance of surviving was much enhanced if it had neighbours feeding nearby. In the case of *Glycaspis* there is no sign of any damage to the leaf where the nymphs have fed. *Cardiaspina* nymphs grow slowly and the species completes three discrete generations each year: in spring, summer, and autumn. Nymphs of *Glycaspis* grow much more quickly and there may be numerous overlapping generations, with all stages present in the time that new growth is available.

So, *Cardiaspina*, the senescence feeder, grows slowly while feeding on the nutrients which are gradually released from the slowly senescing leaf. In addition, it hastens the rate of senescence of the tissues immediately surrounding its feeding site, so increasing the rate of delivery of nutrients into the phloem where it feeds. *Cardiaspina* is not just intercepting the normal flow of nutrients, it is inducing the plant to release them more quickly than it would if the insect were not there. The flush-feeding *Glycaspis*, on the other hand, grows quickly by simply intercepting a superabundant inflow of highly concentrated food along the main delivery channels to the growing leaf. It has no need to hurry this process. Rather, it has a fast, flexible life cycle with which to quickly capitalize on a flush of food whenever it occurs. I mentioned above that there is one exception to *Glycaspis* being confined to soft new growth: it is when its food plant is stressed by a shortage of water. Then nymphs will settle along the veins of mature leaves and grow to adults. While they lack *Cardiaspina*'s capacity to hasten the senescence of leaves, they appear to be able to take advantage of the increased flow of nutrients when it is provided for them by this hastened senescence.

3.1.2 Two African Scale Insects on Californian Ice Plants

The difference between individuals of a "pair" of closely related species, one a flush and the other a senescence feeder, has often gone unremarked, or not been recognized as the probable basis for unexplained differences in the responses of two species to the same treatment. So, it seems, was the case with two African scale insects which had become pests on the introduced ice plants, *Carpobrotus* spp., in California (Washburn and Frankie 1985; Washburn et al. 1987). Individuals of these two scale insects are almost identical morphologically, and they settle and feed together on the same plants. One, however, *Pulvinariella mesembryanthemi*, is a flush feeder, settling preferentially on young leaves and passing through two generations a year, with peaks in abundance in spring and autumn. The other, *Pulvinaria delottoi*, is a senescence feeder. It has but one generation each year and its crawlers, which are produced earlier in the spring than those of *P. mesembryanthemi*, settle mostly on the large older leaves. The rate of growth of *P. delottoi* is only half that of *P. mesembryanthemi*. Of the first instar crawlers of both species, 90% die after they have settled and started to feed. Many are found dead with their stylets still inserted in the leaf, indicating that, like *C. densitexta* (Sect.1.1), they starved while feeding on an abundant but inadequate food. When experimental ice plants were fertilized or given extra water the soluble nitrogen content and turgor of their leaves increased. *P. mesembryanthemi* responded to both treatments with increased passage of sap through the gut, faster growth, and increased fecundity; *P. delottoi* did not.

The difference in the settling preferences and speed of development of the two species suggests that they differ in their capacity to use the same food. The authors could not find any evidence of their feeding from other than the same vascular tissues, and speculated that they may have different endosymbionts. It seems more probable, however, that they are feeding on quite different foods. The sap being exported from the older leaves would provide a different balance and volume of nitrogenous elements, and flow more slowly, than that flowing to new leaves. Furthermore, as has been found with studies of aphids (Sect. 3.5.1), the application of fertilizers is more likely to make it a poorer source of nutrients.

3.1.3 Two Aphids on Scots Pine

Another interesting case of a pair of species, one a flush and one a senescence feeder, on the one host is that of the grey pine aphid, *Schizolachnus pineti*, and the spotted pine aphid, *Eulachnus agilis*, feeding on the needles of Scots pine, *Pinus sylvestris*, in Britain (Kidd et al. 1985). Both aphids have similar life histories and frequently feed on the same needles, but they differ in important ways. The grey aphid feeds in closely packed groups on the outer curved surfaces of current and previous year's needles, rarely on older needles, and never on senescing ones. The spotted aphid feeds on the flat inner surface of needles of much the same age. But they do not aggregate, with two exceptions. First, if there are any yellowing, senescing leaves they strongly prefer to feed on these (up to eight times as many aphids on senescing as on mature needles). Second, in the presence of the grey aphid, they aggregate on the flat surface of a needle *exactly opposite* the *S. pineti* colony on the curved surface.

The authors measured the spotted aphids' success when reared either in isolation or in the presence of grey aphids on otherwise similar needles. More of them survived and they grew more quickly in the presence of the grey aphids. Presumably the grey *S. pineti* is a flush feeder with the ability to induce its host to continue to import nutrients to the needle upon which it is feeding as if that needle were still growing. Colonies of *S. pineti* are acting as nutrient sinks, as do many other species of aphids. *E. agilis*, on the other hand, is clearly a senescence feeder, yet has the capacity to detect and take advantage of the grey aphids' sink.

3.1.4 The Green Spruce Aphid in Scotland

A senescence feeder which shows many similarities to *C. densitexta* in its feeding behaviour and effects upon its host is the green spruce aphid, *Elatobium abietinum*. Parry's (1969, 1971, 1974a, b, 1976, 1979) study of this insect in Scotland revealed how it is firmly "locked in" to the way of life of the senescence feeder, and how the changing availability of soluble nitrogen in its food plant dictates its survival and capacity to breed. In Britain *E. abietinum* causes damage to plantations of both the European Norway spruce, *Picea abies*, and the North American sitka spruce, *P. sitchensis*. It has been especially damaging to the latter; there have been several outbreaks on young trees in Scotland. Each year an upsurge in its numbers started in early autumn and continued to a peak in the following summer. These surges in abundance were characterized by a great increase in the proportion of first instar nymphs in the population. Thereafter, the numbers steadily declined until, by late summer, the animals became quite rare and widely scattered. In years when their numbers became very high they damaged many needles so badly that most of them were shed prematurely. In such years the late summer decline became a sudden crash, often leaving few or no survivors.

Parry soon discovered that predators, parasites, emigration of winged aphids, and cold temperatures had little, if anything, to do with either long-term or annual changes in the abundance of *E. abietinum*. Shortage of food resulting from changes in the quality of the phloem sap seemed a more likely cause.

The aphids settle and feed preferentially upon mature needles. They insert their stylets through the stoma to the phloem. Around each feeding site a yellow spot develops which soon increases to a yellow band. Eventually the whole needle becomes chlorotic. Each aphid induces an irreversible and hastened senescence of the tissues surrounding the phloem from which it draws its food. On Norway spruce this process takes about 12 to 18 days; on sitka about half that time. In the autumn survivors from the late-summer trough in abundance begin to colonize mature needles which flushed in the spring. These insects thrive and multiply to a new peak in the following summer. Then the needles are some 18 months old and becoming progrcssivcly less suitable as a source of food. But the new spring needles are still unacccptable. So, during the middle to late summer, there are few needles which provide an adequate diet for the aphids. Their numbers drop precipitously; the survivors lose weight and do not breed. As autumn approaches, however, the previously unacceptable current year's needles have matured to the point where they are acceptable, and colonization and a renewed build-up in numbers starts.

When Parry examined the level of nitrogen in the needles as they aged he found that new flushing needles contained a very high level of soluble nitrogen, but once the needles were fully expanded this fell rapidly and remained at a low level until late summer. It then rose again (although not to anywhere near the initial level of the new flush) and stayed at higher levels from autumn until the following summer; higher than levels in the current year's needles.

So, the behaviour (and success) of the aphids was closely following the changing pattern of availability of soluble amino acids in the needles. The aphids colonized the needles as soon as the slow process of senescence started breaking down and exporting nutrients from the needles. Additionally, however, they boosted this process by hastening the senescence of the tissues around their feeding sites. But there comes a time when the nitrogen in these needles is exhausted and proteolysis has not yet got under way in the new season's ones. Persistance of the population of the aphids at this time each year seems to depend upon a few individuals (possibly those containing higher than usual fat reserves) chancing upon a few trees or shoots which may still contain higher levels of amino acids than do the majority.

Parry found evidence in the field and in the laboratory which further illustrated this sequence of events, and the dependence of the aphids upon the availability of an adequate supply of soluble nitrogen. When numbers in the trees were so high as to cause premature fall of most of the older needles upon which the aphids were feeding, they moved to the normally unacceptable needles of the current season, but all clustered on the oldest of these needles at the base of the shoots. Most of them nevertheless died, and did so while still in their feeding positions; they had starved to death. He also noticed that instead of producing the typical yellow lesions as they do when feeding on mature needles, they caused the whole needle to go yellow and flaccid and then drop off. Subsequently, aphids which he forced to feed on current needles in the laboratory starved to death at the same rate as unfed aphids – except when he added amino acids to the sap by placing cut stems bearing new needles in solutions containing these amino acids. One summer he discovered a single tree which continued to support high numbers of aphids for 5 weeks longer than the surrounding trees. On analysis he found that the needles of this tree contained 32% more soluble nitrogen than did those of neighbouring trees upon which most aphids had died.

Twenty years later Fisher (1987) experimented with *E. abietinum* in Wales to look at the influence of the chlorosis of the needles around the feeding site upon the aphids themselves. He fed individual aphids on the needles of sitka and Norway spruce in an outside insectary in winter. Half of them he reared from birth to adult on the same needle. The other half he moved after their third moult so that for their last nymphal instar they had to feed on a previously uninfested needle. On both species of spruce those aphids left on the same needle throughout their development were heavier and more fecund than those shifted to new needles. The effect was stronger on the introduced sitka spruce.

In a separate experiment Fisher then caged one branch of each of ten sitka spruce trees and infested them with spruce aphids. He similarly caged, but kept uninfested, one other branch on each of these trees, and one branch on ten other trees which he also kept completely free of aphids. After 30 days he analysed the content of nine essential and nine non-essential amino acids in the needles of all these branches. He found that while the total amount of amino acids was a little higher in the needles from the infested branches, this was not statistically significant. There were, however, significant differences in the concentration of individual amino acids. Some (mostly essential ones) had increased, while some had decreased, relative to those in the two lots of uninfested needles (which did not differ from each other). He concluded that the only effect of feeding by the green spruce aphid was to alter the balance of amino acids in the needles and that this was beneficial to the aphids. In fact, his published figures show that the *total* non-essential amino acids had decreased by only 6% compared with the uninfested needles on the same tree, while the total *essential* amino acids had *increased* by 59%; clearly a considerable improvement in the aphids' food. These results of Fisher's again illustrated how, by hastening the senescence of the tissues surrounding the site where they feed, these senescence-feeding aphids improve the quality of their food. They have induced the plant to release more nitrogen in a soluble form than it would have done if left alone.

3.1.5 Aphids on Sycamore, Apple, Wheat, and Alfalfa

Another aphid, *Periphyllus acericola*, which lives in aggregations and causes the tissues around the feeding site to turn yellow, lives on the undersides of leaves of sycamore, *Acer pseudoplatanus*. When Shearer (1976) confined individuals of this species, or of another aphid species, *Drepanosiphum platanoidis*, on the opposite side of a leaf to one of these aggregations, they grew up to two and a half times heavier than when caged on leaves not infested with *P. acericola*. And in the field in early autumn *D. platanoidis* often gather to feed beside *P. acericola* nymphs, especially when yellowing of the leaf begins around the site of aggregation.

Similarly, Forrest (1971) had found that nymphs of *Aphis fabae*, when fed on apple leaves galled by the apple aphid, *Dysaphis devecta*, grew bigger and more quickly than when fed on ungalled leaves (see Sect. 3.3.2). Furthermore, the presence of a *D. devecta* gall on a nitrogen-deprived plant, which will not support the growth of *A. fabae*, reversed this situation. Such plants supported better growth of *A. fabae* than ungalled control plants. Finally, he found that the honeydew of *A. fabae* nymphs feeding on galled tissue was considerably richer in amino acids, indicating that the phloem from which they were feeding also contained a better supply of these essential nutrients.

A study in Oklahoma of two biotypes of the same species, *Schizaphis graminum*, which lives on winter wheat, *Triticum aestivum*, showed very clearly the added benefits which accrue to a senescence feeder which can hurry along the breakdown of the tissues of the plant (Dorschner et al. 1987). The wheat variety Amigo is resistant to biotype C but not to the virulent biotype E of *S. graminum*. When individual type E aphids were fed on mature leaves of Amigo they caused yellowing and eventual necrosis of the tissues around their feeding sites, and a great increase in the level of free amino acids in these tissues. Biotype C individuals fed on the same leaves did not do this. The biotype E aphids grew faster and bigger, and produced more progeny than did the equivalent type C individuals. Also, their bodies and honeydew contained twice as much nitrogen as those of the biotype C aphids. However, when the avirulent type C aphids were fed on leaf tissue that type E's had previously fed upon, they responded as if they were feeding on a susceptible variety of wheat. They too grew faster and bigger and produced more young. However, other type C individuals, feeding where others of their own biotype had previously fed, did not respond in this way. The virulent type Es are hastening senescence of the plant's tissues (but only immmediately around the site of their feeding; the remainder of the leaf stays green and healthy), and thrives on the resulting enriched diet. Type C individuals, on the other hand, are unable to do this. They thrive only when, like *D. platanoidis* at aggregations of *P. acericola*, and *A. fabae* on galls of *D. devecta*, they are able to feed where others which *can* hasten senescence have already boosted the supply of soluble nitrogen in the tissues for them.

The introduced scourge of alfalfa (lucerne) *Medicago sativa* in North America, and more recently in Australasia, the spotted alfalfa aphid, *Therioaphis maculata*, is a classic senescence feeder. It settles preferentially on older, lower leaves, and causes chlorosis to spread from the feeding site. Aphids grow and reproduce faster than on younger leaves, but do not achieve maximum rates of reproduction until discolouration appears. Recognizing that it is a senescence feeder, Dillwith et al. (1991) at Oklahama State University, have directed their studies towards elucidating the biochemical response of the host plant to the aphids' feeding. They have found that feeding induces many changes in the leaf, including the reduction of protein and fatty acids and increased production of ethylene, which closely resemble those of natural senescence.

3.1.6 Scale Insects on *Euphorbia* and *Euonymus*

A very clearcut demonstration of the dependence of a senescence feeder upon the actual outflow of nitrogenous food from senescing tissues was provided by the study Newbery (1980) did of the coccid *Icerya seychellarum* living on *Euphorbia pyrifolia* on Aldabra atoll. *I. seychellarum* is a polyphagous phloem-feeding coccid which was unintentionally introduced to the atoll in the mid 1960s. It quickly spread, infesting a wide range of hosts. One of these, *E. pyrifolia*, is a deciduous tree which produces leaves continuously through the wet season, but is dormant and leafless during the dry season.

Newbery established that the insects strongly prefer to feed from senescing leaves of *E. pyrifolia*. Infestation was five times greater on senescing than on green leaves, and the trees with the heaviest infestations were those with a much higher ratio of senescing leaf area. When contrasted with leaves kept free of coccids, he found that feeding by these insects increased the level of soluble nitrogen in the senescing leaves. They apparently act as a "sink", as do many other sap-sucking insects. His measurements showed that the success of these insects depended not upon the concentration of soluble nitrogen in those leaves, but upon the amount of nitrogen being *exported* from them and the time for which the flow was *available*. As long as leaves were senescing, coccids were able to grow.

Cockfield et al. (1987) investigated the yellowing that the armoured scale insect, *Unaspis euonymi*, causes when feeding upon its native food plant, the widely distributed ornamental evergreen vine *Euonymus fortunei*. Crawlers settle on either side of a leaf and insert their stylets into the phloem sieve elements of the vascular tissues. As they grow a chlorotic halo develops around the feeding site because the chloroplasts in the pallisade parenchyma cells break down as they do during senescence. This degrades the proteins of the chloroplasts, which comprise a sizeable amount of the protein in a leaf, releasing amino acids into the cells and thus to the phloem, providing a large dietary supplement of nitrogen to the feeding insects.

3.1.7 A Leafhopper with Alternate Generations on Brambles and Oak

The mesophyll-feeding leafhopper *Lindbergina aurovittata* provides another, but somewhat different, example of a senescence feeder. In Wales it completes two generations a year, the first on evergreen brambles (*Rubus* spp.), and the second on a variety of deciduous trees, principally oak, *Quercus robur* (Claridge and Wilson 1978). On brambles the nymphs feed exclusively on old leaves formed in the previous growing season, and their feeding produces extensive yellow patches on these leaves. Even when crowded as final instars they will not move to younger leaves of the current season's growth. When experimentally placed on such young leaves nymphs of all instars quickly moved back to old ones. Yet, when these same nymphs were given a choice – the mature leaves of several late-summer tree hosts or those of bramble – they showed no clearcut preference for bramble. Their preference was always for older leaves, overriding any taxonomic preferences. On the other hand, summer females arising from these nymphs would lay their eggs only on leaves of the trees, not brambles. The females arising at the end of the summer from nymphs of the second generation, however, reject the leaves of the deciduous trees upon which they were raised and return to bramble to lay overwintering eggs.

Throughout its life cycle the animal is a senescence feeder. It tracks a wide range of mature leaves through the summer and then returns to the evergreen primary host. There it lays eggs from which nymphs will hatch next spring to feed upon the leaves which have matured over the winter.

Adaptation to feeding upon either flush or senescing tissues is not confined to insects which suck the sap of plants, however. The great majority of leaf-chewing insects, like the geometrid winter moth *Operophthera brumata* on oak, *Quercus robur*, in England (Varley and Gradwell 1970), are flush feeders. Their life histories are well synchronized to place newly hatched first instars on flush new growth. Old foliage is not part of their environment.

However, some are senescence feeders. A good example is that of another geometrid, *Stathmorrhopa aphotista*, which feeds, occasionally in very high numbers, on eucalypts in Tasmania (Elliott et al. 1980). The moths do not lay their eggs until late autumn and the larvae feed throughout the winter. Pupation is in late spring but the adults do not emerge until the following autumn.

An example of a flush and a senescence chewer on the same host is that of the leaf-tying larvae of two moths from Brazil, one in the family Gelachiidae, and the other in the Oecophoridae (Macedo and Langenheim 1989). Both eat the leaves of the legume *Copaifera langsdorfii*. The former consumes young expanding leaves and grows rapidly to pupate within 3 weeks. The oecophorid eats only fully expanded mature leaves and grows slowly through the rainy season.

3.1.8 Two Species of Caterpillars Eating Oak Leaves

The work of Lawson et al. (1984) illustrated some important and basic differences between flush and senescence feeding leaf chewers. These authors worked in Michigan with the caterpillars of two species of Lepidoptera which live on the leaves of various American species of *Quercus*; *Alsophila pometaria*, a geometrid which feeds on new leaves in spring, and *Anisota senatoria*, a citheroniid which eats the late-season leaves. In the laboratory Lawson and his colleagues fed penultimate instar larvae of both species on the appropriate-aged leaves of six different species of oak, and measured the efficiency with which the caterpillars of each ate and assimilated their respective foods. The young oak foliage contained more nitrogen and less fibre than did the older leaves, and there were some differences between the two in terms of their content of allelochemicals, but the capacity of their total tannin constituents to precipitate dietary proteins was the same.

Both insects have very similar digestive systems, well designed to digest protein, and both appear to be most influenced by the nitrogen content of their diet. On the other hand, both are poorly equipped to exploit the energy and carbon content of the structural polysaccharides of plant tissue. In neither species was their digestion and assimilation of protein significantly influenced by the presence of tannins in the foliage; the high alkalinity of their guts (pH 9.5 to 10.5) tends to dissociate protein-tannin complexes. Nor did the water content of the foliage have any effect upon the efficiency of their feeding.

Yet there were important differences. *A. pometaria* eats all the foliage at roughly the same rate, and grows faster on leaves with higher levels of nitrogen in them. By contrast *A. senatoria* eats leaves of low nitrogen content more quickly than those with more nitrogen in them, but grows at approximately the same rate on either. The flush feeding *A. pometaria* grows faster, and has a much faster rate of passage of food through its gut, than does the senescence feeding *A. senatoria*. The flush feeder has adopted a strategy of a high rate of consumption to achieve a high rate of growth on food rich in nitrogen, albeit at some loss of efficiency, while the senescence feeder relies on a low rate of consumption and long residence time of food in the gut to maximize the digestion of the scarce protein in its food. One favours rate over efficiency; the other efficiency over rate. The end point is that both are equally good at converting ingested nitrogen (*A. senatoria* was, in fact, even better on two species of oak). The higher growth rate of the flush-feeding *A. pomateria* is due entirely to its higher rate of consumption.

This was all revealed by the performance of caterpillars which were near the end of their feeding life. They had successfully survived the much more hazardous early period of establishment as newly hatched first instars. Just how much more vital an adequate supply of nitrogen is at this early stage is illustrated by the behaviour of the gravid females of *A. senatoria*, the senescence feeder. They lay their eggs on the tips of branches where the hatching first instars will have the youngest (and therefore highest nitrogen) leaves to feed upon. Their first instars are flush feeders. Only in later instars do the gregarious caterpillars become truly senescence feeders when they move down the branch consuming all leaves of various ages (Lawson et al. 1982). In this respect they are like many leaf-miners (Sect. 3.2.1) and aphids (Sect. 3.5.1).

So, both species make the best they can of what they have got, the life style forced upon them by natural selection. Each gets one chance (generation) per yearly cycle of its food plant to convert as much of it as it can into the next generation of moths and caterpillars. Both are limited in how successful they are at doing this by the amount of nitrogen that is available to them: how long it is available for the flush feeder and how concentrated it is for the senescence feeder. But neither is more "successful" than the other. Both persist in the environment and convert all the resource that they can into the next generation.

3.1.9 Two Species of Sawflies Mining in Birch Leaves

Fenusa pusilla and *Messa nana* (Tenthredinidae), both introduced to North America from Europe, mine the leaves of native birchs. In Ontario, Canada, DeClerck and Shorthouse (1985) found that both species occur simultaneously in the leaves of white birch, *Betula papyrifera*. Females of both can be seen laying their eggs on the same tree, and the larvae of both mine in and feed upon the palisade cells of the leaves.

There the similarity ends. *F. pusilla* will lay its eggs in only new, soft, expanding leaves, and *M. nana* in only fully expanded mature leaves. *F. pusilla* larvae grow through five instars in only 8 to 13 days. The fourth instar eats spongy mesophyl as well as palisade cells so that by the end of the feeding period nothing is left of the leaf but the epidermis and veins. It completes several generations in a season, new females ovipositing each time in new leaves on the tip of the growing long shoots of the birch. The larvae of *M. nana*, on the other hand, grow much more slowly (approximately 36 days to complete larval development), never eat anything but palisade cells, and at maturity a larva has eaten less than a quarter of the leaf. It completes but one generation each year. *F. pusilla* is a flush feeder, *M. nana* a senescence feeder.

3.1.10 A Chewer and a Sucker on Poplar Leaves

An interesting variation of the flush/senescence feeder story, and one which contrasts a sap sucker and a leaf chewer on the same food plant, is provided by Kearsley and Whitham (1989). They had found that on narrow-leaf cottonwood, *Populus angustifolia*, in Utah the galls of the aphid, *Pemphigus betae*, are 70 times more common on the leaves of mature trees than they are on those of juvenile trees. With transfer experiments using different aged trees from the same clones, they showed that the survival of the "stem mother" aphids which form the galls was 50% higher on mature than on juvenile trees.

The larvae of the beetle *Chrysomela confluens*, which also feeds on the leaves of *P. angustifolia*, were, by contrast, 400 times more abundant on juvenile than on mature trees. Again, using transfer experiments, they demonstrated that the survival of chrysomelids was 50% lower on mature leaves, and they took 25% longer to reach adulthood than they did on juvenile trees.

Now, while *P. betae* stem mothers attack expanding leaves in spring, they and their progeny feed on nutrients being exported from these leaves once they have matured. They are senescence feeders. In contrast, the larvae of *C. confluens* are flush feeders. They skeletonize newly developing leaves of the poplar (mostly the late leaves of this heterophyllous species). However, the leaves on individual trees grow, mature, and senesce each year. The point of Kearsley and Whitham's study, however, was that the insects' preferences are for the age of the *tree*. They were not thinking about insects being flush or senescence feeders. They noted that young trees produced many more late-flushed leaves than do the mature trees, and that these leaves contained more water and less fibre than the late leaves on mature trees. So, they said, there is more food for the chrysomelid larvae on the young trees, and it should be easier to eat.

But what of the quality of the nutrients in these leaves? Shouldn't there be much more soluble nitrogen available to the larvae if there are many more young expanding leaves to eat? Also, what about the aphids feeding on the early leaves which are senescing? Do the leaves of older trees senesce more quickly? Or do they export more nutrients over a given time than those of equivalent age on juvenile trees? Or is the answer simply as follows?

Most leaves on older trees flush early and would senesce sooner than leaves on young trees. Most leaves on young trees flush late and would be importing nutrients after most leaves on older trees have stopped growing. The aphids have evolved to feed on nutrients exported from senescing tissues, and the chrysomelids to eat nutrients imported to growing tissues. Individuals of each species will seek out those tissues to which they are adapted, and on these tissues more of them will survive and grow better.

There are many such unanswered questions and possible explanations about species which feed on either flushing or senescing tissues. The examples I have given should make it clear, however, that these *are* two very different ways of life, and that recognizing the differences is important to a proper understanding of the ecology of a species, whatever other specializations it may have developed [as we shall see with the switch from flush to senescence feeding in

Fig. 2. A Leaf beetle larvae, flush feeders on poplar leaves. More survive, and grow faster on leaves of young trees. **B** Aphid galls, senescence feeders on leaves of the same species of poplar. They are more successful on older trees. Photos courtesy of T. G. Whitham

some leaf miners (Sect. 3.2) and gall formers (Sect. 3.3.4), or the more unusual reverse switch by the caterpillars of some butterflies (Sect. 3.4.3)].

But I now want to discuss how some of these other specializations, like these two ways of life, also lead to better access to readily assimilable nitrogen for the insects' reproduction and growth.

3.2 Leaf-miners

By living as larvae inside the tissues of the plant, and selectively feeding on only some of these tissues, leaf-miners avoid having to ingest structural carbohydrates and most potential allelochemicals. The vulnerable early instars of many species avoid these substances altogether by feeding entirely within the sap of selected cells (Hering 1951). However, like their free-living relatives, they are still susceptible to quite minor changes in the concentration of nitrogen in their food.

3.2.1 The Switch from Flush to Senescence Feeding

Very many leaf miners which differ widely in taxonomy, host plant, and geographical location, seem to have adapted to senescence feeding. Their larvae are growing (albeit more slowly) in ageing leaves. They are dependent upon the slow release of soluble nitrogen over a long period, rather than upon a rapid flood of it in a short time.

However, most have it both ways. They feed on flush tissue as early instars and on senescing tissues in their later instars. The switch from one to the other seems to be associated with the change from a diet of cell sap to one of chewed up tissues. This change of feeding mode is accompanied by a distinct change in the morphology of the head and mouthparts of the larvae: from a head held forward parallel to the line of the body, to one at right angles to the body and with downward biting jaws.

Lithocolletis quercus, which mines the leaves of the evergreen oak *Quercus calliprinos* in Israel, is a good example of this switch (Auerbach and Simberloff 1989). Female moths lay their eggs in freshly flushed leaves in spring. The larvae emerge and immediately start to mine the undersides of the leaves. For the first three instars they feed only one cell deep, consuming a wholly liquid diet of cell sap. The larger fourth and fifth instars deepen the mine, chewing up whole cells. But it takes 10.5 months to reach pupation.

The leaves are very low in nitrogen (less than 2% dry weight) and the authors suggest that this very long development time may be associated with nutritional factors. Yet most of the larvae that die (other than those lost when leaves absciss prematurely) do so as early instars, while still undisturbed in their mines. This suggests – as does the finding of dead psyllids and coccids with their mouthparts still inserted in the phloem (Sects. 3.1.1 and 3.1.2) –

that these larvae starve. Access to adequate nitrogen is more critical for these neonates than it is for slower-growing late instars.

West (1985) investigated 14 species of Lepidoptera which mine the leaves of *Quercus robur* in Oxford's Wytham wood and discovered, as he expressed it, how very close they are to "their basic nutritional threshold". In the first two instars 67% of their larvae died when they were feeding in the sap of epidermal cells, and still within their mines, apparently unharmed. When he cut off the top quarter of a series of leaves, each with a mine of a first instar larva in it, two things happened relative to a similar group of undamaged leaves. The nitrogen content of the damaged leaves dropped by as little as 0.25%, but the number of early instar larvae found dead in their mines in these same leaves increased by 100%. They are, indeed, on a nutritional knife edge when such a small change in the amount of nitrogen in their food means the difference between life and death.

The pergid sawfly, *Phylacteophaga froggatti*, which mines the upper epidermal and palisade cells of the leaves of *Eucalyptus* species in Australia, again illustrates this life style which combines prolonged development in senescing leaves with the need for high levels of nitrogen for the young. It selects only mature leaves, but prefers the juvenile form and epicormic leaves which, apart from having shapes quite distinct from leaves of adult growth, also have a higher nitrogen content (Farrell and New 1980).

3.2.2 Leaf-miners Which Induce "Green Islands" in Leaves

Some leaf miners which have become *very* late senescence feeders (Faeth 1985; Kahn and Cornell 1983; West 1985) have evolved another interesting adaptation. They have had to prolong the period when an adequate flow of highly nutritious sap is still available to them, and have done so by inducing the tissues immediately surrounding their mine to continue actively photosynthesizing after the remainder of the leaf has died. Plant (1984) has suggested that the creation of these "green islands" is an intermediate state between leaf-mining and galling. But it is more likely the reverse. A galler is a flush feeder that prolongs the period of flush growth by creating a gall. It may subsequently extend the availability of good food by inducing that gall to senesce more quickly than the other tissues immediately surrounding it (Sect. 3.3.4). Late leaf miners, on the other hand, are obligate senescence feeders which run the risk of their food supply running out as natural senescence proceeds. They do not hasten that process; they delay it. They induce the leaf to continue active production of nutrients for longer than it would have done in the absence of the miner.

3.2.3 A New Zealand Weevil Mining in Fallen Beech Leaves

An extreme case of senescence feeding, apparently from the first instar, is provided by the larvae of the small leaf-mining weevil *Neomycta pulicaris*. These larvae feed in the leaves of hard beech, *Nothofagus truncata*, in the North Island of New Zealand. The sexually immature adult weevils, which overwinter under dry moss, become active just before the spring flush of new leaves on the beech. They feed on the expanding leaves for 2 or 3 weeks until the leaves have hardened, so they are flush feeders. Only then do they start to lay their eggs in the mid-rib close to the base of each leaf. Attacked leaves, most of which contain only a single egg, are shed by the tree within a few days. Eggs laid early in the spring take up to 4 weeks to hatch; those laid late in the spring take only 8 days. The larvae complete their development in the fallen leaves in about 3 weeks, by which time one mine occupies most of the blade of the leaf. Prepupal larvae move to the litter and pupate in the soil (Hosking and Hutcheson 1986). Apparently this is not a unique case. Hering (1951) reported a similar leaf-mining weevil, *Rhynchites pauxillus*, which induces early abscission by ovipositing in the leaf petiole, the larva completing its entire life cycle in the leaf on the ground.

3.2.4 The American Holly Leaf-miner

A nice study, conducted in a cemetery in Lexington, Kentucky, revealed the native American holly leaf miner, *Phytomyza ilicicola*, as a good general example (Potter and Kimmerer 1986; Kimmerer and Potter 1987). Its larvae live in the leaves of the native holly, *Ilex opaca*, a tree much used as an ornamental in parks, gardens, and cemeteries. The tiny (2 mm long) flies (they are true Diptera) emerge in the spring. The females seek out the soft, new, expanding leaves of the holly, sting them with their ovipositors, and feed on the sap that exudes from the wounds. They also lay their eggs in these developing leaves. All this must be achieved in a brief 2 or 3 weeks, because after that the leaves are fully expanded and very tough.

Larvae hatch from the eggs in May and June and start feeding within the leaf, but their rate of development is notably slow. Not until late October – early November, when winter is fast setting in, do they ecdyse to second instars. And the winter is over by the time they are fully grown third instars and about to pupate in March. In another month the adults emerge to start the next generation on the new flush of leaves. The old leaves, meanwhile, mined or not, and now completely senesced, fall from the tree.

The second noteworthy thing about these larvae is that they are very selective feeders. They feed only in the palisade mesophyll of the upper surface of leaves, and usually only of sun leaves. Furthermore, they eat only the middle layer of cells of this palisade tissue (it is usually three to four cells thick in sun leaves, but only two cells thick in shade leaves). Even as relatively large third instar larvae, when their bodies much expand the mine, they still eat only these middle cells. Not only are the larvae much more abundant in sun leaves, they grow more quickly and survive better in them. (This is reflected in the fact that ornamental holly trees, usually grown in full sunlight, are frequently heavily

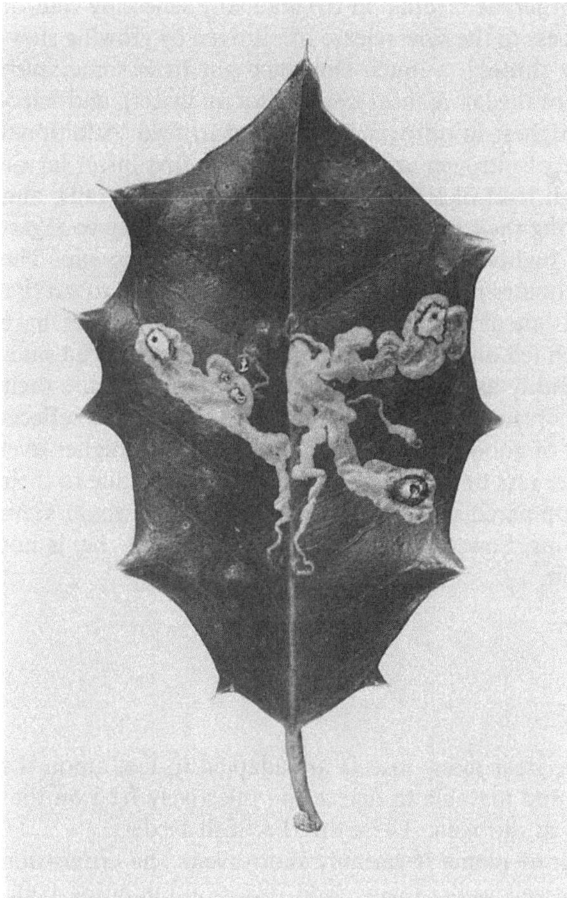

Fig. 3. Leaf miners of American holly hatch in summer in fully-expanded leaves, and feed slowly through winter on senescing tissues, but eat only middle palisade cells with the most nitrogen. Photo courtesy of William Mesner

damaged by the miner, whereas in the forest, where trees grow naturally under partial shade, they are little affected.)

When the holly leaves first expand they have very high levels of both soluble proteins and saponins (the latter are regarded as fairly nasty allelochemicals, possibly functioning as both toxins and inhibitors of digestion). Both protein and saponins decline very quickly once the leaves mature and harden. Shortly after the new larvae hatch and start to feed the soluble protein in the leaves has dropped to 1.2 to 1.5% dry weight, and remains low for the life of the leaves. But when the authors looked specifically at the tissues which the larvae eat – the upper mesophyll cells – they found that these cells had more than twice as much protein as the rest of the leaf, 66% more water, and more than nine times as much saponins.

So the picture emerges of larvae feeding on the gradually senescing cells of the leaf, increasing their access to the slow release of nitrogen by growing slowly and passing food slowly through a much elongated gut (it is some seven times longer than the body of the larva, most unusual for an insect), and selecting the tissues which are highest in nitrogen from which to feed. Additional pointers to the limiting role of nitrogen are the fact that the first instar larvae start feeding *before* the high level of soluble nitrogen in the leaves falls; and that the females, before laying their eggs, feed on the sap of the very youngest leaves which have the very highest levels of soluble protein in their sap. The former suggests that the neonates need access to abundant nitrogen to survive through their early establishment and growth; the latter that females have developed from larvae which fed on a diet of minimum nitrogen, so need similar access to a source of readily assimilable nitrogen in order to mature their eggs. Perhaps, too, the higher survival and faster growth on sun leaves reflects not just the greater amount of good tissue they contain, but both a higher level of total nitrogen and a faster rate of its release as soluble nitrogen due to their faster rate of senescence compared with shade leaves. Finally, it would seem that the presence of saponins, however potentially nasty they may be, is not bothering the larvae one bit.

3.3 Gall Makers

As we have already seen, a great many insects are adapted to feed upon the soft new growth of plants and are able to detect and selectively feed on that part of it which is highest in nitrogen. These are the flush feeders.

However, all new growth of plants is variably short-lived. The proportion of fibrous and indigestible cell walls, toxins, and digestion-inhibiting compounds, increases; that of nutrients, especially nitrogen, decreases. The supply of good food is temporary. A counter to this would be to stimulate the plant to continue to grow new tissue, to enhance the world of the flush feeder and to prolong access to an adequate diet. Many vertebrate animals do this by repeatedly grazing the same plants e.g. geese, ungulates, hares, and turtles. They are discussed in later sections. By removing the new growth they stimulate the plant to mobilize more of its reserves into new tissue. Many animals are too small or too few to do this. Some have achieved the same ends, however, by inducing their food plants to form a gall. The greatest diversity of gall makers has probably occurred in the Arthropoda, especially the insects. Gall makers are found in six orders of insects and include both chewing and sap-sucking forms.

"At the simplest level a gall provides the food for the gall-maker to feed" (Abrahamson and Weis 1987). But it goes further than this. A gall maker is not simply an "interceptor" or a "redirector" of nutrients; it is both these things *and* a "prolonger". The plant is forced to continue to produce new tissue long after its normal flush of growth has finished. To do so it must continue to import high-nitrogen sap to this new tissue. The gall acts as a "nutritional sink" and the supply of good food is prolonged (Larson and Whitham 1991).

3.3.1 Physiological Galls

There are varying degrees of this tactic. Some fall short of producing an actual gall. Thus, Kloft (1957) demonstrated what he so aptly termed "physiological galls" in the phloem of the bark of the European silver fir, *Abies alba*, where two species of *Adelges* fed. At these sites soluble nutrients, especially amino acids and peptides, were found to be concentrated in slightly enlarged parenchyma cells exhibiting hypertrophy of their nuclei. Also, at these sites the development of the adelgids was accelerated. Balch (1952) had found the same phenomenon with *Adelges piceae* feeding on Canadian balsam fir, *Abies balsamea*.

The European beech scale, *Cryptococcus fagisuga*, also feeds in the cells of the bark parenchyma of both *Fagus sylvatica* and the American *F. grandifolia* (Wainhouse et al. 1988). It too forms a zone of proliferating parenchyma cells – an "internal gall" – near to the site of the end of the feeding stylets. This is considered to be a zone of enhanced nutrition probably induced by growth hormones secreted by the scale insect. That this is a site of better nutrition was supported by the experimental demonstration that newly hatched crawlers settled preferentially on sites on the bark previously occupied by scales rather than on surrounding previously uninfested bark. Yet, in spite of this capacity to improve the supply of nitrogen in their food, gaining access to adequate food is still a chancy business. This is shown by the fact that 85 to 90% of crawlers do not survive when they first attempt to settle and feed. Wainhouse and Gate (1988) found that this high mortality was not the result of physical interference or damage by the elements; nor of attack by predators (they have no known parasitoids). But Wargo (1988) showed that nitrogen nutrition is important to the success of *C. fagisuga* attacking American beech. Some trees are more susceptible to attack by the scale than others, and Wargo showed that the more susceptible trees had higher levels of amino acids in their bark. He further showed that these higher levels of amino acids were not an effect of feeding by the scales, but the basis of the difference in suitability of the bark as food for the settling larvae. The level of phenols in the bark appeared to play no part in the degree of infestation by this scale insect. The difference was not due to any effect of the scale, but to differences in susceptibility of the trees. Again, larger trees, which are more susceptible to infection, had higher concentrations of amino acids. Clearly the ability to concentrate nitrogen at the feeding site does not solve all the scale's nutritional problems. It has a better chance if the tree upon which it settles provides a higher level of nitrogen to start with.

3.3.2 Nutritional Benefits of Galling

Forrest (1971) demonstrated not only the nutritional benefit of galling but that this benefit need not be confined to the galler (Sect. 3.1.5). He fed free-living *Aphis fabae* on young apple plants infected with the "pseudo galls" (they are really only leaf rolls) of *Dysaphis devecta*. Compared with those on ungalled plants these *A. fabae* grew faster and bigger, and maintained their larger size over four generations. The controls diminished markedly in size each generation and did not survive beyond the third generation. They suffered the same fate on apple plants experimentally deprived of nitrogen. However, such nitrogen-deficient plants with *D. devecta* galls on them were as nutritionally adequate for *A. fabae* as the galled control plants, and supported better growth of aphids than ungalled control plants. In a separate experiment, the honeydew of *A. fabae* feeding on galled plants was much richer in amino acids than that of aphids feeding on ungalled plants, another reflection of the increased amino acids in the sap of the former plants.

Nor is the nutritional benefit confined to the galled leaves. Aphids feeding on the stems responded in the same way as those feeding on the leaves, so the nutrients must be translocated from the galled leaves to other parts of the plants.

Packham (1982) showed that shoots of the grass *Holcus mollis* infested with the "galls" of the aphid *Holcaphis holci* resembled younger vegetative tillers rather than uninfested shoots of the same age. They were less lignified with more vascular bundles, younger leaves, and much higher levels of soluble nitrogen in all tissues. As she nicely expressed it, the aphids altered the phenology of the plant and lived in a perpetual "spring flush" of nitrogen.

This nutritional benefit of galling is fairly obvious for sap suckers, and there are very many studies which demonstrate this benefit. Sap suckers are directly "plugged in" to the nutritious sap which must continue to flow in order to supply the prolonged growth of new plant tissue. But it is equally true for chewing insects which form galls. They feed on proliferating layers of nutritive cells growing all around them and containing high concentrations of nutrients (Shorthouse 1986). Alternatively, they may be the equivalent of sucking insects, feeding on solutes transferred to them through a few specialized cells from the vascular tissues (Rohfritsch 1987). As with sucking insects, however, it is not necessary to produce a conventional gall in order to induce the plant to proliferate these nutritive cells. Larvae of the New Zealand hepialid moth, *Aenetus virescens*, provide a good example (Grehan 1983). Having spent their early instars in the forest litter feeding on fungi, they then ecdyse to a morphologically and behaviourally distinct "transfer instar". In this form they climb the stem of a living tree, bore a hole in it, and cover the entrance with a webbing of silk and wood chips. But they do not eat the wood. They remove bark from around the entrance to their burrows. This stimulates the bark phloem into a "wound response", proliferating callous tissue around the entrance to the burrow. On this renewing callous tissue the transfer and all subsequent instars graze, just as chewing gall formers graze the nutritive cells lining their gall.

Weis and Kapelinski (1984) demonstrated experimentally that chewing gall formers similarly induce the plant to import additional nutrients from elsewhere in the plant. They worked with the gall midge *Rhabdophaga strobiloides* which forms "pine cone" galls on tips of new growth of *Salix chordata* in Michigan. Twigs bearing a gall achieved a greater diameter than those not galled, in spite of galling causing the early cessation of the production of leaves on the twig. This suggested to them that the presence of the gall caused the importation of nutrients from elsewhere in the plant. They tested the hypothesis by stripping leaves from a series of carefully matched twigs, with and without galls, and compared the subsequent growth of galls and twigs on these stripped twigs and on a matched series of unstripped control twigs. Galls continued to grow, and grew equally well, whether leaves were left intact or removed from the twigs on which they grew (or stripped from the two most adjacent twigs as well). Removal of leaves reduced the diameter growth of both galled and normal twigs, but the effect was slight, and not statistically significant, on the galled twigs. Most striking, however, galling had such a great effect upon the growth of the twigs that the defoliated and galled twigs grew significantly larger than did the fully leaved normal twigs. Clearly, the result of galling is to draw more nutrients from other parts of the plant. Finally, they found that the largest galls grew on the twigs with the greatest diameter, and contained the largest larvae. So, assuming vigour and fecundity of adults are positively influenced by the final size of the larvae, those insects which are best at stimulating the most vigorous growth of their host plant will have the greatest reproductive potential.

Another clear example of this redirection of nutrients came from a study by Abrahamson and McCrea (1986). They set out to investigate the cost to goldenrod, *Solidago altissima*, of galling by a lepidopteran and a dipteran. In their words "...the most striking feature of this study is the marked magnification of nutrient levels going up to the gall maker". They found that nitrogen and phosphorus were concentrated in the galls, in the gall makers, and in their parasitoids and inquilines; but with the concentrations of nitrogen "somewhat more pronounced" than that of phosphorus.

3.3.3 Adaptive Nature of Galls Debated Anew

In spite of what would seem to be overwhelming evidence that the adaptive advantage of galling is nutritional, Peter Price and his colleagues have seen fit to, in their words, "revive the old debate on the adaptive nature of galls" by proposing six hypotheses to explain galling (Price et al. 1987a). They were quickly able to reduce this to three; the "nutrition", "microenvironment", and "enemy" hypotheses. With the last they appear to be on weak ground. Galls are very apparent structures and their occupants, which cannot escape, are, anyhow, as often plagued by predators as their free-living relatives (Hawkins 1988).

It is true that many present-day galls may well provide protection from the elements, especially desiccation in a very dry environment. There are certainly a great variety and abundance of galls in such places, but there are plenty of galls in wet places too. Most larvae living in a gall could not now survive without the protection it affords them. This does not, however, imply that protection in their own microenvironment was the initial advantage which allowed ancestral "proto-gall makers" to escape elimination by natural selection. (An analogy would be to apply similar arguments to present-day cave dwellers.)

However, there would seem to be no exception to the nutritional benefit of galling. Price and his colleagues concede that there is most support for this hypothesis, but with the qualification that the evidence is largely correlational and the nutritional requirements of the galler are largely unknown. It seems unlikely that gall makers have dietary requirements different from those of other herbivores. And, long ago, Kennedy (1951) demonstrated that the major benefit of galling was nutritional rather than protective. He was able to evert the pocket galls formed on the leaves of red-currant by the feeding of the aphid *Cryptomyzus ribis* so that the inner surface of the pocket where the aphids feed was fully exposed and facing up rather than down. When he did this the aphids still preferred to feed on this surface in spite of the loss of any benefit of shelter or shade, and the reversal of their normal orientation to gravity. Furthermore, they grew faster than others which were fed on ungalled leaves (see Sects. 3.1.5 and 3.3.2).

Yet the results of some studies have lead their authors to reject the concept of a nutrient sink. For example, Hawkins and Unruh (1988) did not find there was more soluble protein in or near cecidomyid galls on a Californian species of *Atriplex* than elsewhere in the plant, or that galled plants were different from ungalled ones. Because they found that gall tissue had more water than ungalled stems, and because plants growing near water had a much greater density of galls than plants growing out in the desert, they concluded that, in this arid environment, water was more important than nutrition. But the leaves growing from the galls had considerably less soluble protein in them than did leaves from elsewhere on the galled plants, or leaves on ungalled plants. There was no difference in the water content of different leaves. Also, they had not included the larvae inhabiting the gall in their analyses.

Brewer et al. (1987a) collected mature galls of five species of cecidomyids; four on pine needles and one on nettle leaves. After removing all contained larvae they measured the amount of various chemicals in these galls and in ungalled tissues from the same plants. Many were not significantly different between the two types of tissue, but all galls had less nitrogen than the ungalled tissues. The authors suggested that this was because much of the nitrogen from the galls had already been incorporated into the larvae. Had they included the larvae in their analyses they would have found more nitrogen in the galls. I think this likely, as it would have been in the previous example. Also, had they collected and analysed the galls in the early stages of growth they may have obtained a very different result.

3.3.4 Double-Dipping: Prolonged Growth Plus Hastened Senescence

There is another side to the nutritional advantage of gall making. Some gall makers have tapped the best of both the flush and senescing worlds. First they prolong the import of nutrients to their feeding site by inducing a "continuing juvenility" in the surrounding tissues, leading to the growth of a gall. Then, once the gall is fully formed, they hasten its senescence relative to the rest of the plant. The gall, at first lush and green, begins to discolour, and eventually dies, while surrounding tissues remain green and alive. The galler has induced the release of a further supply of high-nitrogen sap, and continues to feed until shortly before the death of the gall. Often such galls dehisce to release the adult insect.

Whitham (1978) demonstrated this nicely for *Pemphigus populivenae* galling leaves of *Populus fremonti*. The "stem mother" aphid settles when the leaf first starts to expand, forming a gall in which to feed and reproduce. Then the part of the leaf distal to the gall begins to yellow. With galls at the base of the leaf the entire leaf becomes chlorotic; with galls placed more distally the area of leaf that yellows is less. As Whitham says, *P. populivenae* has telescoped the two discontinuous periods of the life of a leaf known to be most favourable to aphids, because of the increased availability of nitrogen (flush and senes-

Fig. 4. "Double dipping". A *Phylloxera* gall on a pecan leaf imports nutrients as it grows, then it and nearby tissues senesce more quickly than the leaf, providing a second supply of food to the aphids within. Photo courtesy of P. C. Andersen

cence), into one. It does this by causing the leaf to advance senescence from autumn to early summer prior to the gall dehiscing and releasing aphids to migrate to their second host.

The story has proved to be the same for *Pemphigus betae* feeding on *Populus angustifolia* (Larson and Whitham 1991). The aphids act as strong "sinks": the colonizing female for storage reserves from the stem of the plant and her progeny in the gall for the resources exported from the galled and neighbouring leaves.

Another example of this double dipping is the study by Andersen and Mizell (1987) of the effects of galling by *Phylloxera notabilis* on the physiology of the leaves of 3-year-old pecan trees, *Carya illinoensis*. The galls are found on the leaf between main veins on each side of the midrib. They mature in 4 or 5 weeks, cracking open to release the aphids. The authors measured the content of 13 elements (including nitrogen, phosphorus and potassium) and the concentration of chlorophyll in mature galls, but before they dehisced. They made the same measurements for leaf tissue immediately surrounding the galls, and for tissue on ungalled control leaves. They also compared the rates of gas exchange of galled and nearby tissues on the same leaf, and of ungalled control leaves. Compared with the controls, galls contained less of all the elements measured (60% less for nitrogen). A ring of tissue 1.5 mm in diameter around galls contained significantly more nitrogen than the galls, but significantly less than tissue 1.5 to 3 cm around galls. This outer ring did not differ from tissue from the control leaves. Similarly, galls had only 4.0% of the chlorophyll of the control leaves; the 1.5 cm ring had 90%. The story was the same for gas exchange. Galled and nearby tissue showed great reductions compared to controls. The authors found no sign of visible necrosis associated with galls. They had, nevertheless, measured the near-end result of the insects hastening the senescence of the galls, and the tissues immediately surrounding the galls, draining the nutrients to their own "sinks".

Furthermore, as we saw with the importing of nutrients to a feeding site (and as I have discussed in earlier sections), it is not necessary to form a gall in order to hasten senescence of tissues surrounding a feeding site, and so increase the supply of good food available at that site.

3.3.5 Selection for High Nitrogen and Survival of the Young

Having evolved the ability to induce a plant to release more nitrogen at times and places where it is most needed does not mean, however, that the gall former has solved all its nutritional problems. Far from it. Establishing and surviving is still a chancy business governed, as always, by being able to gain access to enough nitrogenous food. Whitham's (1978) study of *Pemphigus* aphids galling leaves of *Populus* is a classic example of this. Stem mothers of these aphids hatching from overwintering eggs select newly expanding leaves upon which to settle and form a gall, but they must do so in a period of only

3 days or it is too late. Those forced to move to a second leaf have an even lower chance of survival.

However, not just any leaf will do; size is critical. The larger the leaf the higher the survival and the greater the subsequent weight of the colonizing stem mother, and the greater the number, weight, rate of growth and contained embryos of her progeny. Whitham found that 32% of available leaves were so small that there was an 80% chance that any aphid settling on one would fail to establish. Nearly all these leaves were avoided by the aphids. Only 1.6% of the available leaves were big enough to ensure a 100% probability of success for settling aphids. All of these were colonized. This means that most stem mothers were forced to colonize suboptimal leaves. Even so, the average stem mother colonized a leaf which was 60% larger than the mean size of all leaves on a tree, and this selectivity gave a more than two-fold increase in fitness over that expected from a random settling on leaves.

All this comes about because the amount of nitrogen in leaves of equal age is a function of their size. Yet Whitham's data indicate that no leaf was so large that it provided enough food for maximum reproductive success. That this is so he demonstrated by spraying some leaves with a foliar fertilizer for 6 weeks following bud burst. The number of progeny produced in galls on these fertilized leaves increased by 37%, and their weight by 30%. However, the size of the fertilized leaves did *not* increase.

Further evidence of the critical shortage of nitrogen for these aphids is that the position on the leaf where they form their galls is also crucial. The base of the leaf blade at its junction with the petiole is the site through which all nutrients moving in and out of the leaf must go. Stem mothers in basal galls matured more quickly, were 34% heavier, and produced 90% more progeny than those in distal galls. The more recent studies of Larson and Whitham (1991), using radioactive carbon as a label, have convincingly demonstrated the nutritional importance of the siting of these galls to the success of the growing young. Little wonder that they fight for these prime sites (Sect. 11.2.3).

So it is for all gall makers. Even on the most vigorous and fast-growing tissues, which most easily produce a gall, the young which hatch and attempt to feed in the gall have a low probability of survival.

Sone (1987) monitored several generations of a cecidomyid which forms galls on pine needles in Kyoto, Japan. He estimated mortality and its likely causes at different stages of the life cycle in each generation. He found that 80 to 95% of the total mortality occurred after the eggs had hatched but before galls were formed. Most first instars never even established themselves. Of those which did establish, 35 to 90% died just after forming a gall. Sone demonstrated that predation was not responsible for this early mortality, nor was physical interference between larvae trying to establish their feeding sites. Resinosis was estimated to kill no more than 1 to 7% and desiccation may have been responsible for some few deaths. He was left guessing that unspecified "abiotic factors" must be responsible for this low survival of the very young. I would suggest that there is nothing abiotic about the cause of their death. They starve because of the inadequacy of the supply of food in the plant.

In spite of galling diverting nutrients to the site where the larvae feed, the chance of any one individual getting enough good food will vary. Most will not get enough. Different plants, different parts of one plant, and the same plant at different times will vary in the level of nutrition they can provide.

The work of Sitch et al. (1988) illustrates this point. They showed that the "pea galls" of the cynipid wasp, *Cynips divisa*, on leaves of oak, *Quercus robur*, in Nottingham, England, are far from being randomly distributed on their host. Many trees have no galls at all, or only some branches on a tree may be attacked. On individual leaves galls are formed only on the main veins branching out from the mid-rib, but they occur more frequently on the middle veins than either the distal or proximal ones. On individual veins most galls are closer to the mid-rib than to the margin of the leaf. Where there is more than one gall per vein those nearest the mid-rib are larger, and the first gall on a vein is larger the more companions it has with it on the vein. These authors noted that the latter phenomenon probably means more galls import more nutrients, and the gall which gets first go at the supply grows biggest.

They make no other suggestions, however, to explain these patterns of distribution of successful galls. The results of other studies reported later in this book suggest an adequate and parsimonious explanation. Both the concentration and rate of flow of soluble nitrogen vary from one tree to another, from one branch to another, and from time to time at any one site. The insects can either detect those sites where the supply of nitrogen will be adequate for their needs or, more likely, they survive only at such sites.

Morgan et al. (1983) and Frankie and Morgan (1984) observed a similar situation for another cynipid gall *Disholcaspis cinerosa* on oak trees (*Quercus* spp.) in Texas. A few trees have high numbers of galls, most have low to moderate numbers, and a few have none. What is more, apparently resistant trees still carry few or no galls when growing in close proximity to heavily infested trees. Experimentally exposing resistant and susceptible trees to infestation by the gall wasps showed significantly more galls established on the susceptible ones.

The authors could find no correlation of levels of infestation with the production of acorns or weather patterns; nor did stress from disturbing the roots affect the resistance of two trees so disturbed. There was, however, some indication that as trees grew older they became less susceptible to galling. Only young trees were heavily attacked, and then only in urban environments. Trees in natural habitats never had higher numbers of galls. They suggested that unknown factors of the host plant set the capacity of the trees to resist successful attack by the gall maker, but that predators keep the number of cynipids below a level which the susceptibility of the oaks would permit.

This does not explain why trees become less susceptible as they age, nor why urban trees are more heavily attacked. It is not an adequate explanation to suggest that otherwise efficient parasites take a long time to find newly planted trees when the cynipids have no trouble in quickly locating them. And what about trees which are free of galls growing next to heavily infested trees? Why do the parasites not find the latter? Furthermore, the authors found with caging trials that equal numbers of eggs were laid on susceptible and resistant trees. This suggests to me that, as I said in the case of the pea galls on English oaks, the answers are to be found in the survival of very young cynipids. As Kearsley and Whitham (1989) found (Sect. 3.1.10), the age of the trees, and thus their changing physiology, may have more influence on the chances of those young than their ability to select the best sites.

3.3.6 Selection of Growing Tissues for Proliferation of Galls

Price (1989) and his colleagues (Price and Clancy 1986; Price et al. 1987b) in Flagstaff, Arizona, have made extensive studies of three species of sawflies of the tenthredinid genus *Euura* which form galls on different species of willow (*Salix* spp.). For all three they have found a consistent pattern; there are many more galls formed on younger, more vigorous shoots. Because the number of flush shoots and their vigour decline as a ramet of willow ages, so too does the success of gall makers decline on older plants.

The greater success, and therefore abundance, of all three *Euura* sawflies in these younger, more vigorous tissues is determined by two things: first, the extent to which a female will lay eggs in the shoots; and second, the survival of eggs and (especially) the very early instars in the shoots. At no time are the natural enemies of the gall makers of any importance as an influence on their abundance. They have found that the genotype of the plants can have a considerable influence on the success of gall makers (largely through differences in the length and vigour of shoots with age) and so can the activities of the animals themselves. For example, galling by *E. lasiolepis* stunts and kills shoots of *Salix lasiolepis*. The plant then puts out new flush shoots which are infested by the next generation of sawflies. Or the environmental conditions that the plants experience, by varying the amount and vigour of new growth, may influence the success of gall makers from generation to generation.

Price and Clancy (1986) and Presler and Price (1988) showed that *E. lasiolepis* is less successful when its host plants are experimentally or naturally subjected to drought. As yet they have not looked for the processes which might be responsible for such a dramatic effect on the acceptability of shoots for oviposition, or for their suitability for the survival of the very young larvae in them. (For example, why, even on the "best" plants in Presler and Price's (1988) experiments, did nearly 30% of attempted ovipositions not produce an egg, and 14% of hatched larvae quickly die?)

They (Waring and Price 1988; Price 1989) confuse my hypothesis (White 1969, 1974, 1984) of the influence of stress upon survival of the young through changes in their nutrition, with the situations they found for these gall makers on willow. Gall making is a special case. What is of paramount importance for a gall maker is its ability to get the plant to direct nutrients to the place where it is trying to feed. To do this it must induce the plant to continue to proliferate juvenile tissue – to grow a gall. It can do this only if the plant's tissues are *still vigorously growing*. The less growth there is, and the sooner it stops, the less chance there is that a gall maker will get enough food to survive. Stress of the host plant, sufficient to slow its vigour and growth, will mean that galling is more difficult and less likely to succeed. So, the first essential is a gall. Once this has been initiated the nutritional variables can become predominant. *Then*, the location of the gall; the stage of development and activity of its contained animal; the location, age, and physiological condition of the plant; and the impact of changes in the environment upon the plant's physiology will all contribute to the survival, speed of development, and fecundity of the gall maker.

There is, as Price (1989) suggests, an important dichotomy between herbivores, but it is not one of the difference between those which are "keyed to" (adapted to?) either vigorous or stressed plants. It is one of the differing responses of flush and/or senescence feeders to changes in their host plants.

3.4 Chewing Insects

There is a multitude of insects which feed by chewing the tissues of plants. They include most of those which eat plants propagated for human use. The most destructive and best known of these are the caterpillars of moths and butterflies, and the locusts and grasshoppers, so I have drawn my few examples from among them (but see also the examples in Sect. 3.1).

It is no surprise that most of these herbivores that have been studied in detail are economic pests of our food and fibre crops. But to expect that a study of any of them as pests on crops will necessarily reveal the processes responsible for their original distribution and abundance in nature is risky. It would be akin to trying to explain what influenced the distribution and abundance of ancestral bovids – or even present day bovids living in the wild – by studying the fecundity, survival, and growth rate of a herd of pedigree cattle eating selected species of grasses grown on fertilized and irrigated pastures.

In the case of Lepidoptera which attack fibre crops (except cotton) the contrast is not nearly so great. Most of them are defoliators of trees. For some, like the spruce budworms mentioned later in this section, their ecologies remain essentially unaltered by human activity. Even those that attack trees grown in artificial plantations are far less removed from nature than those which eat agricultural food crops. There have been few studies of herbivores attacking trees to find out what effect nutrition has on the chance of the individual surviving.

There have, on the other hand, been very many detailed studies of the feeding efficiency of lepidopterous larvae. Nearly all have concentrated on economic pests of food crops which have become hugely abundant, and remain so, in a milieu far removed from that in which they evolved. In addition, nearly all such studies have been of penultimate- or final-instar larvae – large, robust animals well past most of the hazards of early life. To use the bovid analogy again, this is equivalent to measuring the feeding efficiency and growth rate of pedigree fat steers, while ignoring the part nutrition plays in the wild in the successful breeding of cows and the survival of their calves.

Furthermore, in most such studies larvae are fed on material previously severed from the plant, on potted plants, or on artificial diets. Also, the chemical analyses of plant tissues have been done on material far removed in time and condition from the tissues the larvae would eat in nature. So the results of most of these studies must be treated with caution if they are to be used to try to understand what it is that most influences the success of herbivores in nature – and before human intervention changed the rules.

Yet, in spite of these handicaps, most studies of feeding by late instar larvae do show a positive response by the larvae to an increase of nitrogen in their diet. How much more critical then is the level of available nitrogen likely to be when they are neonates alone in the wild? There are some studies which give us a pretty good idea.

3.4.1 Creaming-off as a Tactic to Increase Access to Nitrogen: White Butterflies on Crucifers

Slansky and Feeny (1977) reared larvae of the white butterfly, *Pieris rapae*, on several species and varieties of crucifers, both wild and cultivated. These differed from each other in their content of nitrogen either naturally or because they had been fertilized with ammonium nitrate. The authors measured the rates of growth of *Pieris* larvae in their fifth instar on each of these plants. They also measured the rate at which the larvae consumed their food and the efficiency with which they used it.

The growth rates of larvae were as high on plants with as little as 1.5% dry weight of nitrogen in their tissues as on plants with 4.8% nitrogen. This was because the larvae feeding on plants with little nitrogen in them ate their food more quickly than when feeding on plants with high levels of nitrogen. Even though this meant that they used these low-nitrogen tissues less efficiently (they assimilated less of the gross weight of food that they ingested) they nevertheless gained a higher proportion of the total nitrogen in those tissues.

By making large adjustments in the rate at which they feed, these animals can, as the authors express it, be more "powerful" but less efficient at extracting nitrogen from food which contains little nitrogen. They do this by passing several lots of food through their guts in the time that it would take to fully digest all the nitrogen in a single gut-ful. They are, in effect, "creaming off" the portion of the nitrogen which is readily available, thus gaining the same amount of nitrogen that they would have done from one fully digested load. [This adjustment has since been found to be quite common among larvae of the Lepidoptera. They are maximizing the rate at which they can take up nutrients from the gut (Reynolds 1990)].

Slansky and Feeny found that larvae grew exceptionally fast when fed on one of the wild species, *Dentaria diphylla*, and especially on the most heavily fertilized of the highly domesticated collards, *Brassica oleracea* var *acephala*

(6.1% dry weight of nitrogen). On both they ate the tissues of the plants and consumed the nitrogen in them at a rate expected from the total nitrogen content of the plants. But they grew more efficiently and they used the nitrogen they consumed more efficiently than expected. Their better performance stemmed from more efficient use of the food they ate, not from eating more of it. The reason for this lies in the *greater proportion of the total nitrogen* in the tissues of both of these plants that was *in soluble form*: in *Dentaria* because of the naturally more rapid growth of its leaves relative to those of the other plants containing similar levels of total nitrogen; and in collards because of greater flush growth induced by the high levels of nitrogenous fertilizer.

However, they found no correlation between the rate at which the larvae grew and the energy content of the plants, nor the amount or type of glycosinolates in their leaves. There is a superabundance of energy available, and the insects not only tolerate or detoxify the glycosinolates – they are addicted to them. Without glycosinolates females will not lay eggs and larvae will not attempt to feed. The authors concluded that the growth of the larvae of *P. rapae* is limited by the *availability* of nitrogen in all their food plants. It might have seemed at first sight that because the larvae ate less of the high nitrogen plants to maintain a similar rate of growth to that achieved on plants with much less nitrogen, they were therefore *not* limited by nitrogen when feeding on these latter plants. This is not so. The greatly improved performance of larvae feeding on *Dentaria* and fertilized collards gives the clue. If they had tapered their rate of feeding because they had reached the maximum rate of growth of which they were physiologically capable, no matter how much more nitrogen was available, then they would not have grown faster on these two plants.

What is really limiting their rate of growth is the proportion of the total nitrogen in the food which is readily accessible. When there is not much nitrogen in the food, that part of it which is bound into various complexes takes too long to extract in relation to the value gained from it. So it is better to "waste" this bound fraction by passing it quickly through the gut, and in the process pick up much more of the unbound fraction which is quickly and readily available (more nitrogen, in fact, than would have been assimilated by fully digesting a single gut-ful in the same time). But there is a limit to how fast food can be passed through the gut. At this point, confronted with food with even less nitrogen in it, or less of it in soluble form, this strategy will fail and the rate of growth will decline or stop.

At the other end of the spectrum where plants contain more nitrogen (but not more of it in soluble form) there still comes a point where the time taken to extract and digest the bound nitrogen ceases to be profitable. So a balance must be struck between fast passage with quick skimming, and slower passage to extract more of the total nitrogen in the plant, unless more of the nitrogen is in soluble form as was the case in the leaves of *Dentaria* and the fertilized collards.

This is why so many different herbivorous animals seek out and feed on new flush growth. More of the total nitrogen in that tissue is in soluble form and readily available to them. Similarly, an animal feeding on senescing tissue gains access to more of the nitrogen more readily – but more slowly – as it is released by the plant in soluble form. In both cases time and work, which would otherwise be expended on releasing the bound nitrogen, is saved. All this was demonstrated with large and robust fifth instars. They have already completed most of their growth, have a lesser need for nitrogen than they did when they were first instars, and have the size and mobility to cope with considerable changes in the distribution and quality of their food. This is not so for the first instars: quite small changes in the proportion of soluble to bound nitrogen in their food can mean the difference between life and death.

3.4.2 Early Instars Need More Nitrogen: Gypsy Moth on Artificial Diet

Just how much more critical an adequate supply of readily accessible nitrogen is at the early stages in the life of a lepidopterous herbivore was shown by Montgomery (1982). He reared larvae of the gypsy moth, *Lymantria dispar*, individually from hatching to pupation on an artificial diet, and measured their growth and the rate of efficiency with which they consumed and assimilated nitrogen. In the 6 months to pupation the dry weight of female larvae increased 4000 times and that of their assimilated nitrogen 3000 times, both at a constant exponential rate for the first three instars, but slowing considerably thereafter. Similarly, their rate of consumption and assimilation of nitrogen dropped markedly after the third instar. The newly hatched larvae contained about 12% nitrogen but this declined steeply to 10% in the third instar and then more gradually to 9% in the sixth instar. The percentage of nitrogen in the frass of first instars was much higher than in that of the second and third instars (apparently because the small first instars avoided eating the fibrous bran particles in the diet), but steadily increased through the fourth, fifth, and sixth instars. All these measures indicated that the need for nitrogen is very high in the first instar, somewhat less so in the second and third instars, and progressively less pressing in each instar thereafter.

3.4.3 Illustrations from the Life Cycles of Economically Unimportant Butterflies

It is likely that an investigation of economically unimportant species which are still living in a relatively undisturbed relationship with their food plants, and at unremarkable levels of abundance, might more readily reveal the fundamental processes responsible for their distribution and abundance. There are few such studies. One is the long-term investigation of checkerspot butterflies (*Euphydryas* spp.) in California that Paul Ehrlich and his colleagues at Stanford University have been conducting since 1960 (Ehrlich et al. 1975; Singer and Ehrlich 1979).

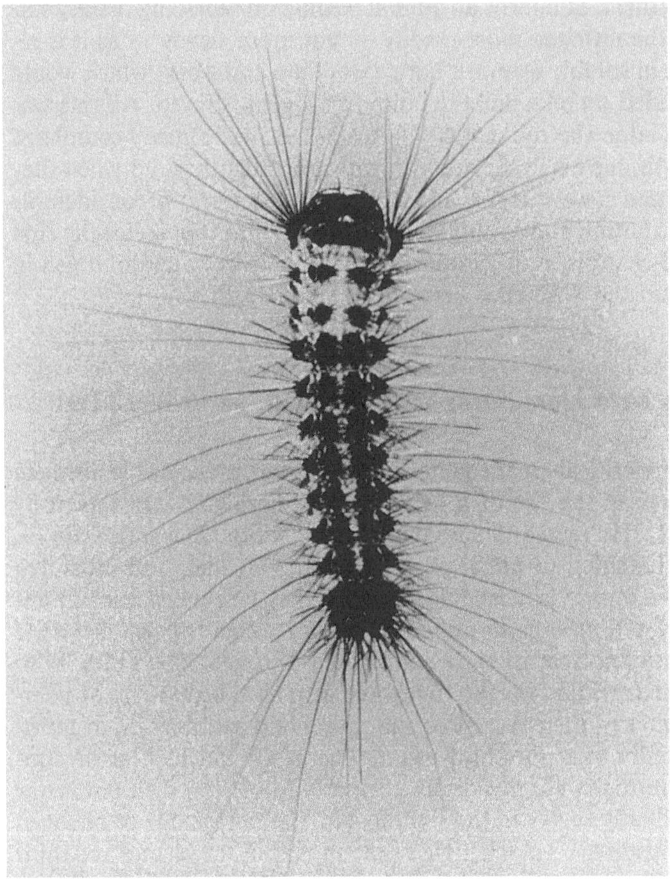

Fig. 5. A first instar gypsy moth. Rearing these on artificial diet has shown that a newly-hatched lepidopterous larva has a greater need for high levels of nitrogen in its food than do larger, older ones. Photo courtesy of R. Zerillo, USDA Forest Service

The life cycle of the first of these butterflies that they studied, *E. editha*, shows an interesting adaptation to the harsh environment in which it lives, and a reversal of the more usual change from flush to senescence feeding seen in some leaf-miners (Sect. 3.2.1) and aphids (Sect. 3.5.1). Rain falls in the 6 months of winter and spring; summer and autumn are periods of intense drought. The caterpillars' food plants (mostly species of Scrophulariaceae) are deciduous or semi-deciduous, flushing new growth in the spring but shedding senesced leaves by mid-summer. Adult butterflies emerge and lay their eggs in the summer not long before leaves are shed as the drought takes hold. The early instars feed on the rapidly deteriorating leaves before entering obligate diapause in the fourth instar, usually only some 10 days after hatching. They are senescence feeders, dependent upon the flow of soluble nitrogen from these leaves for their rapid growth. Small variations in the amount of nitrogen in the leaves, and the rate at which it is broken down and released, will be crucial to their survival.

In the following spring diapause is broken as the plants are flushing new growth. The fourth and fifth instars feed very selectively on this new growth: they are flush feeders, and so much so that if, for any reason, they are unable to feed on the newest growth, they will, after 2 weeks feeding on less-preferred leaves, enter a second diapause which lasts through until the following spring. This process may be repeated over 2 or more years – a wonderful survival strategy in an uncertain environment. However, as Singer and Ehrlich ruefully point out, such behaviour is a terrible frustration to a graduate student with limited time trying to raise and study these animals in the laboratory!

Early in the study of *E. editha* it became clear that neither the death of adult butterflies, the death of diapausing larvae nor parasitism and predation of larvae were playing any significant part in the observed changes in abundance of its populations. The death from starvation of the early pre-diapause larvae was the overriding influence. In some generations up to 99% of these larvae failed to survive long enough to enter diapause because their food supply was inadequate. Only in places where plants are a little later dying back in the dry season could any significant number survive, and this varies with the vagaries of the weather. In some localities with a fairly predictable rainfall numbers do not vary a great deal. In other places where the rainfall is more variable, great changes in abundance occur, from "outbreaks", where all available food is eaten before any larvae can pupate, to local populations dwindling to extinction. The key throughout is not how much nitrogen there is in the environment – there is usually far more than ever gets used – it is how much of that nitrogen becomes *available* as it is shipped out of senescing leaves at the time when the young larvae must have an abundant supply of it.

Another species, *E. chalcedona*, feeds principally on *Diplacus aurantiacus*, a semi-deciduous, drought-adapted, scrophulariaceous shrub (Mooney et al. 1981). Larvae which have been in diapause since the previous summer emerge in the spring and feed preferentially on the new expanding leaves. These expanding leaves, although coated with considerable quantities of phenolic resins (which are said to inhibit the growth of the larvae), also contain their highest-ever levels of nitrogen at this stage. Additionally, more of that nitrogen will be in a soluble and therefore more readily assimilable form than will be the case once the leaves are fully expanded. The amount of resin on the leaves decreases as they expand, but the larvae still keep to the youngest leaves available. They rapidly complete their growth and pupate before the nitrogen in the leaves is drained away to the developing flowers of *Diplacus*. The emerging adults lay eggs soon after flowering starts. First instar larvae hatching from these eggs in mid-summer are faced with having to feed and grow to the fourth instar just as the leaves are finally senescing and dropping from the shrub as the drought takes hold. They have a poor quality food available for only a short time before they go into diapause. The authors noted that "this period may represent the major constraint on the survival and reproductive output" of this butterfly. So, as with *E. editha*, we see this reversal of the usual – the early pre-diapause instars are senescence feeders and the late post-diapause instars are flush feeders.

Lincoln et al. (1982) raised pre-diapause larvae on a series of artificial diets containing levels of nitrogen and phenolic resins which spanned the range found in and on *Diplacus* leaves. The number of larvae which survived, the speed at which they grew, and the size they attained increased as the amount of nitrogen in the diet increased, and decreased as the level of resin increased. Nearly all larvae died in the first few days of feeding. These results supported evidence from the field that the abundance of this butterfly is limited especially by the protein in the food of the pre-diapause larvae. As is the case with *E. editha*, most of the mortality each year is of these pre-diapause larvae, and, again, predation and parasitism were insignificant.

Lincoln (1985) found that by feeding *Diplacus* shrubs with a slow-release NPK fertilizer he could raise the level of nitrogen in their leaves without increasing the amount of resin on them. This had the effect of increasing the rate of growth of pre-diapause larvae feeding on these leaves; a result concordant with that of the laboratory studies using artificial diets. But because the young larvae preferred to feed some nodes back from the tips on older leaves on both fertilized and unfertilized shrubs, and because the leaves at the tips had higher nitrogen and more resin than the preferred older leaves, he concluded that the resin was inhibiting the larvae from feeding on the "optimal" food provided by the youngest leaves. There is, however, an alternative explanation for this feeding preference. As senescence feeders, the young larvae are programmed to feed only upon leaves which are senescing. They are therefore dependent on the rate at which nitrogen becomes available in soluble form as it is translocated out of those leaves. The rate at which nitrogen is released (and possibly its form) is the critical factor, not the total amount present in the leaves. This rate is likely to be greater in the older leaves than in those younger ones at the tip of the branch (and may well be why the older ones had less nitrogen in them when analysed).

E. chalcedona also lives on another shrub, *Scrophularia californica* (Williams et al. 1983a, b). In fact, given a choice in the laboratory, the larvae strongly prefer to feed upon *Scrophularia* leaves rather than upon those of *Diplacus*; but as post-diapause larvae they did not discriminate in favour of younger leaves as they do on *Diplacus*. These feeding preferences were matched by the performance of larvae reared in the laboratory on leaves of the two species of plant. More larvae survived and they grew faster and bigger on leaves of *Scrophularia* than on those of *Diplacus*. Both preference and performance reflected the nitrogenous quality of the food. *Scrophularia* leaves have a higher content of nitrogen than do those of *Diplacus* but, unlike *Diplacus*, mature *Scrophularia* leaves have as much nitrogen as do the young ones.

Finally, the critical role of nitrogen in the successful production of surviving offspring was again emphasized by a study of the diet of adult females of

E. editha (Murphy et al. 1983). Those females which were fed on nectar containing amino acids laid heavier eggs, and the larvae hatching from these eggs were more likely to survive.

Studies of other species of butterfly living in relatively undisturbed and natural association with their food plants have shown a continuation of the common thread: preferential oviposition and feeding on new flush growth or reproductive tissues, and a differential death rate of early larvae linked to the level of available nitrogen in those tissues.

The young larvae of the swallowtail butterfly, *Battus philenor*, are flush feeders. Their survival is dependent upon the availability of new growing leaves of their food plants. Even though larvae must later move from plant to plant, eating leaves of all ages in order to get enough food to survive to pupation, the females lay their eggs on the plants which will provide the necessary high-nitrogen leaves for the first instars. Two species of their food plants, *Aristolochia reticulata* and *A. serpentaria*, grow sympatrically in the area in Texas where Rausher (1980, 1981) studied these butterflies. The former is more abundant, flushes new leaves earlier, and over a shorter period than the latter. The first butterflies that emerge in the spring search out and lay their eggs preferentially on the new growth of *A. reticulata*, but later-emerging females of this brood start to shift their preference to *A. serpentaria*. By the time the second (summer) brood of butterflies has emerged all females strongly prefer *A. serpentaria*, which by then is the only source of new growth which newly hatched larvae must have to survive. Rauscher could find no evidence of any other factor influencing their survival. The content of water and secondary chemicals in the leaves had no influence.

A similar tracking of new growth late in the season (but this time on plants of the same species) is illustrated by *Inachis io* and *Aglais urticae*. The caterpillars of these two species of butterfly eat stinging nettle, *Urtica dioica*, in England. Both prefer to feed on plants of high nutritional quality. Females lay their eggs on young growing plants in the spring and the larvae eat the young leaves at the top of these growing plants. Later in the season the second generation of butterflies will seek out regrowth whenever it occurs on the plants. In a "good" year, when there is a lot of late regrowth, the butterflies may even complete a partial third generation. Pullin (1986, 1987) reared larvae of both species on mature leaves and on regrowth leaves from nettles which he had earlier cut back. These regrowth leaves had nearly twice as much nitrogen in them as the mature leaves, and the proportion of their total nitrogen which was in soluble form was more than three times greater than that in mature leaves. He measured the rate of growth of the larvae, and the efficiency with which they absorbed and digested their food, from the time they became third instars. Not surprisingly, the caterpillars eating regrowth leaves outperformed those fed mature leaves, and it made no difference whether they had been eating mature or regrowth leaves prior to the third instar. They converted the material they ingested and digested more efficiently, and consequently grew faster, pupated earlier, and achieved greater final weights. They also ate less of their food per unit body weight than did those eating mature leaves.

In England, again, the first instar larvae of the orange tip butterfly, *Anthocharis cardamines*, feed exclusively on flower buds and opening flowers of their cruciferous host plants (Courtney and Duggan 1983). Later instars continue the trend, eating developing seedpods and rarely touching the leaves. Yet, still most larvae die as first instars. As in most such cases reported, these tiny animals just disappear. Dempster (1967) had found this for the white butterfly, *Pieris rapae*, and so had Warren et al. (1986) for the wood white butterfly, *Leptidea sinapis*, both in England. In Costa Rica, Young (1978) reported the same thing for caterpillars of *Helconius* butterflies. Like many others working with all sorts of insects other than butterflies, each of these authors attributed the disappearance to predation. Courtney and Duggan, however, could find no evidence in their study that predation was responsible for these early deaths, so they concluded that the food plants must be directly responsible for death – but by way of mechanical or chemical means of defence, not as a result of inadequate nutrition!

Myers (1985) perhaps got closer to the mark. She found that caterpillars of the white butterfly, feeding on plants which had higher levels of nitrogen in them because they had been fertilized, survived better, grew faster and produced heavier pupae; and the female butterflies preferred to lay their eggs on these plants. Tabashnik (1982) found that the fifth instar caterpillars of two species of *Colias* butterflies in Colorado that ate leaves of their food plant with the most nitrogen in them grew faster and bigger. Again, when Finke and Scriber (1988) fed fourth instars of the eastern black swallowtail butterfly, *Papilio polyxenes*, on a number of umbelliferous forbs on which they would normally feed in their natural habitat in Ohio, they found the same thing. The growth rate of the larvae was positively correlated with the amount of nitrogen in the leaves of the forbs, irrespective of the stage of seasonal development the plants had reached.

3.4.4 Further Examples from Forest Defoliators

Some of the most destructive native defoliators of indigenous forests are the North American spruce budworms (*Choristoneura* spp.; Tortricidae) (see Sect. 18.2). Their young larvae feed on the expanding buds and new needles of several species of conifer. By the time the new shoots have fully expanded and are beginning to lignify, the larvae are in their final (sixth) instar prior to pupation (Volney and Cerezke 1992). They are flush feeders, "programmed" to eat only new needles. So, the fact that the level of nitrogen in these needles soon drops as they mature, and that the larvae will mostly die if forced to feed on older needles, is not an issue. What is important is that the quantity and form of nitrogen in the developing needles are sufficient to sustain the early instars. However, the concentration and composition of nitrogenous material in the new needles is not constant. Both vary between species, between individuals of one species at any one time, and in any individual tree at different times (Durzan 1968; Durzan and Lopushanki 1968; Kimmins 1971; Koller and Leonard 1981). And the budworms exhibit appropriately varying feeding preferences, rates of survival and growth, and final size of pupae and adults (Campbell 1989; Wagner et al. 1987; Clancy et al. 1988).

Miller (1963) had shown that even during outbreaks of the eastern spruce budworm, *C. fumiferana*, on balsam fir, *Abies balsamea*, in Canada, 84% of the larvae died before they reached the third instar. Mostly this was as dispersing but non-feeding first instars prior to hibernation over winter; they never found a food plant. But 40% of the surviving second instars died after they had emerged from hibernation and started to feed. Foltz et al. (1972) similarly found that the greatest influence on the numbers in each generation of the jack pine budworm, *C. pinus*, feeding on *Pinus banksiana* in Michigan was deaths between the egg and second instar.

Brewer et al. (1985, 1987b) demonstrated very clearly in two separate sets of experiments with the western spruce budworm, *C. occidentalis*, that the level of nitrogen available in the foliage is a major influence on the success of the larvae that eat that foliage. They grew seedlings of white fir, *Abies concolor*, and Douglas fir, *Pseudotsuga menziesii*, with levels of nitrogen in their needles varying from much lower to much higher than would ever be encountered in nature. They did this by growing the trees in nutrient solutions containing optimum amounts of all nutrients except nitrogen, which they varied from 0 to 800 ppm. They then placed budworm larvae which had been raised on an artificial diet on these trees. They used final (fifth) instars in the first set of experiments, but in the second set they used third instars. (Significantly, they found the very small second instars, which are the first to feed, were not only very difficult to handle, but too many of them died to leave manageable numbers for their experiments.) In both series of experiments the results were the same. At very low and very high levels of nitrogen in the needles the animals fared very badly, but at the intermediate levels, which straddled the concentrations normally found in needles in the field, they showed a graded response. At "optimum" levels of nitrogen (around 2 to 3% dry weight) more larvae survived, they grew faster, and produced heavier pupae from which emerged female moths which lived longer and laid more eggs. Their response beyond this level suggested a "platforming off" once there was an adequate amount of nitrogen available. Their survival and performance did not continue to improve linearly with increased concentrations of nitrogen. In fact, at unnaturally high levels there was a "toxic" effect, their performance, while not as poor as that on nitrogen-deficient plants, fell significantly below those eating needles with intermediate levels of nitrogen. This effect may have been due to a number of factors other than the amount of nitrogen in the needles (for example, plants grown at excessively high levels of nitrogen exhibited aberrant growth, including symptoms similar to those associated with a deficiency of potassium).

Cates et al. (1987) obtained similarly definite responses when they reared post-third instar larvae of *C. occidentalis* on artificial diets. They adjusted the diets to contain high and low levels of nitrogen akin to those found in needles of the larvae's food plants, but adequate in all other constituents. Of those larvae eating the diet with a high level of nitrogen, 20% survived to the adult

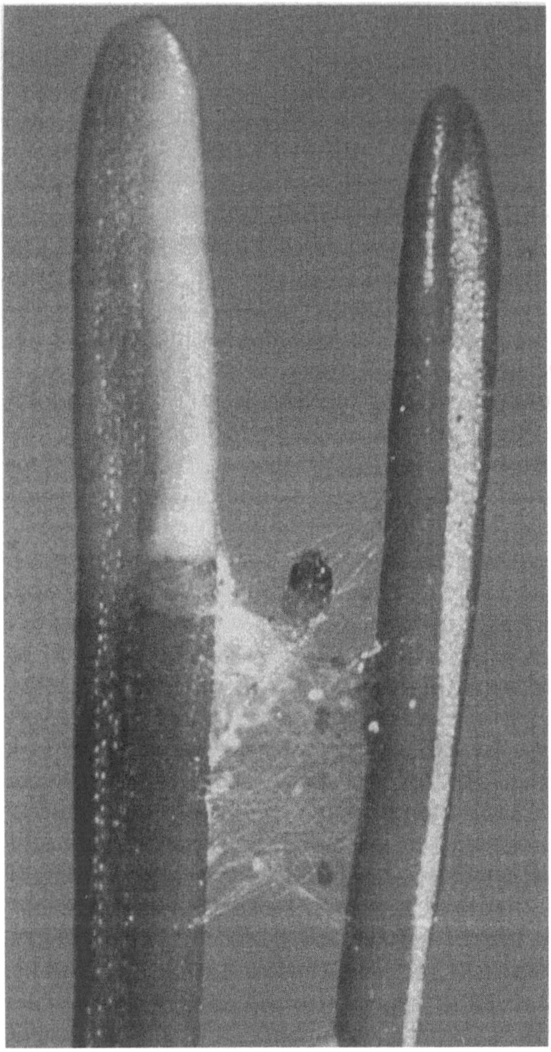

Fig. 6. A second-instar spruce budworm larva must mine and feed in year-old needles of balsam fir until new buds flush. Its chance of survival will be influenced by changes in the chemistry of those needles, wrought by the aging of the trees, and the stressfulness of the site (see also Sect. 18.2). Photo courtesy of Forestry Canada

stage, but only 0.7% of those on the low-nitrogen diet survived – an almost 30-times difference.

Onstad et al. (1986) found that the polyphagous *C. rosaceana* which feeds on apples in Canada, is also a flush feeder, dependent upon the new growth at the beginning of the season. Most of its larvae died early in the first instar, and their deaths were associated with the rapid fall in the content of nitrogen in the maturing leaves.

However, for the forest budworms at least, this is not the whole story. Their overwintering second instar larvae emerge from their hibernaculae *before* the trees' new buds start to swell, and *they mine in the needles of the previous year.*

This means that, like the pre-diapause instars of the checkerspot butterflies discussed in the previous section, they start their feeding life as senescence feeders. Only when, as third instars, they begin to mine in the swelling new buds do they become flush feeders (McGugan 1954; Volney and Cerezke 1992; Shepherd 1992).

Is this the key to the outbreaks of spruce budworms that I discuss in Section 18.2? The rate of senescence of these needles on old trees, especially when stressed in any way, would be hastened, releasing more soluble nitrogen more quickly to the mining larvae, and markedly improving their chance of survival. Certainly this, plus their subsequent dependence on expanding buds, and their greater survival on flower buds, all highlight how essential access to a flow of concentrated soluble nitrogen is for survival of the very young.

There are reports that deaths of the first instar soon after hatching account for by far the greatest losses in populations of a wide diversity of other species of tortricid moths: for example, the light brown apple moth, *Epiphyas postvittana* (Danthanarayana 1983); the Japanese pine shoot moth, *Evetria cristata* (Kanamitsu 1966); the cowpea moth, *Cydia ptychora* (Perrin 1978; Ezueh 1981); and the codling moth, *Cydia pomonella* (Ferro et al. 1975). As with the butterflies, the causes of these deaths were unknown or attributed on indirect evidence to predation. But sometimes the association of these deaths with the nitrogen nutrition of young larvae is recognized (Ezueh 1981).

The same story emerges time and again from studies of many other forest lepidoptera belonging to quite different families. The most significant influence on their abundance, generation after generation, at outbreak levels or endemic levels, is the death of the very young larvae, but rarely is this obvious Achilles heel of so many economically important species investigated. The lymantriid Douglas fir tussock moth, *Orgyia pseudotsuga*, is a good example. Huge amounts of time and money have been expended on the study of this major defoliator of coniferous forests in western USA and Canada (Brooks et al. 1978). Yet there is still no explanation of what causes the abundance of this insect to change, in any one forest, from being so rare as to be almost impossible to find to being so plentiful as to completely defoliate all the trees over hundreds of hectares of land.

Mason (1974) reported that changes in density of small larvae in a population of *O. pseudotsuga* in Arizona were alone largely responsible for changes in the density of that population from year to year. It was not a function of dispersal from areas of high abundance, but rather "differences in natural environmental factors" favouring the survival of the young larvae in the resident population which was responsible. In other studies (Mason 1981; Mason et al. 1983; Mason and Torgersen 1987) covering both outbreak and non-outbreak populations in different localities, the story was the same. The largest source of loss, and therefore the major factor influencing the size of the populations in each generation, was always of the first and second instar larvae. In none of these studies did the author's establish a cause for this disappearance, but they thought predators might be responsible. Nobody, it seems, has actually looked to see what causes so many of the young to perish so soon (see also Part V).

Another lymantriid which has been very extensively studied is the gypsy moth, *Lymantria dispar* (Doane and McManus 1981). A European species, it was accidentally introduced into North America in the 1860s and is now a serious defoliator of forest, ornamental, and fruit trees over most of the northeastern USA – and it is still spreading. Again, an understanding of what generates the spasmodic outbreaks of this moth remains elusive. Montgomery's (1982) detailed study of the growth and assimilation of nitrogen by gypsy moth larvae which I discussed in Section 3.4.2, clearly illustrates the enormous demand for nitrogen in the early instars. Also it is known that the survival of first to third instar larvae is much higher in populations which increase to outbreak levels than it is in contemporary populations which do not reach great abundance (Campbell 1976). The suitability of different species of plants as hosts seems to be largely a function of the time that they have soft expanding leaves available for the first instars to eat during the relatively short period when gypsy moth eggs are hatching. Raupp et al. (1988) found that the difference in the survival of the neonates was the major cause of differential success of the gypsy moth on different hosts. On inadequate hosts between 50 and 100% of neonates died within 10 days of hatching (and they don't feed for the first 2 days). A good host like white oak, *Quercus alba*, flushes new leaves over a 2-week period, supporting 90% survival during that time, but as soon as all its leaves were fully expanded survival dropped precipitously. A poor host like the beech, *Fagus grandifolia*, on the other hand, has a much shorter flushing time. Survival of neonates on its leaves dropped from 73 to 0% after only 6 days. Unfortunately the authors did not look at what chemical changes took place in the leaves as they expanded. However, it is a fairly safe prediction that the situation is not really any different from, for example, that of the expanding tips of heather for red grouse (Sect. 7.1.2), or flush new grass for rabbits (Sect. 6.2.5.2).

Another family of Lepidoptera which includes many important defoliators of both natural forests and artificial plantations of trees in many parts of the world is the Geometridae, the looper caterpillars. Most are fairly polyphagous but, nevertheless, prefer to feed on some species more than others, and on some individual trees more than others. Usually their preferences have been found to be directed towards obtaining the tissues with the greatest availability of nitrogen (Reader 1979; Werner 1979; Ayres et al. 1987). Predators have nearly always been found to be insignificant as an influence on changes in abundance. Most deaths in each generation are of first instars, and usually these have remained unexplained (Danthanarayana and Kathiravetpillai 1969; Varley and Gradwell 1970; Elliott et al. 1980; Barbour 1988). I have previously discussed how a shortage of available nitrogen may have been the major factor influencing the survival of these neonates, and thus the changes in abundance of a number of species of geometrids (White 1974; see also Part V).

3.4.5 Pests of Crops Also Reveal the Need for Nitrogen

As I have earlier pointed out, it is risky to try to interpret what processes might have influenced the distribution and abundance in nature of "pest" species which have had their evolved relationships with their food plants radically altered by agriculture. But if, as I am contending, these processes are the same for all herbivores, then it is a fairly safe bet that these are "pests" because we have vastly increased the quantity and accessibility of nitrogenous food in their environment (see Sect. 18.6). There are many pointers to this being so. Not least of these is their possession of the same sorts of behavioural and physiological traits which other herbivores have evolved to enhance their access to usable nitrogen in the environment.

Many major agricultural pests which belong in this category are noctuid moths, of which the corn earworm, *Helicoverpa zea*, and a number of closely related species are good examples. They preferentially eat flowers and *developing* seeds whenever these are available. When they are not available they seek out the newest leaves available – always maximizing their access to nitrogen in its most readily assimilable form (Hardwick 1965). Denied access to high-protein food, most of the early instars die, and those that survive grow more slowly and become less fecund adults.

Just how critical it is for neonates to have access to nitrogen that is flowing to new growth was shown for a group of 15 species of noctuid moths from the northeastern USA. They came from eight genera of one subfamily and were feeding on 14 different species of host plants, ranging from hardwood trees to herbs. Schweitzer (1979) conducted 26 tests in each of which he raised two groups of larvae of one of these species. One group was placed as newly hatched larvae on partially opened leaf buds of its normal food plant and the other (usually from eggs of the same female) was placed 1–2 weeks later on expanded leaves of the same plant. The early groups fared much better than the late groups. Survival ranged from 50 to 90% in the early groups and from 10 to 0% (this latter in 11 of the 26 trials) in the late groups. In those trials where some larvae in the late groups survived, they weighed much less as fully grown larvae than those in the comparable early group.

Al-Zubaida and Capinera (1984) demonstrated the dependence on the level of nitrogen available in the food for the larvae of another notorious noctuid, *Spodoptera exigua*, the beet armyworm. They fed larvae on the leaves of sugar beet, *Beta vulgaris*, and two weeds which commonly grow in the beet fields and are used by the larvae as alternative food plants, *Amaranthus retroflexus* and *Chenopodium album*. Plants of all three species were grown in nutrient solutions at different levels of nitrogen. On the leaves with more nitrogen more larvae survived and they developed more quickly. As female moths they produced more eggs. There was some variation in the efficiency with which the larvae were able to use the leaves of the three species as food but, in general, foliage with more nitrogen in it was more digestible and the growing larvae used it more efficiently.

Al-Zubaida and Capinera also demonstrated that the less nitrogen there was in the leaves they fed to *Spodoptera* larvae the more likely those larvae were to eat conspecific pupae. Noctuids are renowned for their propensity for carnivory in general, and for cannibalism in particular (Gould et al. 1980). This has become particularly well known from the widespread laboratory rearing of pest species. And it is not confined to noctuids. In my experience just about any lepidopterous larva, given half a chance, will eat newly pupated siblings, but the noctuids seem to have a greater penchant than most. *H. zea* is especially renowned for its aggressiveness, even to the extent of biting the hand of any unwary entomologist! Cannibalism is not, however, an aberration deriving from crowding in laboratory cultures, but a well documented natural phenomenon (Semlitsch and West 1988; see also Sect. 3.4.7 and Chap. 12).

Another family of Lepidoptera which includes several major agricultural pests is the Pyralidae. Early studies of some of these revealed that for these moths as well survival of first instars, and dependence upon a sufficient level of available nitrogen in their food, were the crucial factors determining the species' success.

Bottger (1951) found that the larvae of the European corn borer, *Pyrausta nubilalis*, when fed on the leaves of sweetcorn (which are relatively high in nitrogen but low in sugar) survived better but were of lower final weight than those fed on the plants' internodes (which have less nitrogen but more sugar than the leaves). Beck (1956), using artificial diets, demonstrated that the first instars of *P. nubilalis* need high levels of nitrogen in their diet to survive but have a negligible need for sugars, whereas the late instars need relatively little nitrogen but a lot of sugars. Cannon and Ortega (1966) experimented with the effect of fertilizing corn plants on the survival of *P. nubilalis* larvae which ate them. They found that when they increased the level of nitrogen from 10 to 200 ppm in the nutrient solutions in which they grew susceptible varieties of corn, there was a ten-fold increase in the survival of *Pyrausta* larvae eating those plants. (There was a similar but much less marked response on resistant varieties.)

The growth of the larvae of another pyralid of great economic importance, *Chilo suppressalis*, on rice was shown by Hirano (1964) to be closely connected to variations in the amount of nitrogen in the plants. Kalode and Pant (1967) demonstrated the same relationship for larvae of *C. zonellus* eating varieties of maize and sorghum. Susceptible varieties had more amino acids and higher total nitrogen than resistant varieties.

More recently, Padhi and Chatterji (1986) found the same story for varieties of rice plants resistant and susceptible to the yellow stem borer, *Scirpophaga incertulas*. The susceptible varieties contained more nitrogen than the resistant ones; more *S. incertulas* larvae survived on them and attained greater final weights than those eating resistant plants.

A more unusual case is that of the greater wax moth, *Galleria mellonella*. The larvae of this pyralid can apparently survive on a diet of nothing but beeswax with the help of intestinal microorganisms (Waterhouse 1957); however, Pant and Kapoor (1964) found that there was more to it than that. They fed newly hatched larvae in the laboratory on fresh combs, old beeswax taken from used combs, and wax that had been refined ready for use as foundation for new comb. Only the old beeswax supported normal survival and growth. Test insects fed and tunnelled in both the new comb and refined wax but were all dead within 7 days. When Pant and Kapoor analysed the three lots of wax they found that the old wax contained all the essential amino acids except threonine, but the other two contained no amino acids at all.

3.4.6 Examples from Biological Control of Weeds

Attempts to use herbivorous insects for the biological control of weeds have provided some good examples of the importance of nitrogen to the success of pyralid moths. One such programme, which has in recent years been remarkably successful in achieving control of a weed in several countries, was developed and implemented by a group of ecologists based at the Brisbane laboratories of Australia's Commonwealth Scientific and Industrial Research Organisation (MFJ Taylor 1984, 1988; Taylor and Forno 1987; Room et al. 1989). It represents a situation in which the only human interference with an evolved relationship between a plant and its herbivore has been the moving of first the plant and then the herbivore to a new location. Furthermore, the plant, *Salvinia molesta*, is sterile, reproducing entirely vegetatively. It is an aquatic fern from Brazil which has become a major pest in many tropical regions of the world. It forms thick mats floating on the surface of lakes and slow-flowing rivers. One of two insects imported from Brazil to northeast Australia and Papua New Guinea as a potential agent of biological control of this weed was a pyralid moth, *Samea multiplicatis*. Initial attempts to establish this insect in the field met with varying success. Extremes of temperature set geographical limits to where this moth could persist, but within this constraint whether or not the animal multiplied and thrived was found to depend largely on the amount of nitrogen in the fronds of the fern. This differed in ferns growing in different localities. At concentrations of nitrogen of 1% dry weight or less the moth may not become established, or may persist at numbers so low as to go undetected by normal sampling methods. However, increasing the level of nitrogen in the fern to only 1.3% dry weight by simply adding urea fertilizer to the water can cause an explosive increase in the abundance of the moth and severe damage to the plants. Numbers of the moth remain low where levels of nitrogen in the food of the larvae is low because most of the larvae die in their first or second instar. Raising the level of available nitrogen in their food dramatically increases the survival of these neonates. It also increases the rate at which they grow. The few larvae which manage to survive on plants low in nitrogen can spend up to 70% longer growing to pupation, and pass through an extra (sixth) instar in the process. Significantly, this retarded rate of growth stems almost entirely from the poor nutrition experienced by those first two instars. Providing food with higher levels of nitrogen to surviving larvae in their later instars made little difference to their rate of growth, but females derived from larvae given late access to more nitrogen produced more eggs. All females preferred to lay their eggs on plants with the highest level of nitrogen in their fronds. Predators, including diseases, were shown to have no significant influence on the observed changes in abundance of the moth. The other insect from Brazil, a weevil, introduced to Australia to combat *Salvinia*, is also limited by the availabilty of nitrogen in the plant as food for its very young larvae.

This same story had already emerged in Australia nearly 60 years ago when another pyralid moth, *Cactoblastis cactorum*, was introduced to Queensland and northern New South Wales to combat prickly pears (*Opuntia* spp.). Some 26 species of this genus had become established as feral populations. The first had been introduced by Captain Arthur Phillip in 1788 in an attempt to establish a cochineal industry so that he could "dye his soldiers coats red". Later ones were from unrecorded imports of garden and hothouse plants. By 1925 some 12 to 24 *million* hectares of farmland were literally clogged up beneath dense and impenetrable stands of cacti. The dramatic destruction of most of these stands in a few short years by *Cactoblastis* is frequently quoted as the greatest success story in the history of the biological control of weeds (Dodd 1940).

Forgotten, however, is the fact that some stands of *Opuntia* were not destroyed. Dodd identified two sorts of cacti which were resistant to attacks by the moth. The first comprised plants which had an abnormally high carbohydrate content. *Cactoblastis* larvae could survive in the young growth of such plants but died when they had to feed in the starchy, low-nitrogen, older segments. This, however, was the least common and least important of the two, as repeated attacks eventually killed it. After each onslaught the larvae died once they had destroyed all new growth, but the plants quickly produced even greater amounts of new growth, giving rise to a greater number of insects until, after several repititions, the plants disintegrated and died.

The second sort of resistant plant Dodd called "yellow pear". These were chlorotic, slow-growing and stunted plants which bore few fruit and which had less than half the nitrogen content of normal green plants. They grew on patches of poor soil and on hilly wooded ridge country. Small patches of these nitrogen-deficient plants among large expanses of green plants were no problem. They were soon swamped and killed by overflow populations from the more nutritious plants all around them. But in some areas where the chlorotic plants predominated the moth did not make any headway. Dodd demonstrated with field experiments that most young larvae died on these chlorotic plants; survivors grew more slowly, moths were smaller, and they laid fewer eggs. A moderate application of nitrogen fertilizer – enough to turn plants green but not enough to stimulate new growth – dramatically reversed the situation. It produced a five-fold improvement in the population of *Cactoblastis* and rapid destruction of the plants. Subsequent studies (Myers et al. 1981) have confirmed that the level of nitrogen in *Opuntia* plants is critical to the success of *Cactoblastis*. Females are attracted to and lay more eggs on plants with higher levels of nitrogen in their cladodes, and more larvae survive on them.

A final example from the field of biological control of weeds (although this time not with a pyralid but with an arctiid moth) again illustrates this same "resistance" – this immunity – of plants to serious damage by their herbivores when the level of nitrogen in their tissues is below a critical minimum for the survival of young growing larvae. The tansy ragwort, *Senecio jacobaea*, is a native of Europe which has become established in many parts of the world as an aggressive weed of pastures. The cinnabar moth, *Tyria jacobaeae*, has been introduced as a potential agent of biological control of this weed, but with mixed success. Basically, as with all herbivores, the problem is, as Myers (1980) expressed it, that the plant, and not the herbivore, is the driving force of the system. *Tyria* has very little real influence on the abundance of its food plant except at odd times and in odd places when the plants are disadvantaged by a change in their environment. Generally it is changes in the nutritive value of the plant, influenced largely by changes in the weather, which have the major influence on the success of the moth. Other studies (Dempster and Lakhani 1979; Van der Meijden 1979) support this conclusion. Myers and Post (1981) found that the larvae of *T. jacobaeae* survived better and produced larger moths (which laid more eggs) when they ate plants that contained more nitrogen, either naturally or because they had been fertilized with urea. When this happened the number of early instars which survived could be great enough to completely defoliate all plants in a locality so that many late-instar larvae starved. The local population of the moth would then "crash". A mini local outbreak had been generated, different only in extent and duration from outbreaks of forest defoliators like the spruce budworm (Chap. 18). The authors noted that this response by *T. jacobaeae* to increased nitrogen in its food plant is akin to that of *Cactoblastis* on *Opuntia*. The close dependence on the level of nitrogen in the food has been further illustrated by *Tyria* larvae growing faster and larger and producing more fecund females when they are fed on flowers rather than on leaves of ragwort; the flowers contain higher levels of nitrogen than the leaves (Myers 1979; Dempster 1982). The concentration of nitrogen in the food plant generally plays a predominant role in the nutritional ecology of the species (Soldaat and Vrieling 1992). And, as always, most larvae die as early first or second instars. Dempster (1971) had previously recorded this for successive generations of cinnabar moths, but could find no cause for these early deaths – most just disappeared. So he attributed them (on very indirect evidence) to the attacks of generalist predators. Today's evidence from the similar disappearance of the very young of so many herbivores would suggest that they starved.

These dramatic responses to manipulations of small changes in the amount of nitrogen in the food of herbivores nicely illustrate how plants may be enormously plentiful, and yet severely limit the abundance of the herbivores because they are an inadequate source of nitrogen for the herbivores' young.

3.4.7 Locusts and Grasshoppers

The family Acrididae includes some of the most destructive pests on Earth – the legendary plague locusts – and a great number of other lesser but still formidable destroyers of our crops.

I have previously presented evidence (White 1976) that a change in the supply of nitrogenous food could be the basic cause of the changes in abundance of a number of species of acridids which, from time to time, explode to plague proportions in various parts of the world. This comes about because the quantity and quality of the plants that locusts eat change as a consequence of the weather acting directly upon those plants. This all operates through variations of the usually very high mortality of newly hatched nymphs. Most quickly starve; occasionally many more survive. Predators have no influence upon these huge changes in abundance.

Farrow's (1982) studies have since confirmed this association for the Australian plague locust, *Chortoicetes terminifera*, in central New South Wales. The major influence on changes in their abundance in each generation was the amount of rainfall acting through its influence upon the amount of green feed for the hopper stage of the life cycle (all instars prior to the winged adult). This was when most locusts died in all generations that Farrow monitored. Mortality averaged 89% and was never less than 66%. The greatest proportion of these deaths was always in the first instar, soon after hatching. These early deaths were fewest when eggs hatched in recently flushed green pastures, but were as great as any recorded during droughts if the eggs hatched in mature but still green pasture. So access to *currently growing* grass, high in soluble nitrogen, is of major importance to the survival of these neonates. Even then, when conditions seemed ideal, 60% of them disappeared without trace! At best survival was a chancy business – mostly less than 10% of the population ever get past the first few hours of life; they starve. But a small difference in the stage of growth of the plants they eat may increase to 40% the proportion of these neonates that survive – enough to generate a massive increase in the population.

Clearly *C. terminifera* is a flush feeder, on a par with kangaroos and feral rabbits (Sects. 6.5.4 and 6.2.5.2 and Part V) in its ability to respond to unpredictable flushes of good food following rain; so are many other locusts and grasshoppers. But some are either senescence feeders or, like many aphids, can track the best source of available nitrogen wherever and whenever it occurs in the plant. Some American grasshoppers fall into this latter category.

Lewis (1984) had observed that the nymphs of several species of Texan grasshoppers appeared to prefer to eat parts of plants which were senescing at an accelerated rate. She counted many more of them feeding upon tissues which had previously been damaged by other chewing insects, had been infected by rust fungi, or had wilted following attack by stem-girdling beetles. On the chewed leaves the grasshoppers confined their feeding to tissues made necrotic by the earlier damage, and on the diseased leaves they ate only the rust and the tissues immediately surrounding it. Lewis tested these observations with one of these grasshoppers, the generalist-feeding *Melanoplus differentialis*, on

Fig. 7. *Melanoplus* grasshopper nymphs have a fairly catholic taste, but strongly prefer, and survive and grow better, on plants which have senesced faster as a result of being wilted. Photo courtesy of A. C. Lewis

the leaves of the native sunflower, *Helianthus annuus*. When she gave nymphs the choice between leaves infected with *Puccinia helianthii* and clean leaves, they preferred the former. Adults chose leaves she had wilted by damaging their petioles, ahead of turgid ones.

In the field she had noted that grasshoppers preferred to eat the inflorescences and senescing leaves of otherwise healthy sunflower plants. They would, however, choose turgid leaves of naturally *less* vigorous plants ahead of those of vigorous ones, and of those plants which had *not* been fertilized ahead of those which had. They also preferred, ahead of leaves which had never been wilted, leaves which had wilted on water-stressed plants and subsequently recovered full turgidity. These leaves did not need to have been wilted for very long and needed to have lost as little as 5% of their water to be preferred. She also observed that early instars were much more likely to choose wilted tissue than were older insects.

I suspect it is very likely that these last three preferences, especially that for previously wilted leaves which had regained full turgidity, would more often than not go unnoticed during normal field observations. How often have such important influences on the success of a herbivore gone undetected in the past? Note too the similarities with Dixon's aphids (Sect. 3.5.1) and McNeill's mirids (Sect. 3.5.4).

Lewis found that *M. differentialis* showed the same preferences for reproductive, senescing and damaged tissues on 21 different species of plants. She found also that the quality of the plant's tissues would override a grasshopper's usual choice between species that it would eat. In the laboratory she could reverse the first instars' choice between two species by wilting the less preferred one. Throughout, all indications were that chemical changes in the plants rather than a simple lack of water were responsible for the observed responses of the insects. All these preferences clearly indicated a seeking out by the insects of parts of a plant where more of its nitrogen is in soluble form. That this, in turn, leads to improved chances of surviving and reproducing was shown when Lewis raised grasshoppers from hatchlings to adults on two different diets. One batch was given a mixture of wilted and turgid sunflower leaves. They were free to eat whichever sort of leaf they chose; a situation, Lewis said, more akin to what they would find in nature than being confined to wilted tissue. A second batch of siblings of the first lot was fed only turgid leaves. More of those which were able to eat wilted leaves survived, they grew faster, and produced more and heavier eggs than did their siblings raised on turgid leaves alone.

Other workers have reported a preference by various acridids for the tissues of water-stressed plants (Boutton et al. 1980) and Lewis has found that the wilted tissues of many species of plants are more palatable to the African desert locust, *Schistocerca gregaria* (Bernays and Lewis 1986). So this is not an isolated case. And Haglund's (1980) discovery that several species of grasshoppers on Montana grasslands show a preference for grass which he had treated with proline and valine suggests that acridids are well able to detect the amino acids released by stress-induced breakdown of leaf proteins. Furthermore, Simpson and Abisgold (1985) showed with laboratory-reared African *Locusta migratoria* that acridids are well able to distinguish the relative amounts of protein and carbohydrate in their food. When the proportions are varied, they regulate their intake with respect to the protein and not to the digestible carbohydrate. There is also evidence from the field and laboratory of the crucial importance to the success of acridids of the level of protein in their diet. For example, Grayson and Hassall (1985) measured the effect of excluding high numbers of rabbits from a Norfolk heath on the success of the grasshopper *Chorthippus brunneus*. The heath is a closely cropped mixed turf of grasses (principally *Festuca ovina*), numerous dicotyledons, mosses, and lichens. Within 5 months of excluding the rabbits Grayson and Hassall found marked and continuing changes. The biomass of dicotyledons increased throughout the 2 years of exclusion. Cessation of grazing by the rabbits resulted in a dramatic three- to four-fold increase in flowering in the turf. In spring, during the critical period of growth of grasshopper nymphs, the nitrogen content of grass leaves from within the exclosure was higher than those from the open heath (but in the summer it was higher in the grass kept close cropped by the rabbits). The grasshoppers responded to these changes within the exclosure with denser populations in which more nymphs survived and grew faster and larger to become more fecund adults which laid more eggs. In the second year following exclusion of rabbits these responses were even more marked. In the laboratory grasshoppers raised on *F. ovina* leaves from the exclosure grew faster and larger than those fed on leaves from the open heath. In other words, more new growth of both dicotyledons and grasses, and the production of more flowers all meant more mobilized soluble nitrogen available in the young nymphs' food.

Access to tissues in which the nutrients are being actively transported in soluble form is as important to acridids as it is to other herbivores. This was nicely highlighted by McCaffery (1975) when investigating the influence of the quality of the food on the production of eggs by *Locusta migratoria migratorioides* raised in the laboratory in London. In a series of experiments newly emerged female locusts were fed daily upon freshly cut plants of the grass *Agropyron repens*, either on new flush plants, or on older ones still with green stems but with leaves starting to go brown. The former had a mean dry matter content of 12% and 22.4% crude protein; the latter had 40% dry matter and only 10.7% protein. Females eating ad lib. amounts of the good food prospered. They gained weight, mated and laid many eggs. Females fed with the poor quality plants fared badly however. They eventually lost weight and some of them died. None showed any sexual activity and, of course, they produced no eggs. Switching another lot of females from good food to bad after they had completed somatic growth, but before they had mated, caused them to stop putting on weight and to lay practically no eggs. The reverse switch caused females to start gaining weight and, although they mated and started to lay later than the first group, they gradually increased their production of eggs. So, having survived to adulthood, a female locust's chances of reproducing depends almost entirely upon her gaining access to newly flushing plants. While these plants had more than twice as much protein as the older plants, they would also have had a much greater proportion of that nitrogen present in soluble form; a situation clearly analogous to that of various vertebrate females discussed in later chapters.

Tests with some generalist feeding acridids have lent support to the hypothesis of Caswell et al. (1973) that C_4 plants are likely to be a poorer source of food for herbivores than C_3 plants, and thus avoided by them. The two sorts of plants differ fundamentally from each other in the biochemistry of their photosynthetic pathways of fixing CO_2 (Bidwell 1983). The new C_4 pathway was only discovered in the 1960s (Hatch and Slack 1970). On average C_4 plants have approximately only half the protein of C_3 plants, and more carbohydrate. In particular they have much more structural material; bundle sheath cells comprise a high proportion of all cells present. These cells are especially difficult to digest because of their thick walls, and they contain the majority of the leaves' protein and starch. There are, of course, many insects which eat either C_4 or C_3 plants with equal facility. Others are specialists on C_4 plants and show expected adaptations (such as the morphology of their mouthparts) associated with eating such plants (Boutton et al. 1978; Bernays and Barbehenn 1987). But when acridids which will eat a wide variety of plants are confronted with a choice they seem to favour C_3 plants, taking them in greater proportion than would be expected from their abundance in the habitat. Also, females show a much stronger preference for C_3 plants than do males. These responses would be expected to maximize the grasshoppers' intake of protein, the greater bias of females reflecting their greater need of protein for reproduction (Boutton et al. 1980).

Caswell and Reed (1976) did a simple experiment which dramatically illustrated why C_4 plants are not favoured by generalist herbivores which lack the specialized adaptations for feeding upon them. They fed fresh, actively growing C_4 grasses to grasshoppers belonging to ten North American species, and then examined their faeces. These were mostly bright green and contained large numbers of unbroken sheath cells with their contents still intact. So, not only is there less protein to start with in C_4 plants, the nutrients in their cells are not accessible to these grasshoppers.

Where the problem has become acute, as with acridids which eat only harsh grasses, the insects' mandibles have become modified to crush as well as bite (Bernays and Barbehenn 1987). Apparently the acridids have not developed the capacity to digest the contents of intact cells that leaf-chewing lepidopterous larvae have done (Barbehenn 1992). This means that quite small changes in the concentration and composition of the nitrogen in the sap of the cells that are ruptured could mean the difference between life and death, especially for a neonate. The total nitrogen in the plant is irrelevant. How much is accessible for quick absorption from the gut is the crucial factor.

Finally, for acridids, what part might animal protein play in the nutrition of these herbivores? There is no evidence that they benefit in any way from symbiotic microorganisms, but many orthopterans, not just acridids, otherwise seemingly strict herbivores, will readily and commonly eat dead insects, including their own species. For example, 21% of the diet of the mormon cricket, *Anabrus simplex*, a major pest of crops in parts of the USA, is made up of arthropods, and the diets of several species of Colorado grasshoppers include arthropods. For some of them arthropods comprise a major source of food (Hansen and Ueckert 1970; Ueckert and Hansen 1970, 1971). Farrow (1982) occasionally observed cannibalism at hatching in egg beds of the Australian plague locust, even in the presence of fresh green food. This seeking after animal food is particularly common among the acridids, and often involves eating conspecifics. Lockwood's (1989a, b) investigation of cannibalism (or, more strictly, "necrophagy", the eating of cadavers) by a complex of grasshoppers species on Colorado grasslands illustrates this. He placed dead grasshoppers at 15 m intervals in a field and recorded what happened with a video camera. Most cadavers were found in less than 10 minutes and often had attracted six or more grasshoppers within 20 minutes. Feeding was actively contested, with frequent jostling and kicking. Clearly grasshoppers eagerly and actively seek out animal protein, including the bodies of their own kind. Of added significance, Lockwood found that the behaviour is most common in adults, especially in females, where it coincides with oogenesis, mating, and the deterioration of their food plants. Might careful observation reveal that cannibalism and scavenging for dead animals is more common among acridids than was realized? It seems unlikely that eating animal tissue is a usual or necessary component of their nutrition, especially when they are in plague proportions on cultivated crops. But it may play a crucial role in maintaining a small breeding population when good plant food is scarce, a time of great importance to the persistance of a natural population (Sect. 3.4.5 and Chap. 12).

3.5 Sap-sucking Insects

3.5.1 Aphids

Aphids, like the Lepidoptera, are a group of insects of great economic importance as pests of agriculture, horticulture and forestry. The same sort of caveat that I gave about extrapolating from studies of pest Lepidoptera to explain their distribution and abundance in nature holds for aphids which are pests of cultivated plants. They are pests because they quickly multiply to enormous numbers in response to humans increasing both the quantity and quality of food in their environment. But again, like the Lepidoptera, the ecology of aphids which feed on trees is likely to be less removed from its evolved state than is, for example, that of aphids attacking cruciferous vegetables. Yet there are very many studies, some now more than 50 years old (e.g. Evans 1938), of these latter sorts of aphids which have clearly shown that, in spite of the benefits bestowed by these manipulations, the successful breeding and abundance of aphids is still largely determined by the availability of nitrogen in the form of amino acids in their diet (Auclair and Maltais 1950; Kennedy 1958; Mittler 1967; Jansson and Smilowitz 1986). As with other sorts of herbivores this is especially crucial for the survival of the very young. First instar nymphs can take four times as much nitrogen per unit weight as fourth instars, and five and a half times more than adults (Kennedy and Stroyan 1959). This comes about because parthenogenetic aphids invest in both somatic and gonadal growth during their larval development, unlike most animals which grow first and then develop reproductive tissues. As gonads grow exponentially, while somatic tissues grow logistically, the ratio of soma to gonad changes, so that a large percentage of an adult aphid's body is made up of gonads (Brough et al. 1990).

Nearly all aphids feed directly on the phloem sap of their food plants. This is high in sugars and low in nitrogen. The major nitrogenous compounds in the phloem are amino acids, usually of a grossly unbalanced composition. So small changes in the concentration of nitrogen in the phloem can have a major effect on the aphids' survival. They must therefore ingest much more carbohydrate than they need for their energy budget. Most of this excess carbohydrate is passed in their liquid faeces (honeydew) either unaltered or converted to more complex oligosaccharides (Auclair 1964; White 1974; Fisher et al. 1984). This point was well illustrated by a study of the energy and nitrogen budgets of *Macrosiphum liriodendri* living on tulip trees, *Liriodendron tulipifera*, in a

Tennessee mixed hardwood forest. Van Hook et al. (1980) found that while these aphids consumed about 400% of their dry weight each day, they removed a mere 1% of the trees' annual production of photosynthate. However, this represented the equivalent of 17% of the annual standing crop of nitrogen in the trees' leaves. They remove nearly all this in the short time following budburst, when amino acids are being mobilized from storage into the production of new leaves (they remove a smaller proportion in early autumn when amino acids are being transported from leaves to storage prior to the abscission of the leaves). However, while 65% of the carbohydrate the aphids ingested was passed in their faeces, the authors could find only 2% of the ingested nitrogen in their honeydew.

Many studies have demonstrated that the observed performance of aphids in the field depends on an adequate supply of soluble nitrogen in their food. There are several examples in my earlier discussion of flush and senescence feeders, like Parry's study of the green spruce aphid. Many aphids, however, rather than being specialized to feed on flushing or senescing tissues, have evolved to seek out and feed upon *both* sorts of tissue. But they too are usually short of enough nitrogen to survive and breed.

Dixon's (1963, 1966, 1970, 1976) elegant long-term study of the sycamore aphid, *Drepanosiphum platanoidis*, in Scotland is a good example. *D. platanoidis* is specific to sycamore, *Acer pseudoplatanus*, throughout the British Isles, but Dixon studied populations living in just a few trees in and near Glasgow, and on seedlings maintained in the laboratory. He found that many and various factors in the environment played some part in determining how many aphids there were living on a sycamore tree during each season, and from year to year. Above all, however, it is access to assimilable nitrogen in its food which drives the ecology of this insect. *D. platanoidis* extends the time over which it can gain access to an adequate diet by tracking the source of the production of soluble nitrogen in the plant. In the spring it reproduces while feeding on the high-amino-nitrogen sap flowing to the actively growing leaves, and in the autumn on the outflow of nutrients from the senescing leaves. In the summmer, when the leaves are mature and growth has ceased, it breeds little or not at all. Individual aphids partially compensate for the low level of amino-nitrogen in the sap of the mature leaves by feeding more quickly and efficiently; they use all the amino acids that they ingest and pass none out in the honeydew. But they nevertheless enter a reproductive diapause. So, when the concentrated flow of soluble amino acids in the phloem sap ceases, aphids can maintain themselves (with some extra effort) but they cannot reproduce, nor can their young grow.

Dixon made some astute observations and did some simple experiments which highlighted and confirmed the aphids' dependence on this supply of soluble nitrogen. In the spring the aphids feed as close to the greatest concentration of food being imported to the new leaves as the length of their stylets permits – on the petioles and stems. When growth ceases in summer they move out over the lamina of the leaf and feed from the veins closest to the site of synthesis of nutrients. Here the amino acids being exported to the rest of the

Fig. 8. The sycamore aphid is a flush and a senescence feeder. Nutrients flowing into growing leaves support its spring reproduction, those flowing from senescing leaves its autumn reproduction. Aphids can survive on summer leaves, but cannot breed or grow. Photo courtesy of A. F. G. Dixon

plant will be most concentrated, and increasingly so as the leaves senesce. (Compare this change of feeding behaviour with the two distinct behaviours of the flush-feeding and senescence-feeding Australian psyllids which I discussed in Sect. 3.1.1.)

Dixon cut off the lamina of several mature summer leaves with the result that their petioles senesced ahead of those of the surrounding entire leaves. These senescing petioles immediately became attractive to the aphids, which moved on to them and bred, while the average number of aphids on the laminae of whole leaves continued to decline.

An odd twig or branch in the crown of a sycamore tree will occasionally become stressed, and the leaves on it will senesce prematurely, eventually turning yellow ahead of the rest of the foliage. Several times Dixon observed that when this happened there were more aphids on these leaves in the summer, and that they had come out of their reproductive aestivation and were breeding while those feeding on surrounding mature leaves had not. He collected both sorts of leaves and measured the amount of nitrogen in them. The prematurely senescing ones had exported much more of their nitrogen than had the still-green ones.

Finally, he found that depriving potted sycamore saplings of nitrogen had no effect on the spring populations feeding on the new growth. However, it did result in the reproductive rate of the autumn aphids feeding on the senescing leaves being nearly twice as great as that of aphids feeding on the control plants. The reason for this was that, before leaf-fall, the nitrogen-starved plants had salvaged nearly twice as much nitrogen from their senescing tissues as had the control plants. The translocation of this extra nitrogen back into the stem of the plant had therefore resulted in a highly nutritious food becoming available to the aphids. Since then Shearer (1976; see Sect. 3.1.5) has shown that in the autumn *D. platinoidis* will congregate and feed alongside aggregations of the senescence-inducing aphid *P. acericola*. Again, the aphids are tracking a source of enrichment of the amount of nitrogen being transported in the phloem.

There are numerous other species of aphids which track the supply of soluble nitrogen in this way. Their growth and reproduction varies throughout the season, and from season to season, according to the quality and amount of nitrogen available. The birch aphid, *Euceraphis punctipennis*, is one such species (Wratten 1974). In the spring they are evenly distributed on the leaves of *Betula pubescens*, but as most leaves approach maturity they concentrate on the young terminal leaves. Still their rate of reproduction falls in summer as the supply of soluble nitrogen declines. During summer any leaves which are injured by chewing insects senesce prematurely, and there are more aphids on such leaves, as there are in late summer on the first of the undamaged leaves which start to senesce. *Cinara pini* feeding on the stem and branch internodes of Scots pine, *Pinus sylvestris*, behaves in the same way (Larsson 1985). It moves to different parts of the tree to feed where seasonal changes in the flow of soluble nitrogen in the phloem are generated by the changing photosynthetic rate and ageing of needles produced in different years.

Schaefers and Montgomery (1973) at the New York State Agricultural Station produced another compelling demonstration of the dependence of aphids upon the level of available soluble nitrogen, irrespective of the age of the tissues which provide their food. They worked with *Chaetosiphon fragaefolii* feeding upon the leaves of strawberries, *Fragaria vesca*. When they reared these aphids on the tips of growing runners of strawberry plants they grew to an average length of 1.51 mm, on young growing leaflets to 1.41 mm (6.6% smaller), and on mature to overmature leaflets to only 1.08 mm (28.5% smaller). These results suggested that aphids feeding on growing tissues received better nutrition than those feeding on mature tissues. Yet when they grew aphids on leaf discs floated on distilled water and cut from leaflets opposite those mature and overmature ones on which the aphids had grown least of all, the aphids grew to a mean length of 1.55 mm – larger than those on the runners. Excising mature leaves hastened the senescence of their tissues and caused more soluble nitrogen to be released more quickly from the breakdown of leaf proteins. Discs which Schaefers and Montgomery cut from leaves and floated for 5 days on distilled water contained 58% more soluble nitrogen than uncut leaves. They further demonstrated that it is the release of additional nitrogen from the accelerated senescence of excised tissues which improved the nutrition, and thus the growth, of the aphids, by growing aphids on discs floated on solutions of the cytokinin N^6benzyladenine (N^6B). Cytokinins inhibit senescence in excised leaves. After 5 days floating on a 10.0 ppm solution of N^6B leaf discs contained 29% less soluble nitrogen than did the uncut leaves, and 55% less than the discs floating on water. Newly born aphids raised on leaf discs on N^6B solutions grew to adults which, relative to those fed on discs on distilled water were, respectively 4, 12, 30 and 37% smaller on 0.1, 1.0, 10.0 and 20.0 ppm N^6B solutions. The authors demonstrated that this response was not the result of a direct influence of cytokinin on the aphids, but an indirect one via the physiology of the leaf. *C. fragaefolii* does not grow well on artificial diets so they fed *Myzus persicae* on artificial diets with and without the addition of double the maximum concentration of N^6B fed to the strawberry aphids. This had no effect upon either the rate at which the aphids fed, or their growth.

3.5.2 Psyllids

The psyllids are another group of sap suckers closely related to the aphids. They also feed exclusively from phloem, so are confronted by the same grossly imbalanced diet – a large surplus of sugars and a shortage of nitrogen. They too must pass much of the carbohydrate they ingest through their alimentary canal. They too must attempt to feed where and when the nitrogen is most concentrated, or manipulate the plant's physiology to prolong and concentrate its availability. My own work with psyllids living on eucalypts in South Australia (White 1969, 1970a, b) showed that their abundance was determined largely by the quality of their food which is, in turn, a function of the changing physiology of the food plants. The trees' physiology changes during the seasonal phenology of their growth and reproduction, and in response to a variety of factors – but especially the weather – which impinge upon them (see also Part V).

Later work on a total of five species of lerp-forming psyllids (three senescence feeders; two flush feeders) feeding upon the leaves of *Eucalyptus blakelyi* in the Australian Capital Territory confirmed and extended my conclusions (Clark and Dallwitz 1974, 1975; Journet 1980). Variations in the abundance of all five of these psyllids was a function of the favourability of the host plant as a source of food for them. The amount and quality of this food is influenced by many factors in the environment, such as the amount and timing of growth of new leaves; rainfall; the age, vigour and location of individual trees; and the feeding behaviour of the insects (i.e. the senescence feeders "plug in" to small ultimate vascular bundles, the flush feeders into large ones: some of the former have the capacity, like *C. densitexta*, to hasten the senescence of the leaf tissue, while some do not). The major proximate influence, however, was the physiological changes in the leaves affecting the quality of the sap the insects ate. These changes were associated with the growth and senescence of the leaves and, especially, with the changing patterns of the flowering of the trees. As I had found, parasites and predators played no significant part in how many insects survived in each generation. It depended on there being an adequate supply of soluble nitrogen in their diet, and, as with all animals, this is most critical for the neonates. In each generation most individuals of all five species died as first instars. These tiny beasts were usually found safely beneath their lerps, mouthparts inserted into the phloem, unmarked, but dead.

All these Australian psyllids are indigenous insects living on the species of plant with which they evolved. Human interference has been minimal, usually only that associated with clearing and farming, or the planting of eucalypts as urban trees. The relationships and responses of the psyllids to their hosts seem to be undisturbed by such interference, and are the same as those on trees in uncleared bush. So they present few, if any, of the problems alluded to before which are associated with studying pests of crops.

Sutton (1984) also found that changes in the concentration of soluble nitrogen associated with changes in the flowering of trees can have a considerable influence upon the success of psyllids. *Psylla perigrina* and *P. subferruginea* feed upon the new growth and inflorescences of hawthorn, *Crataegus monogyna*, in England. The distribution on the trees of the growing nymphs of these two flush feeders closely tracks the availability of soluble nitrogen in the infloresences and growing tips throughout the season. But individual hawthorn trees have a biennial flowering pattern, and their tissues have higher concentrations of soluble nitrogen in their flowering years than in their non-flowering years. Sutton found that the psyllids grew to be larger adults on flowering than on non-flowering trees, and that their size was closely correlated with the flowering patterns of individual trees.

A study by Webb and Moran (1978) at the University of Grahamstown in South Africa of populations of the psyllid *Acizzia russellae* was another that had none of the possible drawbacks of studying pests of cultivated plants. Their study also showed quite clearly that an adequate supply of nitrogenous food was the only significant factor influencing the success of the psyllids. *A. russellae* is an "innocuous non-pest insect" which feeds from the phloem of young flush leaves of only one species of plant, the thorn tree, *Acacia karroo*. Both insect and plant are indigenous to southern Africa. Webb and Moran chose to study this psyllid specifically because it provided an example of an undisturbed and ancient evolved association between a herbivore and its food plant.

The insect is never very abundant. Predictably, populations increase to a peak each spring when the thorn trees flush, but its numbers vary from tree to tree and from year to year. A small and patchy population persists throughout the year because there is always some flush available on some trees. Previous work had shown that the level of abundance of *A. russellae* was not being materially influenced by the weather, nor by parasites and predators. From laboratory rearings and from measurements in the field Webb and Moran showed that fewer psyllids survived, they grew more slowly, and were smaller on low-nitrogen plants. In the field their numbers were correlated with the amount of nitrogen in the foliage from which they fed.

The authors could, however, create local "mini-outbreaks" of the psyllid simply by cutting a number of thorn trees back to ground level in the spring. The trees responded by vigorous regrowth, producing much more flush foliage much more rapidly than uncut trees. On the leaves of this regrowth the number of psyllids exploded to reach mean densities per sample shoot 20 to 40 times greater than those ever achieved on unpruned trees. The authors found that there was no difference in the level of nutrients (or potential feeding deterrents) in the flush leaves of pruned and unpruned trees. The *quality* of the food was the same in both. The normal flush of these trees could provide this good phloem only patchily distributed and in small quantities over the entire period of spring growth. The number of psyllids was able to explode because the regrowth provided the same high grade phloem, but in much larger amounts, and all at the one time and place.

An example from a situation even less disturbed by human activity is that of *Psylla palmeni* and *P. phlebophyllae* which live on several species of willows (*Salix*) in arctic Alaska. Hodkinson et al. (1979) found that in the short time available each season in this very harsh environment, these insects had evolved to concentrate their entire life cycle on the earliest appearing and most nutritious part of their food plant – the female catkins. They must complete their development in the time the catkins take to mature and burst, about 40 days. Female psyllids will lay their eggs only on catkins, and virtually all nymphs complete their development on them. The female catkins of arctic willows are nutrient sinks with a level of soluble nitrogen higher than that of any other plant component growing on the Alaskan tundra.

There are several species of psyllids living under conditions which are the reverse of these examples. These psyllids attack fruit trees that are intensively bred, cloned, fertilized, irrigated, and propagated where they would never have grown naturally. Yet they all evolved on wild ancestral trees in the face of the same shortage of nitrogenous food. Today such psyllids are so successful because we now provide them with that which they never encountered in nature – vast amounts of phloem containing elevated levels of soluble nitrogen, and available over periods extending far beyond natural flushes of new growth.

The pear psyllid, *Psylla pyricola*, is one of these. It is a major pest of pears, *Pyrus communis*. It is a flush feeder seeking out the young new growth and water shoots of pear trees. Pfeiffer and Burts (1983, 1984) showed that in Washington State females lay their eggs preferentially, and more nymphs survive and grow faster to produce denser populations, on trees which had higher levels of amino acids in the phloem. Also Westigard et al. (1980) in Oregon found that when they reduced the amount of new flush on pear trees by applying the plant growth regulator daminozide to them, they reduced the population of pear psylla on their leaves to only one-seventh of that on untreated trees.

Another flush-feeding psyllid which is a major pest of an intensively cultivated crop is *Trioza erytreae* (Catling 1971). It is the principal vector of the viral "greening disease" of irrigated citrus in South Africa. Temperatures are warm enough in the orchards in the northeast of the country where Catling worked that citrus may be forced to flush at almost any time by manipulating the water supply. So there is some new growth on some trees at all times of the year, but the major flush of new growth in all areas is still in the spring. Catling found that parasites, predators, and the weather had only minimal influence on the survival and rate of growth of populations of *T. erytreae*. Their abundance was determined almost entirely by the amount, distribution, and nitrogen content of new growth on the trees. As with all plants, the nitrogen content of the flush growth of citrus is very high, but soon declines as the leaves mature. Young trees flush for longer and have higher levels of nitrogen in their new growth than the new growth of mature trees. Catling discovered, however, that on any given tree, the spring flush contained higher levels of nitrogen than did new growth produced at other times of the year, especially in winter. Populations of *T. erytreae* complete six or seven generations throughout the year, living only on flush tissue (they will not even lay their eggs on mature leaves). But they attain maximum survival and growth, and increase to peak density only when they can feed on the spring flush.

This dependence upon a high level of nitrogen in flush tissue was serendipitously highlighted one spring in one orchard. Catling found abnormally high mortality of early instars, prolonged development of survivors, and much smaller adults in a population which was feeding on succulent but chlorotic flush foliage. Analysis revealed that this foliage had only half the nitrogen content of normal green trees in a nearby grove where the psyllids had developed normally.

3.5.3 Scale Insects

Many of the Coccidae, the scale insects, are major sap-sucking pests of our cultivated plants. McClure's (1980) studies of the elongate hemlock scale, *Fiorinia externa*, is a clear example that they too depend on the level of nitrogen that is available in their food. This armoured diaspidid scale was introduced from Japan early this century and is now a major pest of eastern hemlock, *Tsuga canadensis*, in northeastern USA. Like all scales the insect disperses and finds new food plants as first instar nymphs. These "crawlers" of *F. externa* settle preferentially on the new growing needles of hemlock. McClure demonstrated that the survival, growth rate and subsequent fecundity – and hence the density – of this flush-feeder depends upon the concentration of nitrogen in those needles. Fertilizing trees with ammonium nitrate in spring prior to flush of new needles raised the concentration of nitrogen in the subsequent young needles and increased the survival, rate of development, and fecundity of scales settling on those needles. He also found that the level of infestation reached by *F. externa* on 14 different species (in five genera) of native and introduced conifers growing in Connecticut was influenced by the phenology of each of these species in relation to the time of dispersal of the crawlers. The older the new needles, the lower their nitrogen content, and the lower the survival, rate of development, fecundity and density of *F. externa* on them. As McClure concluded: "Nymphs on trees with foliage providing a poor source of nitrogen suffered greater mortality, developed at a slower rate and ultimately produced fewer progeny than did nymphs feeding on trees with high-nitrogen foliage."

Sadof and Raupp (1991) recognized that McClure's results represent the response of the scale insects to changes in the *form* and *mobility* of nitrogen in the plant in the same way as did Newbery's work with *Icerya* on Aldabra atoll (see Sect. 3.1.6). For both insects it was access to the flow of soluble nitrogen – in for the flush-feeding *F. externa* and out for the senescence-feeding *Icerya* – which was important, not the total amount of nitrogen present in those tissues at any one time. Anything which increased the concentration and rate of that flow benefits the herbivore. These authors nicely illustrated this benefit for another scale insect and an aphid. The fecundity of both *Unaspis euonymi* and *Aphis fabae* increased when they were fed on the variegated leaves of *Euonymus japonica* var. *aureus* rather than on the green leaves of non-variegated plants of this shrub. This came about because the variegated shrubs compensate for the loss of photosynthetic capacity by increasing their leaf area. To do this they had to mobilize more nitrogen into those leaves. The insects feeding from the consequent increased inflow in the phloem benefited accordingly. There was also a behavioural response to this improved flow of good food. Adult *A. fabae* fed more often from the veins of variegated than green leaves.

3.5.4 Planthoppers, Leafhoppers, and a Mirid

Moving to a quite different family of sap suckers, the cicadellids, the Japanese green rice leafhopper, *Nephotettix cincticeps*, like many of the aphids, has the capacity to opportunistically track the best source of nitrogen wherever it may occur in the plant (Andow 1984). Basically, the hoppers have two preferred feeding sites on rice plants; new leaves and senescing leaves. The former are most preferred when still expanding (i.e. at the peak of import of nutrients) and the latter when first becoming chlorotic (i.e. at the peak of export of nutrients). Previous workers had demonstrated that *N. cincticeps* survived better, grew faster, and reproduced more on well-fertilized plants, seeking out the growing tips where the concentration of soluble nitrogen was greatest. When Andow varied the level of nitrogen in rice plants by differentially fertilizing them, he found that the hoppers preferred the growing tillers of normally fertilized plants, but the senescing leaves of *underfertilized* ones!

The most parsimonious explanation of this unexpected result is again that the insects were seeking out the best source of the most readily assimilable nitrogen in the plant. They concentrated on the senescing leaves of the underfertilized plants for the same reason that Dixon's sycamore aphids bred more profusely on nitrogen-starved plants. Such plants salvage much more of the nitrogen in their senescing leaves, thus increasing the supply of soluble nitrogen available to insects feeding from the phloem sap of those leaves.

For planthoppers in general Denno and Roderick (1990) have summed up their dependence upon nitrogen in their diet, and the adaptations they have evolved to obtain sufficient of it to survive and breed. Like other sap suckers which I have been discussing, they depend principally on their capacity to select the plants, and parts of plants, which have the most nitrogen in their sap, and the time of year (spring and autumn) when the flow of nitrogen in that sap is greatest. They can also increase their rate of feeding, and some of them can modify the physiology of the plant so that more nitrogen is released into the phloem. Finally, the authors point out that the endosymbionts which all leafhoppers contain may be synthesizing essential nutrients for them.

Endosymbionts are mostly prokaryotes, probably reduced forms of bacteria. A few are yeasts. They are contained *within* specially modified cells (mycetocytes) of the host's body (as distinct from the typical gut microbes which live in the lumen of the digestive tract). Again, unlike ordinary gut microbes, they are passed from generation to generation via the ovaries and eggs. They are widespread among insects, particularly those which eat a nutritionally poor or unbalanced diet, and are universally present in those which feed on phloem sap. They are essential to the survival of their hosts, but their role is not well understood. What evidence there is, however, strongly suggests that they are concerned with the nitrogen nutrition of their hosts, and especially that of actively reproducing females, most probably synthesizing essential amino acids which the host cannot obtain from its food nor synthesize for itself.

On the other hand, not all insects which feed on phloem have extracellular bacteria in their guts. However, those that do apparently can grow but little, and survive for only a short time, without them (Douglas 1988, 1989). It is a fairly safe bet that, as has been shown for many other vertebrate and invertebrate herbivores, these gut microbes, when present, provide an essential supplement of nitrogen to their hosts' otherwise inadequate diet.

A study by McNeill (1973) in England of an insect from yet another family of sap suckers, the mirid bug *Leptopterna dolobrata*, yielded another picture typical of the flush feeders. It is an insect with a wide range of host plants, but commonly feeds on the grass *Holcus mollis*. Eggs hatch and first instar nymphs start to feed on newly growing leaves with high concentrations of nitrogen. Nevertheless the deaths of these and the second instars constitute the greatest loss of individuals during the life cycle. McNeill could not explain this, but noted that more nymphs died when there was more rain (he thought they might drown). Parasites were not a significant influence on the abundance of the population. Later in the life cycle he noted an important change in the feeding behaviour of the nymphs that survived this early mortality. As their growth rate approached its maximum and their gonads started to form, they moved from the leaves to the flowers and started to feed upon the developing seeds.

To see how important this shift of diet might be, McNeill reared nymphs in the laboratory either just on *Holcus* leaves or with flowers as well. Practically none of the nymphs denied access to flowers survived. The few which did reach adulthood were very weak, their gonads had not developed, and they died in a day or so. Many more of those allowed to feed on *developing* seeds in the flowers survived to adulthood. They also lived longer and the females amongst them laid eggs. So, at this time in the life cycle when their need for nitrogen in the diet was at a peak, the nymphs abandoned the maturing leaves in which the level of nitrogen was rapidly falling, and moved to feed from the concentrated inflow of soluble nitrogen to the developing seeds.

As McNeill's experiment showed, access to this nitrogenous food is especially critical for the females if they are to be able to reproduce. As we shall see, access to the nutrients translocated to developing seeds can be crucial to the successful breeding of other females (Sects. 6.2.2.1 and 7.3.4). The explosive growth of the neonates provides the other peak demand for nitrogen in the life cycle. I would hazard the guess that the reason that so many of them die is that most are unable to get enough nitrogen, even from fast growing new leaves.

Prestidge and McNeill (1983) have illustrated that the abundance of phloem feeders is not simply a matter of the total amount of soluble nitrogen present in the phloem. From their studies of a complex of species of homopteran leafhoppers which feed upon species of *Holcus* grasses in England, they discovered that an "amino acid quality index" was more important. This is the ratio in the phloem of a group of essential amino acids, or their precursors, to a group of non-essential (or possibly toxic) ones. The amino acids in the non-essential group stay at relatively low and constant levels from season to season and from year to year. But the essential ones peak during the flush of nutrients associated with growth, reproduction, and senescence of the plant each year, and vary greatly from year to year. This means that in years when the total levels of amino acids are low, the potentially harmful group is relatively much more important, and can form almost half of the total pool of amino acids available to the insects. So, not surprisingly, Prestidge and McNeill found that hoppers were much more abundant at times when there were more of the essential amino acids in the phloem and the potentially harmful group madc up less than 25% of the total.

3.5.5 Xylem Feeders

There is a much smaller number of species of sap-sucking insects which have evolved to feed entirely from the xylem of plants (Raven 1983). These insects are confronted with an even more impoverished diet than that provided by the phloem. Xylem contains orders of magnitude less carbon and nitrogen than phloem; it is greater than 98% water, by far the most dilute food encountered by herbivores. However, its nitrogen to carbon ratio is much higher than that of most other plant tissues – up to 98% of all organic matter in it may consist of amino acids (Wiegart 1964). This means that insects which feed exclusively on xylem must use nitrogen as a source of energy as well as for maintenance and growth. But the problem is compounded because often the little nitrogen the xylem does contain is even more nutritionally unbalanced than that in the phloem. There is an excess of non-essential amino acids. So xylem feeders are not so much faced with a shortage of nitrogen in a sea of carbohydrate, like their phloem-feeding cousins, but with a shortage of nutritionally unbalanced nitrogen in a sea of water!

In the light of all this it is no surprise to find that xylem feeders depend for their success on being able to seek out xylem which contains the highest level of nitrogen, and to have the capacity to use virtually all and any of that small amount of nitrogen. Neither is it surprising to find that in spite of such adaptations they ingest sap at a far greater rate than do phloem feeders, produce greater quantities of very watery faeces, and grow much more slowly. The periodical cicada, *Magicicada cassini*, which takes 17 years to grow from egg to adult, feeding on the xylem of the roots of hardwood trees, is an extreme example of this slow growth (White and Strehl 1978).

Horsfield (1977, 1978) demonstrated these special features of xylem feeders through his studies of the spittlebug, *Philaenus spumarius*. Feeding on a range of host species its nymphs can ingest 150 to 250 times their own body weight in 24 hours (phloem-feeding aphids can achieve 8 to 32 times) and take 10 weeks to complete their development, something most aphids do in less than a week under similar circumstances. In the laboratory he allowed *P. spumarius* nymphs to choose where they would settle and feed on plants of the burweed, *Xanthium strumarium*. They chose the upper younger leaves which contained the highest concentrations of amino acids. Given a choice between mature leaves (which they would normally reject) with the cut petioles in either distilled water or a solution of amino acids, they chose the latter. When caged for 7 days on leaves of all ages on several plants, up to 90% of those on the oldest leaves with the lowest level of amino acids died. By comparison only 11% of those confined on young leaves with high concentrations of amino acids in the xylem died.

Working with adult insects of the cicadellid *Homalodisca coagulata*, Andersen et al. (1989) were able to demonstrate the remarkable efficiency with which xylem feeders can metabolize all forms of nitrogen present in their food. Brodbeck et al. (1990) found that the distinct seasonal patterns of selection of different species of host plants by these adult *H. coagulata*, and their abundance on any one species at different times of the year, were closely correlated with the concentration of amino acids in the xylem of those plants. By grafting scions of a low-nitrogen, non-preferred variety of *Prunus* on to rootstocks of both high-nitrogen, preferred and low-nitrogen, non-preferred varieties, they demonstated that this correlation represents the adult *H. coagulata* seeking out xylem with the greatest amounts of amino acids. Scions retained the morphology and phenology of the non-preferred parent plant, but the level of amino acids in their xylem was dictated by the variety of rootstock. Concentrations of amino acids were two-fold greater in scions grafted on the preferred, high-nitrogen rootstocks, and so was the insects' preference for feeding upon them.

The same constraints are found to apply at the scale of daily feeding upon the one plant (Andersen et al. 1992; Brodbeck et al. 1993). The feeding rate of *H. coagulata* (and of two other xylem-feeding leafhoppers) peaks at the time of day when nutrients in the xylem are at their maximum concentration, even when, as happens with some host species, this is at night.

Also, like the *P. spumarius* nymphs, these adult insects ingest enormous amounts of food (up to 300 times their dry body weight per day). Additionally, however, they exhibit a sexual difference which highlights the need of access to nitrogen for reproduction. Males not only consume up to four times less fluid than the females, they retain much less of the nitrogen that they ingest. The females, which lay over 1000 eggs each, retain more than five times as much nitrogen as do the males.

3.6 Fruit Flies

The tephritid fruit flies (including the economically important Queensland fruit fly, *Bactrocera tryoni*, and the Mediterranean fruit fly, *Ceratitis capitata*) are herbivorous insects worth noting as a separate group. This is because some, but not all (Howard and Bush 1989; Daser and Brandl 1992), have adopted a strategy for circumventing the acute shortage of nitrogen in their food (ripe fruit), a strategy which is only indifferently developed among sap suckers and apparently totally absent in most chewing insects. This strategy is an association with cellulose-digesting bacteria. They seem to have overcome the problem, however, in a more direct manner than other herbivores which have adopted this strategy. The relationship is apparently not one of gut symbiont and host. They eat bacteria which eat the fruit (Drew et al. 1983; Courtice and Drew 1984; Drew and Lloyd 1987, 1991; Drew 1988).

Female fruit flies emerge from their pupae with various species of enterobacteria contained within a special oesophageal bulb. These bacteria proliferate throughout the fruit flies' digestive tract and are regurgitated on the surface of ripening fruit to multiply on the nitrogen-rich leachates from the fruit. The flies ingest and digest these bacteria as necessary protein food for the maturation of their eggs. *B. tryoni* fed pure cultures of these bacteria laid four times as many eggs as females eating a conventional laboratory diet of autolysed brewers' yeast, and 97% of their eggs hatched compared with only 75% of those from the females on the laboratory diet. Before they lay their eggs the females again regurgitate on the surface of the ripe fruit. They then oviposit through the drop of "spittle" into the fruit. Larvae that hatch from these eggs immediately tunnel to the centre of the fruit, spreading bacteria as they go. They then feed on their way back towards the outside, but eat the bacterial "soup" which has developed in the fruit and not the fruit itself. Analyses of separate samples of the flesh of fruit protected from attack by fruit flies and of infected fruit revealed that amino acids had increased by 40 to 100% in infected pulp. Clean fruit contained 4.8% protein and infected pulp 8.5%. Eating the "soup" provided a diet nearly twice as rich in protein as that available in the absence of bacteria, and in a much more readily assimilable form.

The work of others confirms the essential contribution of these bacteria to the nitrogen nutrition of fruit fly larvae. Lambrou and Tzanakakis (1978) found that the larvae of the olive fruit fly, *B. oleae*, died as first instars in green olives if their mothers had been treated with streptomycin to kill their bacteria. But in ripe olives, which contain much higher levels of amino acids, they could persist without the bacteria. Also Fitt and O'Brien (1985) showed that the larvae of *B. jarvisi* could not develop in an artificial medium containing unhydrolysed protein and no bacteria, but they developed normally when *Enterobacter cloacae*, isolated from the flies, was added.

It seems that there is no close symbiotic dependence between fly and bacteria, but a fairly loose association wherein a number of species of bacteria, capable of growing independently outside the bodies of the flies, are cultivated and eaten. A measure of the importance of these bacteria to the flies, however, is the fact that they are passed to the next generation via the oesophageal pouch of the pupae and are not left to the chancy business of being picked up afresh from the environment.

The Mediterranean fruit fly also exhibits the typical herbivore's nitrogen hunger. It is routinely reared in the laboratory on artificial diets where females with immature eggs will choose a diet rich in protein, but prefer one high in carbohydrate once their eggs are mature. Zucoloto (1987) experimented with the larvae of *C. capitata* to see if they too can distinguish between different diets on the basis of their nutritional value. In all but one of numerous choice tests which he conducted, using different combinations and deletions of ingredients, the larvae always chose the diet with the most protein in it. The exception was with full- grown larvae shortly before they pupated. They chose a high-energy diet ahead of a high-protein one. It seems that their adults in the wild are similarly drawn to the sources of high nitrogen where there will be a rich bacterial flora. Hendrichs and Hendrichs (1990) recorded them in an Egyptian orchard seeking out and feeding upon bird droppings on the leaves of their food plants.

Another well known group of insects, the common vinegar flies (*Drosophila* spp.), have evolved a similar relationship to that of the tephritids. Their larvae also feed in the flesh of ripe and decaying fruit, but are nutritionally dependent upon bacteria and yeasts which first grow in the fruit (Starmer et al. 1986).

3.7 Wood-eating Insects

As is so often the case, the best illustration of what is limiting natural populations can be found by looking at extreme situations. Insects which eat wood are a good example. They have to cope with a food which is impossibly low in nitrogen, between 0.03 and 0.2% (Martin and Kukor 1984). But in doing so they very clearly reveal the key role that a lack of nitrogen plays in the ecology of herbivorous insects, and illustrate some of the adaptations that they have evolved to cope with this shortage.

3.7.1 The Key Role of Fungi: Increasing Nitrogen in Wood

Wood is such an uncompromising source of nitrogen nutrition that only very few organisms have evolved the capacity to exploit it in its *unaltered* state as a food. Essentially only the basidiomycete fungi have succeeded (Swift and Boddy 1984). No animal, vertebrate or invertebrate, even if it had evolved the ability to digest all the structural carbohydrates of wood, could possibly survive on a diet so dilute in nitrogen. As it is, only relatively few animals have been able to adapt to surviving on a diet of wood *plus* fungus. Most of these are insects. All of them in fact eat the microorganisms, or the products of their metabolism, as well as the wood. They depend upon the microorganisms to supply the protein component of their diet. Even those insects which employ other microorganisms to digest wood in their gut, depend on fungi to at least start to break down the raw wood before they ingest it.

Fungi, by converting wood to carbon dioxide, water, and fungal tissue (mycelium), reduce the carbon content of the wood by some two thirds. They concentrate a *very* dilute source of nitrogen (C:N ratios of from 100–500:1) to a *relatively* dilute source (C:N ratios of 40–90:1) plus a *moderately* dilute source (mycelium with C:N ratio of about 35:1). Insects' bodies have a C:N ratio of approximately 6:1 so they still have a way to go to concentrate the available nitrogen sufficiently to support their survival and growth. To do so they all depend upon other microorganisms which live in their intestines. These organisms, yeasts, bacteria, and protozoa, further convert the indigestible fractions to digestible microbial tissues. They also produce byproducts like essential amino acids; they recycle metabolic excretory nitrogen; and some fix atmospheric nitrogen.

The level of nitrogen in wood, although very low, docs vary from species to species, from time to time, and in different parts of the wood of a single tree. In a living tree the cambium has by far the highest level of nitrogen. The amount declines steeply in the sapwood, still further in the immature xylem, and in the heartwood there is very little nitrogen at all (Merrill and Cowling 1966a). The success of a fungus growing in wood, and the amount of nitrogen in its mycelium, varies according to the level of nitrogen in that wood. Merrill and Cowling (1966b) showed how fungi can adapt the level of nitrogen in their mycelium to that of the substrate on which they grow. The authors found that when they grew *Polyporus versicola* on a medium with a C:N ratio of 8000:1 it contained a mere 0.18% nitrogen, but had 5.0% when grown at a C:N ratio of 10:1: more than a 27-fold increase. When they grew the same fungus on poplar wood it had a nitrogen content of 0.5 to 1.0%, but *Fomes applanatus* grown on the same wood attained 2–3% nitrogen in its tissues. Many fungi, when growing on a substrate which is especially deficient in nitrogen, can preferentially allocate what little nitrogen *is* available to essentials like nucleic acids within their mycelium (Levi and Cowling 1969). Also, they can retrieve nitrogen from old hyphae and concentrate it in currently growing parts, as well as transporting it from the soil.

So the fungi too are limited by the amount of nitrogen available to them. They will be most successful in wood which provides nitrogen at the highest level and, of course, will be the best source of nitrogen nutrition for insects which feed on them.

3.7.2 Termites and Woodroaches: Gut Fauna, Coprophagy, and Recycled Nitrogen

The termites (Isoptera) are probably the insects best adapted to living on a diet of nothing but wood; however, they cannot live on wood alone. Even if they could digest the structural carbohydrates which comprise 99% of wood, there is not enough nitrogen in wood to support their survival and growth. They must rely on their associated microorganisms to convert, concentrate, recycle, conserve, and supplement the meagre amounts of nitrogen in the wood; otherwise they could never gain enough, quickly enough, to grow and reproduce (Breznak 1984; Collins 1983). Yet, in spite of the many sophisticated interactions with microorganisms which make possible the upgrading of an otherwise totally inadequate food, they must still seek out those foods in their environment which contain the greatest concentrations of nitrogen. This selective feeding on high-nitrogen food is widespread in termites, including those species which eat dead plant litter other than wood (Prestwich et al. 1980).

Termites *can* persist on intact wood – wood which has not previously been degraded by microorganisms – but will preferentially feed on wood which has already been attacked by fungi. And the more it is decayed the better they thrive. Some have gone even further (Wood and Thomas 1989). They cultivate specific fungi on beds of their faeces (dead plant material which is only partly digested by their gut microfauna in one quick passage through the gut). They reingest faeces plus fungus some 5 to 8 weeks later. They also feed directly on the tissues of the fungus, in particular on groups of protein-rich (C:N ratio 7:1) asexual spores called mycotetes, which grow on the surface of the fungal comb. The net effect is that these termites can convert dead vegetable matter into termite tissue more efficiently than can other termites, but only with the aid of their special fungus. To ensure that each new colony has the proper fungus their winged reproductives ingest a bolus of spores before they leave the nest.

All termites have a gut fauna of bacteria; some also harbour other microbes, particularly protozoans. All these symbiotic microorganisms are housed in the enlarged "paunch" of the multi-chambered hindgut of the termite, and digest the structural carbohydrates of wood. In addition, the bacteria in the gut of many – possibly all – species of termite can fix atmospheric nitrogen. The rate at which they do this and the amount of nitrogen fixed seem to vary inversely with the amount of nitrogen already in the food the termites are eating. Experimentally, a significant change in the rate of fixation has been detected within 5 hours of nitrogen being added to a diet of filter paper; up to a 200-fold variation in the rate of fixation has been recorded. Also, rates of fixation have been measured which would have contributed 50% of all the nitrogen needs of a colony, or, in the case of one species, *Nasutitermes corniger*, at a rate which would have allowed a doubling of the biomass in a nest in less than a year (Bentley 1984).

As with other insects, the nitrogenous waste of termites is uric acid. It is transported from its site of synthesis and of storage in the fat bodies via the Malpighian tubes to the hindgut, where it is anaerobically broken down by bacteria for recycling. This recycling may be so significant in terms of its contribution to the nitrogen economy of some termites that many believe uric acid functions in termites as an important form of stored nitrogen rather than simply as a waste product [it has been calculated that the process could liberate enough nitrogen to support a biomass of termites equivalent to 30% of the total biomass of a colony (Breznak 1984)].

In addition to this recycling of uric acid there is a great deal of recycling of nitrogen in other forms within termite colonies. Caste skins and dead termites are eaten by their nest mates. At times when nitrogen in the food is low (for example, when termites are fed experimentally on sound wood or diets of pure cellulose) or prior to the production of reproductives, cannibalism becomes commonplace in a colony. Also, supernumerary reproductives are usually eaten. The incidence of cannibalism can be reduced or eliminated by experimentally adding a source of nitrogen such as casein to their diet.

There is disagreement as to the extent to which termites can digest the bodies of microbes in their hindgut, but many species achieve this indirectly by active coprophagy. The benefit that the growing young of wood-eating insects can gain from the recycling of cellulolytic microorganisms via coprophagy is well illustrated by the woodroach, *Cryptocercus punctulatus*, (Nalepa 1984, 1990). These roaches are thought to be closely related to ancestral termites and, like present day primitive termites, they carry symbiotic protozoans, and nitrogen-fixing and uric acid-recycling bacteria in their hindgut paunches. Their newly hatched young remain together in a "family" with both parents and feed on anal fluids from the adults (usually the female). They thus acquire an inoculation of the microorganisms essential to their gaining sufficient protein to survive and grow. Unlike the termites, their symbionts, once established, are retained encysted in the wall of the hindgut when the exoskeleton is shed at each ecdysis. But this is not achieved until the third instar, some 1 to 3 months after hatching. After that they can survive if removed from their parents. Yet they stay with them and continue with proctodaeal feeding for 3 years or more! The most likely explanation for this prolonged association is that the parent roaches provide their young with a supplementary supply of nitrogen during the period of greatest growth. This is still a knife-edge existence for the neo-

Fig. 9. Newly-hatched nymphs of the primitive woodroach cannot survive without feeding upon their parents' anal fluids, and continue to benefit from this high-protein food for 3 years or more, long after innoculation with gut microbes. Photo courtesy of C. A. Nalepa

nates. In spite of this parental care and nurturing nearly half die in the first few instars, and only one-third survive to the following spring. From the evidence it seemed improbable that predators or parental cannibalism could account for such large losses. Starvation of first instars which do not get enough proctodaeal fluids is a much more likely cause.

Increased access to and conservation of nitrogen is, as with other herbivores, of greatest importance to termites for the production and growth of their young. Just how vital – and how physiological and behavioural adaptations enhance the availability of nitrogen for breeding females and growing young – has recently and succinctly been demonstrated by Shellman-Reeve (1990). She reared founding pairs of the Californian dampwood termite, *Zootermopsis nevadensis*, on both sterile filter paper (0.03% N) and filter paper with uric acid added (0.15% N), as well as on their natural food of *Pinus radiata* wood (0.03 to 0.18% N). Those pairs that she raised on wood showed the greatest variability in the success of their breeding – from no production of young to the highest production – but had a 36% average probability of reproducing (defined as producing at least one offspring). For those raised on

Fig. 10. Adult Californian dampwood termites produce special faecal pellets of concentrated gut microbes, with 1400 times more protein than their ordinary faeces. The male feeds his to the female before she lays her eggs; both feed them to the hatching young. They do not eat their ordinary faeces. Photo courtesy of J. S. Shellman-Reeves

filter paper plus uric acid, however, this average probability was 73%, and for those on sterile filter paper it was only 23%.

She found that the bodies of reproductives of these termites contained by far the greatest reserves of uric acid (twice as much as first and sccond instar nymphs, and 40 times more than later instars). Their hindgut protozoans metabolize the uric acid, synthesizing amino acids. The protozoans are subsequently eaten and digested by the adult termites in the form of special high-protein proctodaeal pellets. These pellets contain 1400 times more protein than the ordinary faeces which the termites do not eat. However, the consumption of these pellets was not equal. Females ate more of them than did the males, and the males transferred far more of their own pellets to the females than did the females to the males; that is until the eggs had hatched. Then both sexes equally fed their proctodaeal pellets to the young. So, not only is there a concentration of available nitrogen, first to the production of eggs and then to the growth of young larvae, there is also a clear paternal investment in both processes.

3.7.3 Furniture and Longhorn Beetles

The larvae of a great variety of beetles (principally cerambycids, buprestids and anobiids) which eat the wood of dead or dying trees rely on fungi to first digest the wood for them, or they eat the fungi themselves. They also rely on more specialized fungi, their endosymbiotic yeasts, to further digest cellulose and to recycle waste metabolic uric acid. This further conserves precious nitrogen (Kukor et al. 1988). Commonly these yeasts reside in special caecal pouches in the midgut, and are attached to the shells of eggs when laid in or on the wood by the female. The hatching larvae eat the egg shells and are inoculated with the yeast.

Kelsey (1958) prevented hatching larvae of the common furniture beetle, *Anobium punctatum*, from acquiring their yeasts in this way. Few of them survived more than a short time in the wood and those few that did survive grew much more slowly than larvae which had been allowed to eat their egg shells when they hatched. Bletchly (1969) demonstrated that *A. punctatum* preferred the outer sapwood of softwood timbers which is higher in nitrogen than the inner sapwood, and grew much more quickly when feeding in it. Baker et al. (1970) showed that *Anobium* larvae are able to use the small amounts of soluble nitrogen in the wood far more completely than the insoluble fraction. Even so the total amount of nitrogen in the wood was less than that taken up by the growing larvae. Between 40 and 60% of the nitrogen used for their growth must have come from a source other than the wood. At the time these authors could only speculate that this extra nitrogen might have come from intestinal microorganisms fixing atmospheric nitrogen. Now it is known that such bacteria are widespread in various animals which feed upon wood.

Another anobiid, the deathwatch beetle, *Xestobium rufovillosum*, is notorious in England for the damage it does to timbers (especially of oak, *Quercus robur*) in buildings. It is totally dependent on basidiomycetes first infecting the timber. Fisher (1941) was able to show that the speed with which the larvae grew in the wood, and the amount of destruction they caused, was directly related to the amount of fungal decay of the wood. In undecayed wood nearly all larvae died before or soon after they entered the timber (presumably they starved), and those few that did survive grew very slowly (as little as 3 mm long after 10 years). In well decayed wood, on the other hand, larvae could complete their life cycle and pupate in as little as 10 months. Intermediate levels of decay produced intermediate rates of developement. So, again, not only is the thinly spread nitrogen in the wood physically concentrated by the action of the fungi, it is made physiologically more available as fungal protoplasm which the larvae can digest.

The larvae of longhorn beetles (Cerambycidae) which feed in the wood of dying or recently dead trees always start feeding in the inner bark and cambium rather than in the wood itself. Only when the inner bark is exhausted or dried out will they move into the sapwood. As the cambium and inner bark are the site of growth and of translocation of nutrients, and therefore a much richer food than the wood, this is not really surprising. Just how dramatic the difference is was demonstrated by Hosking and Hutcheson (1979) working with the cerambycid *Arhopalus ferus* living in logs of *Pinus radiata* in New Zealand. They had previously established that the larvae strongly preferred to feed in the inner bark rather than in the sapwood. So they reared two lots of larvae in the laboratory in either bark (comprising phloem, cambium and some cortex) or outer sapwood sandwiched between sheets of plexiglass. When they introduced small larvae into these sandwiches the bark contained almost ten times more nitrogen than the sapwood, and after 48 days still contained six times more nitrogen. The larvae feeding in the bark grew four times as fast as those in the sapwood. The larvae in the bark always fed against the plexiglass on the cambial side of the sandwich, indicating a preference for the tissues with the highest concentration of nitrogen in them. However, two of the larvae in the sapwood sandwiches had gained twice the average weight of the group, and were intermediate in weight between the remainder feeding on sapwood and the group feeding on inner bark sandwiches. Inspection revealed that both these larvae had been feeding in a band of resin-impregnated wood which they had completely consumed before feeding on the remaining resin-free sapwood. The nitrogen content of wood impregnated with resin was almost twice that of plain sapwood.

There are a number of species of cerambycids which, either as adults or as larvae, girdle twigs and branches of living trees. These twigs subsequently fall to the ground or, if they stay on the tree, die and the larvae complete their development in the dead tissues. Forcella (1982) neatly demonstrated for the hickory twig-girdling beetle *Oncideres singulata* in Oklahoma, just how this behaviour increases the chances of such larvae surviving and growing to maturity. Adult *O. singulata* emerge in late summer, about a month before the leaves of hickory, *Carya texana*, begin to fall. The female beetles immediately start girdling twigs and laying their eggs in the bark above the girdles. The girdles are deep enough to cut through the phloem but leave the xylem intact. So, water and solutes can continue to be transported to the leaves but nutrients, especially the nitrogen which is being mobilized and transported out of the senescing leaves at this time, are blocked and accumulate above the girdle. This greatly enriches the food of a larva hatching from an egg deposited above the girdle.

Forcella did three things. He collected a lot of the previous season's girdled twigs from the ground, both those from which adults had emerged and those from which they had failed to emerge, and measured the nitrogen in their tissues. He collected neighbouring, living, girdled and ungirdled twigs from trees each week throughout the autumn and measured the amount of nitrogen in the leaves, in their bark plus phloem, and in their xylem. He detached two lots of girdled twigs from the same trees; one lot as soon as they had been girdled by the beetles, the other lot not until all their leaves had senesced and fallen. He tagged both lots of twigs and left them on the forest floor until the following summer when he inspected them for the presence of live larvae.

The results were unequivocal. There was significantly more nitrogen in the twigs from which beetles had successfully emerged the previous season than in those which had failed to support the development of a beetle. The level of nitrogen in the leaves of all twigs declined steadily through the autumn. In the phloem above the girdle of girdled twigs it increased. Below the girdles, and in ungirdled twigs, there was no increase of nitrogen in the phloem, nor was there an increase in the xylem of any twigs. Next summer there were far fewer living larvae in the twigs picked immediately after they had been girdled than there were in those left until all the leaves had senesced and fallen. It is evident that without the supplementation of extra nitrogen in this manner the twigs of hickory are usually not an adequate food for the larvae of *O. singulata.*

Incidentally, I think that Forcella's acknowledgement that he made these investigations after showing what he thought was the work of clever nut-gathering squirrels to a colleague, who politely suggested that what he had found was the work of an insect, is a nice example of the serendipity of scientific advance!

3.7.4 Woodwasps

The larvae of the woodwasps (*Sirex* spp.; Hymenoptera) also depend upon fungi to survive and grow on a diet of wood (Gilbertson 1984). They, like beetles that live on wood, carry their own symbiotic fungus – species of *Amylostereum* – with them from generation to generation. The females have special pouches set at the base of their ovipositors which contain spores of these fungi. They lay their eggs deep in the wood with their long ovipositor and a few oidia of the fungus are transferred with each egg. The survival and subsequent growth of the larvae hatching from these eggs depends upon the fungus having first grown in the wood. The fungus both digests the wood and provides a source of protein for the larvae. Enzymes capable of digesting fungal and plant polysaccharides are produced by the fungus and ingested by the larvae, where they further digest wood in the larval midguts (Kukor and Martin 1983).

In the specific case of *S. noctilio,* which kills many trees in plantations of unthinned and drought-stressed *Pinus radiata* in New Zealand and Australia, Titze and Turnbull (1970) demonstrated that the secretions of mucus from the females' club glands stimulate the fungus *A. areolatum* to grow more quickly and vigorously. Also, Titze (1970) found that *A. areolatum* produced metabolites which, when extracted and applied to seedlings of *P. radiata,* caused them to wilt. In the field the larvae of *S. noctilio* survive and grow only when the trees which they attack quickly wilt and die.

3.7.5 Borers Which Do Not Ingest Wood

There are insects which, although apparently wood-eating, are not. The shothole borer beetles (Platypodidae) which, like *Sirex,* attack badly stressed or dying trees, do not actually ingest wood at all. The shavings from the tunnels that they rapidly chew in the wood are ejected from those tunnels, but the beetles carry with them the spores of specific fungi which grow through the wood creating the familiar stain which surrounds their tunnels. The adult beetles and their subsequent broods of larvae live entirely on the tissues of the fungus which grow from the walls of the tunnels.

The larvae of the hepialid moth *Aenetus virescens* are another example. They bore into the stems of living trees but do not kill them or eat the wood, nor do they eat fungus. They eat the callus tissue which proliferates from the cambium around the entrance to their tunnels (see Sect. 3.3.2).

Another group of insects which bore into living trees are the bark beetles (Scolytidae). Outbreaks of these in forests of senescent or stressed trees are of major economic importance, but their larvae feed entirely in the living phloem, not in the wood. They are thus more akin to aphids and others that depend for their reproductive success upon the nitrogenous nutrients in the phloem, than they are to woodborers.

But, as with all things in nature, ultimately there are no absolute boundaries. What is the difference between eating wood plus fungus to just fungus? The essential food for the insect is the fungus which must first "eat" the wood. And what is the difference between eating living or dead cambium and sapwood? In both the quality of the food is dependent upon the amount of nutrients available in the phloem.

Chapter 4 Crustaceans

4.1 Microcrustaceans

In most marine environments the dominant members of the zooplankton are microcrustacea. We can consider these animals as "herbivores" in so far as they feed on photosynthesizing primary producing organisms, the phytoplankton. Yet it is impossible to make precise separations, as most zooplankton will be omnivorous, eating whatever small organic particles they are able to ingest; and many phytoplankton are, at times, carnivorous. But it is undeniable that, as in the terrestrial world, the distribution and abundance of non-photosynthetic life in the sea depends on the quantity and quality of photosynthetic organisms available as food either living or as detritus recycled by microorganisms.

4.1.1 Distribution and Abundance of Food Limited by Nitrogen

As Conover (1968) said, the supply of food for zooplankton is essentially limitless were it not for the fact that it is usually very thinly or patchily dispersed. Hence the problem of any individual microcrustacean getting enough to eat is formidable. Be they herbivore, carnivore or omnivore, they all suffer a relative shortage of food. They all live in a nutritionally dilute environment.

However, the dilution is not uniform. The seemingly homogeneous aquatic environment is as heterogeneous in terms of the availability of good food as is any terrestrial one. The patchiness will, of course, vary both in extent and intensity. Checkley (1985) and others have demonstrated this relationship, and Checkley was able to measure for the copepod *Paracalanus parvus*, living off the coast of California, just how variable an adequate supply of food is both in time and in distance from the shore. As Huntley and Boyd (1984) concluded, herbivorous zooplankton are less likely to experience a shortage of food in coastal waters than they are in the open ocean, although in both environments there will be a shortage whenever conditions arise which dilute the phytoplankton.

The "condition" most likely to cause the density and distribution of the phytoplankton to vary is the amount and location of inorganic nutrients in the water. More specifically, the growth and abundance of phytoplankton, as with all organisms, are largely dependent upon a minimum supply of nitrogen in the environment.

In the sea the primary source of nitrogen is inorganic nitrate in the upwelling of cold, nutrient-laden waters, supplemented by gaseous nitrogen from the atmosphere. A recent report (Carpenter and Romans 1991) has it that large, but previously overlooked nitrogen-fixing *Trichodesmium* bacteria are, at least in most tropical seas, the *primary* source of this new nitrogen.

As well as new nitrogen there is a constant recycling of old nitrogen (mostly as ammonia and urea) derived from the metabolic activity, death and decay of all organisms in the illuminated zone of biological activity in the surface waters of the ocean. The fact that phytoplankton are often themselves heterotrophic further accentuates this dependence on scarce supplies of nitrogen (Porter 1988; Tranvik et al. 1989). The ingestion of bacteria, very small algae, and particles of organic matter by photosynthetic phytoplankton is quite widespread and common; but apparently varies according to the amount of light available for photosynthesis and the concentration of soluble nutrients in the water. It seems that this carnivory is a response to inadequate levels of nutrients, especially nitrogen, rather than a shortage of carbon for energy.

That the supply of inorganic nitrogen is usually inadequate is emphasized by the capacity of many aquatic organisms to directly absorb free amino acids dissolved in the water, even when these are present in the quite infinitesimal quantities found in the ocean depths (Morita 1980). Many algae (North and Stephens 1971), and bacteria (Keil and Kirchman 1991), and a wide variety of marine invertebrates, especially their non-feeding larvae, embryos and eggs, can readily take up these simple molecules present in the water as dissolved organic matter (Stephens and Schinske 1961; Jaeckle and Manahan 1989; Thomas 1990; Poulet et al. 1991). Amino acids in solution make a significant contribution to the nutrition of many of these organisms, especially at times when food is scarce or of low quality (Gorham 1990). Some can increase their biomass in this manner in the absence of particulate foods (Manahan 1990). A specific example of the significance of this adaptation to a shortage of nitrogen was demonstrated by North and Stevens (1971) when they grew the photosynthetic alga *Platymonas* in a culture lacking inorganic nitrogen (normally its sole source of nitrogen). The alga's uptake of amino acids increased by as much as ten-fold. It seems, however, that crustaceans are unable to take up nutrients in this way (Gorham 1990).

Another group of organisms recently revealed as of probably equal importance as the algae, and similarly dependent upon the supply of nitrogen in the marine environment, are the bacteria. They too comprise a very considerable source of food for zooplankton (Ducklow 1983; Cole et al. 1988; Fuhrman et al. 1989; Sherr and Sherr 1991).

So the size and chemical composition of individual algae and bacteria, the number and proportion of species present, and their abundance are all limited by the relative paucity of nitrogen available in the water. In turn the distribution and abundance of the zooplankton will be influenced by the range, quality and abundance of their food (Eppley and Peterson 1979).

4.1.2 Nitrogen Content of Food also Important

It is not simply a matter of there being a sufficient number of phytoplankton and bacteria present and available as food particles in a given time. The *quality* of those particles, especially their nitrogen content, is just as vital. Zooplankton can adjust the rate at which they filter particles from the water to compensate for there being too few of them available, or for individual but abundant particles being of low quality. But, as with all other animals, there is a maximum bulk of food that they can process through their gut in a given time. At this point, no matter how dense the food particles, if the nitrogen they contain is too "dilute" they will be an inadequate food for the zooplankton eating them. Like other herbivores faced with the same situation, the zooplankton will suffer a relative shortage of nitrogen in an apparently abundant supply of food.

The importance of this nitrogenous quality of the food ingested by crustacean zooplankton was nicely demonstrated by Guidi (1984). He experimented with a benthic marine copepod *Tisbe cucumariae.* He said it could hardly be classed as a herbivore because it will attempt to eat just about anything presented to it which is even remotely edible. He fed laboratory populations with seven carefully prepared pure diets which ranged from polychaete worms to low-nitrogen marsh grass, and an even poorer red macroalga. He uniformly ground up all diets so as to minimize any possible differences due to the size of the food particles. He found that both the time these animals took to develop and their survival were significantly influenced by the level of total nitrogen and protein, and the C:N ratios of these foods, but not by their carbon or caloric values. On all but the two very poorest sources of nitrogen their speed of development particularly, and to a lesser extent their survival, tended to platform off, suggesting a threshold level of nitrogen in the food. On the poorest diet of all (the red alga) they had still not completed their life cycles when Guidi stopped the experiment after 22 days, whereas on all other diets they matured in 8 days. At the other extreme the diet of polychaete worms produced by far the greatest number of mature females bearing eggs.

The story is the same in fresh water. Byron et al.'s (1984) sampling of freshwater cladocerans and copepods in an oligotrophic lake in California is an example. They found that the phytoplanktonic algal food of these zooplankton is commonly limited by the availability of inorganic nitrogen in the water, and the abundance of these herbivorous crustacea is, in turn, limited by the abundance and nitrogen content of their algal food.

Species of the cladoceran genus *Daphnia* are a major component of the freshwater zooplankton throughout the world. Hebert (1978) presented evidence that the populations of these usually parthenogenetic crustaceans are limited by their food supplies. Again, it is stretching a point to call them herbivores; they eat algae, bacteria, fungi, protozoa, and general organic debris. But when populations of *Daphnia* are kept short of food in the laboratory the adult females do not produce any eggs. These non-reproductive adults frequently predominate in quite dense natural populations, which are also found to have much higher mortality rates than the food-limited laboratory populations – largely as a result of starvation of the young. In nature the particulate food of *Daphnia* is both variable in size and patchily distributed. Under such conditions some adults in a population will survive and produce eggs, but the young hatching from these eggs will have a low probability of surviving; as neonates they are much more likely to die of starvation. Additionally, their longevity in the absence of food will depend upon the nutrition of the females that produced them – in turn influenced by the quality of *their* food (Cowgill et al. 1984). Furthermore, the evidence of similar death rates in populations of *Daphnia* with and without predators present, suggests that predation is not a significant influence.

Supporting this contention was the study Vanni (1987) carried out in a shallow freshwater lake in Illinois, specifically to test the influence of predation and food on the abundance of the zooplankton in the lake. By setting up enclosures in the lake he was able to monitor the effects of the presence or absence of fish known to feed on the zooplankton in the lake. Also, by adding nitrogen and phosphorus fertilizer to the water, he could measure the influence of an increase in abundance of the zooplankton's algal food. His results showed unequivocally that increasing the availability of food had much more effect than did predation on the density of individual species of the zooplankton, and on the total abundance of zooplankton.

However, it is when these animals are forced to live as strict herbivores in the laboratory, feeding on only one or two species of algae, that the importance of access to nitrogen becomes clearly apparent. Two recent studies highlight this.

Groeger et al. (1991) conducted long-term feeding experiments of laboratory-reared *Daphnia pulex* fed on the alga *Scenedesmus obliquus* which had been grown in either a complete or nitrogen-deficient solution. On a diet of the latter the *Daphnia* grew more slowly, were smaller, and produced fewer eggs than those eating the former diet. The fact that they were always fed algae in excess of their capacity to ingest them, and that they always had full guts, showed that they were not limited by the quantity of food available. And the high content of lipid in their bodies belied any suggestion that a lack of energy might have been limiting their growth or reproduction; rather, it was the lack of amino acids and protein in the algae grown on the nitrogen-deficient medium.

Giani (1991) fed cultures of *Daphnia hyalina* and *D. galeata* on either a *Rhodomonas* species of alga, which has high levels of nitrogen and protein, or a diatom, *Stephanodiscus hantzschii*, which has a much lower concentration. Both cladocerans grew and reproduced equally well on both foods so long as there was an abundant supply to ingest. However, reducing the density of both algae in the water revealed the importance of nitrogen in the food at densities more likely to be encountered in nature. The growth and reproduction of both *Daphnia* species were severely reduced on low numbers of the nutritious *Rhodomonas*, but on the same density of the poorer quality *Stephanodiscus* they all died as quickly as animals starved in filtered sterile water.

This dependence upon an adequate level of nitrogen in their food was further confirmed when Giani fed the *Daphnia* on the more nutritious *Rhodomonas* when it had been cultured either in the normal nitrogen-rich medium used for the previous experiments, or in a low-nitrogen one. Again, both species did well on high densities of normally grown *Rhodomonas*. On the nitrogen-limited ones, however, they not only fared equally badly at either high or low densities of algae; at both they grew and reproduced less than they had on the *lowest* density of the algae from the nitrogen-rich medium.

4.1.3 Microcrustaceans Feed Selectively for Nitrogen

It is similarly under controlled conditions of experimental feeding that the need for these animals to be selective, and to select for a maximum intake of nitrogen, is revealed.

Richman and Dodson (1983) used feeding trials with a series of carefully structured mixes of high- and low-quality algae to show that both the quantity and quality of the phytoplankton in lakes can have a major influence on the abundance of cladoceran (eg. *Daphnia* spp.) and calanoid (e.g. *Diaptomus* spp.) zooplankton, and that these animals are capable of quite sophisticated preferential selection of the food they ingest.

Other recent work (still mostly with copepods and cladocerans) has provided abundant evidence that both marine and freshwater zooplankton are very often limited by their food. Fecundity, reproductive rate, survival, and abundance all vary with the quantity and quality of their food. Most populations of zooplankton are limited by a shortage of food at all times other than during short spring flushes (Lampert and Muck 1985). Predation is only of any moment in reducing their abundance when food is in short supply and their birth rates are low. Once the production of phytoplankton increases so does the birth rate of the zooplankton, more than compensating for the losses from predation (Muck and Lampert 1980; Lampert and Muck 1985). Truly, control from the "bottom up" rather than the "top down".

Even more sophisticated studies (including techniques such as microcinematography) have shown that both copepods and cladocerans are very selective feeders, differentiating not only between different species of algae and different sizes and shapes of particles, but also responding to physical and chemical cues from individual cells. Butler et al. (1989) have taken the investigation of these feeding preferences a step further. They looked at the capacity of a calenoid copepod, *Diaptomus kenai*, and a cladoceran, *Daphnia rosea*, to differentiate between single cells of one species of green alga, *Selenastrum minutum*, which had been grown in cultures containing different levels of nitrogen. All these organisms are common in North American freshwater lakes and the work was done at the University of British Columbia in Vancouver.

Butler and her colleagues selected a particular clone of *Selenastrum* which has the unusual feature of not changing the shape or size of individual cells when grown at different rates on different concentrations of nitrogen. Thus, any selection on the basis of quality could not be confounded by physical differences between cells. They also presented the test animals with concentrations of cells in the water which exceeded saturation feeding levels so that, without active selection, the crustaceans should ingest equal quantities of high- or low-nitrogen cells, whether these were presented to them separately or mixed together. The animals quite conclusively rejected the null hypothesis for them, ingesting high quality cells one and a half times faster than the low quality ones when these cells were presented to them separately, and up to three times faster when presented in a mixture. They clearly could detect and preferred the more nutritious high-nitrogen cells. As the authors concluded, because both marine and freshwater phytoplankton are often limited by the availability of nitrogen in the water, variations in the feeding of zooplankton in response to changes in the nutritional status of the phytoplankton could be an important element in the ecology of both.

At about the same time, but on the other side of the continent in Massachusetts, another group was carrying out a similar investigation. Cowles et al. (1988) experimented with the feeding response of the marine copepod *Acartia tonsa* to populations of the single cell diatom *Thalassiosira weissflogii* grown in a nitrogen-rich and in a nitrogen-poor medium. There were no size or other morphological differences which could have influenced the copepods' selection of one or other type of cell. The faster growing cells on the rich medium contained more total protein, more dissolved free amino acids, more chlorophyll, and a lower C:N ratio than the slower growing cells. During 5-hour feeding experiments the copepods ingested the high-nitrogen cells nearly twice as fast as the low-nitrogen ones, regardless of the concentration of the cells in the filtered sea water, or whether they had been fed for 2 days prior to the trials on low- or high-quality cells.

When a cell-free solution containing exudates from the algae was added to test cultures of *Acartia* this "scent" significantly increased the rate at which the copepods ingested cells. When cells of the same size, but of either high or low nitrogen content, were mixed in different proportions, *A. tonsa* always ingested the high-nitrogen cells out of proportion to their abundance in the mixture. But they ingested cells at a slower rate in mixtures than in pure cultures of either high- or low-quality cells, presumably because selecting for individual cells increased their handling time in the mixed populations.

Houde and Roman (1987) found the same sort of response by *A. tonsa* in Chesapeake Bay, Maryland. Not only was the protein and nitrogen content of their algal food the most important cue regulating their rate of ingestion, but they probably assess its quality, before ingesting it, from the amino acids leached into the water by the algae.

All this provides a clear demonstration that these copepods can distinguish between individual algal cells on the basis of their quality as nitrogenous food. They can detect this difference chemically before physically contacting a cell; and they use this capability to select cells which will maximize their intake of protein. Some would not agree. Turner and Tester (1989) working in the field with populations of *A. tonsa* and of several other species of copepods in the

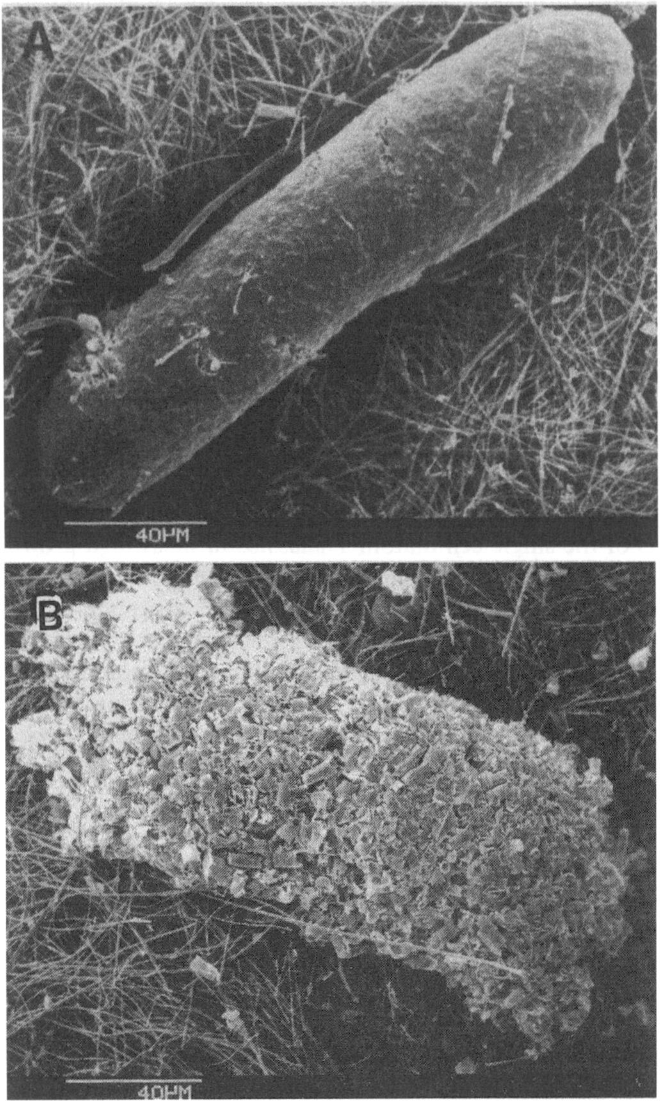

Fig. 11 A, B. Scanning electron micrographs of copepods' faceal pellets incubated for 24 h in their absence (**A**), and in their presence (**B**). In **B** the copepods have eaten the pellet's peritrophic membrane on which bacteria grow, but in **A** the membrane remains intact. Photos courtesy of T. Noji

plume of the Mississippi river, could find no evidence that any of these animals were selective in their grazing. They were critical of conclusions derived from controlled laboratory experiments having any relevance to the real world. But it seems unlikely to me that such definite and well-developed behaviours would be found in animals which do not make use of them in nature.

Kiørboe (1989), working in Denmark, noted that there was a number of instances recorded where zooplankton had increased their feeding, fecundity, growth rate, and density in response to an increase in the biomass of phytoplankton following an upwelling, a tidal front, or a storm, bringing more nutrients into the surface waters of the ocean. He also noted, however, that there were reports of these sorts of responses by zooplankton without there having been any concomitant increase in the biomass of the phytoplankton. He therefore conducted some controlled experiments to see what effect fertilizing the water in which the plankton live might have. He also used the copepod *A. tonsa* and fed it on the same diatom, *Thalassiosira weissflogii*, which he had cultured with different levels of nitrogen in the water. He found that the copepods feeding on the diatoms with the highest levels of nitrogen in their tissues ingested about six times more nitrogen (but no more carbon) than those copepods feeding on diatoms with the lowest levels of nitrogen. They also produced six times as many eggs. And this with a relatively inconsequential increase in the *number* of diatoms that they ate. It seems that it is the level of protein in the food, not simply the numerical abundance of the phytoplankton, which is limiting the capacity of *Acartia* to breed and increase in abundance. This in turn demonstrates how it is that copepods can be seen to respond to increased nutrients in the water without there having been any increase in the biomass of the phytoplankton which they eat.

4.1.4 Coprorhexy

Even though phytoplankton may provide a food which has a considerably higher proportion of nitrogen to carbon than do terrestrial plants, the zooplankton feeding on them still need to have access to other more concentrated forms of protein. Lampitt et al. (1990) while working at the University of Kiel, in Germany, nicely demonstrated one way in which they do this. Copepods eject their faecal pellets wrapped in a skin of peritrophic membrane (unlike that of the insects this is secreted only in the posterior half of the midgut). In the sea this skin is rapidly colonized by bacteria in the process of penetrating to the algal material within. With the aid of algae labelled with radioactive carbon, and some very beautiful scanning electron micrographs, Lampitt and his colleagues were able to show that the copepod *Centropages hamatus* is highly efficient at breaking down its own faecal pellets, selectively removing and ingesting just the membrane and its attached bacteria. While the membrane itself is of little nutritive value, the concentrated bacteria provide a significant supplement of high protein food. The authors have dubbed this form of selective coprophagy "coprorhexy" from the Greek kopros = faeces and rhexis = to break or crumble. An indirect benefit of this selective removal of the membrane from the faecal pellet is that it hastens the dispersal of the contained particles of algal material. These are then more accessible for degradation by the bacteria in the surface waters where the copepods feed, further increasing the supply of bacteria for the copepods. On the broader front this behaviour serves, as does coprophagy in other marine herbivores (Sect. 9.4), to salvage as much nitrogen as possible before it sinks beyond reach (Noji 1991). Finally, the authors noted that copepods sometimes pass empty capsules of peritrophic membrane which they may subsequently ingest. Could this, they ask, be a device which further aids the concentration of bacterial food for the copepods?

4.2 Macrocrustaceans: Land Crabs, Lobsters, and Shrimps

When we turn to the macrocrustacea we find that they all have pelagic larvae which are carnivorous, feeding on zooplankton. So none are true herbivores when they are young. But even as post-larval juveniles and adults very few, if any, are true herbivores. They all reveal the same dependence upon extra nitrogen in their food.

Wolcott and Wolcott (1984, 1987) worked with two species of Caribbean land crabs, *Gecarcinus lateralis* and *Cardisoma guanhumi*, both of which are largely herbivorous, eating living and dead leaves of plants. They found that both species are limited by a relative shortage of nitrogen. *C. guanhumi* eats a greater proportion of leaf litter, so is more nearly a detritivore than *G. lateralis*; as such it consumes a staple diet some two to four times lower in nitrogen than that of *G. lateralis*. Both have rapid transit times of food through the gut but are nevertheless restricted by the bulk of plant food that they can process. There is no evidence of any microbial fermentation in their guts. Both species are, however, opportunistic carnivores; they eat insects, small frogs, lizards, fiddler crabs and carrion. Both are cannibalistic on their own young (see Sect. 12.1). Nevertheless, they normally do not get enough protein. When the Wolcotts fed captive adult crabs on plant food supplemented with casein or soybeans, they found the crabs grew faster and were less prone to cannibalize their young than crabs fed on plants alone.

Freshwater crayfish are mostly omnivores or detritivores, but as Momot (1984) found, their success is largely determined by variations in the level of survival of juveniles in response to the availability of food. France (1985) in a specific study of *Orconectes virilis* living in freshwater lakes in Ontario, found that the size, growth rate, reproductive success, and abundance of these crayfish in different lakes was directly correlated with the production of phytoplankton in each lake. This, in turn, was a reflection of the availability of basic nutrients for the growth of the phytoplankton.

Joll and Phillips (1984) found that the growth of juvenile marine rock lobsters, *Panulirus cygnus*, in Western Australia was far greater at a location where they ate predominantly animal food (molluscs) than at one where their diet was dominated by coralline algae. This was after they had settled following some 9 to 12 months spent as larvae in the plankton. Those living in the poorer algal-dominated site also moulted less frequently and had larger volumes of food in their foregut than those at the site rich in animal food.

Work in Texas with the juveniles of two common penaeid shrimps from the Gulf of Mexico revealed in more detail the importance of animal food for post-larval crustaceans. The post-larvae of both species migrate to saltmarsh estuaries where the juveniles become benthic, browsing omnivores. *Penaeus aztecus* is predominantly carnivorous, but eats algae; *P. setiferus* eats mostly micro-algae. While a diet of algae would provide a better source of nitrogen than would a diet of vascular plants, it is nevertheless inadequate. McTigue and Zimmerman (1991) demonstrated this in laboratory trials in which juveniles of the two species were fed a variety of mixed or pure diets of algal or animal food. Both shrimps grew best on mixed diets, and animal food proved to be mandatory for the survival of the carnivorous *P. aztecus*. The more nearly herbivorous *P. setiferus* could survive on some diets without animal food, but with little or no growth. On a diet of diatoms alone, however, while they grew for a while, none survived.

Similar stories are found for a wide range of generally omnivorous decapods. The death of the very young is the major loss from each cohort in a population. The nitrogenous quality of their food is the principal factor limiting their success (Bages and Sloane 1981) and they will select their food to maximize their intake of protein (Moriarty and Barclay 1981).

Some, in addition, have a gut flora of cellulolytic bacteria to aid in digesting and extracting nitrogen from their food (Fair et al. 1980).

4.3 Terrestrial Crustaceans – The Isopods

Another interesting group of crustaceans, which once again reveal the nitrogen-hunger of all those which depend on a diet of plant tissues, are the terrestrial isopods: the slaters, woodlice, sowbugs, or pillbugs. Common in most gardens, but usually paid little attention, they are essentially detritivores and scavengers, but lean variably towards eating living plant material.

4.3.1 The Case of a Common Woodlouse

Armadillidium vulgare is a widespread and common isopod of Mediterranean origin. It is an omnivore and scavenger, but mostly eats dead leaves. So, technically, it is a detritivore rather than a herbivore. It does, however, eat living plant tissue. In California, Paris and Sikora (1965) demonstrated with radiotracer labelling that in spring these isopods eat growing forbs. Also, Paris (1963) found that they will eat carrion and fungi, and can be cannibalistic. In Texas, Brody and Lawlor (1984) fed female *A. vulgare* in the laboratory on either unlimited (7 days a week) or restricted (2 days a week) supplies of the animals' natural food of soil, litter, and dead leaves, supplemented with fresh carrots. The reproductive capacity of those females on the restricted diet was severely reduced, and reflected the seasonal changes in the success of their breeding in the field in response to the availability of their food.

But, as was the case with their distant relatives the cladocerans and copepods, experimentally restricting them to a diet of nothing but plant material or artificial foods revealed the importance of nitrogen to their successful breeding and survival. When Merriam (1971) fed laboratory cultures of *A. vulgare* on Purina rabbit chow (15% protein) they grew faster and bigger and attained sexual maturity in half the time of those he fed on Cerophil, a compound based on ground cereal leaves with a protein content of only 2.5%. The difference in their survival and reproduction was even more dramatic. After 5 months 69% of the original population on the high nitrogen diet were still alive, but 90% of those on the low nitrogen diet were dead. And while sexually mature females bred freely and successfully on the diet of Purina chow, on the low nitrogen diet practically no young were produced at all.

Rushton and Hassall (1983a, b, 1987) worked with field populations and laboratory cultures of the same isopod in England. They fed animals in the laboratory with diets consisting of either standing dead (still attached to the plant) or decayed leaves of one of several species of dicotyledons and monocotyledons. They measured the animals' preferences for these different leaves, and the rate at which they ate and absorbed them. Freshly fallen dicotyledonous leaves provided the best-quality food for these grassland isopods. The animals preferred to feed on them, ate more of them, and absorbed more from them in a given time. They grew faster, reproduced better, and more of them survived than when they fed on the litter of the much more abundant grasses, or on decayed dicotyledonous leaves. The best growth and highest survival were attained by isopods fed on freshly fallen leaves of a legume which had a much higher total nitrogen content than the leaves of the other dicotyledons.

Interestingly, however, results from all the diets revealed that the animals were not simply responding to total nitrogen in the food, but to how readily it could be digested and absorbed across the gut wall. It was the *form* and *availability* of nitrogen in the food which was critical.

The differences in all responses to variations in the availability of nitrogen were greatest for juveniles. The growth rates of juveniles fed diets in which different proportions (0 to 100%) of freshly fallen dicotyledonous leaves were mixed with decayed grass leaves increased exponentially with increasing proportions of the high quality dicotyledonous food in the diet. On a diet with less than 10% dicotyledonous litter they hardly grew at all, even though there was an excess of grass litter available to them. Similarly, the rate of mortality of these juvenile isopods declined steeply as the proportion of good food in their diet increased. All this reflected the common fate of newborn *A. vulgare* in the English grasslands – up to 80% of them are dead within 2 to 3 weeks (Al-Dabbagh and Block 1981). Paris and Pitelka (1962) had found that the fate of the very young was similarly of major importance to the success of populations of *A. vulgare* in Californian grasslands. Most of the young died in the first few weeks of life, and only 18% lived to 6 months. Mortality of terrestrial isopods in general occurs soon after they are released from the brood pouch (Warburg et al. 1984).

That an adequate source of protein in its food is important to *A. vulgare* in nature was further illustrated by the grassland population studied by Rushton and Hassall. Its numbers declined more than six times over just 2 years as an increasing density of the grass sward reduced the number of dicotyledons growing in the grassland. As the good food became more thinly spread among the bulk of inadequate food, the isopods experienced a growing relative shortage of protein. They were *not*, as many would have it (e.g. Warburg et al. 1984), surrounded by an abundance of good food and controlled by temperature and humidity. They were living in an inadequate environment where most of what they could eat contained too little nitrogen to sustain their reproduction and growth. And all this was exacerbated because these good-quality leaves fall in the autumn after the isopods have finished breeding, and the leaves are much decayed by the following spring, so even less of this already limited supply of protein is available to them when they most need it.

Hassall and Dangerfield (1990) further experimented with manipulation of the food supply of populations of *A. vulgare* and showed how important a change in the availability of litter from dicotyledons, rather than grasses, is to the success of these populations. They excluded rabbits from areas of the grass heath where these isopods live. This caused a dramatic increase in the amount of leaf litter derived from dicotyledonous plants which had previously been heavily and preferentially grazed by the rabbits. More isopods survived in these exclosures, they grew larger and more fecund, and in 5 years their density increased by 90% compared with the population in the open grass sward.

Even with additional supplies of dicotyledonous leaf litter these animals were still constrained by a shortage of nitrogen in their food. The authors demonstrated this by adding powdered high-protein rabbit pellets to enclosed areas of the sward. In these enclosures young isopods in their first 4 months of life grew to double the weight of those in control enclosures. They maintained this difference without further supplementation and grew to maturity much more quickly than the controls. The major factor influencing the abundance of *A. vulgare* in these populations was clearly the supply of high quality food – the amount of nitrogen it contained. It influenced the early survival of the young, the rate at which they grew, and the eventual size (and thus fecundity) of adult females. An example, the authors concluded, of limitation from "below" by food rather than from "above" by predators.

4.3.2 The Role of Microorganisms and Coprophagy

In view of all this it is not surprising that these animals do not depend on leaf litter alone for their food. Nor are they just omnivores; they are predators (Edney et al. 1974) and coprophagous (Carefoot 1984; Hassall and Rushton 1985). Both these behaviours have been ascribed in the past to a lack of copper in their diet. However, the most parsimonious explanation of predation would be that it supplements the isopods' plant food, which is low in protein, with animal food which is high in protein.

Hassall and Rushton's study also revealed how the strategy of coprophagy may be employed by these animals to increase the protein in their diet. They fed a laboratory culture of a different species of terrestrial isopod, *Porcellio scaber*, on its preferred food of decayed birch, *Betula pendula*, leaves. They collected faeces from this culture and "aged" them for from 1 to 50 days in a solution derived from decaying birch leaves. Then they tested animals to see at what age their own faeces were most palatable to them. The isopods showed little interest in reingesting 1-day-old faeces, but a significant overall increase in preference with age up to a peak at 2 to 3 weeks old, followed by a decline beyond that age. These results, and the fact that the isopods have hindguts which are especially efficient at absorbing nutrients, enabled Hassall and Rushton to reject the hypothesis that the behaviour is "refection" of nutrients which would otherwise be lost in the faeces (see Sect. 6.2.5.1). They were left with what they aptly called the "external rumen" hypothesis; that the faeces are improved as a source of food by the action of microorganisms on those faeces. Measurements of bacteria and fungi in faeces of different ages showed that they always contained one to two orders of magnitude more microorganisms than uneaten leaf litter which had been aged for the same time under identical conditions. So the faeces were always a better source of protein than leaf litter. However, the number of bacteria and of fungal propagules in the faeces declined steadily over the first 3 weeks. It therefore seemed unlikely that they were the cause of the preference for 3-week-old faeces. On the other hand, the total biomass of fungal hyphae in the faeces reached a very clear peak in 20- to 30-day-old faeces. The authors cautiously point out that there are a number of possible explanations as to why isopods might prefer to reingest their faeces when they contain the greatest amount of fungal hyphae. But I would think that the fact that the fungal hyphae are a rich source of protein for animals living on plant tissues which are low in protein and high in structural carbohydrates would seem the best explanation. Also, if other processes are operating – such as the acquisition of enzymes from the fungi with which to break down cell wall material, or the microorganisms degrading polyphenols in the faeces – these would further enhance the availability of the nitrogen "locked up" in the plant tissues.

Other isopods are known to actively "graze" fungi growing on decaying leaf litter, and Gunnarsson's (1987) investigation of *Oniscus asellus* feeding on dried maple, *Acer platanoides*, leaves in Sweden supports the concept that the fungi are providing isopods with a richer source of protein than would otherwise be available to them. Using time lapse photography Gunnarsson showed that *O. asellus* eat the patches of the maple leaves which are infested with the tar spot fungus, *Rhytisma acerinum*, before they touch any other part of the leaf. Next they eat the tissue immediately surrounding the fungal colonies they had already eaten, and only then do they feed on the remaining areas of the leaf blade before finally starting to eat the fine leaf veins.

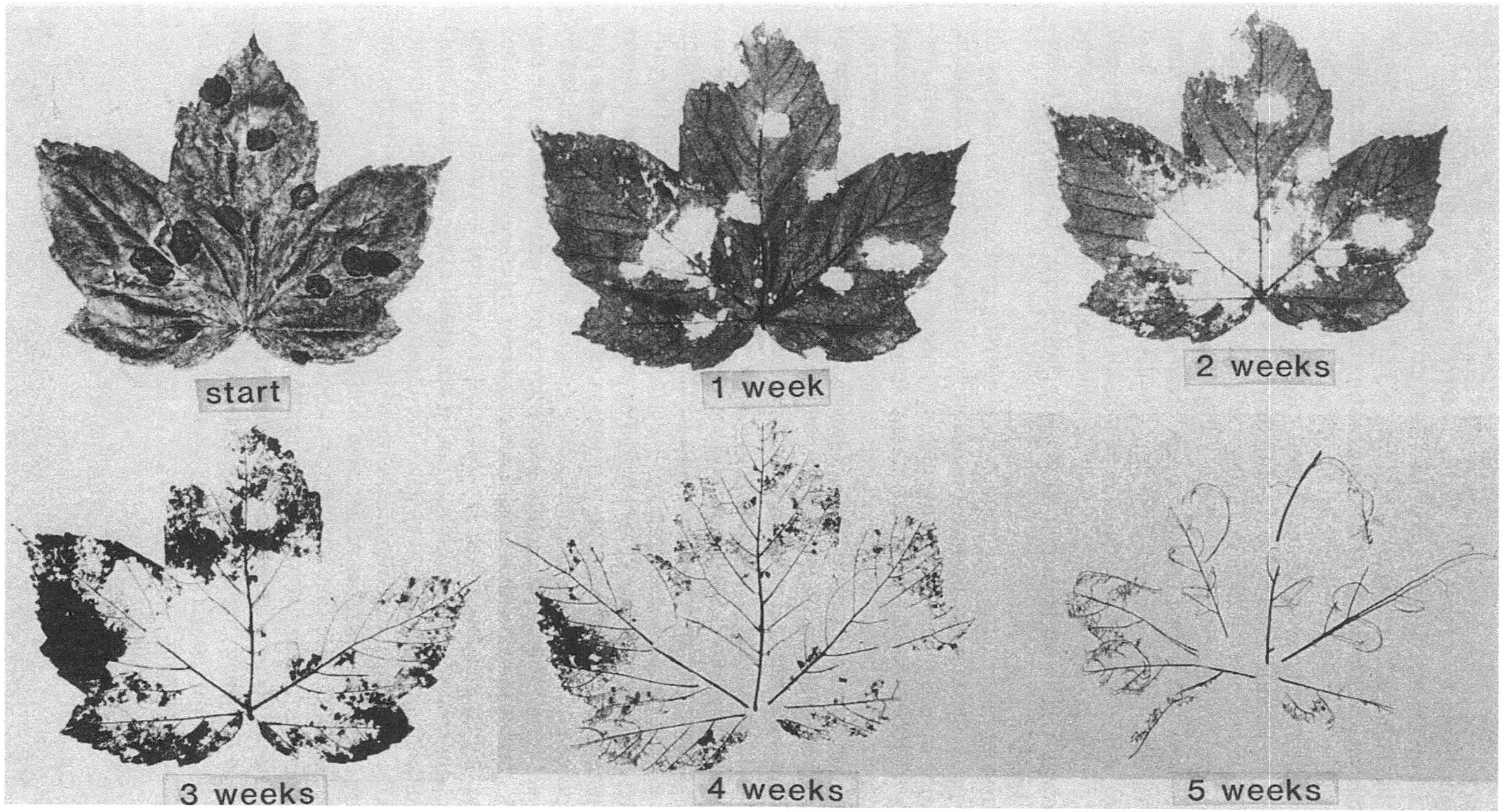

Fig. 12. Time-lapse photographs of a dry maple leaf being eaten by isopods. Patches of fungus contain most nitrogen and are eaten first; veins, with the highest C:N ratio, last. Photo courtesy of T. Gunnarsson

Gunnarsson found that these highly preferred fungal colonies had more nitrogen in them and a lower C:N ratio than uninfected parts of the leaf, and that the tissues left until last – the veins – had the highest C:N ratio. While alternative explanations for this strong feeding preference have not been eliminated, it is clear that the presence of microorganisms guides the choice, and the weight of evidence suggests it is their value as protein food which does this. Similar preferential feeding on rust-infected leaves has been demonstrated for a millipede, for several species of slugs and snails (Ramsell and Paul 1990) and for insects (White 1984; see also Sect. 3.4.7). It would seem too that, like other detritivores and wood-eating insects, terrestrial isopods benefit from cellulolytic enzymes derived from fungi which they ingest with their food (Kukor and Martin 1986). Some desert isopods have a gut biota of "bacteria and fungi, and perhaps protozoa as well, which contribute significantly to the hydrolysis of cellulose and other recalcitrant compounds" that they eat (Crawford 1988).

However, many terrestrial isopods show a strong preference for feeding on litter partly decomposed by microorganisms before they ingest it. A fascinating variation on this theme of making maximum use of the activity of these free-living microbes was found by Hassall and Jennings (1975). They investigated the structure of the gut and the digestive physiology of *Philoscia muscorum*, a common terrestrial isopod of dune grasslands in Yorkshire. The hindgut of this animal is almost as long as its body and is divided into the anterior, papillate, sphincter, and rectal regions. The anterior region is by far the largest and is further divided longitudinally to a large ventral sac with a narrow dorsal typhlosole or channel. Solid ingested food passes quickly into this ventral sac of the anterior region where it remains for 20 to 24 hours. There its digestion is continued by the microorganisms ingested with it and which had already started the process before it was eaten. The products of digestion of cellulose by the microbes are absorbed in this section of the hindgut. The residue plus microorganisms then pass into the papillate region where they are attacked by enzymes from the mid-gut caeca; these enzymes are discharged along the dorsal channel of the anterior region. The products of this digestion are absorbed in the papillate region. Thus, *P. muscorum* has evolved a system somewhat analogous to that of the ruminants (and a step ahead of the "external rumen" of *Porcellio scaber*) which segregates the digestion of cellulose by microbes from the subsequent digestion of the microorganisms by its own enzymes. But it uses free-living microbes rather than obligate internal ones.

There is still much that is not known about the extent and variation of the dependence of terrestrial isopods on microorganisms for the extraction of an adequate supply of protein from their impoverished food. However, it is clear from the foregoing that a variety of strategies of varying sophistication have proved effective enough for the animals possessing them to survive the rigours of natural selection.

Chapter 5 Molluscs

Most molluscs are not herbivores. Apart from the many clearly obligate carnivores, the myriad of aquatic filter and detritus feeders eat bacteria and algae as phyto- and zooplankton, or growing on the surface of rocks, aquatic plants, or organic and mineral particles in sediments. There are few apparently true herbivores, those that eat only the tissues of vascular plants. Of these the terrestrial gastropods, snails and slugs, are probably the best known. They eat two sorts of tissues. When it is available they selectively eat flush new growth, especially seedlings (with devastating effect as any gardener knows), but for most of the time such food is not available. Then they select senescing or dead parts of plants (Grime and Blythe 1969). Others eat rotting wood, fungi, soil detritus, or carrion (Mason 1970). All are saprophagous to a greater or lesser degree, and as such their diet is variably supplemented by microbial or animal protein.

Whatever their diet, however, the common theme recurs. Protein food is either constantly or spasmodically in short supply, limiting their abundance by reducing their reproductive capacity and the viability of their young. This shortage has led to the evolution of the same sorts of adaptations as it has in other herbivorous animals. They feed selectively for protein, supplement vascular plant tissues with microorganisms (or eat nothing but microorganisms, often via the practice of coprophagy), practise cannibalism, and harbour nitrogen-fixing bacteria.

5.1 Some Examples of Freshwater Snails

Eisenberg (1966, 1970) manipulated the number of adults, the presence of predators, and the amount of food of a freshwater snail, *Lymnaea elodes*, in a small permanent pond in southern Michigan. He did this with populations inside snail-proof pens constructed in the pond. Increasing or decreasing the density of adults at the beginning of the breeding season (by as much as five times the initial densities), or adding or removing predators had no measurable effect on the density of populations at the end of the season. But in pens to which he added food in the form of frozen spinach leaves, adult fecundity increased by 25 times and the number of young by nine times. Eisenberg considered this dramatic effect strong evidence that food was limiting the abundance of these animals. Yet because there was an abundance of vegetation and coarse debris in the pens, he concluded that the shortage must be a relative one; that the lack is of high quality food for growth and reproduction. Concurrent studies of the general population in the pond showed that the young snails were growing at only a fraction of the rate of those he fed with superabundant food in the laboratory. And 93 to 98% of them died. Again, he attributed this to a shortage of good food in the pond.

He carried out a further series of trials in snail-proof pens, but this time supplying additional food over a longer period and manipulating snails of different ages. The snails showed a dramatic response. Fed for 6 weeks with a continuing supply of frozen spinach leaves, they were one and a half times larger, twice as heavy, produced egg masses with twice as many eggs in them, and produced 85 times more eggs than did snails that did not get extra food. Without added food snails did not grow particularly well, and hardly reproduced at all. But they did not die; given more food they were not only able to grow, they reproduced abundantly. Eisenberg considered that the extra food added per snail was so small that it must have been supplying some specific limiting component of the diet rather than just relieving a gross shortage of food. Unfortunately he did not look at what this limiting component might be. Nor did he measure the influence of added food on the survival and subsequent growth of the young snails hatching from all the extra eggs produced. I suspect the key component added was microorganisms growing on the spinach leaves. They would have supplied a protein boost to the already abundant carbohydrate in the snails' diet.

That this was probably the case was demonstrated in a more recent study of aquatic snails, also in southern Michigan. Osenberg (1989) experimented with a community of three prosobranch and five pulmonate snails living in a lake with a littoral zone dominated by stands of *Scirpus*. The snails fed primarily on epiphytic micro-algae and bacteria growing on the *Scirpus* leaves. Small blue-green algae and diatoms made up over 90% of the biomass of these epiphytes. Osenberg produced an obvious increase in abundance of epiphytes after only 1 week by adding phosphorous fertilizer to plots set out in the beds of *Scirpus*. After a month there were 13 to 500 times more epiphytes in his fertilized plots than in control ones, giving an approximate 20-times overall increase in abundance. In this 1-month-long experiment the snails responded in the same way as the epiphytes – but with important differences. Those species with young that had hatched *before* Osenberg added the fertilizer did not increase in density (but individual snails grew larger than those in the control plots). Those species with young which hatched *during* the experiment, however, were 15 times more abundant at the end of the month. The time scale of this experiment in relation to the biology of the snails was such that this result could not have come about through the influence of improved nutrition on the fecundity of the adults. Osenberg had previously, and in this experiment, demonstrated that predators had little, if any, effect on the abundance of the snails. The dramatic increase in numbers could have come about only through increased survival of the newly hatched young.

O'Keeffe (1985a, b) obtained very similar results from his populations of the gastropod *Bulinus globosus*, the intermediate host for bilharzia schistosomes in coastal ponds in Kenya. Initial investigations led him to conclude that the high temperature of the water was keeping these populations at low levels. He also observed that reproductive peaks of *Bulinus* occurred after heavy rains, especially after unseasonal rains in the cool months. The influence of rain could have been direct or indirect. One possibility was that it altered the quantity and proportion of microorganisms on which the snails fed.

O'Keeffe had also noted that the performance of his experimental snails was highly dependent on the type of food he provided in the enclosures. So he conducted a second series of trials with known numbers of snails held in netted plastic enclosures suspended in his study ponds. To each of three different densities of snails he added leaves of either the succulent water lily or the coarse sedge, both of which grow in the ponds. Snails with a supplement of lily leaves added to their enclosure grew faster, had higher rates of reproduction, and lived longer than those supplied with only coarse vegetation. And the snails starting at low densities performed better than those at higher densities. At the same time, the growth, reproduction, and survival of the snails living free in the pond were as poor as those for the *least* successful feeding regime in captivity. Clearly these ponds are suboptimal habitats for the snails.

Like Eisenberg's snails, O'Keeffe's snails could, for most of the time, stay alive on a strictly maintenance level of nutrition, but could not reproduce. O'Keeffe concluded that the most likely reason for the bursts of reproduction following heavy rains was runoff adding nutrients to the ponds. These nutrients would have produced increased growth of microorganisms on which snails fed, providing them with a flush of high-quality food. His feeding trials indicated that algae, not the vascular plants he added, were the important limiting food for the snails.

5.2 Marine Limpets and the Flow of Nitrogen Through the Food Chain

Similar limitations seem to hold for marine molluscs. Of the many published studies demonstrating this, Quinn's (1988a, b) more recent study will serve as an example and reference to earlier studies. Working with a pulmonate limpet, *Siphonaria diemenensis*, on the rocky intertidal shore of southeast Australia, Quinn investigated the influence of the availability of food on two populations living in distinct tidal zones. In both zones the animals grazed encrusting brown algae. In the higher zone the supply of algae is seasonal; in the lower zone there is a steady and fairly constant supply all year round. Limpets on the higher zone grew more slowly and did not grow as big as those in the lower zone with a continuous supply of food. They produced many fewer egg masses, and many more of them starved to death. As others had found in similar studies (Underwood 1978), predators played no significant part in limiting the abundance of these limpets.

The story of another limpet, *Patella granularis*, living on the rocky shores of the southwest coast of South Africa (Bosman and Hockey 1986, 1988a, b), nicely illustrates not only how these herbivorous molluscs are limited by their food, but that nitrogen flowing up through the trophic levels is the major factor limiting abundance at each level. This study emphasizes how, for most of the time and in most places, there is an acute shortage of nitrogen in the environment. When this relative shortage is relieved by a supply of available nitrogen being concentrated at one place, organisms at all trophic levels also concentrate at that place and quickly make use of this increased supply to increase their rate of reproduction.

On the offshore islands in Saldanha Bay there are very large permanent colonies of seabirds (gannets, penguins, and cormorants) with consequent great deposits of guano. The seawater washing the intertidal zone of these islands is therefore much enriched. The dominant soluble organic compound is uric acid which is rapidly converted to ammonia by aerobic and anaerobic microor-

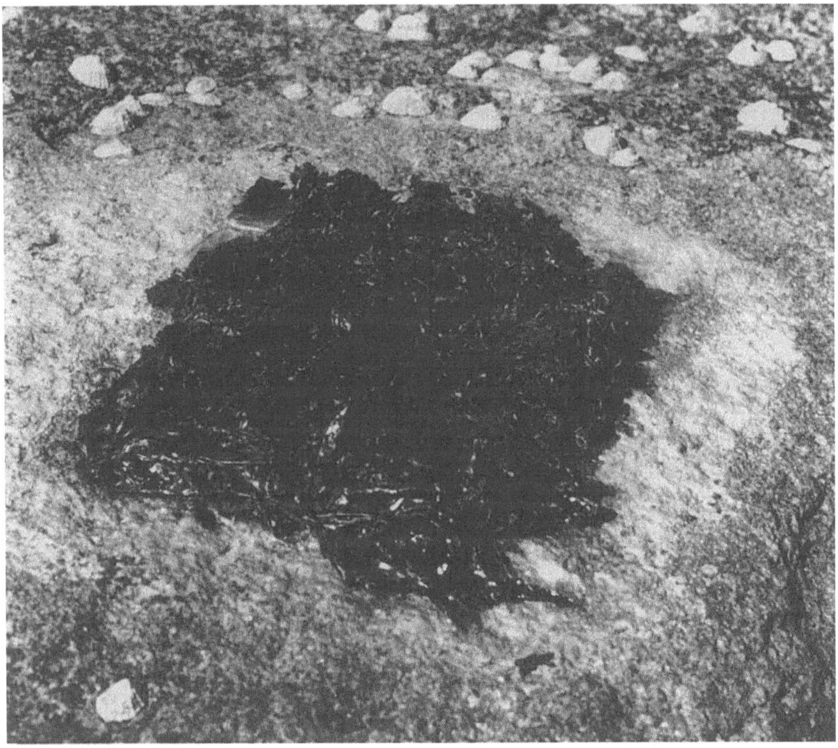

Fig. 13. South African limpets grow to great densities where the algae they eat (foreground with exclosure cage removed) grow profusely in seawater enriched by guano from seabirds (see also Sect. 16.4). Photo courtesy of P. A. R. Hockey

ganisms, and then to nitrite and nitrate by nitrifying bacteria. In these enriched waters dense beds of foliose algae grow, and the major herbivore feeding on these algae is *P. granularis*. This limpet is a generalist forager eating any available microalgae, algal sporelings, and prostrate and creeping algae. In turn, small *P. granularis* are the major food that the African black oystercatcher, *Haematopus moquini*, feeds to its young.

Bosman and Hockey found that the algal mat grew fastest and thickest on the enriched island shores. On the mainland shores the growth of the foliose algae is slow and sparse, never forming mats, even when limpets were excluded from these sites. But when the authors artificially enriched a section of mainland shore with guano, the production of algae at that site was higher than on the adjacent unenriched shore.

On the naturally enriched island sites the authors found that although there were *fewer* limpets than on the impoverished mainland shores, these island limpets grew faster and bigger, and produced, in total, many more young than did the mainland ones. Limpets which they transplanted between island and mainland sites soon showed growth rates similar to those of the local limpets.

The lower numbers of limpets on the shores with abundant algae arose because of the oystercatchers. They throng to these island shores and feast on the enormous numbers of young limpets growing there (an estimated 1.1 million per kilometre of enriched coast are removed each year). On the mainland sites there are few, if any, oystercatchers. Clearly, many more settling larval limpets can survive, and grow quickly, on the dense growth of algae in the nitrogen-rich waters of the guano-supplemented shores. The difference in survival and growth is so great that – *in spite* of supporting so many oystercatchers – the enriched island population still has a greater residual biomass and produces *ten times* more gametes than the mainland populations. In turn, the oystercatchers fed almost exclusively where the supply of their protein food is abundant, were themselves much more abundant, and, no doubt, produced many more young than did the rare birds found on the mainland coast.

As a result of this study Bosman and Hockey concluded that, perhaps, in the past marine ecologists had been too preoccupied with the role of predators, rather than primary production, in seeking to understand the organization of intertidal communities. When one views what happens at these locally enriched sites as but one small facet of the situation throughout the populations of both the limpets and the oystercatchers, it is clear that this is so. I discuss this more fully in Section 16.4.

5.3 Death of the Young, Selective Feeding, and Animal Protein in the Diet

There are many studies reporting the death of the very young as the main factor influencing the abundance of snails, whether they be terrestrial or aquatic, herbivorous or carnivorous. Usually a shortage of food has been implicated in their early demise. For example, McQuaid (1983) found that a decrease in the numbers of the herbivorous winkle, *Oxystele variegata*, feeding on diatoms in South Africa was related to high juvenile mortality associated with a shortage of food. Seager (1982) working with the opisthobranch *Philine gibba* in marine sediments of the antarctic island of South Georgia recorded very high mortality of juveniles. He considered that food was so short that most juveniles starved. Spight (1975, 1982) demonstrated that the average hatchling of several species of carnivorous *Thais* snails, living on the coast of Washington, had a less than 1% chance of surviving and that most of the 99% that died each year did so shortly after hatching. In some years, when their major food (barnacles) was more abundant, more juvenile snails survived.

Very pronounced selection of food items is commonly recorded for both grazing molluscs and for those which feed on detritus. The fact that a mollusc can feed solely on algae, even though they may be better food than vascular plants, does not preclude it from the necessity of selecting those species from which it can most readily obtain the highest possible level of protein. For most (and for many other aquatic animals) diatoms are the preferred food (Calow 1973; Underwood 1978). Apparently, the contents of diatoms are far more readily accessible in spite of their silica shells. On the other hand, green algae, with their cellulose coat, are usually shunned, but when they are preferred, as in the case of the common European periwinkle, *Littorina littorea*, species of *Ulva* and *Enteromorpha* are chosen (Watson and Norton 1985).

Nicotri (1977) neatly demonstrated that although intertidal gastropods can survive on a diet of nothing but microalgae, they are nevertheless highly selective as to which ones they will eat. He maintained populations of three species of limpets and one species of periwinkle on large clay pipes set in the mudflats of San Juan Island, Washington. They could not escape and could eat nothing other than the algae which grew on the surface of the pipes. All four species of gastropod strongly selected just three species of diatoms from some 130 species of microalgae which grew there.

A study of the omnivorous marine snail *Nassarius pauperatus* in South Australia suggested that a diet of algae may not provide sufficient protein for adequate growth and reproduction. It feeds on algae, which are ubiquitous but spaced over the mudflats, and carrion, which is scarce, unpredictable and highly clumped. McKillup and Butler (1983) found that snails grew much more rapidly, and to a greater adult size, when they were fed unlimited amounts of carrion in the laboratory. In the field they found that the size of snails was related to the amount of carrion available in the habitat. It would seem that these snails are surrounded by algal food which is inadequate in either quantity or quality as a source of nitrogenous food. So the extent to which individuals are able to survive, grow, and reproduce depends largely on the amount of carrion that becomes available in their habitat as a supplement of animal protein in their diet.

5.4 Detritus Feeders Feed Selectively, and Depend upon Microorganisms and Coprophagy

For filter and detritus feeders, selection is more a matter of sifting and concentrating the microbial bodies in the water or sediment (Calow 1974). Connor and Edgar (1982) demonstrated that the mud snail, *Ilyanassa obsoleta*, an Atlantic detritivore in the USA, feeds selectively from the material available to it on the mud and sand flats where it grazes. Carbon, nitrogen and algae in the sediments were concentrated 20 to 40 times in their guts. Living diatoms were most preferred, but the snails will also scavenge the remains of dead animals. So, clearly, they are omnivores rather than herbivores. Nevertheless, they are selecting a diet which is high in nitrogen. This, it seems, is the case for terrestrial as well as aquatic snails which feed on detritus. The bacteria and algae which make up the bulk of their diet, especially when they are very small, are high in nitrogen and low in structural carbon by comparison with vascular plants.

Valiela and Rietsma (1984) investigated the chemical cues used by detritivores when feeding on a Massachusetts salt marsh to find out whether or not these explained their feeding preferences. They used several different sorts of detritus-feeding animals in their experiments, but the principal one was a snail, *Melampus bidentatus*. They conducted all the trials by allowing experimental animals to graze in the laboratory on agar suspensions of ground detritus with or without supplementary substances. They experimented with the palatability of four different cues, both singly and in combination: pH, salinity, nitrogen, and ferulic acid (the last because it is the most abundant phenolic acid in *Spartina alterniflora*, the main source of detritus on these marshes). Salinity and pH proved to be relatively unimportant alongside the other two, and the presence of sufficient available nitrogen overwhelmed any response to ferulic acid.

Because nitrogen is known to increase in the litter with both age and eutrophication, the authors tested the palatability of agar preparations of detritus which had been aged in the marsh with or without added fertilizer. The snails preferred the higher nitrogen detritus from the fertilized plots. Contrary to expectations, however, with detritus from both fertilized and control plots, they nearly always preferred younger litter, with a lower content of total nitrogen, to older litter. Valiela and Rietsma's explanation for these apparently contradictory results is that younger litter, although lower in *total* nitrogen (which they were using as a feeding cue), contains more *soluble* nitrogen than older litter. This soluble nitrogen, or some fraction of it, may be the feeding cue to the only nitrogen that is available to these detritivores. The older detritus may contain nitrogen largely bound in an unusable (and possibly undetectable) form. On the other hand, it may be a matter of what proportion of the nitrogen in the litter is contained in living microorganisms feeding on the litter. Their abundance would be affected in turn by the extent to which the litter, as it ages, is exhausted of organic material by the bacteria feeding on that organic matter. Older detritus would then carry a lower population of microorganisms.

Twenty years earlier, Newell (1965) had carried out an elegant study of the role of detritus in the nutrition of two marine deposit-feeding molluscs in the Thames estuary in England. One was a prosobranch, *Hydrobia ulvae*, and the other a bivalve, *Macoma balthica*. Both are common: abundant where the detrital deposits were fine and sparse in coarse deposits. Newell was able to show, by careful laboratory experiments, that bacteria were also much more abundant in the fine deposits, and that it is these bacteria which the molluscs are eating. By measuring the carbon and nitrogen content of the deposits that the molluscs ate, and the content of their faeces, he found that only the nitrogen fraction of the detritus was being removed. This nitrogen fraction is made up entirely of the bodies of the bacteria. Once discharged back into the sediment, the faeces quickly increased in nitrogen content as bacteria grew on them. Then they were reingested by the molluscs.

By feeding these animals on their own faeces in the laboratory Newell showed that they extracted the microbial nitrogen, leaving the carboniferous organic matter largely untouched. So, there is a nice system of recycling indigestible organic debris by "creaming off" the almost pure protein of the crop of bacteria which grows on the debris each time it returns to the sediment. This form of coprophagy is quite common among various aquatic animals which feed in detrital sediments, and in some it may be further assisted by the presence of a bacterial fauna in the gut (Johannes and Satomi 1966).

In the light of his results Newell proposed that the terms "detritus" and "detritus feeders" were misnomers. Rather, there is organic debris on which microorganisms grow, and animals which eat the microbes. This is likely to be equally true for other deposit feeders and filter feeders – and for many which are classed as grazers or herbivores. For example, *H. ulvae*, which he used in these experiments, has a varied diet including green algae, blue green algae, diatoms, and detritus. It may well be that bacterial protein is an essential component ingested with all these foods.

5.5 Terrestrial Snails Live with the Same Constraints

It seems that aquatic molluscs feed mostly upon algae – a very different food from vascular plants. Terrestrial snails, however, more commonly eat vascular plants, but they too must often supplement their vegetable diet with microorganisms or animal tissue.

Williamson and Cameron (1976) studied the population dynamics of the land snail, *Cepaea nemoralis*, living on grassland in West Sussex, England. They found that, like other terrestrial snails, *C. nemoralis* is primarily a detritus feeder and showed a strong preference for feeding on dead or senescent herbs. They made a detailed analysis of the faeces of these snails in order to better understand their diet. This technique has the same disadvantage that it does with other animals. It reveals only what the snail has *not* digested, leaving unknown that which is ingested and completely digested. So their conclusion that the approximately 10% of faecal samples comprising a variety of animal remains represented only accidental ingestion is probably wrong. The remains (largely of insects) could indicate that animal protein is a significant component of the diet. And what chance is there of detecting microorganisms in the diet? The results of their experimental rearing of juvenile snails, captured in the field and fed in the laboratory on a variety of purely plant diets, would tend to confirm this oversight. None of them grew as well as those left in the field. Something important was missing from their diet. The authors did not investigate what this might be, but confirmed with further studies (Williamson et al. 1977) that high mortality of juveniles was the chief cause of changes in abundance of *Cepaea* on these grasslands. Predation was unimportant.

Wolda and Kreulen (1973), in a separate study of *C. nemoralis* in Holland, established that the key factor influencing its abundance there was, without a doubt, the survival of the young. More than 20 times as many young survived their first year in an expanding population as in a declining one. As well, in the expanding population there were twice as many eggs produced, hatchlings had a 25% better survival, and three times more juveniles were produced per adult than in the declining population. But, as the authors said, these latter differences, important as they may seem when considered alone, are of only minor importance when compared with the massive difference in survival of the young. They also noticed that in the declining populations of *Cepaea* the size of egg clutches declined through the summer; but in the laboratory this did not happen. Nor did they have the massive loss of juveniles there. They were able to eliminate all factors likely to have been responsible for the decline except food. There was no drop in the quantity of food in the field, so they decided it must have been the quality of the food which had declined. The most likely cause of this drop in quality, they concluded, was a decrease in digestible proteins in the food.

But neither group of workers attributed the enormous and usually sporadic changes in survival of the young *Cepaea* to change in the availability or quality of their food. Wolda and Kreulen considered drought and predation the most likely causes of death of the young, while Williamson and his colleagues considered it to be unspecified "density-dependent control".

5.6 Cannibalism by Young Snails Illustrates Shortage of Nitrogen

Land snails provide a further indication of how critically short nitrogen is in the food of herbivorous (or near-herbivorous) snails, especially the very young; this indication is cannibalism by newly hatched snails.

The Roman or edible snail, *Helix pomatia*, is thought to be entirely herbivorous, feeding only on live green plant tissue in fresh condition (Pollard 1975).

For obvious reasons its young eat their calcareous egg shells, but they also frequently eat unhatched eggs, and occasionally other young snails.

Similar behaviour has more recently been recorded in detail for another terrestrial pulmonate, *Arianta arbustorum,* in Switzerland and Sweden (Baur and Baur 1986). This snail is deemed to be a "food generalist", eating many species of live plants but also dead plant material. Newly hatched snails readily eat fresh green plants, but they are also cannibals – and the degree of cannibalism is related to their nutrition. The Baurs found by experimental manipulation that the degree of cannibalism was not influenced by the size of the egg clutch in which the young hatched, the density of eggs in the habitat, nor the presence of plant food (fresh lettuce leaves). Normally all eggs in a clutch hatched more or less simultaneously and the only eggs eaten were either infertile or incompletely developed. Lettuce leaves were eaten only after all unhatched eggs had been devoured. However, when the Baurs extended the time over which the eggs from one clutch hatched, by refrigerating some of the eggs for a few days, the first snails to hatch quickly devoured *all* the unhatched eggs.

Nor did the early feeding experience of the hatchlings make any difference to their propensity for cannibalism (Baur 1987). Hatchlings which were fed solely upon either conspecific eggs or fresh lettuce when they first hatched subsequently chose eggs when offered a choice. But by the time they were 4 days old they ate eggs and lettuce in equal proportions. By 16 days of age they would eat only the lettuce, whether or not they had previously eaten eggs. Snails older than 4 weeks, having eaten all the lettuce offered, refused to eat any eggs and starved. Baur (1988b) also found the same initial exclusive preference for eggs by hatchling *H. pomatia*, but again this quickly decreased with age.

The Baurs demonstrated that snail eggs are a rich source of protein by raising hatchlings on either conspecific eggs or lettuce. Hatchlings eating conspecific eggs increased their wet weight by 500% in the first 15 days of life; those on lettuce increased by only 50%. And while only one-third of those on a diet of lettuce survived to adulthood, two-thirds of those on eggs did so – and they matured earlier (Baur 1990). In another experiment with both *A. arbustorum* and *H. pomatia,* hatchlings fed on conspecific eggs doubled their weight in their first 6 days while those on vegetable food only increased their weight by 18% in the same time (Baur 1988a).

The experiments also revealed that the young hatching from the eggs of snails collected from a midlands meadow at 580 m above sea level in Switzerland, were twice as cannibalistic as those from snails collected on a mountain pasture at 2010 m. Snails from the mountains are smaller than those from the midlands, take longer to mature, and lay fewer, smaller eggs. But relative to their adult size, the eggs are larger than those of the more cannibalistic midland snails (0.5 and 0.3%, respectively, of the mother's weight). So, the Baurs hypothesized, mountain snails emerge better nourished from their relatively larger and better provisioned eggs than do the midlands ones; therefore a greater proportion of the eggs of the larger clutches of the midland snails will end up serving as food for the young that hatch from a small proportion of

Fig. 14. Hatchling snails eating unhatched conspecifics redistribute limited protein to fewer individuals. More survive, and grow faster, than without this cannibalism (see also Sect. 12.2). Photo courtesy of B. Baur

those eggs. This is another good example of cannibalism serving as a strategy to make more effective use of a limited supply of protein by redistributing it to fewer individuals (see Chap. 12).

It seems, however, that the behaviour is limited to very recently hatched snails eating unhatched eggs. Juveniles and adults have never been seen to eat eggs or hatchlings. It is also species specific. Even when the authors exposed hatchling *A. arbustorum* and *H. pomatia* to a choice of their own and the other's eggs in the laboratory, they each ate only a very occasional egg of the other species, but quickly ate all of their own (Baur 1988a).

A more specialized form of cannibalism of eggs is seen in the evolution of "nurse eggs" or "trophic eggs". This is an adaptation found in several taxa, but which is common and well studied in prosobranch gastropods (Rivest 1983; Baur 1992; see also Sect. 12.2).

5.7 Teredo Shipworms Depend upon Microorganisms Which Fix Atmospheric Nitrogen

A fairly extreme case of herbivory among molluscs is that of the marine shipworms (Teredinidae). These bivalves are adapted to a way of life akin to that of the termites (Sect. 3.7.2) and are well known because of the damage they do to any wood immersed in the sea. Their larvae are pelagic for up to 100 hours before they must settle in a piece of wood. In this time they feed upon plankton (Lane 1959), so are initially carnivorous or omnivorous. Once established, they can and do normally continue to feed upon plankton filtered from the sea with their gills. But they have been reared to sexual maturity in the laboratory on wood in seawater filtered free of plankton.

So, once settled, they can subsist and grow on wood alone. They are able to do this because they can obtain nitrogen from microorganisms. Like the wood-boring crustaceans (*Limnoria* spp.) they probably gain protein from the mycelia of marine fungi which quickly infest any unprotected wood in the sea (Lane 1959).

Also, they have special glands in their gills, unique to the family Teredinidae, that harbour intracellular bacteria which digest cellulose, synthesize essential amino acids, and fix nitrogen (Trytek and Allan 1980; Waterbury et al. 1983). Carpenter and Culliney (1975), in their laboratory at Woods Hole, Massachusetts, had demonstrated the ability in three coastal species of teredo worms and one from the Sargasso sea to fix nitrogen. In the species from the Sargasso sea, *Teredora malleolus*, the bacteria fixed nitrogen at a rate which doubled cellular nitrogen in 1.5 days. This is nearly 20 times greater than the rates they measured for the coastal species, *Psiloteredo megotara, Lyrodus pedicellatus,* and *Teredo navalis*. Juvenile shipworms had higher rates of fixing nitrogen than had adults. These differences between species and ages in the capacity to fix nitrogen illustrate how fine is the line between sufficiency and deficiency of nitrogen in the food of these animals, and the vital role the nitrogen-fixing bacteria must play in keeping them in a positive nitrogen balance. The C:N ratio of the wood was about 300 – 500:1 and that of phytoplankton 5 – 10:1. So, for either a young teredo unable to filter enough phytoplankton fast enough to match its rate of growth (they have been measured increasing their length 30 times in a month), or for a worm of any age living in the Sargasso sea, where the density of phytoplankton is notoriously low, the nitrogen supplement provided by their gut bacteria becomes crucial. Yet, apparently, these microbial sources of protein may still not be enough. Without a supplement of plankton in their diet many species will grow but are unable to reproduce (Trytek and Allan 1980).

Chapter 6 Mammals

6.1 Large Mammals

Thirty years ago Lauckhart (1962) tried to convince his fellow wildlife managers in North America to "back off a great distance...and...view the whole picture of animal populations". He then presented them with essentially the same picture which I am trying to present – that most individual animals die when young in a habitat which is saturated. They are passively limited by the inadequacy of their environment. As a consequence, their populations are *constantly* pressing hard against the carrying capacity of their habitat; and that capacity is set by the quality of their plant food. Plants, he said, have been able "to retreat *below the animals' threshold of malnutrition* in order to escape total destruction by animals" (my emphasis), and it is this malnutrition, not predation (including human hunting) which determines the abundance of herbivorous mammals and birds. Judging by the slightly condescending questions published with his presentation, and the lack of notice taken of his views since, he was not too successful! Present day managers and conservationists would benefit from reading his simple but astute assessment of how the real world of nature works, and how human activity impinges on it. So too would my fellow ecologists!

There are very many studies which confirm the general inadequacy of plants as food – and specifically as a source of protein – for large herbivorous mammals (Sinclair 1975; McNaughton and Georgiadis 1986; Boutton et al. 1988), and that it is the very young which bear the brunt of this inadequacy (Caughley 1966; Seegmiller and Ohmart 1981). Yet, in spite of recognizing the importance of the role of nutrition in the reproduction of wild mammals, most attempts to understand this have been couched in terms of energy (Sadleir 1969).

Mammals supply their neonates with a rich supply of animal protein in milk. From weaning they all depend on intestinal microorganisms for much of the protein in their diet. They exhibit various territorial and social dominance behaviours which restrict access to the best food to only some of the many. They can adjust the time spent feeding and the rate of passage of food through the gut. They can move from place to place or change from grazing to browsing to follow the availability of food. Different species, differing in size, have adapted to feed on different species and ages of vegetation; and this is reflected in differences in the size and complexity of their rumens or caeca. They must still, however, spend most of their time feeding, and do so in such a way as to select, diligently, those parts of their food plants which will return the greatest quantity of digestible protein per volume of vegetation ingested.

In spite of all these adaptations they still experience a chronic shortage of available nitrogen in their food. For many, lack of enough protein for basic maintenance metabolism is a regular seasonal event. For all, sufficient protein to sustain females through pregnancy and lactation, and the young through growth after weaning is the major factor limiting their abundance. It is true that energy, water, and various nutrients or micronutrients, may, for some species in some situations and at some times, be limiting (McNaughton 1988). It is, however, obvious that the general and crucially limiting resource generating all these adaptations is food which contains sufficient protein for successful reproduction.

6.1.1 Feral Donkeys in Australia

An unequivocal demonstration of the abundance of a large mammalian herbivore being limited by the abundance of its food, operating via the nutrition of the females and the survival of their neonates, was provided by a field experiment with two discrete populations of feral donkeys, *Equus asinus*, in Australia's Northern Territory (Choquenot 1991). For more than a century these introduced ungulates have roamed over hundreds of square kilometres of savannah country in this tropical monsoonal area. Major equid diseases are absent from Australia and there is no evidence that the only potential predator, the dingo, *Canis familiaris dingo*, attacks them. Yet their numbers are stable, although high, at ten per square kilometre. One of the two populations was culled by shooting more than 80% of the animals over an area of 225 km^2 three years before sampling began. The control population, living in a similarly sized area more than 100 km away, was left untouched until sampling began.

The culled population immediately rebounded, and was growing at more than 20% per year when sampled. Animals were growing faster and reaching sexual maturity earlier than those in the dense but stable control population. But the most dramatic change was in the survival of the neonates. While equal numbers of females were pregnant in both populations, in the unculled one 79% of all foals were dead within 6 months of birth, compared with only 38% in the growing population. Also the survivors were in poorer condition and grew more slowly than those in the reduced population. Foaling and lactation in both populations of donkeys is synchronous with the flush of high quality new grasses which grow at the beginning of the wet season. Females in the culled population were found to be eating a wider range of species with more nitrogen and minerals and less fibre than were those in the dense control population; and they were in better condition and better able to feed their young.

By removing some animals from a stable population Choquenot demonstrated the same thing that providing one population with additional protein food for females and young would have done. Like all organisms, the donkeys constantly produce as many young as they can. Once there are more young be-

Fig. 15. Red deer on the Isle of Rhum. The chance of a fawn surviving is enhanced if its mother eats plants containing more nitrogen as a result of being fertilized by guano from nesting seagulls. Photo courtesy of T. H. Clutton-Brock

ing produced than the food can support, the "surplus" quickly die. This apparently wasteful tactic ensures that the population uses all the food resource that is available to produce as many donkeys as possible in that environment – that it is constantly pressing hard against the limit set by the food that is accessible to individual donkeys. But, at the same time, it ensures that any increase in the availablity of food – no matter how temporary – is quickly used to produce still more donkeys.

6.1.2 Red Deer in Scotland

T. H. Clutton-Brock and his colleagues for many years maintained an extensive study of the ecology of a population of red deer, *Cervus elaphus*, living within the 12 km^2 North Block of the Isle of Rhum in Scotland's Hebridean Islands. Female deer live in matrilineal groups occupying distinct feeding areas within this Block. These studies have served to illustrate that microbial assistance with the digestion of food has not solved the problem of large mammalian ungulates gaining access to sufficient protein from their food. These deer live in an environment in which food is, at best, only marginally adequate in terms of nitrogen available for successful reproduction, and they will quickly seize upon any improvement in this supply. Moss et al.'s (1981) report of red deer selectively feeding on heather with the highest concentration of nitrogen in the Dee Valley of Scotland supports this generalization.

Analysis of 13 years of data from their studies (Clutton-Brock et al. 1986, 1987) showed clearly that death of calves in the first year was the major influence on changes in abundance of the population. Lightest and latest-born calves had the least chance of survival in both summer and winter. Most calves that died in summer did so within a few days of birth. Even those heavier calves that survived their first year had a lower chance of surviving as yearlings if they had been born late. Further analyses of these same data (Albon et al. 1987) revealed that the birth weight of calves is influenced by the effects of weather on the quality and availability of food for both females and their young. In particular, warmer springs produced heavier calves. This is a consequence of the positive influence of an earlier flush of growth on the nutrition of pregnant hinds. Warm dry summers, which reduce the amount of nutritious vegetation the hinds can eat (Albon et al. 1983), produce females which subsequently bear lighter calves than would have been expected from their own birth weights.

Generally, their studies revealed a picture of the abundance of a herbivore being primarily under the influence of the survival of the young which, in turn, is largely influenced by the nutrition of the females, which is dependent upon the level of protein in their food. The system is so finely balanced that what seem to be quite small changes in the weather from season to season can have a major impact on the size of the population.

Just how closely these deer are restricted in numbers by the limits set by the level of available nitrogen in their food is further illustrated by Iason et al.'s (1986) study of the same northern Rhum population. They identified three feeding areas around the coast occupied by three different matrilineal groups. One of these areas, sandwiched between the other two, has nesting colonies of herring gulls, *Larus argentatus*, which deposit large quantities of guano each summer. There was little, if any, increase in the biomass of vegetation growing on the areas the gulls used most, but the nitrogen content of the vegetation on these guano-enriched sites was significantly higher. The authors found that the hinds occupying this feeding area showed two important responses to this increased quality of their food plants. First, they selectively grazed the vegetation on those places which were most heavily fertilized by the gulls. Second, they had a higher lifetime reproductive success than hinds in the two neighbouring groups. Their response is the same as that of the limpets in Section 5.2, but less dramatic.

6.1.3 Antelope, Giraffe, and Greater Kudu in Africa

Rowe-Rowe and Scotcher (1986) investigated the carrying capacity of a game reserve in Natal, South Africa, for the five most abundant antelope in the reserve. While each ate a different range of plant species, all fed selectively on particular species, parts, and growth stages. In late winter this selectivity became increasingly critical as the crude protein content of herbage in the reserve dropped to 1–2%. They recorded 64% of deaths of antelope in the 3 months of late winter-early spring and nearly all of these were young animals less than 1 year old.

Of the five antelopes, the eland, *Taurotragus oryx*, was the least selective, feeding mainly on green grass during the growing season, but being selective between species, and browsing woody plants and eating forbs during the winter. In spite of its large body size it has a relatively small rumen, which restricts its ability to compensate for low-quality food by simply eating more. Instead, its strategy is to select for a protein-rich diet. But this necessitates its moving over great distances or from one vegetation type to another.

The point of the study was to check earlier suggestions that wild herbivores could be stocked on natural veld at higher rates than those for domesticated cattle. The authors state that this is not so because of the restricted preferences for habitat and extreme selectivity in feeding by the native antelope compared with the bulk feeding of cattle. Also, the estimates of the carrying capacity of the latter have been based on rotational grazing, the use of cultivated pastures, and supplementary feeding of concentrates. The native antelopes, on the other hand, must subsist on low-protein native vegetation, without supplements, and with only their capacity to feed selectively to aid in concentrating more nitrogen in their diet.

Pellew (1983, 1984a, b) studied an expanding population of the selectively browsing ungulate, *Giraffa camelopardalis tippelskirchi*, living in Tanzania's Serengeti National Park. The expansion in the number of giraffes was a response to an increase in the supply of its food generated by the conversion of mature *Acacia* woodland to a more open habitat with increased amounts of regenerating *Acacia*. The population was increasing at about 5–6% per annum. By comparison with other populations of giraffe, known to be stable or declining, this population showed a greater total density per square kilometre. While mortality of neonates was still high (some 22% in the first month) approximately 46% of calves survived the first year (compared with 29% surviving to 1 year in an apparently stable population in Nairobi National Park). And they grew faster. Females reached sexual maturity at an earlier age and calved more frequently than those in other populations; and adult mortalities were lower.

As browsers, giraffes have access to a substantially more nutritious diet than that of grazing ungulates, where the crude protein content may regularly fall below the threshold for maintenance metabolism. But they are still highly selective in their feeding. In particular, they browse the very young growth of the *Acacia* which make up the bulk of their diet. These new flush tips make up 80% of the annual diet of adult females (which eat a substantially higher nitrogen diet than the males). These tips have very high levels of crude protein, but this falls rapidly after only a few days. Shoots that survive for a week are subsequently ignored by the giraffes. However, their browsing also has a compounding effect. It produces patches of highly palatable regrowth on the *Acacia* trees, a situation analogous to the grazing lawns produced by many large grazers (du

Toit et al. 1990) and by other herbivores like geese (Sect. 7.1.1) and turtles (Sect. 8.2). The intensity of this selection for flush tissue with its high levels of soluble nitrogen is highlighted by the seasonal behaviour of these animals. As the availability and quality of food declines as the dry season advances, their feeding time increases and they move down to the wetter woodlands in the valley bottoms. Here, the production of new shoots is sustained throughout the dry season.

Other studies of giraffes have highlighted the importance of a high-protein diet, and the extent to which the availability of this protein influences their abundance by way of their capacity to breed and produce surviving young. For example, Hall-Martin and Basson (1975) analysed rumen samples taken from giraffe freshly killed by lions or culled by shooting in a South African nature reserve. They found that the protein content of these samples was highest (22%) during the wet season when the animals had an abundance of new growth of deciduous trees to feed on, and fell to its lowest (12%) at the end of the dry season when the amount of leaf material available had become drastically reduced. The giraffes changed to eating leaves of evergreen and small deciduous plants and eventually mostly woody material. That the late dry season is a time when giraffes are severely and critically limited by the lack of both the quantity and quality of their food was emphasized by their extremely poor condition at this time and their greatly increased mortality due to malnutrition, starvation and predation.

Sauer et al.'s (1982) analysis of 54 plants eaten by giraffes in South African nature reserves clearly showed that they consistently select throughout the year both the species and age of tissues which provide the maximum possible content of protein in their diet. A population of giraffes in a national park in Kenya was also recorded concentrating their feeding areas near rivers during the dry season and dispersing into deciduous woodlands when rain had fallen; thereby following the availability of flush growth (Leuthold and Leuthold 1978). This park lies in an area of relatively low rainfall where the supply of food is unpredictable. The result is a very low mean density and large home range of giraffe in the park. The authors also noted the ability of giraffe, apparently shared by elephants, to detect isolated rainstorms over considerable distances. These rainstorms create local flushes of green browse, a vital resource in this unpredictably poor environment. The poor quality of the habitat is also reflected in poorer reproductive performance compared with the Serengeti population studied by Pellew. Of the only 0.3 estimated births per female per year, approximately 50% of the young were never seen and were presumed to have died before 1 month. Only about 15% survived to 1 year.

The way in which the amount of nitrogen in the diet dictates the nutritional well-being of breeding females, and thus the viability of their young, has again been clearly shown by Owen-Smith (1990) for the greater kudu, *Tragelaphus strepsiceros*, in South Africa's Kruger National Park. Because he could identify each individual kudu from its distinctive markings, he was able, over a 10-year period, to build up a precise record of the age, breeding success, and survival of all females and their offspring living in a number of discrete social units. His study demonstated that it is the survival of the very young which is the key factor influencing changes in kudu abundance, and that the whole process is driven by rainfall through its effect on the supply of high-nitrogen food in the habitat. Kudus preferentially feed upon the new flush growth of trees and shrubs, and on herbaceous forbs. The proportion of these high-protein foods present in the apparent superabundance of edible plants varies with the rainfall. In good years there is sufficient for females to produce strong calves and plenty of milk, and there is abundant high-protein vegetation for the newly weaned calves to eat. Survival is high and numbers increase. In poor years of low rainfall the reverse is true.

6.1.4 Deer in North America

Klein (1970) reviewed the ecological importance of food and its quality for two species of North American deer, *Odocoileus virginianus* and *O. hemionus*, and their several subspecies. They are flush feeders, adapted to early successional stages of vegetation. They are selective feeders, choosing between species, individual plants, and parts of plants for the diet highest in protein. The major influence on their abundance is change in the availability of food with a sufficiently high content of nitrogen. Rises and falls in this availability, and thus in their numbers, are reflected in rises and falls in the fecundity of the females, the survival of young calves, and the age of sexual maturity. The quality of their food limits the number of young which survive both directly and through maternal nutrition.

Records from a herd of rocky mountain mule deer, *O. hemionus*, kept in captivity for 10 years in Denver, Colorado, highlighted this chronic shortage of protein. Robinette et al. (1973) fed the herd a diet supplement which substantially increased the deers' protein consumption. As a result, the animals increased in weight and productivity, they had larger antlers, the females bred and fawned earlier, and the mortality of fawns in their first week declined. The overall productivity of the herd (in terms of fawns per doe) surpassed anything ever recorded for wild populations in spite of the animals being kept at a density of 20 to 30 per hectare – far higher than any density recorded in nature.

Hanley et al. (1987), working in southeast Alaska, found that individual sitka black-tailed deer, *Odocoileus hemionus sitkensis*, consistently preferred to eat the leaves of their favoured browse, blueberry and bunchberry, where these had been grown in the shade of the forest rather than in the open. Analyses showed that plants from the forest contained more than twice as much digestible proteins as those growing in full light, while the content of digestible carbohydrate was the same in plants from both sites. The level of protein in the shade plants would be sufficient to meet the protein needs of a lactating doe, but those from the plants growing in the open would meet only about half her requirements. Animals browsing in the open could, therefore, obtain sufficient nitrogen for their needs only by doubling their intake of food – something which the rate at which they can process food through the rumen may not allow.

Van Horne et al. (1988) found a similar situation when they extended these analyses to include three more species of plants considered to be important browse for black-tailed deer. These deer reach the northern limit for members of this genus in southern Alaska. Here, large decreases in their numbers have been recorded from time to time. These declines have especially been associated with severe winters, and it was thought that the availability of digestible energy during winter was the critical factor responsible for the declines. But, as a result of these studies, Hanley and his associates concluded that the nitrogen economy of these deer is likely to be a more important influence on their abundance.

6.1.5 The Giant Panda in China

The giant panda, *Ailuropoda melanoleuca*, is an interesting, if aberrant, example of how, once a species has been "launched" along a particular evolutionary pathway, it manages to persist by means of whatever adaptations happen to enable it to persist – but not necessarily by the "optimum" solution of human perception. The panda is a member of the order Carnivora and as such possesses a simple gastrointestinal tract that is anatomically similar to that of other carnivores. It has a simple stomach, no caecum and a short undifferentiated colon. Yet it is virtually entirely herbivorous and, what is more, specialized to live on nothing but bamboo.

Feeding trials with a captive male and female panda in the USA (Dierenfeld et al. 1982) showed how, without the benefit of either fore- or hindgut bacterial fermentation, giant pandas are able to achieve an intake of protein from a purely vegetable diet which is adequate for day-to-day maintenance. To do this they have to be highly selective, eating first the leaves and then the central pith of the bamboo. The former contains four times as much crude protein as the rest of the plant and much less fibrous cell wall material; the latter is made up of the vascular tissue which carries concentrated nutrients to the growing leaves. All else is discarded. Pandas have specialized teeth which finely crush and comminute the fibrous food, giving increased access to cell contents. They also eat very quickly and consume large quantities of bamboo (up to 6% of their body weight in dry matter daily), and spend a major part of the day feeding. The food passes very rapidly through their gut (complete passage in a mean of 8 hours). They have been forced to adopt the "creaming-off" strategy found in other herbivores living on a strictly vegetable diet without the benefit of microbial digestion of cell walls (e.g. see Sects. 3.4.1 and 7.1.1).

Yet, from field studies in their native China (Schaller et al. 1985; Reid and Hu 1991), it seems that this exclusive diet of bamboo is one of necessity, not choice – they have access to little else, especially in winter. For most of the year they eat one species, *Bashania fangiana*, but in spring prefer the protein-rich shoots of the lowland *Fargesia robusta*.

Fig. 16. The giant panda has a carnivore's digestive tract, yet eats little other than bamboo. The consequence of becoming painted into this evolutionary corner, in spite of adaptations which enhance its intake and use of nitrogen, is life on a knife edge of inadequacy of protein in its diet. Photo courtesy of Don Reid

They *are* able to eat small quantities of more than 25 other species of plants, including more nutritious berries, nuts, and forbs – but mostly do so only after the massive die-off of bamboo following flowering. In normal times they ignore them, just as senescence-feeding insects ignore nearby and more nutritious new growth (Sect. 3.1). They have become narrowly specialized, both anatomically and behaviourally, to feed on bamboo. More optimal food is no longer available to them.

Much more significantly, however, animal remains – rodent, monkey, and musk deer – have been found in their faeces; they have been seen capturing rats; show "obvious predilection" for meat in traps; and, in captivity, will eagerly eat meat offered to them. It seems, however, that their large size and sluggishness make it unlikely they could often or regularly catch animals. So, while remaining essentially omnivores fully capable of assimilating animal foods, they are confined to being herbivores. And, it would seem, at considerable cost. Their nutritional margin of safety is quite narrow. While a diet of nothing but bamboo will usually get them by on a day-to-day basis, it seems doubtful that it would be adequate for successful reproduction. They cannot accumulate fat reserves on their poor-quality diet, so must depend upon their large size, minimal physical activity, and minimal nutritional input by the females to the production of young, in order to make do with bamboo. It would be interesting to know something of the influence of nitrogen nutrition on the capacity of females to breed, and of the young to survive – especially in the light of their suspected poor pregnancy rates (Reid pers comm. 1992), and low survival of neonates in the wild (Wei et al. 1989).

6.1.6 Domestic Stock

Finally, it is interesting to note that domestication is no guarantee of escape from the rigours imposed by an inadequate environment; and that it is the death of the neonates due to inadequate nutrition of their mothers, or their own starvation, not the depredation of predators, which is the principal cause of mortality in domestic stock. They too need an assured supply of protein in their diet. The ability to detect and select plants and parts of plants which will maximize the intake of nitrogen in the food remains highly developed in domestic animals (Bath et al. 1956; Cable and Shumway 1966). Even patches of grass which have a higher content of nitrogen due to the deposition of urine are detected and preferentially grazed by domestic steers (Jaramillo and Detling 1992). They are responding in just the same way as do wild animals like limpets (Sect. 5.5.2), deer (Sect.6.1.2), hares (Sect. 6.2.5.1), geese (Sect. 7.1.1), and grouse (Sect. 7.1.2) to natural or artificial enrichment of their food plants.

Predatory birds are commonly blamed (and usually severely persecuted) by farmers for killing young lambs. Rowley (1970) investigated this supposed predation by Australian crows (*Corvus* spp.) and Brooker and Ridpath (1980) studied that by the Australian wedgetailed eagle, *Aquila audax*. Both found that neonatal mortality was largely due to starvation of unviable lambs shortly after birth, or other complications related to inadequate maternal nutrition. A similar investigation by Houston (1977) of hooded crows, *Corvus cornix*, attacking hill sheep in Scotland demonstrated the same thing. Most pregnant ewes were undernourished, and nearly all the lambs that died did so in the first week of life and from "uncomplicated starvation". In all three studies predation by the birds was not having any significant effect on the survival of the young sheep.

6.2 Rodents

6.2.1 Squirrels

The Sciuridae (squirrels, chipmunks, ground squirrels, prairie dogs and marmots) are a group of mostly herbivorous rodents, many of which have specialized diets of seeds. Many hoard these seeds as winter food; others hibernate. They are particularly abundant and diverse in North America, where there have been many detailed studies of their ecology.

6.2.1.1 True Squirrels

The squirrels (*Sciurus* and *Tamiasciurus* spp.) eat the seeds of coniferous and hardwood trees. Tree mast is their year-round staple food. As soon as new seeds start to form the squirrels feed on them and continue to do so, storing mature seed in autumn to provide what is almost their only food over the winter months. *Sciurus aberti* is an exception. It feeds on ovulate cones of *Pinus ponderosa* throughout the summer, but in winter it feeds on the inner bark of the pine's terminal twigs. Then it shows the same selective behaviour as some tetraonids (Sects. 7.1.2 and 7.1.3), feeding preferentially on twigs of only a few of the trees available to it.

Seeds represent one of the best sources of concentrated protein that can be obtained from a vegetable diet. But, as with other seed-eaters, specializing on a diet of seeds has not been sufficient to release squirrels from a chronic state of incipient deficiency of protein for successful breeding.

Young squirrels are not weaned until they are fully able to fend for themselves. A large caecal microflora aids the digestion and supplements the protein of their fibrous diet [but limited evidence suggests that sciurids are generally not coprophagic (Kenagy and Hoyt 1980)].

Squirrels feed selectively on the most protein-rich seeds available, and their numbers are highest in habitats where these are most abundant (Smith 1968; Nixon et al. 1975). In these more favoured habitats there are not only more ani-

Fig. 17. The red squirrel overwinters on its staple diet of seeds, but to breed successfully in the spring needs access to flower buds and *ripening* seeds and/or animal protein. Photo courtesy of T. P. Sullivan

mals; territories are smaller, breeding is more successful, mortality is lower, numbers are more stable from year to year, and there are fewer juveniles in the population. Each year juveniles are expelled and move into the less preferred habitats where most of them die before winter is over (Rusch and Reeder 1978; Gorman and Roth 1989).

Each spring and summer when they are breeding, the squirrels' diet of seeds is extensively supplemented by a variety of foods which are high not just in protein, but in more readily assimilable soluble nitrogen. In particular, they seek out new reproductive buds, inflorescences, pollen, and fruit of both conifers and angiosperms. Their two seasons of birth, in spring and in summer, coincide with, respectively, the growth of new buds and flowers and the *ripening* seed crop (Koprowski 1991). They feed on the inner bark of trees, especially once nutrients start to flow to new spring growth, and around sites of fungal cankers where the cambium is thicker (Sullivan and Sullivan 1982b). They also eat large quantities of fungal fruiting bodies, selecting the spore-bearing gill tissues, and cache many in trees where they provide a winter supplement to seeds (Smith 1968). All these foods, not just tree seeds, have been found to be more abundant in their most preferred habitats (Rusch and Reeder 1978). And they miss no chance to eat animal protein, probably far more generally than is realized. Feeding on insects would go largely unnoticed unless consciously looked for. Even then, as with other mammals and birds, readily identifiable remains would soon vanish in the gut. There are a number of records of squirrels eating lepidopterous caterpillars, especially if they are abundant (Jennings and Crawford 1989), and Nixon (1970) observed juvenile gray squirrels, *Sciurus carolinensis*, foraging under bark and eating insects that they found. He said that insects are frequently eaten by gray squirrels, especially in late spring when the young from the first litter are being weaned. At that time he found significantly more insects in the stomachs of juveniles than of adults. It has long been known that squirrels will chew bones, eat aphids, regularly take nestling birds, and eat lizards and carrion as opportunity affords (they are even recorded as eating soil soaked in human urine!) In captivity they will readily eat meat. Significantly, such behaviour has been observed in pregnant and lactating females and juveniles, but not in males (Keith 1965; Smith 1968). Short (1976) showed that such supplementation of a vegetarian diet is more than just a casual matter. When he fed captive Arkansas fox squirrels, *Sciurus niger*, on a diet of nothing but their usual staple of acorns, these were not sufficient for normal maintenance of body nitrogen of adult squirrels. For captive fox squirrels to breed successfully they needed about 4% of their diet as a combination of vertebrate and invertebrate animal matter.

Yet, in nature many breeding females and their pre- and post-weaning young still die each year. Spring-born young generally survive better than the summer-born litters. Breeding is more successful when initiated early in the season (there is a strong correlation between the survival of young and the time of snow-melt).

There is also a strong link for many populations between the density of the population, the success of breeding and the size of the crop of whichever species of seed forms the squirrels' staple diet. In years when a seed crop fails as few as 10% of females may produce a spring litter, and none of them a summer one. Even if there are some young produced in summer, their survival is drastically reduced in years when seed crops are low. Territorial species expand their territories and much previously occupied habitat is abandoned by all species, territorial or not. Juveniles that do survive in remaining occupied habitat are quickly expelled to disperse and mostly die.

The reverse is also true. In response to a good seed crop more females breed, more and larger litters are produced, and many more young survive from both spring and summer litters. These young quickly disperse and reoccupy previously abandoned habitat (Smith 1968; Nixon et al. 1975; Gorman and Roth 1989).

Changes in the number of squirrels in the environment would seem, then, to depend on changes in abundance of their staple tree seeds (Gurnell 1983). This is normally considered to be a matter of how much seed the squirrels have to eat during the winter. Predation (including in many cases heavy pressure of human hunting) is unimportant. Social and territorial behaviours are determining *which* individuals get what food is available in the habitat, not *how many* can live in the habitat (Sects. 11.2.1 and 11.3). Space is not limiting, nor, apparently, are any other resources such as nesting sites.

But is this the whole story? Clearly a minimum supply of food in winter is necessary if animals are to survive until spring, but it may not be sufficient. The success of breeding may well be influenced more by the presence of spring and summer foods, which have been shown to provide the additional protein which pregnant and lactating females and their young must have. And this holds whether there was a shortage or a surplus of winter food. The report by Koprowski (1991) of a season of exceptionally high mortality of fox and gray squirrels in Kansas illustrates this dependence upon spring foods. An unusual period of severe frost in early spring killed most of the buds and developing fruit of their major food plants. This had no measurable effect on the survival of adult squirrels, but severely reduced that of the juveniles. Those young which survived to early summer had lower body weights and much higher mortality than in other years.

There are a number of other relevant facts here. The initiation of breeding is tied to snow-melt and, thus, the beginning of new lush growth. Supplementing food in *summer* dramatically increases breeding and abundance (Sullivan and Sullivan 1982a). Survival through the winter of the young from summer litters is usually much less than that of spring-born young. The observations of Kemp and Keith (1970) are particularly pertinent. They found that numbers in a population of the American red squirrel, *Tamiasciurus hudsonicus*, increased when the crop of spruce cones was large. This, in turn, was influenced by dry summers producing increased seeding, and was reflected in increased survival of young squirrels. Nothing very unusual about all that. Except that the improvement in the reproductive performance of the squirrels was happening in the years *before* heavy cone crops; in fact in the spring of years of *poor* cone crops. Kemp and Keith suggested that this was because in the spring following a winter of poor supplies of seed, hungry squirrels would be "forced" to feed on the newly developing flower buds of the spruce. These buds are nutritionally superior to spruce seed and would therefore stimulate good breeding in that season. I doubt the squirrels would need forcing. There is plenty of evidence that they feed avidly on such good food whenever it is available. The point is, I think, that spruce cones are initiated in the previous season's flowering, so the year before a bumper crop would be a year of great abundance of flower buds and forming fruit, both of which are, indeed, a particularly rich source of soluble nitrogen.

It may be, therefore, that the spring and summer diet of females and their young is more important than their subsequent winter diet. Without an abundance of rich food in spring and summer, fewer young will be produced, and the survivors will enter the winter with fewer reserves to carry them through to the spring – even when there is enough seed to last the winter. Conversely, well fed and strong juveniles may much better survive winter even when stores of food for maintenance are inadequate and they lose condition. For these reasons the emphasis that is commonly given to the influence of winter diets upon the success of many herbivores, not just squirrels, living in harsh winter climates may be misplaced.

6.2.1.2 Chipmunks and Ground Squirrels

Another group of sciurids usually regarded as totally herbivorous are the chipmunks (*Tamias* and *Eutamias* spp.). They, like the squirrels, eat seed mast throughout the year and hoard it underground as their sole winter food. However, they also regularly eat significant amounts of animal food – insects, molluscs, annelids, frogs, mice, birds, and snakes – and in the spring concentrate also on bulbils, flowers, small seeds, and fungi. They feed selectively from among each of the plant, fungal, and insect foods available in their habitat (Forbes 1966; Wrazen and Svendsen 1978). Breeding females and young eat more invertebrates than do males, which eat more plant material, highlighting the separate needs of protein for the former and energy for the latter. The females especially exploited the abundant lepidopteran caterpillars to support the growth of their spring litters. All the young ate more invertebrates than did adult females.

Similar stories can be found for other sciurids. A few appear to be wholly herbivorous once weaned, all feed selectively for readily assimilable nitrogen, and many include animal food in their diet. For example, Karasov (1982) showed for the ground squirrel, *Ammospermophilus leucurus*, that if they consumed only leaves they needed to eat tissues which contained at least 13 to 16 mg nitrogen per gram dry mass to maintain a positive nitrogen balance. Selective feeding helps them to achieve this, but they eat small quantities of arthropods and vertebrate flesh throughout the year, ensuring that they remain in nitrogen balance even when plant tissues fall below the critical minimum level. Beldings ground squirrel, *Spermophilus beldingi*, in "cafeteria" feeding trials, selected species and parts of plants (mostly flowers and then fresh leaves) for their high protein content, but there was no significant difference in the content of energy of those plants they selected and those they rejected (Eshelman and Jenkins 1989). The pocket gopher, *Thomomys monticola*, in similar trials selected high-protein plant food, low in fibre (Jenkins and Bollinger 1989). Also, it would seem that ground squirrels in general may quite regularly eat animal food. Hansen and Ueckert (1970) recorded arthropods and flesh in the diet of *Spermophilus richardsonii* and commonly saw them eating road-killed squirrels, while Boonstra et al. (1990) have discovered that the arctic ground squirrel, *Spermophilus parryii*, is a regular predator of the collared lemming, *Dicrostonyx kilangmiutak*, in Canada's Northwest Territories.

Most sciurids feed quite extensively on the fruiting bodies of fungi, usually late in the summer when they are abundant. One such species which does this is the golden-mantled ground squirrel, *Spermophilus saturatus*. These animals live in ponderosa pine-Douglas fir forests in Washington State and spend more than 6 months of each year hibernating. In their short above-ground season they must produce and wean a litter in time for the juveniles to gain enough weight to survive their first winter in hibernation. Most of their food is green herbaceous plants which begin to grow as the snow melts in the spring. This flush tissue remains the major food for most of the time that the females are lactating; the animals track later-growing plants in the shade as those growing in the open mature and die-off. But by late summer more than 50% of their diet is fungi, mostly one species of truffle which they dig up (Kenagy et al. 1989). For these, and other sciurids which eat a lot of fungi, it has been assumed that the fungi were a source of high-protein food, but recent work by Cork and Kenagy (1989) suggests that this might not be especially true. They found that when they fed individual *S. saturatus* on a diet of nothing but fungi the animals were unable to maintain body weight. Mainly, it seems, this is because the spores, which contain most of the protein, are totally indigestible. Rather, the authors suggest, the fungi are taken because they are abundant, easy to find, and they provide just sufficient nitrogen as a base which can be "topped up" by small amounts of harder-to-find foods which are higher in protein. The squirrels are making the best of a situation where an adequate supply of protein is rapidly declining as the season progresses.

There are many small mammals like these ground squirrels which, rather than store food to tide them over winter, hibernate. It is generally accepted that this is an adaptation to conserve energy when food is short and is not simply a response to cold temperatures. Torpor can be induced experimentally at any time without reducing the ambient temperature by depriving an animal of food. It would seem, however, that once again energy is not the critical factor. Montoya et al. (1979) experimentally induced torpor in the European dormouse, *Eliomys quercinus*, by feeding captive animals a diet deficient in protein but which ensured that the energy requirements of the animals are amply satisfied. This, combined with the fact that hibernating rodents, like others given a low-protein diet (Kennedy and Hume 1978; Rubsamen et al. 1982), recycle a much greater proportion of their endogenous urea to their caecal bacteria, suggests that shortage of protein drives this adaptation. Recycling of urea by gut microbes is common in mammals and birds. The small rodents are no exception (Steffen et al. 1980). The increase in this recycling in response to a shortage of protein in the diet is not confined to hibernating animals (Bintz and Torgerson 1981).

The importance of adequate protein nutrition for the survival of these animals which *do not* feed in winter also highlights the point I have been stressing about those animals which *do* feed in winter. It is the spring and summer diet of the females and their young which is vital to their surviving the winter, not their subsequent winter diet.

6.2.2 Rats and Mice

The same theme can be followed in studies of species which belong to other families of small rodents. Few are totally herbivorous, most incorporate more or less animal food (usually insects) in their diet, especially during the breeding season. All feed selectively for plant tissues with high concentrations of digestible nitrogen, and live in places where such foods are most abundant. Caecal fermentation with associated coprophagy and microbial recycling of metabolic urea aid in conserving scarce nitrogen. In all, the impact of changes in available nitrogen on year-to-year changes in their abundance is through its effect on the success of breeding females and survival of their young.

At one extreme are such ubiquitous, aggressive, and adaptable animals as the black or ships rat, *Rattus rattus*, and the house mouse, *Mus musculus*. In and around human habitation these two species probably rarely experience any shortage of good food, but in the wild it is a different story. Gaining enough protein becomes paramount. Both are opportunistic, yet highly selective, omnivores choosing to eat only some species of plants, selecting new growth, flowers and seeds when available, and consuming a wide range of animal food. Thus, Clark (1980) on the Galapagos Islands, and Gales (1982) in a New Zealand forest, found males and females, and young and old *Rattus* choosing different foods. Females, and particularly actively breeding ones, eat mostly animal food while males eat more fruit and seeds. Growing youngsters of both sexes eat far more animal food than do adult females. In the case of the Galapagos population, young rats additionally ate only the seeds of a particular fruit while the adults ate only the flesh.

There are accounts scattered through the literature of the numbers of these and other small rodents rapidly increasing in response to a sudden abundance of high quality food in their environment. As in the case of the feral rabbits in Australia (Sects. 6.2.5.2 and 16.2), these responses are more dramatic in harsh environments. One moment the animals are rare and almost impossible to find, the next they are literally everywhere. For example, Pefaur et al. (1979) reported the irruption of populations of *M. musculus* and *R. rattus* plus four species of native rodents over wide areas of the semi-arid region of Chile. This outbreak followed the sudden mass germination of grasses in the region which was, in turn, generated by the exceptional El Niño rains of 1971–1972.

Newsome and Corbett's (1975) analysis of plagues of *M. musculus* and three species of native rodents in the arid and semi-arid inland of Australia shows the same processes operating. The plagues of all these animals are responses to flushes of vegetation over large areas of usually barren land after unpredictable heavy rains, or floodwaters from distant heavy rains.

In all such cases animals live for most of the time in low numbers, surviving where chance conditions provide scattered, often ephemeral, and shifting refuges throughout wide areas of otherwise totally inhospitable terrain. In these refuges there is enough food for a few individuals to breed successfully. With the sudden advent of widespread abundance of new growth these few individuals have the reproductive capacity to quickly "explode" out from these refuges, tracking the increasing supply of food. The abundance of soluble nitrogen in the growing plants and their developing seeds enables females to breed repeatedly and successfully and ensures that their young nearly all survive, grow quickly, and in turn reproduce successfully. But such profusion is short-lived in these sorts of environments. As the vegetation dies off, so too do most of the rodents. Only those lucky enough to find themselves in refuges survive to provide the nuclei for the next irruption (see also Chaps. 16 and 18).

6.2.2.1 The House Mouse

Pennycuik and Reisner (1989) kept populations of house mice, *Mus musculus*, in pens with abundant space and shelter but limited food. Once numbers exceeded the food supply they stopped increasing because there were fewer pregnant females, most nestlings starved to death in the first 2 weeks from an inadequate supply of milk from their mothers, and at weaning older mice pre-empted all the food. Their experiments demonstrated that "when food is in short supply the major mechanism limiting growth in populations is high mortality among pups".

An especially revealing study of the importance of the form and availabilty of nitrogen in the food is that by Bomford (1987a, b) of populations of the house mouse living in irrigated cereal fields of the Murrumbidgee Irrigation Region at Griffith in New South Wales, Australia. She found that in this habitat the mice were essentially granivorous, although approximately 9% of their diet was invertebrates (mostly lepidopterous larvae and earthworms). Grain spilt during harvest (up to 1 t per hectare!) provided their staple food during the summer and winter non-breeding period, although by the end of winter this food was usually becoming scarce.

Irrigation began in the late winter and early spring in the wheat and barley fields, producing good growth of weed grasses. Early in September these grasses started to produce milk-ripe seeds (i.e. full-sized seeds with starch granules in the endosperm, but still green, soft and exuding a "milky" juice if squeezed). The mice immediately switched to eating these seeds and started breeding 1 month later. When the rice and sorghum fields were irrigated about 3 months later another lot of grass weeds grew and set seeds. Again, the mice fed on these and bred 1 month later. But this was up to 5 months after the mice in

Fig. 18. The breeding of house mice which live in cereal-growing areas in Australia is triggered by a diet of *ripening* seeds into which concentrated soluble nutrients are flowing. Photo courtesy of Ederic Slater

the wheat and barley crops had started to breed, and long after they had stopped breeding. In all fields the picture was the same. Mice started breeding only after they had access to milk-ripe seeds of the grass weeds, peaked as fresh milk-ripe cereal grain became available, declined as the crop ripened, and ceased after harvest. Spilt grain was abundant, as was ripe seed of the weeds on embankments around the irrigated fields, so the mice were not short of food. But they stopped breeding.

When Bomford fed mice in the laboratory on nothing but rice grain they bred poorly or not at all. However, if she supplemented the rice with casein or with sprouted rice grain, or fed the mice on high-protein pellets, mealworms, or sunflower seeds and oats, many more females bred. The response to these different high-protein diets indicated that lack of protein was inhibiting breeding, and supplementing the rice grain with casein indicated that it was lack of protein which was preventing the mice from breeding on a diet of ripe rice grain alone. She tested this latter conclusion in the field (Bomford and Redhead 1987). In the autumn, when there was still an abundant supply of rice grain in the field (and thus no shortage of calories), she put out ad lib. supplements of oats and sunflower seeds. These she had found in the laboratory to be an acceptable diet which supported good breeding, having a protein content of 11.6 and 22% respectively, compared with the 7 to 9% of rice grain. The provision of this high quality food reversed the decline of breeding which usually occurs in rice fields in autumn.

However, Bomford was unclear as to why the sprouted grain should have stimulated breeding. She observed that the mice ate all the endosperm left in the grain and the cotyledons, but hardly touched the shoot or root tissue. There was no evidence of a chemical stimulant like 6-methoxybenzoxazolinone (6-MBOA; which had already been shown not to have any effect on the breeding of house mice). She concluded that during germination either a nutrient that was limiting in dormant grain became more available, or the concentration of proteinase inhibitors decreased allowing the mice to absorb more of the protein in the seed than they could from dormant grain. Also, her feeding trials in the laboratory did not provide any evidence to explain why milk-ripe seeds should stimulate breeding in the field.

I think the answers to the response to both milk-ripe seeds and germinating seeds are one and the same. In the field the mice bred only when milk-ripe seed, weed or crop, was available. As these seeds ripened, breeding declined. Once they were fully ripe (but still abundant) the mice stopped breeding. In the laboratory they bred on very high protein diets, or when germinating seeds were added to dormant seeds. However, in Bomford's field trials the mice in the control populations continued to breed at a much higher level than is usual in autumn (but still at a level significantly below that of the supplemented populations). It happened to be a wet autumn, and better breeding had been previously reported for wet autumns. One suggested possibility for this unexpected continuation of breeding was that sufficient spilt grain had sprouted in the wet conditions to provide the nutritional boost necessary to keep some of the females breeding. As Bomfield noted, there was no shortage of energy at the time the mice stopped breeding. Nor would there have been a shortage in mice fed on the ad lib. diets in the laboratory. And there was no evidence of a lack of a necessary chemical stimulant, nor of the presence of any inhibitor. The seeds of the weed grasses had a much higher content of protein than did the cereal grains (14 to 21% compared with 7 to 13% in Bomford's study). So they might have provided the higher level of nitrogen necessary for breeding. But the peak of breeding was on the flush of milk-ripe seed of the crop plants.

The key is in the active import or export of nitrogen to and from the seed; when it is being formed and when it is germinating. The formation and filling of seed is a time during which a complex process of import and synthesis of amino acids is proceeding apace prior to and as part of the deposition of protein in the grain (Ugalde and Jenner 1990). The process is reversed when the stored protein is being mobilized to fuel the growth of the new plant; the endosperm and the cotyledons are sites of concentration of the soluble amino acids being released. Only at these times is the limited amount of nitrogen in rice grain converted to forms which are readily enough assimilated to provide a diet adequate for breeding and lactating females.

A more recent study of *Mus* in wheat fields in Victoria has lent strong support to the evidence from Bomford's studies. Tann et al. (1991) monitored the breeding of mice in relation to their food by analysing the stomach contents and breeding condition of animals trapped every 6 weeks over 2 years. Again, seeds, mostly from the wheat crop, were the staple food of the mice, but the onset of breeding each year followed an increase of insects and *green* grass seeds in their diet. The peak of breeding coincided with the availability of *fresh* cereal grain. Breeding ceased when there was still a great abundance of seed, but it was all mature cereal grain. Also, breeding females ate far more insects and green plant tissue than did non-breeding ones.

This story is strikingly similar to that of the galah in Western Australia (Sect. 7.3.4), and essentially the same as for other seed-eaters like the goldfinch (Sect. 7.3.2), the squirrels (Sect. 6.2.1.1), some mirids (Sect. 3.5.4) and tephritids (Sect. 1.4.2), and for folivorous voles, like *M. montanus* (sect. 6.2.3), the rabbit on flush grass (Sect. 6.2.5.2), and the red grouse on growing shoots of heather (Sect. 7.1.2). All are seeking access to *nitrogen in its most available form*. And where the level of nitrogen in their food is normally below a level which will support breeding, such access will be sufficient to tip the scales and trigger widespread and successful breeding so long as, but only so long as, the supply lasts.

6.2.2.2 The Australian Smokey Mouse

In contrast to those species with wildly fluctuating numbers, there are those which live cryptic lives and change little in average abundance from year to year. Yet the struggle to survive – to continue to gain enough protein from the environment to be able to persist in it – is just as harsh.

Cockburn (1981a, b) was able to demonstrate this for one such animal, the Australian smokey mouse, *Pseudomys fumeus*, living in the subalpine climate of western Victoria. The strong preference these mice have for only some parts of the habitat, where they spend most of their lives, has nothing to do with competitive exclusion by other species of small mammals that live in the same locality (Sect. 1.4.2). Rather, they are tracking food which will maximize their intake of protein. This is especially true of the females during the breeding season. They seek out localities where fungal fruiting bodies, moss capsules, and insects are most abundant. Their staple diet is seeds but, again, they have a strong preference for leguminous ones, high in nitrogen.

Pseudomys, like many species of small rodents, experiences a severe decline in numbers each spring just before breeding starts. At this time a large proportion of the males disappears each year, but so do many females. In the population which Cockburn studied, living in a preferred habitat was essential if a mouse was to survive this crash and successfully breed. Those animals occupying preferred habitat did so with little sign of aggressive behaviour. But those living in non-preferred sites suffered abrupt mortality accompanied by a great deal of fighting.

It seems the trigger for this was a "nutritional crisis". On all but the favoured sites supplies of fungi run out, and/or supplies of seeds and insects do not become available early enough, to provide animals with sufficient protein for them to survive the stresses of the breeding season, let alone produce viable young. Those animals relegated to the less favourable sites are analogous to other examples of "doomed surpluses" (Sect. 11.3). They provide for the population as a whole seeking out and using all possible parts of the habitat which contains enough adequate food, and give it the capacity each year to absorb any increase in the amount of food in the habitat.

The population is effectively pressing continually against the limits set by the supply of protein in the environment that is attainable by individual mice; and this is at the expense of all but those favoured few that find and occupy those parts of the environment where there is an adequate supply. If that supply changes but little from year to year the numbers in the population will not vary very much. But if, as happens in the desert, the supply changes suddenly, then the numbers in the population will do the same, and in proportion to the change in the supply of food (see also Chaps. 11 and 16).

6.2.2.3 The American White-Footed Mouse

Just how critical an adequate supply of protein is for successful breeding of small rodents was demonstrated by Merson and Kirkpatrick (1981) for the North American white footed mouse, *Peromyscus leucopus*. They fed captive and individually segregated female mice on a diet of 100, 90, 80 and 70% of their mean daily consumption of an ad lib. supply of high-protein laboratory chow. After only 3 weeks on these diets the weight of the ovaries of all mice were the same, but those females on 80 or 70% of the ad lib. diet had significantly lighter uteri. All the mice on the 100% diet exhibited oestrus during the last week of the trial but only 10% of those on restricted diets did so. And this occurred whether they had suffered as little as 10% or as much as 30% restriction of their intake of chow. However, none of the mice had lost weight nor was the level of their carcass fat reduced.

When similarly food-restricted females were caged with males, their breeding success was much lowered. After 30 days half of the females on the 100% diet were either pregnant or had already had a litter, in contrast to 29, 13 and 0% of those mice on, respectively, 90, 80, and 70% of the ad lib. diet. Again, there was no difference in the weights or amount of body fat of females on the different diets.

So an individual female mouse's capacity to breed is immediately and drastically reduced when her supply of food is reduced by so small an amount, or for so short a time, that it has no detectable influence on her physical condition.

In nature there is no steady supply of a carefully balanced diet of laboratory chow, but only unpredictably scattered items of food with variable, and usually low, nitrogen content. How, then, is she likely to fare there? Little better than her Australian cousin *Pseudomys*, judging from Krohne's (1989) study of a population of *Peromyscus leucopus* in Indiana. There, mice living in restricted parts of the habitat are able to breed each year, but their young must disperse to what are clearly suboptimal areas able to support but a few animals. In these "dispersal sinks" practically none of the young *Peromyscus* survive (see Chap. 11).

6.2.2.4 American Woodrats

A group of rodents usually considered to be strict herbivores, and extremely narrow specialists to boot, are several species of American woodrats of the genus *Neotoma* (Atsatt and Ingram 1983; Vaughan and Czaplewski 1985). The staple diet of many of them consists of the leaves of but one species of tree. What is more, they seem to be rare examples of senecsence-feeding mammals (see Sect. 3.1). They largely reject young leaves and feed instead on fibrous, yellowing, senescing leaves low in nitrogen and loaded with tannins and other polyphenolics.

Woodrats have developed a number of specialized (yet familiar) strategies which enable them to maintain an adequate nitrogen balance on such a daunting diet. First, they are not such strict herbivores. Like so many others they eat insects, and they eat other high protein foods such as flowers, pollen, fruit, seeds, fungi, and green leaves of shrubs, especially in the spring. Their staple diet, however, is limited to the leaves of one or a few species, the phenolics of which they can degrade or avoid.

They very selectively feed on the foliage of only some individual trees. This is thought to be the avoidance of high levels of phenolics, but may well be a response to the level of protein in the leaves, as seems to be the case with some other folivorous specialists like the capercaillie (see Sect. 7.1.2).

They process (with the aid of caecal bacteria) very large volumes of low grade foliage. They reingest the caecal protein via coprophagy (including the ingestion of adults' pellets by the young).

But they still pay a heavy price for having gone down this evolutionary bottleneck. Females are slow to reach reproductive maturity. They lose a great deal of weight while lactating, in spite of greatly increasing their intake of food, and they are slow to regain sufficient weight to be able to produce their next litter. Many die at this time. Their litters are small (generally only one pup), the young grow slowly and are not weaned until well after they have started eating solid food. Often they starve to death, especially if there are more than one in a litter; and those in second and third litters grow more slowly and more often die of starvation than those in first litters.

6.2.3 Voles

The microtines, the voles and lemmings, are probably the most famed among all the small rodents because of the spectacular and regular rises and falls in abundance of many of them, and of the mythology about Norwegian lemmings marching to mass suicide in the sea. In the northern latitudes of the Northern Hemisphere the numbers of many different species peak every 3 or 4 years, often synchronously, over wide areas of land. A great many studies of microtines have been, and are being, conducted, the thrust of most of them is aimed at explaining the enigma of these periodic eruptions. These studies have generated a huge bank of published literature and many hypotheses to explain the cycles of abundance. Here I can but touch on this literature, and do so only so far as it is necessary to show that the common thread of dependence on an inadequate supply of nitrogen continues through this group, and that we find the same strategies being adopted to attempt to overcome this inadequacy (see also Chap. 17 for further discussion of their cyclic eruptions).

Cockburn (1988), in reviewing the diet of voles and lemmings, points out that although all 110 described species of these rodents are herbivores, they range from extreme specialists, feeding on only one species of plant, to omnivorous generalists. He and dozens of other authors dealing with the feeding and nutrition of many different species of vole, make no mention of their eating anything other than plant food. If this were so, microtines would be the only small rodents, other than the lagomorphs, which are complete vegetarians once weaned. But they are not – or at least the two common European species, *Microtus agrestis* and *Clethrionomys glareolus*, are not. Hansson (1971) made an intensive study of the diet of these two voles and of that of the woodmouse, *Apodemus sylvaticus*. The mouse is an omnivore, its diet throughout the year comprising mostly insects with seeds and herbaceous foliage being eaten as available. *M. agrestis*, however, is considered an "extreme herbivore", feeding almost exclusively on grasses, and *C. glareolus* is a seed-eater. Yet Hansson found that in the spring and early summer approximately 10% of the diet of *M. agrestis* and 30% of the diet of *C. glareolus* was insects, mostly soft-bodied larvae. And while the staple for *M. agrestis* was grasses and for *C. glareolus* was seeds, in spring and early summer herbs equalled or exceeded grasses or seeds, respectively, in their diets. Clearly, both voles depend on flush dicotyledonous tissues and animal protein in the breeding season.

It may well be – I expect it is so – that many, perhaps most, microtines eat insects in the spring plus a good deal of flush herbaceous growth. If so, workers concentrating on the obvious staple foods and on feeding in winter, when so many animals disappear, have largely ignored or overlooked the importance of these components of the diet in ensuring successful breeding and survival through the subsequent winter. Hahus and Smith (1990) have shown that perhaps careful searching for such feeding by microtines, especially in the breeding season, may prove rewarding. They discovered that prairie voles, *Microtus ochrogaster*, ate periodical cicadas (*Magicicada* spp.) during a mass emergence of these insects. These apparently strictly herbivorous microtines obviously do take animal food when the opportunity offers. Why not others?

Cockburn discusses the problem faced by these small mammals in coping with their fibrous diet, consisting largely of indigestible carbohydrates, so as to extract sufficient nutrients and energy from it. What he does not say is that the problem is largely one of extracting sufficient *protein* from that diet, not energy. The adaptations which he lists as having been evolved by microtines to overcome their dietary difficulties are the same as those found in other herbivores, and as in those other herbivores, they operate primarily to increase the intake and efficient use of the limited amount of nitrogen available in their environment.

Microtines spend most of their time feeding, yet all are strongly selective of what they will eat. They seek out those species, individual plants and specific tissues which contain the greatest concentration of available nitrogen (e.g. leguminous plants, new growth generally, flower primordia, seeds, and fungal fruiting bodies) and they live in those parts of their habitat where these sorts of food are most abundant and readily available (Batzli and Lesieutre 1991). They exhibit a strong change in their diet during the breeding season, showing first preference for flush and reproductive tissues. This switch is particularly pronounced in the more specialist feeders. Cockburn and Lidicker's (1983) study of *Microtus californicus* emphasized the importance of the variability of the habitat as an influence on the dynamics of a population of voles. Nearly all successful breeding and survival in the population they studied was confined to quite specific patches of habitat. The vegetation of these patches was quite distinctive, and provided green vegetation further into the summer than did other areas. The voles selectively fed on high-protein legumes and broad-leaved forbs, and increases in the number of resident females on these preferred habitats occurred at the time of spring flush of the vegetation in them.

Like all rodents microtines are hindgut fermenters with a highly modified proximal colon and a specialized caecum housing a bacterial microflora. Linked with this microbial fermentation is a specialized dentition, enabling very fine comminution of their food, and a longer gut to allow for more efficient absorption of nutrients. They also have a remarkable mechanism for separating most of the bacteria in the colon for recycling through the caecum. Unlike the lagomorphs and the caecal-fermenting birds (Sects. 6.2.5 and 7.1.5) this mechanism separates just the bacteria from the particulate food. The proximal colon is greatly modified to form a double spiral within which bacteria, separated in the distal end of the colon, are returned – against the normal peristaltic flow of the outer channel (Sperber et al. 1983). Also, they are coprophagic, selectively reingesting soft pellets of their own faeces which have originated in the caecum. These soft pellets have the same energy content as the hard pellets which are not reingested, but contain significantly more nitrogen. While this adaptation is common and widespread in small rodents, it is interesting that apparently only in the microtines is it as well developed as in the lagomorphs (Kenagy and Hoyt 1980). For example, *M. californicus* reingests 25% of its faeces as caecal pellets, while many omnivorous and seed-eating species from other families take in much smaller amounts. The sciurids may not be coprophagous at all. Apparently the less animal food there is in a small rodent's diet, the more dependent it is on microbial protein produced in its gut from the breakdown of cell wall material in its food.

Just how significant coprophagy is to the nutrition of voles was illustrated by Cranford and Johnson (1989). They maintained two species of voles,

Microtus pennsylvanicus and *M. pinetorum*, in the laboratory on a diet of either high protein (24% crude protein) laboratory rodent chow or low protein (10% crude protein) horse chow. Half the animals on each diet were fitted with collars which prevented them from eating their own faeces. All the expected things happened. On the high protein diet all animals gained weight, but those allowed to be coprophagic gained the most. On the low-protein diet all animals lost weight, but the non-coprophagic ones lost the most. The coprophagic voles on the low diet ate more of their faeces than did those on the high diet. The non-coprophagic ones ate more food than the coprophagic ones, thus maintaining their intake of energy. The authors therefore concluded that coprophagy was not significant to the voles' energy budget (anyhow, the faeces contain less digestible energy than the food) but that the caecal faeces supply some essential "specific dietary nutrients". I would suggest that digestible nitrogen is likely to be the nutrient in question.

The literature is replete with studies which collectively reveal that, in spite of all these adaptations, the abundance of microtines is tightly and precariously linked to their food supply; not just its quantity, but its quality. These studies also reveal that this quality of diet influences changes in their abundance via the effect it has on how successfully the females breed, and how well their offspring survive and grow after they are weaned. Not so clearly revealed, but beginning to emerge in some more recent studies, is the fact that the essential quality of the food is the usable nitrogen it contains, not its energy value.

For example, Myllymaki (1977) showed that nutritional factors play a major role in the decline of populations of the European field vole, *Microtus agrestis*. The decline usually starts gradually, often right at the end of the breeding season in the peak year, when few of the overwintering cohort and its offspring survive. But the size of litters and prenatal mortality do not change throughout a cycle. Growth and sexual maturity of the females coincides closely with the start of growth of their food plants. Even in winter a mild spell of weather may induce breeding activity. Such early attempts to reproduce were always elicited and maintained by newly grown grass. In fact, maturation of overwintering females seemed possible whenever such high-quality food became abundant.

More recently, Boonstra and Boag (1987) conclusively demonstrated for a population of *M. pennsylvanicus* that maternal and environmental effects, not genetic differences, were of overriding importance in determining the weight, rate of growth, speed of attaining sexual maturity, and survival of young voles. This was most likely a function of the quality of food available to the breeding females and its effect on the quantity and quality of their milk.

Also working with populations of *M. pennsylvanicus*, Bergeron and Jodoin (1987, 1989) and Belanger and Bergeron (1987) have demonstrated that these voles choose to eat only a few of the many species of plants in their habitat, and that these are the ones with high protein but low levels of fibre and digestion inhibitors, particularly total phenolics. Neither the availability of these plants, their energy content, nor their content of allelochemicals influenced selection; and there appeared to be a far greater biomass of these preferred species in the habitat than the voles could use. But the protein content of these

Fig. 19. The success of meadow voles fluctuates with availability of nitrogenous food, in spite of selective feeding on high-protein plants, hindgut fermentation, and eating their own caecal faeces. Photo courtesy of J.M. Bergeron

preferred plants decreased markedly during the summer. In the summer following a peak in the number of voles the proportion of the total biomass of plants made up of the preferred species declined steeply, as did the quality of the plants. This decline in the availability of high-protein food was reflected not just in a considerable drop in numbers of voles but in widespread physiological symptoms of malnutrition in the surviving animals. These changes had a greater impact on the females than the males.

Bergeron and Jodoin say that there is still no clear understanding of what constitutes high-quality food for the voles, nor whether all voles in a population have equal access to it. However, their studies, and the others referred to here, would seem to point clearly to the importance of the level of protein being of paramount importance, and that it is access to this food by breeding females and their young which is the key to the number of voles that the habitat can support.

Their work also shows how the amount of high-protein plant food which is available in the habitat may change dramatically without there being any obvious change in the *quantity* of vegetation in the habitat. They suggest that this may be the reason why many ecologists working in the field with voles have failed to see or accept how these animals could be affected by their food supply, even when their numbers were high. I would agree, but add that this misconception is widespread among ecologists studying all kinds of herbivores.

These authors also draw attention to the fact that it is the females which are most affected by changes in the availability of high-quality food, and as nobody denies the importance of maternal influences on the viability of their offspring, perhaps this is "...the link that ecologists look for to explain differential survival rates among fluctuating populations of voles". I couldn't agree more! That is my message for those who study *all* herbivores.

It is interesting that these relationships of voles (and of most other small rodents) to their food have been elicited by the use of what are really very blunt instruments: grid trapping, chemical analyses of oven-dried plants, and counts of remnants of plant tissues in small samples of faeces or stomach contents. The very young, the survival of which is so crucial in deciding abundance from generation to generation, are not caught in traps. The chemistry of a plant's tissues, once detached, change very rapidly. In particular, the nitrogen component degenerates, masking its likely nutritional value prior to detachment. Also, it is well established that many items of food are very soon undetectable in samples from the alimentary canal of birds (Sect. 7.1.1 and 7.3.4). Frase and Armitage's (1989) study of the feeding of the yellow-bellied marmot, *Marmota flaviventris*, in Colorado has demonstrated that this is equally true for faecal samples of small mammals. They took the trouble to feed known species of plants to their animals over several days in the laboratory, and then search their faeces for remnants of these plants. Not surprisingly, they could find no trace of many of them. This was especially so for several forbs. The authors knew from field observations that these forbs were highly prized by the marmots, and that their high content of nitrogen makes them critical to the maintenance of an adequate level of protein in the marmots' diet.

How much better a picture is likely to emerge, then, if studies were directed towards observing what parts of which plants females and their young actually eat – and how much and when they eat it – and analysing not just gross protein content of the plants, but the concentration and the form of the nitrogen in the tissues they eat at the time when they eat them. Beyond this, however, what part is played by insects in their diet? How much less likely is it that a soft-bodied caterpillar, rather than a piece of a plant's cell wall, might survive the digestive process? I have drawn attention to the possibility that this aspect of the nutrition of small rodents, other mammals, and invertebrates may often have been overlooked, and will do so again for birds, reptiles, and fish. Might not careful investigation reveal that access to animal food in spring and summer could make the difference between indifferent or outstanding success of breeding, and of the viability of juveniles entering their first winter? And might the oft-recorded increase in abundance of a population correlated with a bumper seed crop owe its success as much, or more, to access to the fruiting buds and developing seeds which precede the mast crop or to concomitant changes, such as increases in herbaceous plants and insects in the spring, as to extra staple in the winter? Others (e.g. Kemp and Keith 1970; King 1983; Hansson 1987) have questioned whether the obvious correlation is necessarily always cause and effect.

6.2.4 Supplemental Feeding of Small Rodents

Dobson and Kjelgaard (1985) were able to provide a convincing overview and explanation of the way in which supplying supplemental food to small herbivorous mammals reveals the ubiquitous shortage of protein food for their breeding females and young. These authors did so using techniques which avoided the shortcomings just discussed. They worked with populations of the Columbian ground squirrel, *Spermophilus columbianus*, in the Rocky Mountains of southeast Alberta. They captured and marked virtually all the animals in these populations; the adults as they emerged from hibernation and the young from their natal burrows. Over 3 years they were thus able to conduct intensive direct observations and counts of identified individuals. Being able to do this meant they did not have to rely on estimates of numbers based on live trapping; they could mark and identify complete litters as they were produced, and they could observe directly what the animals ate.

Just after the first young of the year emerged from the breeding burrows at the beginning of their study, Dobson and Kjelgaard started supplemental feeding. They had earlier, and on a separate population, found that a high-protein diet of mixed grains was readily eaten by the squirrels. They provided this diet continuously and in excess to their experimental populations throughout two subsequent spring emergences of young.

The results were dramatic. Numbers in the supplemented populations increased by 85% annually and the biomass by 94% annually. There was an immediate immigration of all age classes into the supplemented sites. This immigration continued but decreased in relative importance in the two subsequent years. Many more females reproduced and they produced larger litters (ground squirrels produce but one litter a year). Some females bred as yearlings instead of as 2-year-olds. And many more young and yearlings survived. Other factors which might have contributed to the limitation of the size of these populations were clearly of only secondary, if any, consequence. Such a dramatic and immediate response to the added food could not have happened if space, burrows, hibernacula, or weather were in any way directly limiting the numbers that could live in the habitat. Even an increase of predation by badgers in one of the supplemented populations did no more than dampen the increase that was underway. The evidence was unambiguous. An increase in protein food *at the appropriate time* in their life cycle greatly increased the number of young the females could produce, and the number of those offspring which subsequently survived to breed.

From this study, and from data from 20 previous studies where supplementary food had been supplied to populations of *Mus, Peromyscus, Clethrionomys, Apodemus, Microtus, Tamias, Eutamias, Tamiasciurus* and *Lepus*, they concluded that "increased survival of resident young appears to be the only general response of a population. . .to supplemental food", and that food is a "critical resource" that females needed for reproduction. Subsequent studies (e.g. Gregory et al. 1988) have supported this interpretation and at the same time emphasized the different responses of males and females to increased food in the environment. The females' breeding success and their density is set by the level and timing of the availability of appropriate food, and the males' by the availability of females (Galloway and Boonstra 1989; see also Sect. 11.2.1).

Other studies (e.g. Young and Stout 1986) have served to emphasize how the response of a population to the provision of additional food will vary according to the sort of food supplied, and the relationship of that food to the natural diet and feeding behaviour of the species being studied. The animals' responses will also vary, of course, depending upon when their diet is supplemented, and for how long.

In this context one of the most significant experiments with supplemental feeding of small rodents was that of Negus and Berger (1977). They simply placed freshly sprouted wheat shoots alongside the runways of the grass-feeding *Microtus montanus* in the depth of winter. In as little as 2 weeks all the females in the population were pregnant, at a time when temperatures and food supply were both quite unfavourable for reproduction and survival of young voles. A specific chemical in flushing grasses, 6-methoxybenzoxazolinone (6-MBOA), has since been shown to act, in the absence of green food, as the stimulus to reproduction by species of *Microtus* (Berger et al. 1981; Sanders et al. 1981; Korn and Taitt 1987). The importance of this result, however, is that it demonstrates that these animals have evolved to start breeding as soon as, but only when, new flush tissue high in soluble nitrogen becomes available. The cue and the resource are inextricably linked.

Batzli (1986) has pointed out that while there is considerable evidence that green forage stimulates reproduction by small mammalian herbivores, there are species of voles which breed in winter when there is no new green growth. However, such winter breeding is recognized as being exceptional, found only in populations that are increasing in abundance, and confined to the exceptionally heavy individuals in such populations (Keller and Krebs 1970). Winter breeding seems to be a "carry-over" effect. In particularly good spring-summer seasons many more better-fed, and thus bigger and heavier, animals would be sufficiently well nourished and strong to not just survive the season, but continue breeding longer than would normally be the case.

I would expect this type of evolved response to flush new growth to be general among those small mammals which are essentially dependent on a folivorous diet, but it may not be so pronounced and widespread, nor obligatory, among the more omnivorous and carnivorous species. Korn (1989) found that 6-MBOA did not have any triggering effect on the breeding of *Peromyscus maniculatus*. Several other similar studies which he reviewed in that paper showed that while species feeding on grasses responded positively, there had been a mixed response from other species tested.

Yet, Reichman and van de Graaff (1975) showed that for species of *Dipodomys* and *Perognathus* reproducing animals, and more especially the females, ate far more green vegetation than at other times. These are all primarily sead-eating heteromyids with insects comprising about 10% of their diets. Green flush of any species of plant is always a rich source of highly digestible nitrogen, and seems to be sought after like this in the breeding season by all but obligate carnivores. So it is likely that there will be various specific chemical cues to which different herbivores will respond.

All the foregoing discussions about small herbivorous rodents would seem to suggest, fairly conclusively, that changes from one year to the next in the abundance of these animals are determined by the number of young that survive each year to breed in the following spring. The number and viability of the young would seem to be largely determined by the quality of their and their mothers' protein nutrition, which, in turn, is dependent upon the availability of high-protein food in spring and early summer, rather than upon the amount of staple in winter. If this is so then a logical following assumption would be that the reason why so many of their populations fluctuate so violently in abundance must have to do with equally violent changes in the availability of that protein food, and vice versa where numbers remain relatively stable. If that can be demonstrated, the next logical step is to look for what it is that changes the supply of that good food.

The examples of the rabbits and rats and mice living in arid places would seem to be straightforward and obvious enough. The physical *quantity* of their food can be *seen* to change violently as the plants bloom after rain and then wither as the earth dries out. But what is not immediately obvious is that this is a great (but even more short-lived) increase in the quantity of a special sort of food – soluble nitrogen flowing to growing leaves and developing seeds. This food is finished long before the plants wither. For other species which live in more equable habitats there is little observable change in the amount of vegetation in their habitat, let alone its quality, from one year to the next, or over several years. At very high densities some populations sometimes visibly deplete the amount of vegetation locally, but this quickly grows again. Most of the time most species seem to be living in the midst of a superabundance of unused food.

Of the many hypotheses proposed in the past to explain cycles of abundance of microtines, most have, at least tacitly, been discarded to leave a "short list" of food or predators or, perhaps, a combination of the two (Hansson 1987), just as has happened with studies of the snowshoe hare (Sect. 6.2.5.3). I will deal more fully with this problem of the cause of cycles in abundance in Chapter 17.

6.2.5 Rabbits and Hares

Lagomorphs are another group of rodents the abundance of which is once again largely influenced by the nitrogenous quality of the food available to breeding females and young weanlings. Like all herbivores they are highly selective feeders, seeking out and eating only the species and parts of plants which provide the maximum possible availability of protein in their diet. Again like other herbivores, this is not sufficient. There is no evidence that they eat any form of invertebrate or vertebrate protein other than milk, but all lagomorphs are hindgut fermenters with a specialized caecum housing a large microflora. And they practise coprophagy, referred to as "refection". Soft caecal faeces, produced when the animal is resting and not feeding, are taken directly from the anus and eaten. They thus reingest and digest the microbial protein which grew on the cellulose in the caecum. When active and feeding they produce normal hard faecal pellets which are not derived from the caecum, and are not eaten.

6.2.5.1 The European Mountain Hare

There are many studies of many species of rabbits and hares in many parts of the world which illustrate these facts. The study of the European mountain hare, *Lepus timidus*, in Scotland is a good example (Miller 1968; Hewson 1970, 1976; Moss et al. 1981; Moss and Hewson 1985). Heather, *Calluna vulgaris*, constitutes 90% of the their winter diet and 50% of their summer diet. Hares can quickly detect plots of fertilized heather as small as 1 m^2 set in the middle of large areas of moorland. In these they feed preferentially (along with the rabbit, *Oryctolagus cuniculus*) on the foliage which has a higher nitrogen content than that of the surrounding unfertilized plants. Like the red grouse living in the same area (Sect. 7.1.2), they are more abundant, breed earlier, raise bigger litters, and their leverets survive better on the base-rich moors where the nitrogen content of the heather is higher. Their populations do not increase after predators have been removed. Their breeding success is highest when their numbers are increasing from low densities. On many moors the hares maintain and heavily graze isolated patches of heather no more than 1 m in diameter. As a result, the heather in these patches retains permanent and dense juvenile foliage which is much higher in nitrogen than is mature foliage. It is not known why these small plots are chosen by the hares. The heather on them may have been initially richer in nitrogen by virtue of small local differences in soil or topography. Or it may be that, by chance, they concentrate their browsing to one spot, as this is more economic than browsing widely. Once started, however, their constant grazing acts to maintain new growth with high levels of soluble nitrogen in the same way as it does for geese (Sect. 7.1.1) and many other grazers (Sects. 8.1 and 8.2). This behaviour has a similar effect, on a small scale, to the management practice of continuous rotational burning of heather moors. The burnt heather produces new green shoots, high in soluble nitrogen, which are heavily and preferentially grazed by sheep, grouse, and hares alike.

In subarctic Sweden and Finland *L. timidus* feeds mostly on deciduous plants like birch, willow, and poplar. In summer hares eat the leaves and young shoots plus herbaceous material and blueberries. In winter they eat the twigs and bark plus, when desperate, pine foliage (Pulliainen 1972, 1982; Lindlof et al. 1974a, b; Pulliainen and Tunkkari 1987; Rousi et al. 1987). Again, they are strongly selective in their feeding, taking only the most nutritious plants and parts of plants. This includes feeding on trees fertilized with nitrogen and on grafted trees in poor condition. And they seek out certain habitats where the food is better. Before a crash in their numbers they have been observed to increase their use of least-favoured foods. After a crash in abundance their populations are confined to refuges provided by the patches of most-preferred habitat.

6.2.5.2 The European Rabbit

Probably some of the most detailed studies of a lagomorph are those made of the European rabbit, *Oryctolagus cuniculus*, as a feral pest in Australia. These studies have resulted in a fairly complete understanding of the ecology of the animal. Andrewartha and Birch (1984) give a cogent review of that ecology, but the essential points for the present discussion are as follows: The rabbit can breed only when there is fresh new herbage to eat, a source of highly concentrated and readily digestible protein. A flush of growth following rain has animals quickly coming into breeding condition, and they will continue to breed so long as a supply of new green growth is available. Normally, in the Mediterranean climate of southeastern Australia, this is only in winter following autumn rains. How long they breed and how successfully depends on how long it is before the hot dry summer kills off all the new growth. But in the mostly dry interior of Australia rain is little, infrequent and unpredictable. Here, the rabbit is an opportunistic breeder in response to any rain which is sufficient to cause herbs and grasses to germinate. Occasionally there are widespread and heavy rains in the interior which set water courses flowing, flooding vast areas of land, and filling the beds of normally dry salt lakes. Then the desert blooms green and there are enormous and widespread outbreaks of rabbits; their numbers build up continuously for as long as 3 to 5 years in favourable periods. Equally spectacular crashes of numbers occur when the inevitable drought finally eliminates all green feed. There is selective death of the young as the food disappears (Richardson and Wood 1982). The youngest die first, but any rabbit which is not fully grown has little chance of surviving without a diet of new growth. Ultimately, only remnant populations persist in refuges where some new herbage can still grow. There the rabbits' complex social behaviour of separate male and female hierarchies and defence of a territory containing a breeding warren, ensures maximum efficiency in converting whatever adequate food is available to surviving young. These young are forced out as they mature and must seek a new home. Mostly they will not find enough to eat, and will quickly die as part of the "doomed surplus". But should there have been any rain, and some more of the habitat again flushes green, they will quickly start to breed (see also Chap. 11).

The rabbit, then, is an animal superbly and finely tuned both physiologically and behaviourally to keep as close as possible to the limits of the available food in its environment. But only when human intervention transplanted it to a land where the amount of adequate food fluctuates wildly from next to nothing to superabundance was this potential revealed. The rabbit is just like the native Australian locusts (Sect. 3.4.7) and kangaroos (Sect. 6.5.4) which now share their arid environment with the rabbit. All are adapted to "boom and bust" in response to the unpredictable availability of food of sufficient quality to support the production and growth of young. The experimental removal of predators of the rabbit in Australia by Newsome et al. (1989), discussed in Section 16.2, dramatically illustrated the powerlessness of predators to influence either the increase or the decrease of a population which is responding to changes in the availability of its food in this way.

By way of contrast, rabbits were introduced by nineteenth-century sealers to subantarctic Maquarie Island, an environment about as different from the hot dry interior of Australia as it is possible to find. There, in a bleak climate varying little throughout the year from a mean temperature of 4 to 5 °C, with less than 2 hours sunshine a day, and with rainfall spread evenly over all seasons, they flourish for nearly 9 months of the year. Only in the depth of winter, when there is no new green grass growing, do they enter a short anoestrus (Shipp et al. 1963).

Wallage-Drees and Michielsen (1989) came to the same conclusions for populations of rabbits living in a habitat which is more nearly like that of Maquarie Island than the interior of Australia: coastal dunes of The Netherlands. They found that while predation could at times slow down the rate of increase in the number of rabbits, it was never strong enough to keep densities below the level permitted by the food supply. It was the quality and quantity of food, varying stochastically with the weather, and influencing the success of breeding and survival of the young, which decided the abundance of rabbits.

6.2.5.3 The North American Snowshoe Hare

Another lagomorph which has been the subject of long and detailed studies is the North American snowshoe hare, *Lepus americanus*. It is renowned for its 8- to 11-year cycles of abundance in Alaska and northern Canada. Attempts to explain these cycles (and their absence in populations near the southern limit of the hares' distribution) have been going on for most of this century (Green and Evans 1940; Dolbeer and Clark 1975) and continue to this day (see Chap. 17).

Such things as blueberries, cranberries and *Equisetum* feature high in the spring diet of snowshoe hares, along with new growth of grasses and succulent herbs. In summer the leaves of deciduous trees and shrubs, especially willow and birch, predominate. In winter there is little choice. Conifer needles and the bark and twig-ends of deciduous trees which remain above the snow line are all that is available (Wolff 1978). Within these seasonal constraints snowshoe hares feed selectively to maximize their intake of nutrients, particularly protein, and rely on coprophagy to recycle precious nitrogen (Pease et al. 1979; Sinclair et al. 1982).

They are not territorial but maintain individual home ranges, which may overlap considerably, and they move from winter to summer ranges in response to the availability of different foods (Wolff 1980). There is a social hierarchial status but this has only been observed to affect the acquisition of food when it is limiting. Then it does not determine how many the food can support, only which hares obtain the food (Sinclair et al. 1988).

During the lows of cyclic populations, and at all times in the non-cyclic populations near the southern limits of the hares' range high in the Rocky Mountains, only scattered and isolated patches of the habitat are occupied. These patches may comprise as little as 10% of the apparently available range. Presumably they are those places – like the refuges of the rabbits in Australia – where enough adequate food persists to maintain at least a few breeding females (Wolff 1980; Dolbeer and Clark 1975). In cyclic populations numbers increase as more and more of the range becomes reinhabited by young breeding females. In the Rocky Mountains this apparently never happens. Juveniles which emigrate from refuges each summer never survive for long. This would suggest that the habitat changes in the north and is quickly tracked by the hares, but that in the south the habitat is more nearly constant.

Breeding in all populations of snowshoe hares starts each spring as soon as new growth of forbs and grasses appear following snow-melt. There appears to be no real difference in reproductive potential of females from cyclic and non-cyclic populations; in both it is formidable. Each female will produce two or three, and sometimes four, litters before breeding ceases in late summer. Each litter will have from three to six young. Juveniles from the first litters of the season survive better than those from subsequent ones, but in all cases very few young survive until the following spring. This loss of young has been shown to be the greatest source of mortality in all populations that have been studied.

A group of biologists from the University of Wisconsin, led by Lloyd Keith, conducted a 17 year study of snowshoe hares within the southern edge of the boreal forest in Alberta (Keith 1983). In addition to intensive monitoring of wild populations over this period, they conducted field experiments with populations of known high and low densities established on small islands and in large (approximately 4 hectare) enclosures of areas of natural habitat. They supplemented both the winter and summer diets of some of these experimental populations with pellets of commercial rabbit food and matched their subsequent performance against neighbouring populations of the same initial densities.

Artificially high summer populations soon exhausted the hares' normal summer food of succulent herbaceous plants, and forced them to eat woody browse, something they would normally do only in the winter when there is no choice. These populations declined sharply because of a high loss of young early in life. In particular, they found there was a much lower survival of later litters after the herbaceous food had largely been consumed. Providing commercial pellets to populations of similar density from the beginning of summer until freeze-up in late autumn dramatically changed the outcome. While the hares still grazed out their summer food and ate much of the browse, they increased in weight and their numbers dropped hardly at all. Matched unfed populations did not produce *any* surviving young, while those with supplemental food produced third litters with survival rates equalling those of unmanipulated populations which had only about a tenth of their density (Windberg and Keith 1976).

So, high numbers per se were not affecting survival providing there was abundant summer food. All Windberg and Keith's evidence pointed to mortality of the young from a lack of summer food as the dominant factor in declining populations. They saw two critical times for the young: weaning, and the autumn switch from a herbaceous to a woody diet.

Monitoring of natural populations (Keith and Windberg 1978) revealed that the number of young surviving the breeding season starts to drop in the summer of the peak year of abundance. It drops further during the first two or three summers of the decline, but starts to recover before the low. This lower survival of young animals persisted well into their second year of life. Years of high reproduction corresponded with years of population increase, low reproduction with decline, and intermediate reproduction with peak years. During years of high population the hares became much more widely distributed, but during low years only "disjunct sections of hare habitat" continued to hold hares.

Populations which were given supplemental food in winter (actually from November until April, 1 month into the breeding season) showed little difference in survival of adults (Vaughan and Keith 1981), but the reproductive capacity of these hares increased dramatically. They bred for longer and produced 54% more young than did hares short of winter food. Their young grew faster, survived better, and some first litter juveniles even bred in the season of their birth.

The outcome of this work led Keith to the conclusion that lack of food in the winter was chiefly responsible for peak populations of hares declining through starvation. Predators contributed to and accelerated the decline, and subsequently kept hares scarce, allowing vegetation to regrow. With the return of abundant food and the decline in abundance of their predators, the hares again began to increase towards another peak.

Others working on *L. americanus* in Alaska (Bryant and Kuropat 1980; Fox and Bryant 1984), however, concluded that browsing by the hares caused cyclic variations in the amount of deterrent and toxic chemicals in their food plants, and in the ratio of palatable to unpalatable twigs on these plants. These changes in the food plants in turn resulted in cyclic mortality and reproduction of the hares.

Charles Krebs and his colleagues at the University of British Columbia in Canada set out to test these hypotheses on populations of snowshoe hares in the southern Yukon of Canada (Krebs et al. 1986a, b; Sinclair et al. 1988; Smith et al. 1988). In doing so they identified five hypotheses to explain the cycles in abundance of hares. These were: (1) a simple shortage of food for peak populations caused the declines; (2) increases in secondary compounds in the hares' food plants, as a result of browsing by the hares, were responsible; (3) the proportion of unpalatable juvenile twigs increased with browsing and forced the decline; (4) a shortage of food followed by increased predation was the cause; and (5) predation alone decreased the numbers.

From 1977 they have monitored control and experimental populations by live trapping, and provided additional food over winter to several experimental populations. Mostly the additional food was pellets of commercial rabbit chow, but they provided one declining population with extra natural food by placing freshly cut preferred food plants on the experimental site throughout one winter. Supplying artificial food maintained experimental populations during the increase and peak of a cycle of abundance at up to three times the density of the control populations, and delayed the decline by 12 to 18 months. But it did not stop the decline. Nor did supplementation with natural food. It did stop the loss of body weight of fed hares, but did not increase their survival or reproduction. Even at peak densities hares in the control populations did not eat all the natural food available to them. The nitrogen content of faecal pellets suggested that hares rarely suffered a shortage of protein, and the workers saw little sign of starving animals. So they concluded that there was no evidence to support the first hypothesis. The number of hares continued to decline even in the presence of an apparent superabundance of food.

They monitored the amount of resins and phenolics in the most preferred winter food plants and the relative proportions of juvenile to mature twigs on these plants to see if these changed so as to influence the abundance of the hares, as predicted by the second and third hypotheses. They did not. Rather, changes in both the form and chemistry of the plants appeared to be following the hare cycle, not causing it.

This left them with the fourth and fifth hypotheses. Is it food plus predators, or predators alone that drives these cycles? They concluded that further work should be directed towards distinguishing between these two.

During the course of these studies in the Yukon winters many other facts came to light. A number of these point towards a sixth hypothesis which should be tested.

There was better (and longer) breeding by heavier animals in years of increasing and early peak numbers; shorter and less successful breeding in late peak years; and shorter still and even less successful breeding through the decline, but picking up in the years just before the start of the next increase. The rate of recruitment of juveniles into the population was clearly the major force behind the cycles. This recruitment was highly correlated with the annual rate of population change and was not affected by the addition of extra food in winter. There were signs that supplemental feeding in winter improved the condition of hares prior to the breeding season. Such improved winter nutrition may be necessary for females to breed successfully, but it is clearly not sufficient, especially for the production of second and third litters in the summer.

The UBC team thought that the poor rate of recruitment of juveniles at the end of the peak was most strongly affected by the survival in summer of the very small hares during their first 8 weeks of life – before they can be caught in the live traps from which most of the data are derived. They do not know the cause of this mortality, but they did note that this poor survival of young may be influenced by the effect of summer food on the behaviour or lactation of the females. More than 50 years ago Green and Evans (1940) found that the most important factor causing cycles of snowshoe hares in Minnesota was the survival of young *in the first few months of life*. Keith's group had found the same thing for populations in Alberta (Meslow and Keith 1968).

Here, then, is the sixth hypothesis: Changes in the abundance of the snowshoe hare are driven by changes in the survival of the very young which, in turn, is driven by changes in the nitrogen nutrition of breeding and lactating females and of the young themselves, at and immediately following, weaning.

The plane of nutrition of mother and young will determine not only how many young survive to the end of the summer, but how viable they are as adults. As Keith and Windberg (1978) had found for years of poor survival of young, "the lower survival of younger animals persists well into their second year of life". There is a lag effect. The observed winter loss of juveniles stems from their being weaker, as a result of poorer summer nutrition and/or poorer foetal growth, plus a lowered quality and quantity of their mothers' milk. There is up to a 2 year lag effect. It has been *assumed* that it was the lack of winter food which was the proximate cause of juveniles dying in the winter. Krebs et al. found that this food was *not* exhausted prior to a decline in the abundance of the population. And there will be an equal but opposite lag effect for the young born in years when their survival is high.

If both the quantity and quality of food in spring and summer is high, and it is available for a prolonged season, then females can produce more litters in a season and more young in each litter. More of these young will survive, be better able to survive in winter, and be stronger, better breeding animals in future seasons. The population will explode, and vice versa.

It matters not at this stage that the change in abundance is cyclic. First find out what generates changes in the success with which the hares breed from one year to the next. Pin that down and then look to see why *it* changes from year to year and place to place. Then we may have an explanation for the cycles – and the lack of them in populations further south (Dolbeer and Clark 1975). Certainly we will be much closer to knowing (see also Chap. 17).

All the recent work has concentrated on winter food and nutrition. I have found very little record of summer food and nutrition of breeding females (other than the studies of Windberg and Keith mentioned earlier) and none on that of the young.

The assessment of what causes animals to die has also concentrated on winter time, and on the death of grown hares from predation, not on the fate of the young between conception and weaning.

Both the Alberta and Yukon studies found that predation of grown hares is considerable, especially in a declining population. At this time the predators, which have themselves increased in abundance as *their* food supply increased, are pressing hard on a shrinking resource, exacerbating and speeding the decline in the number of hares. But the point is that they are overriding an already declining population; they are *not* the *cause* of that decline. The experimental removal of predators of the rabbit by Newsome et al. (1989), discussed in Section 16.2, nicely confirmed Keith's point that predators will further depress a declining population and prolong its subsequent period of low abundance.

6.3 Primates

6.3.1 Colobine Monkeys

Most primates are omnivores, but with the bulk of their diet comprising a staple of plant food. A few are apparently truly herbivorous. Some of the leaf monkeys would seem to be among the latter. Davies et al. (1988) and Waterman et al. (1988) studied the chemistry and ecology of the food and feeding of two colobine monkeys (the red and banded leaf monkeys, *Presbytis rubicunda* and *P. melalophos*) in Malaysia; and nicely compared and synthesized their findings with earlier studies of other colobines. Their review covers eight different species of *Presbytis* and *Colobus* in five widely separated localities (two in Africa, one in India and two in Malaysia) providing habitats which differ considerably in the composition of their flora and in their carrying capacity for these herbivorous primates.

Colobines are recorded feeding almost entirely on the foliage of trees, but supplementing their diet with some fruits and seeds when they are available. They apparently eat practically nothing else, and are even differentiated as a group from other herbivorous monkeys by the lack of insects in their diet. They are able to be more folivorous than most primates because they all have sacculated forestomachs specialized for anaerobic bacterial fermentation of cellulose. It is a system analogous to that of small ruminants, producing fatty acids which, along with the microbes themselves, pass to the stomach for digestion. In the caecum and colon microbes again ferment the food.

Yet, in spite of this microbial assistance, colobine monkeys are highly selective in their choice of which leaves they will eat. Without exception they select young leaves in preference to mature ones, and then from only some of the many species of trees available to them. This choice is for nitrogen. All the studies that these authors reviewed demonstrated that colobines select foliage to maximize the intake of protein and minimize the intake of fibre. Toxins such as alkaloids do not seem to influence their feeding, probably because the foregut bacteria can detoxify them. Nor do tannins seem to have any consistent negative effect on their choice of food. What makes the similarity of these feeding preferences even more interesting is that they occur in species which are not closely related, and which live in different habitats with widely differing food plants. At all sites fibre was lower and nitrogen higher in the young foliage that the monkeys ate, than in the average young leaves present; this dichotomy was greatest for the two species living in the least hospitable forests with the lowest quality foliage. Furthermore, in the impoverished habitat the average size of the groups in which *P. rubicunda* lived was smaller (six), and their home ranges much larger (84 hectares), than the 17-member group and 30-hectare home range of *P. melalophus* living in the richer habitat.

Finally, these authors demonstrated that the higher the ratio of protein to fibre (or fibre plus condensed tannins) in the leaves, the better the food quality of the leaves. The overall biomass of colobines at any of the sites was closely and positively correlated with the protein to fibre ratio of *mature* leaves at each site. There was no evidence of disease, predation, or human hunting limiting the density of any of their populations. However, the mature leaves formed a relatively small part of the diet of all colobines studied except for those at the richest site. All species fed mostly on young leaves which have more protein and less fibre than the mature leaves. The authors concluded that because the young leaves are available only seasonally for limited times, "bottlenecks" may be created when the monkeys are forced to subsist on mature leaves. The scarcity and lack of nutritional quality of these may directly limit the population density and biomass of the colobines.

However, this may not be the case. Compare this concept with that discussed in the previous section, of winter food limiting herbivores in colder climes. Which young adult monkey is more likely to survive the season when good food is scarcest or absent? One nursed by a well-fed mother, and itself fed as a post-weaning youngster on a high-protein diet, or one that just managed to survive in a state of chronic malnutrition?

A more recent analysis (Oates et al. 1990) of the likely causes of observed densities of primates – especially colobines – in tropical forests confirms that they are most abundant in places where the foliage they eat is highest in proteins and tannins, and lowest in fibre. These sites are, however, those with the most acid and nutrient-poor soils. This supports the idea that such sites have plants with the highest concentrations of secondary chemicals. But it flies in the face of the conclusion that these plants will therefore support the fewest herbivores.

It is on these poorest sites, however, that the greatest proportion of legumes grow – plants with particularly protein-rich foliage and seeds. And it is these same legumes which are the monkeys' favoured food, especially the seeds which, it now seems, form a large part of their diet.

All this points to the quality of the food of breeding females and their young, and not to the supply of food in lean times or toxic chemicals in the food, being responsible for the success of these herbivores. It would be interesting to know what part access to *developing* seeds plays in the quality of that food (see for example Sects. 3.5.4, 6.2.2.1, 6.3.4 and 7.3).

More recently Ganzhorn (1992) has confirmed the correlation of biomass of primates with the protein content of the foliage for 28 species of Madagascan lemurs – and that these animals, too, feed selectively for a high concentration of easily extractable protein, or low concentration of fibre, or both.

Finally, are these primates as herbivorous as was thought? There are scattered records of individuals of several species of *Presbytis* eating insects, and Srivastava's (1991) observations confirm that when such protein is attainable it is readily and eagerly consumed. A troop of *P. entellus*, which he had under observation in Jodhpur, devoted a fifth of their feeding time to eating large quantities of scale insects while they were in great abundance in the locality.

6.3.2 Cercopithecid Monkeys

Whatever the quality and quantity of foliage available, few herbivores can persist on such a diet with nothing other than microbial protein as a nitrogen supplement. Most have had to include significant amounts of animal food in their diets. Cords' (1986) study in western Kenya of the diet of *Cercopithecus ascanius* and *C. mitis* showed the importance of the animal protein component of the diet to the reproduction and growth of these much more omnivorous primates. In both species adult females ate more insects and leaves than did the males, which ate more fruit; and pregnant and lactating females ate less fruit and up to three times more insects than did other females. Very young monkeys, before they were weaned, ate more fruit and less insects than did larger juveniles. The larger juveniles, however, when going through puberty and growing faster than the younger ones, ate more young leaves and insects than did the adults.

Beeson (1989) reported that the blue monkey, *Cercopithecus mitis*, of the montane forests of Malawi, lives on a diet consisting essentially of leaves (and especially the high-nitrogen petioles of some species) with small amounts of fruit and flowers; except during most of the short wet season when fruit predominates. These monkeys select items positively for their protein content but experience a "strong dip in protein intake per dietary item" in October, at the end of the 8-month-long dry season. At this time their diet is most varied and contains the least amount of fruit, even although this is the time of year when most trees are flowering and fruiting. The monkeys range further to find food then, and it is the only time of year when they have been observed to leave the forest to feed in open ground. It is also the time of year when they eat most arthropods. The author concluded that the poor quality of their diet at this time of year is related to "increased chemical defences in edible plant parts". Yet analyses of samples of the items of food they ate showed total phenolics to be at their lowest for any month, fibre much the same, and only condensed tannins higher. It seems much more likely that the plant food available to them at this time is simply a much poorer source of nitrogen, forcing them to search more widely, select more carefully, and seek more animal protein in order to maintain sufficient nitrogen in their diet.

6.3.3 Howler Monkeys

Milton (1979) records that the New World howler monkey (*Alouatta palliata*) on Barro Colorado Island, Panama, has a strong preference for feeding on young, rather than mature leaves, although these are far less abundant and more patchily distributed in the habitat than are the mature leaves. Analysis of leaves selected or ignored by these animals revealed that the young leaves they preferred had higher protein and lower fibre content than had mature leaves, and less non-structural carbohydrates. But they had higher levels of phenolics and equal or higher levels of condensed tannins. She concluded that high protein and low fibre are the major factors determining the feeding preferences of howler monkeys, and that secondary "toxins" are not.

These small (for a mammalian herbivore) animals, which lack extensive areas for fermentation in the foregut or hindgut, must be very dependent on a fast passage of food through the alimentary tract and on a high level of available protein in their food to maintain nitrogen balance. A combination of high protein and less cell wall material in their food therefore becomes a major factor in selecting food. Milton (1981) went on to show that, by comparison with the sympatric fruit-eating spider monkeys, *Ateles geoffroyi*, howlers have a long retention time of food in the alimentary tract and a large caecum. This, she said, enables them to obtain sufficient energy from fermentation products of leaves when fruit and flowers are in short supply, but does not relieve them of the necessity to be highly selective for high protein foods. On the other hand, they could not get enough protein from a diet of fruit with the slow time of passage of food through the gut. The spider monkeys, however, with a much faster rate of passage, can; but they could never obtain sufficient energy from a diet of leaves. In effect, each of these primate species is living on a diet that is not tenable for the other, a good example of how evolved behaviours, morphologies, and physiologies isolate animals from an apparently readily available source of food. What seems available is not. It is not part of that animal's environment (Sect. 1.4.2).

Estrada and Coates-Estrada (1986) studied the same species of howler monkey in Los Tuxtlas, Mexico. They found them to be highly selective, feeding on only 34 species (14%) of the trees in the forest. Only five of these were used consistently, but far more than would be expected from their density. The monkeys often travelled considerable distances to follow the pattern of flushing of the preferred new growth of these species. Analyses showed that their strong preference for young, rather than mature, leaves was reflected in the higher protein and lower fibre content of the young leaves. Yet, when they contrasted these preferred leaves with similar ones not used by the monkeys, they found no difference in the protein content, but the preferred leaves had significantly less fibre in them. Again, this influence on the ratio of fibre to protein in the selection of food illustrates the importance of the *availability* of food to an animal. If there is a lot of cell wall material which cannot be broken down during its passage through the gut, as far as the animal eating those tissues is concerned, the high-nitrogen cell sap may as well not exist. It is also not simply a matter of the *total* of nitrogen in the leaves, but the proportion of it which *is* in soluble form in the cell sap.

6.3.4 The Gorilla

The largest of the primates, the gorilla, has been considered the most herbivorous of the hominoids. Calvert's (1985) study of food selection by the West African *Gorilla gorilla gorilla* in Cameroon, showed that they seek out foods containing less lignin and more protein, and which are most digestible. She

A

B

found no evidence throughout her study of them eating animal matter. The gorilla is a non-ruminant and, unlike other primates, the bulk of its diet is made up of shoots and stems [they do, however, have capacious colons, which house cellulose-digesting ciliates, and have been recorded practising coprophagy (Rogers et al. 1990)]. This preference and other puzzling ones (such as eating mature, rather than young, foliage of one particular species, and leaving banana fruit when consuming the core of the stems) were all shown by chemical analysis to be caused by the gorillas selecting for higher protein and digestibility, and avoiding lignin and condensed tannins. Calvert pointed out that maximizing the intake of energy would not adequately explain the gorillas' choice of foods. They are limited by the time they can hold food in the gut for digestion, so any improvement in digestibility will be beneficial. And given the apparent lack of animal matter in the diet, their strong selection for high protein is not surprising.

More recently a 6-year study of these western lowland gorillas, carried out in central Gabon by a group of women scientists, has produced a much clearer picture of their diet and feeding tactics (Tutin and Fernandez 1983; Rogers et al. 1990; Williamson et al. 1990). Their staple diet throughout the year consists of leaves and pith of mostly herbaceous, but some woody, plants. They select the youngest of those leaves which are highest in protein and lowest in fibre compared with the general vegetation. Also, they choose between individual plants and parts of plants. Furthermore, they commonly strip foliage leaving the mid-ribs attached, and discard petioles and leaf bases. The fleshy stems of herbs are split and peeled; only the inner pith is eaten. When such preferred young growing tissue is scarce, they turn to proteinaceous bark and mature leaves. They also eat considerable quantities of fleshy fruit whenever this is available. The ripe seeds are either discarded or pass undigested through their gut, but they frequently extract and eat *immature* seeds (see, for example, Sects 3.5.4, 6.2.2.1 and 7.3.3 for the possible significance of this).

They also eat insects; what sort depends upon location and availability, but principally weaver ants, *Oecophylla longinoda*, and termites, *Cubitermes* spp. The former bind leaves together to form compact nests, each containing approximately 5 g of concentrated protein – mostly soft eggs, larvae, and pupae – high in amino acids. The gorillas deliberately and regularly eat these whole like a sandwich! The termites build nests on the sides of trees, and these are broken open and the inhabitants eaten. They are also recorded as eating other sorts of insects, mostly ants, but also caterpillars. In all, insects were estimated to comprise only 3 to 4% of the animals' diet, and a third of all faeces sampled included insects. The authors do note, however, that their major sampling technique of faecal analysis means that most soft tissues may not be detected. Other recent work (Nishihara and Kuroda 1991; Yamagiwa et al. 1991) would seem to confirm and reinforce these findings. Insects, especially ants and ter-

Fig. 20. A The gorilla is considered the most herbivorous of all hominoids, but eats significant numbers of insects. **B** A nest of weaver ants, a 5-g package of concentrated protein which gorillas eat with relish like sandwiches. Photos courtesy of M.E. Rogers

mites, are probably a regular and important component of the diet of gorillas wherever they live, in spite of some authors' scepticism that they eat enough of them to be nutritionally important (Watts 1989).

So, it seems that those primates which are solely or predominantly herbivorous are no different from other herbivores. They too must seek out the youngest growing leaves, flowers, and seeds – especially unripe ones. It is from these tissues that they can get the maximum proportion of the nitrogen in their food in a soluble and most readily assimilable form. But still, most must also supplement this meagre source of nitrogen with either or both microbial and animal protein. The extent of the latter in their diet is probably far greater than is usually recognised, especially in the diet of lactating females and newly weaned young. For example, much eating of fruits and seeds may, in fact, be directed towards obtaining the insects in them (Redford et al. 1984). Such patterns of feeding to enhance the intake of protein are undoubtedly quite general in primates, as recent observations of the Brazilian fruit sakis, *Chiropotes satanas*, are revealing. These animals are known to selectively eat immature seeds, and it is now emerging that they also eat significant quantities of soft larval insects (Frazao 1991).

6.4 Fruit and Flower Bats

A special group of mammalian herbivores are those bats which eat a diet consisting mostly of fruit. Fruits can be classed as nutrient-poor, containing less than 0.5% protein, or nutritious, containing up to 6–7% protein and with a high lipid content. The former are among the most protein-poor of all plant tissues (Thomas 1984). Frugivorous bats of both the Old World (Pteropodidae) and New World (Phyllostomidae) feed largely on nutrient-poor fruits. The New World phyllostomids supplement their diet of fruit with insects. They are thus similar to fruit-eating monkeys and birds (Sect. 7.2) which supplement their diet with arthropods and feed their young on them.

Herbst (1986) investigated the role of nitrogen from five species of fruit in the nutrition of the phyllostomid bat *Carollia perspicillata* in Costa Rica. The five fruits he selected were reported to represent about half of the species eaten by this bat, and to comprise 60% of their faecal contents. He found that all but *Ficus* fruit could apparently provide an adequate source of nitrogen for day-to-day maintenance of these bats without the need for them to eat insects. But the fruit of only one species, *Piper amalago*, contained sufficient nitrogen to meet the need of lactating females for nitrogen. (Herbst calculated that a 20-g bat would secrete about one fifth of its weight in milk each day.) However, by comparison with the known requirements of the laboratory rat, all five fruits were deficient in one or more of the essential amino acids. If, therefore, these bats have to eat more fruit to compensate for these deficiencies, not even *Piper* fruit would be adequate for lactating females, and none of the other four, on their own, would be sufficient for maintenance metabolism. Not surprisingly, *C. perspicillata* shows a marked feeding preference for the fruit of *Piper* (Heithaus and Fleming 1978); and more insects are found in the stomachs of these bats when the number of lactating females is at a peak (Fleming et al. 1972).

So, it would seem that maintaining a positive nitrogen balance on a diet of fruit alone is, at best, a chancy business; females must have access to animal protein in their diet if they are to successfully produce and raise their young.

Morrison (1980) had concluded that a diet of figs was an inadequate source of nitrogen for maintenance metabolism of another phyllostomid, *Artibeus jamaicensis*, on Barro Colorado Island, Panama. Thomas (1984) compared the results of his experimental studies of two West African pteropodid bats, *Epomops buettikoferi* and *Micropteropus pusillus*, with Morrison's results. Unlike the phyllostomids, the Old World pteropodids apparently do not supplement their diet with animal protein, and are remarkable among mammals for their ability to rely on a low-protein diet. Instead they have evolved the capacity to ingest large quantities of food in a short time (up to two and a half times their body weight of fruit per night) and to regulate the amount of fruit they ingest each day according to its protein and not its energy content. Furthermore, they can pass food through the alimentary tract in as little as 20 minutes. A major factor contributing to this ability to process so much food in so short a time is their habit of carefully masticating the fruit and spitting out pellets of fibre – they live on a liquid diet of fruit juice. This behaviour is also used by other pteropodid bats which eat leaves rather than fruit. Lowry (1989) has observed two species of fruit bats in Australia, *Pteropus alecto* and *P. poliocephalus*, feeding on leaves of species of *Albizia*, leguminous plants noted for the unusually high digestibility and high concentration of their leaf proteins. The bats carefully chew the leaf into a bolus which is discarded while the liquid fraction is swallowed. Lowry demonstrated that for *P. alecto* the extract that they ingested contained about 51% of the crude protein of the leaf and is itself 36% protein.

P. poliocephalus is also recorded as feeding preferentially on nectar and pollen rather than fruit (Parry-Jones and Augee 1991). They take flowers all through the year, chewing and spitting out whatever they choose not to swallow. They digest large masses of pollen. Many of their droppings are almost entirely composed of large numbers of pollen grains, most of which had had their nitrogen-rich contents removed (see below). So pronounced is this preference that they are a problem to commercial fruit growers only in those wet seasons when flowering of the native eucalypts is poor. They will even leave their normal haunts to look for preferred food, leaving behind untouched and abundant commercial crops.

This lack of interest in commercial fruit is not too surprising in the light of Steller's (1986) analysis of the energy and nitrogen requirements of *P. poliocephalus* on a diet of native figs and commercial apples. He discovered that apples are a grossly deficient source of nitrogen for even resting, nonbreeding individuals. So, in the field, they would eat them only as a last resort when their normal food was not available. His study also points out the value of the capacity these fruit bats have of raising their body temperature when foraging at night (Noll 1979), as this highlights that it is nitrogen, not energy, which is limiting. They thus use up excess energy from a diet with a low protein to energy ratio so that they can meet their nitrogen requirements. By burning up energy, rather than laying it down as adipose tissue, they increase their capacity to make long foraging flights in search of food.

Another group of even more specialized herbivores are the nectivorous or flower bats. Until quite recently they were thought to be able to exist on a diet of "virtually pure carbohydrate", but Howell (1974) neatly dispelled this myth. He sampled the faeces and stomach contents of species of the genus *Leptonycteris* feeding in the flowers of cacti from Arizona to Mexico, and experimented in the laboratory with individuals of *L. sanborni*. These small bats (15 to 20 g) exhibit a strong preference for feeding in the flowers of only a few of the many species available. For *L. sanborni* in Arizona the preferred flowers were those of *Agave palmeri* and *Carnegeia gigantea*.

Howell's field collections revealed traces of insect cuticle in less than 5% of stomachs, but the cuticle was from small flower-dwelling thrips, and the dentition and sonar characteristics of *L. sanborni* are not those of an insectivore. Occasionally the bats eat the fruit of cacti which can yield 16% protein. However, these are seasonal and could, at best, provide only a bridge between peaks of flowering. He could find no evidence of symbiotic organisms in the gut.

He did find, however, that pollen comprised 97.8% of the faecal pellets and 25% of the stomach contents collected in the field. Analysis showed that the pollen from the two preferred flowers contained 22.9 and 43.7% protein, respectively, and has an amino acid composition equivalent to whole egg and bee protein; more than adequate, he calculated, to supply the gross protein requirements of these bats. These protein contents of the "bat-favoured" species are considerably higher than the 9 to 15% protein recorded for the closely related "non-bat" species.

The extreme resistance of the exine of pollen grains to either abrasive or chemical attack (such as boiling in hydrofluoric acid!) is renowned. Yet the grains will germinate in a warm sugar solution, and 0.001 to 0.1*N* HCl, or weak solutions of urea, will extract protein from pollen grains. The stomachs of these flower bats obviously provide the warm sugar solution, and Howell found that the concentration of HCl in the stomach of *L. samborni* was 0.1*N*. Furthermore, these bats exhibit an unusual but quite definite behaviour. They ingest some of their urine. There are several possible explanations for this, but it seems likely that the behaviour could be enhancing access to protein by urea degradation of pollen grains.

In the laboratory Howell fed five lots of ten *L. sanborni* (maintained in individual cages) on five different diets for 10 days. Those fed on synthetic nectar plus pollen of either *Agave* or *Carnegeia*, and synthetic nectar plus whole-egg protein, lost less than 5% of their body weight (those on the totally unnatural diet lost a little more), remained in positive nitrogen balance, and were in good health at the end of the experimental period. Those fed on either synthetic or natural nectar alone (containing no protein), however, quickly went into negative nitrogen balance and rapidly lost weight. In 5 days three were dead, eight were dead by the 8th day, and all by the 10th day.

In Australia, recent work (Law 1992a, b) has demonstrated that the small megachiropteran Queensland blossom bat, *Syconycteris australis*, is similarly dependent upon pollen as its source of protein. It has the undifferentiated gut and fast rate of passage of food through it that is typical of other pollen feeders (see also Sects. 6.5.2 and 7.2).

6.5 Marsupials

The herbivorous marsupials of Australia are no different from their eutherian cousins when it comes to having to eke out an existence on a purely vegetable diet. They too are hard pressed to gain enough protein from their food to support successful reproduction and survival of their young.

6.5.1 The Koala

The koala, *Phascolarctos cinereus,* is the largest of the folivorous marsupials in Australia (average weight 6 to 10 kg) and possibly the most narrowly specialized. It eats nothing but the leaves of the genus *Eucalyptus*, a fibrous and abrasive diet low in nitrogen and high in cell wall fibre, lignin, tannins, and natural oils. So, it is not surprising that koalas have been found to have a diversity of adaptations which both increase the availability of nitrogen in their diet and improve the efficiency with which they absorb and digest it, particularly when they are young and growing fast.

The metabolic rate of marsupials generally has been found to be lower than that of eutherian mammals and they require less nitrogen to maintain nitrogen equilibrium than do the eutherians. The metabolic rate of the koala is only 70% of the marsupial average (Degabriele and Dawson 1979), in itself an important adaptation to a diet low in protein (Cork 1986).

They are hindgut fermenters par excellence with a proximal colon and caecum which, at more than four times the body length, is the greatest caecal development recorded for any mammal (Cork and Hume 1983; Cork et al. 1983). And they have a mean retention time of food in the gut longer than most [213 hours for solutes; and 100 hours for particulate matter (Cork and Warner 1983)]. In addition, the fine particles and solute in the food are retained in the caecum and colon for twice as long as the coarse particles. This is an adaptation shared with other small- to moderate-sized mammals (discussed in previous sections) and with many birds (e.g. Sect. 7.1.5), and which increases the utilization of the cell contents of their food. The koala's reliance on gaining access to the cell contents of its food is further reflected in its dentition. The teeth are specialized to grind leaves to fine particles (Lanyon and Sanson

1986). (It is interesting to contrast this combination of fine grinding and long retention time with the fruit bats' initial separation of cell contents from cell walls and fast passage of contents through the gut, as discussed in the last section.)

The young feed solely upon milk until they are about 5 months old. Weaning is gradual over the next 6 to 8 months. At 5 months of age the young first put their head out of the mother's pouch. They are then fed on "pap", soft faeces produced by the mother and presumed to be concentrated caecal material (Smith 1979; Martin and Lee 1984). The young continue to feed upon pap for up to 6 weeks, during which time they enter a phase of exponential growth. The first teeth erupt at this time, after which the young start to emerge from the pouch and begin to feed on leaves. Eating pap would clearly serve to inoculate the young with the microorganisms necessary for caecal fermentation (Chilcott and Hume 1985) but it would seem unnecessary to maintain this special diet for 6 weeks to achieve that inoculation. It must also provide a valuable high-protein supplement to the mother's milk at a time when the young are most in need of protein for growth. Adult koalas are apparently not coprophagic.

Koalas are very selective feeders. They are recorded eating the leaves of 35 different species of eucalypt (Degabriele 1981, 1983) but normally browse upon only one or a few preferred species of those available to them in any particular locality (Hindell and Lee 1987). More importantly, they eat mostly soft young growth, turning to older leaves only when there are no new leaves available. This commonly happens in winter, and they must compensate for more fibre and less nitrogen in the older leaves by eating more of them (Cork 1986). A koala's preferred browse has been shown to be significantly higher in crude protein and lower in fibre, lignin, and gross energy (Ullrey et al. 1981a, b; Martin 1985).

Koalas are aggressive animals with marked social dominance and territorial behaviours (Degabriele 1981). Dominant males defend several females. All 2- to 3-year-olds are excluded from territories held by adults and join nomadic non-territorial groups of koalas. Members of these groups have a low probability of survival compared with those holding territories. They are excluded from access to the best food source, which is sequestered for the exclusive use of a few (see also Chap. 11).

Numbers of koalas fluctuate quite markedly, both in nature and in populations artificially maintained in reserves. In the latter, great increases in abundance have been the result of provision of year-round supplies of flush growth from the coppicing of their preferred species. In the wild, increases are likely to be caused by increased and prolonged availability of new growth; for example, in patches reshooting after fire, or when the weather results in better and more frequent new growth of trees. Similarly, decreases are likely at times when the supply of adequate food declines, such as during a severe drought. Then there is no new growth and many trees may lose some or all of their leaves. Then koalas starve. They become dehydrated, in poor condition, anaemic, carry a much increased load of ticks, and many (mostly the young) die. And this

Fig. 21. A koala, in spite of adaptations to increase access to nitrogen in *Eucalyptus* leaves, suffers "wasting disease" – starving to death with a full stomach – if it cannot eat enough new flush foliage. Photo courtesy of R. Degabriele

may happen at times when, a few kilometres away, koalas living on healthy trees by waterholes remain in good condition and suffer little mortality (Gordon et al. 1988). It can also happen when increasing numbers, unable to disperse from isolated populations or on islands, defoliate preferred trees (Martin 1985a, b; see also Sect. 14.5).

In recent years there has been much public concern about the sickness and death of many koalas in Australian reserves and overseas zoos from causes apparently unrelated to the quality of their diet. Commonly the animals suffer from a chlamidial infection which reduces their fertility, and/or from "wasting disease". In the latter case koalas become emaciated and die in a coma with a full stomach, and for no apparent reason. It is most prevalent among the very young and very old, especially during winter or following a drought.

Degabriele (1989) has provided a succinct explanation for these deaths in terms of the nutrition of the koalas. He points out that a koala in its prime, and with all the adaptations for the extraction of nitrogen from its food functioning at maximum efficiency, may still starve if it is forced to eat nothing but old fibrous leaves with little protein in them. For the very young, with the cutting edges of their teeth not fully developed, and for the very old, with those edges worn away, such a diet could be quickly fatal. The "cure" for wasting disease is to provide more nutritious leaves.

6.5.2 Possums and Gliders

The common ringtail possum, *Pseudocheirus peregrinus*, is the smallest arboreal folivore (600 to 1000 g) in Australia. Pahl (1987a) found that it feeds primarily on eucalypt foliage but also eats small amounts of (usually) other Myrtaceae depending on their relative abundance and, very occasionally, fruit. It practises food selection at four levels: species, tree, leaf, and part of leaf. Of the several species of *Eucalyptus* available the ringtail prefers to feed in only three of these, and within a species it shows preference for individual trees [all very reminiscent of the behaviour of many small rodents (Sect. 6.2) and some folivorous birds (Sects. 7.1.2 and 7.1.3)]. Young leaves predominate in the diet in spring and early summer when they are most abundant. These, and some mature leaves, are eaten, petiole and all, but often the mid-rib and petiole of larger tougher leaves are left. The choice of diet (including the non-eucalypt component) maximizes their intake of nitrogen and minimizes their intake of cell walls (Cork and Pahl 1984).

There is evidence that young leaves are more important to lactating females. In a "normal" year about two-thirds of the diet of both males and females was young leaves in the spring. This is also the time when females are lactating. In a drought year the situation was dramatically different. There would have been much less young growth in the spring, reflected in the great decrease in the proportion of young leaves in the diet of both males and females at that time. But, whereas the males managed to obtain only 10% of their diet as young foliage, the females' diet was 25% young leaves; when there was a shortage of good food the lactating females got the lion's share of it. That this was still inadequate for successful breeding is reflected in the changes in survival of the young that Pahl (1987b) recorded. In the "normal" years with average rainfall, survival to weaning at age 6 months varied from 26 to 80%, depending on the locality, and to sexual maturity at 1 year it averaged 30%. In the following drought year, however, survival to weaning was as low as 6%, and only 2% survived to 1 year old.

Chilcott and Hume (1984a, b, 1985) have studied the common ringtail when it was fed nothing but the foliage of *Eucalyptus andrewsii* and given access to drinking water. It is a hindgut fermenter, and while young expanding leaves are its preferred food, the very low nitrogen and high fibre content of these leaves create problems for so small an animal with its consequent high mass-specific energy requirements and small gut volume. The efficiency with which it can extract nitrogen from its food is therefore likely to be critical to its survival and reproduction. They demonstrated that the common ringtail has a low maintenance requirement for nitrogen and recycles a high proportion of its endogenous urea to the gut (some 11% of its intake of truly digestible nitrogen) where it is available for protein synthesis by caecal microorganisms. Thus, these small animals are able to maintain a positive nitrogen balance on a diet containing only 1.1% nitrogen.

Like the koala, the common ringtail's retention time of the fluid digesta is significantly longer than that of the particulate matter, although nowhere near as long as the times recorded for the koala. But, unlike the koala, adult common ringtails are coprophagic, or, more correctly, caecotrophic. During the night, when they are active and feeding, they produce hard faeces which have about the same content of nitrogen, but a higher fibre content than the ingested leaves. However, during the daytime when they are resting they produce, and eat, soft faeces. These are produced when the usual anti-peristaltic movements of the proximal colon, which are responsible for the long retention time of the liquid digesta, cease. Consequently, the soft faeces are virtually pure caecal contents, high in nitrogen and low in fibre compared with the hard faeces. They are ingested regularly and daily. The benefits of this are two-fold: it increases the time available for microbial fermentation of fibre in the caecum and recycles microbial protein, which would otherwise be lost to the animal, for digestion. The authors calculated that the soft faeces contributed some 56% of the animals' energy intake. But, of much greater significance, without caecotrophy the animals' estimated maintenance requirements for nitrogen would be doubled. So, the ringtail exhibits similar, but different, adaptations to those of its much larger cousin, the koala [but nearly identical to its very distant relative, the hare (Sect. 6.2.5.)], enabling it too to succeed on the same low nitrogen diet of eucalypt leaves.

Slightly larger than the common ringtail possum, the greater glider, *Petauroides volans* (1000 to 1200 g) is also a strict herbivore. It eats the foliage of several species of eucalypts, including the occasional bud and flower, but apparently not that of non-eucalypt species (Hume et al. 1984). Like the ringtail it is a hindgut fermenter with similar adaptations of low field metabolic rate, slow passage of digesta, and selective retention of fine particulate matter in the caecum. It differs, however, in important ways. Both are highly selective as to the species of eucalypt they will eat – but not to the same species. Its mean retention time of digesta is not as long, nor does it have as marked a difference between the retention of fine and coarse particulate matter. It does not appear to practise caecotrophy, and its maintenance nitrogen requirement is nearly double that of the common ringtail. All this shows in the food on which the two species can subsist. Hume et al. (1984) could maintain both animals for up to 3 years on a sole diet of *Eucalyptus* foliage. The common ringtail survives happily on *E. andrewsii*, which has a nitrogen content of only 1.1%, but the greater glider needed the 1.9% nitrogen content of *E. radiata* foliage.

Other Australian arboreal marsupials, while folivorous, often predominantly so, and normally specialising on eucalypt leaves, eat a variety of other plant products as well. Some eat nectar and pollen (Turner 1984a, b; Goldingay 1987; Wooller et al. 1988), and others plant exudates (Smith 1982). The yellow-bellied glider, *Petaurus australis*, makes incisions in the bark of selected trees and feeds upon the phloem sap that exudes from the wounds (Goldingay 1987). All exhibit selective behaviour in their feeding which would maximize

the amount of nitrogen they obtained from these foods (Goldingay 1987; Landsberg 1987).

A common thread between them all is that they supplement their diets of plant material with significant quantities of insects, especially during the breeding season and when females are lactating (Smith 1982; Henry and Craig 1984; Wooller et al. 1984; Goldingay 1986). They are not true herbivores; except one, the honey possum, *Tarsipes rostratus*. Turner (1984a, b) investigated the diet of the honey possum and of two other small nectivorous marsupials, the eastern pygmy possum, *Cercartetus nanus*, and the feathertail glider, *Acrobates pygmaeus*. All three ingest and digest large quantities of pollen in addition to nectar. *C. nanus* certainly and *A. pygmaeus* probably also eat significant quantities of insects. But the honey possum is apparently the only non-flying mammal that is totally dependent on nectar and pollen for its food. This tiny beast (11 to 22 g) is highly specialized for feeding at flowers, especially those of several species of *Banksia* (Proteaceae). Turner was able to show from inspection of faecal samples and by experimentally feeding a captive animal, that the honey possum can, with a rate of passage through the gut of as little as 1.5 hours, digest the contents of 100% of the *Banksia* pollen it ingests. She calculated that it could readily meet its 0.11 g minimum daily protein requirement from the pollen it ate. Richardson et al. (1986) confirmed *Tarsipe*'s dependence on pollen as its source of protein. They also demonstrated the specialized morphological adaptations of its tongue and palate for collecting pollen, and the digestion of pollen grains in the intestine, probably by means of direct enzymatic action through the pores of the exine coat of the grains, a similar adaptation to that found in nectivorous bats (Sect. 6.4).

6.5.3 The Habitat of Possums and Gliders

Braithwaite (1983, 1984) and his colleagues (Braithwaite et al. 1983, 1984) adopted a novel strategy to attempt to discover the relationship of the distribution and abundance of arboreal possums to variables in the environment. They solicited information on the precise location of all these animals seen by members of 36 logging crews felling eucalypt forest in the Eden area of southern New South Wales over an 18-month period. This gave them a record of the location of 930 individuals of seven species over 2890 hectares of forest for which floristic, soil and geological information was available.

The species included two strict herbivores, the greater glider, *Petauroides volans*, and the ringtail possum, *Pseudocheirus peregrinus*; and five omnivores, the sugar glider, *Petaurus breviceps*, the yellow-bellied glider, *P. australis*, the feathertail glider, *Acrobates pygmaeus*, the pygmy possum, *Cercartetus nanus*, and the brushtail possum, *Trichosurus vulpecula*.

Fig. 22. The pygmy possum was thought to ingest pollen only accidentally when feeding on nectar. While it commonly eats insects, it can satisfy its minimum daily protein requirement from pollen alone. Photo courtesy of B. Fuhrer

No animals were reported from 52% of this area, and 63% of all those found came from only 9% of the total area. All the species, and a majority of the individuals of each species, tended to occur in the same areas. Of the 22 vegetation communities identified in the forest nine of them contained all the species of possums recorded and the maximum densities of each of them. These nine communities were distinguished by the presence of one or more of only a few of the many species of *Eucalyptus* in the forests, and one or more of the three peppermint eucalypts, *E. radiata, E. dives*, and *E. elata*. All, but especially the three peppermints, were found to have high levels of nutrients (especially nitrogen) in their foliage. These communities grew mostly on soils of high fertility derived from one geological formation, Devonian intrusives.

So, not surprisingly, all these marsupials – the two strict herbivores and the five omnivores – are found concentrated in habitats where the richest food is to be found. Elsewhere they are rare or absent. The authors note that the five omnivores eat considerable quantities of invertebrates, so for them the relationship may be once removed. They are concentrated where the high nitrogen foliage produces higher densities of herbivorous insects. Subsequent work with one of these species, the yellow-bellied glider, has confirmed this dependence upon arthropods. Goldingay and Kavanagh (1990) found that late lactation and weaning (the time when the protein requirements of mother and young are greatest) coincides with the highly predictable peak of abundance of arthropods in the habitat.

Yet for all of them the environment is largely inadequate. Only small patches of what seems like a great abundance of apparently similar *Eucalyptus* forest provides food from which they can gain enough protein. So most of the forest is a desert where they cannot survive.

That this link between the distribution and abundance of different species of marsupials and that of the species of eucalypt that they eat is based upon the protein needs of the animals, has since been clearly demonstrated for one of them, the greater glider, *Petauroides volans* (Kavanagh and Lambert 1990). Extensive observations, spread over 27 months near Bombala in southern New South Wales, have shown that which trees these gliders are found in, and what parts of those trees they eat, are almost entirely dictated by the availability of nitrogen in their food. Like the koala, whenever possible these gliders eat only young leaves. And, like the koala, they select the young leaves of one particular species, *E. viminalis*, in preference to those of any other eucalypt in the forest. Only when there is no new growth on *E. viminalis* will they feed in trees of other species, and then only on their young leaves. The young leaves of *E. viminalis* have a much higher concentration of nitrogen (around 2.5% dry weight) and a nitrogen to fibre ratio nearly three times greater than that of any other species present in the forest. The other less preferred species had about 1.5 to 2.0% nitrogen, while the new growth of species which predominate in parts of the forest where gliders are hardly ever found had 1% or less.

Interestingly, the leaves of these species also have very low levels of fibre, which points to nitrogen rather than fibre as the important determinant of the quality of the gliders' food. Nor could the authors find a pattern of levels of polyphenols or essential oils in the leaves of the different species of eucalypts. Often they were as high in the leaves of *E. viminalis* as in those of the least preferred species.

Recent work by McArthur and Sanson (1991) demonstrates, as one might expect, that such "deterrents" are nothing of the kind for adapted animals. They have shown that the three strict folivores, the ringtail possum, the koala, and the greater glider, have evolved mechanisms involving energetic, rather than nitrogen, costs. They excrete tannin or tannin breakdown products rather than tannin-protein complexes in their faeces – an appropriate pathway for a diet which is consistently low in nitrogen.

6.5.4 Kangaroos and Wallabies

The other major group of apparently truly herbivorous Australian marsupials is the Macropodidae, the kangaroos and wallabies. Their physiological and anatomical adaptations to cope with a supply of food which is usually low in nitrogen and often ephemeral, have been extensively discussed by Hume (1982). His and his colleagues' physiological studies have supplemented and confirmed earlier field studies.

Macropods share with their folivorous relatives many similar traits for extracting as much nitrogen as they can from their food, but there are many differences. In contrast to the folivorous marsupials, macropods are foregut fermenters, and grazers, not browsers. The food of folivores – eucalypt leaves – is continuously abundant and widespread, but nearly always low in nitrogen. They must cope with this chronic shortage only variably mitigated by seasonal flushes. The food of macropods, on the other hand – mostly grasses – fluctuates violently both in abundance and in nitrogen content, seasonally as it senesces, and from one year to another in response to infrequent and unpredictable rainfall. The macropods are exquisitely adapted to rapidly tracking and capitalizing on such flushes of abundance, be they seasonal or spasmodic. Their reproductive strategy of embryonic diapause ensures that females can produce young immediately there is an increase in good food, and, virtually without pause, continue to do so as long as the food lasts. There is always a partially developed young in the uterus, ready to be born the moment its older sibling vacates the mother's pouch – and before it has been weaned. Nevertheless, they must, like all other herbivores, seek out and concentrate the nitrogen in their food. Thus, R. J Taylor (1984) found on the New England tablelands of New South Wales that the eastern grey kangaroo, *Macropus giganteus*, and the wallaroo, *M. robustus robustus*, have a strong preference for feeding upon leaves of grasses which are high in protein. On fertilized pastures sown with introduced high-protein grasses among the native coarse tussock grasses, their day-to-day distribution, and time spent foraging, were closely correlated with the presence of these introduced grasses. This correlation held even in winter when the good food was much less abundant and they were forced to eat greater quantities of low-protein tussock grasses. Osawa's (1990) analyses of the stomach contents of the swamp wallabies, *Wallabia bicolor*, killed on the road in Queensland, revealed that, throughout the year, they selected those plants with the highest level of nitrogen. In so doing, this often meant that they were also eating plants with the highest level of fibre.

The quokka, *Setonix brachyurus*, is a small wallaby that lives on 2000-hectare Rottnest Island off the coast of Western Australia. There it is protected from hunting and has no other predators. It is strictly nocturnal and eats a variety of shrubs that grow on the island, but is highly selective for high protein (Main et al. 1959).

In the mild, winter wet season new growth on plants provides protein at an adequate level for growth and reproduction. But in the hot dry summers the vegetation is dormant, mostly sclerophyllous, and with protein levels at their lowest. Then the quokkas are constrained by time and the capacity of their stomachs, and cannot eat sufficient to meet their needs. Like their cousins the koala, they starve, becoming anaemic and debilitated. By late summer many die, but it is mostly the juveniles that succumb. How many will depend largely on their condition at the beginning of this annual period of inadequacy, which in turn depends on the outcome of the quality of food available to them and their mothers in the growing season.

On the wider and more long term scale, Newsome (1975) reviewed the ecology of two major macropods of the dry hot interior of Australia, the hills euro, *Macropus robustus*, and the plains red kangaroo, *M. rufa*. While they differ in many aspects of their ecology – the places they prefer to live, the plants they eat, and the minimum level of nitrogen upon which they can subsist – they are both limited in their distribution and abundance by the supply of protein-rich food. The changes wrought by European settlement nicely illustrate this. The changes in the vegetation, caused by the feeding of ruminant domestic sheep and cattle, the introduction of high-protein exotic pasture grasses, and the provision of artificial waterholes greatly increased the range and abundance of both species of macropods. Yet they still remain at the mercy of the weather. They explode in abundance, feeding on the flush new growth following good rains, and die in their millions when drought again intervenes. That this is so became even more abundantly clear during the great drought associated with the 1982–1983 El Niño. Then, across all of eastern and central Australia, some 40% of all species of kangaroo died in one season (Caughley et al. 1985). It is this weather-generated change in the supply of good food, and this alone, which determines the number of kangaroos (Caughley 1987; see also Part V).

But, as in the case of several other apparently "strict" herbivores, is this the whole story? Is their repertoire of physiological and behavioural adaptations sufficient to always get them by on a purely vegetable diet? One recent report would suggest perhaps not. As a result of patient and careful observations of the tammar wallaby, *Macropus eugeneii*, on South Australia's Kangaroo Island, McKelvey (pers. comm. 1992) has discovered that these herbivores are, in fact, not inconsiderable opportunistic carnivores. He commonly saw them eating insects which they encountered while foraging. He also discovered that

Fig. 23. A Kangaroo Island tammar wallaby eating a house mouse. These "strict" herbivores also eat any small birds or lizards they can catch, and actively search for and dig up insects to eat. Photo courtesy of M. W. McKelvey

they actively listen for and dig up insects in the litter and humus, readily and repeatedly take a variety of small lizards whenever they encounter and can capture them, and eat nestling mice and birds. He also observed them feasting upon the large numbers of cossid and hepialid moths when they are synchronously emerging from the soil. How many other macropods might be found to eat animal food if someone looks? Is this a "one off" adaptation peculiar to the Kangaroo Island wallaby, or is it a usual part of the feeding repertoire of macropods? Or might it be a "survival strategy" to enable them to breed at times when flushing new vegetation is scarce or absent?

Chapter 7 Birds

An acceptable starting point for any discussion about the importance of food for birds is the work of Lack (1954). He maintained that they are limited by their food. A great deal of work has since been directed towards testing this hypothesis, and today most would accept that Lack was right – about carnivorous birds. The problems arise with herbivorous birds. Many would not accept that they are limited by their food. One of the reasons for this has been that most studies look only at the quantity of food, with no thought for its quality, and consider only the energy needs of the animals.

There is now much evidence, however, which makes it clear that it is not just the amount of food that is important, nor is it the energy content – it is the level of nitrogen in the food (Robbins 1981). But many still believe the case is "not proven". Newton (1980) points out that a great deal of evidence about the feeding ecology of birds is indirect or circumstantial, and often not amenable to experimental testing. The need for experimental testability is, of course, paramount, but a true understanding will be long coming if the wrong hypothesis is being tested. For example, manipulation of the winter staple will be of little avail if it is the diet in spring and summer which is limiting. Nor is measuring the consumption of energy in relation to breeding success of much use if the animals are short of protein.

Martin's (1987) review has brought the focus clearly back to the key role of food – and of spring and summer food rather than winter food. Numerous studies of both carnivores and herbivores clearly show it as the major influence on the success of all stages of breeding: from the quality of the females' eggs to the continuing diet of the growing young. Yet, while protein is increasingly being recognized as the key component of food for successful breeding, most of the emphasis in present-day studies remains firmly upon measuring the supply of energy. I hope that what follows will help to rectify this.

7.1 Birds Eating Green Leaves

7.1.1 Geese in Europe and North America

Geese would seem to be truly herbivorous animals (Owen 1980), and incapable of digesting cellulose (Mattocks 1971). They are large, strong fliers, so are constrained by the need to trade weight against flight. This precludes holding a large bulk of vegetation in sacculated sections of the gut for digestion by microorganisms. The gastrointestinal tract of most geese apparently lacks any obvious modifications for the fermentation of cellulose. Some species do have enlarged caeca, but seem capable of supporting only limited microbial fermentation (Buchsbaum et al. 1986).

However, recent work reveals that they have a more efficient system of microbial digestion than was supposed. Prop and Vulink (1992) have found that barnacle geese, *Branta leucopsis*, in Holland have bacteria in the caeca and colon which, digestion trials have confirmed, ferment hemicellulose and cellulose. But the system comes into its own only in the summer when barnacle geese, along with several sympatric species, must subsist on poor quality plants (mostly mosses). Then the time that they retain food in their gut increases two to four-fold, the interval between their defaecations increases from 3.5 to 15 minutes, and bacteria become much more concentrated and active in the caeca and colon. At other times the bacteria are much less abundant and dispersed throughout the digestive tract.

Geese also have two other important digestive adaptations: they can separate small particles of food in the proximal colon, concentrating them in the caeca, and they can transport 20 to 30% of their urine from the cloacal region to the caeca (Bjornhag and Sperber 1977). Both mechanisms would contribute significantly to improved utilization of scarce nitrogen, recycling much of it for use by the caecal bacteria.

In the spring and summer, however, geese depend on processing food through the alimentary canal in as little as 30 minutes (Owen 1975) and continuously harvesting large quantities of forage. They can consume more than 25% of their body weight in fresh grass each day (Ebbinge et al. 1975) and return to defaecating every 3.5 minutes while grazing (Owen 1971).

However, there are costs to such a strategy. They cannot lyse cell walls to gain access to the contents, and the rapid passage through the alimentary tract is not conducive to fully absorbing all the nutrients released. So, all the protein is not available to them. The quality of the food they eat is therefore of extreme importance to geese. This is especially true in spring when they must lay down reserves of fat and protein to allow reproduction in the arctic environment where they arrive to breed before food is commonly available. The timing and success of breeding is dependent upon this accumulation – they must produce and incubate their eggs largely on reserves formed during or before migration.

Although they are largish birds, geese are relatively small animals, so their nitrogen requirement per unit body weight is high, needing a greater intake of protein for maintenance at the expense of fibre. Their demand for a diet high in protein during spring, when they must accumulate these substantial reserves in addition to maintenance, is, therefore, extraordinarily high.

Apart from their digestive adaptations, they have two major behavioural strategies which help meet this need. First, they forage selectively for the species of plants which have highest levels of nitrogen, and take from them only fresh green growth which has the highest protein to fibre ratio, and the highest levels of soluble nitrogen. Second, they repeatedly graze the same plants, harvesting the highly nutritious regrowth which this grazing stimulates.

Studies of several species of geese, mostly in Alaska, Canada and Holland, have served to highlight the way in which access to enough nitrogen dominates the ecologies of these birds.

Ydenberg and Prins (1981) found that grazing by barnacle geese, *Branta leucopsis*, on their main spring food, *Festuca rubra*, on a salt marsh in The Netherlands, where they overwinter, was concentrated where plants grow fastest. The protein content of plants on grazed sites was significantly higher than of those on ungrazed ones. Furthermore, this high level of protein in their food plants was a direct result of repeated grazing by the geese and caused the consequent sustained regeneration of young, protein-rich tissues. Once the geese departed for their arctic breeding grounds in mid-April the protein content of the grass on the grazed areas fell to a level equal to that in experimental exclosures not grazed by geese. However, during the winter these geese foraged on heavily fertilized dairy pastures adjacent to the marsh, making only short, sporadic excursions into the marsh. The level of protein available in the pasture grasses is higher than that in the *Festuca* on the marsh for most of the year, and markedly so during winter. But in the early spring the *Festuca* undergoes a burst of growth ahead of the pasture grasses. Then, for a short time, it is higher in nitrogen than the pasture grasses. At the precise time when the availability of dietary protein is equal in both pasture and marsh, the geese shift to the marsh. There they remain until they depart for their breeding grounds in 6 to 8 weeks time even though the pasture grasses soon surge past the *Festuca* and are again a richer source of nitrogen. Prins and Ydenberg (1985) suggested that the salt marsh is a more preferred habitat but that its low level of dietary protein prevents the geese using it during winter. A more positive interpretation might be that the geese simply desert their winter feeding ground, to which they are adapted, for the much richer one provided by human manipulation of the habitat. [Other species of geese are known to seek out and feed preferentially upon plants that have been artificially fertilized (Owen 1975)]. By doing this they illustrate that their acute need for high-protein food for successful breeding is often not, or only marginally, met by their natural feeding grounds. The conversion of much salt marsh to dairy pasture must have measurably improved the success rate of their breeding. But, not surprisingly, the geese would otherwise prefer to be in their native marsh. They constantly return to "sample" it during winter, and return permanently the moment it provides food as good as that of the pasture. The more puzzling fact is that the geese stay in the marsh in the spring after the pasture is again a source of more protein than the marsh plants. Perhaps, in spring, the level of protein in the marsh plants is high enough that it ceases to be the limiting factor, so the birds stay where they are "happier".

A more recent study casts doubt on this idea. Bazely et al. (1991) have found that the barnacle geese, like the red deer on Rhum (Sect. 6.1.2), spend significantly more time feeding on sites where seagulls nest. And at these sites the *Festuca* plants have some 25% more nitrogen than those growing on nearby sites where the gulls do not breed. Clearly a higher level of nitrogen in their food is still important to these geese. It may be that once nitrogen in the *Festuca* rises above a critical minimum, it is as Ydenberg and Prins suggest: the lower levels of disturbance in their preferred habitat allows the geese to graze more slowly, and thus more thoroughly use their food plants.

The same group of researchers (Prins et al. 1980) investigated the interaction between the brent goose, *Branta bernicla*, and its food plants. This species also overwinters in The Netherlands and prior to departing to breed in the Arctic, large numbers of them gather in the spring in the same salt marshes where the authors studied the barnacle geese. There the brent geese graze on *Plantago maritima* and *Puccinellia maritima* rather than *Festuca.* Every 4 days they regraze the same plants, feeding almost exclusively on the youngest leaves. This regime ensures that the geese harvest the highest possible proportion of new plant tissue, and, by grazing as a flock, function as "an integrated harvester" maximizing a consistent return to the individual of food of the highest quality protein, rather than of biomass (McNaughton 1979).

Cargill and Jefferies (1984) demonstrated the same tactics by the lesser snow goose, *Chen caerulescens caerulescens*, feeding in the salt marshes of Canada's subarctic Hudson Bay. They highlighted the importance of nitrogen to successful breeding, and how the behaviours of selective feeding and grazing increased the availabilty of nitrogen in their diet. The plants they grazed had significantly higher concentrations of total nitrogen until late in the growing season. And, as Kotanen and Jefferies (1987) showed, they graze leaves that provide them with the maximum biomass of the most nutritious tissue per bite. They ate few very young and older leaves, and most fully expanded but recently mature leaves. It is the latter which have the highest protein and nitrogen content. The supply of available nitrogen is further enhanced by the fact that their favoured food plants in the salt marshes are the halophytes which accumulate high levels of soluble amino acids as a result of salt stress (Jefferies 1980). The feeding preference of the closely related greater snow goose, *C. c. atlantica*, is similarly influenced by the protein content of its food plants (Gauthier and Bedard 1990).

Like other geese snow geese process large quantities of food very rapidly through their gut. The goslings, during their period of exponential growth, retain a larger percentage of ingested nitrogen than do adults. The growth and quality of their primary food plants are, in turn, limited by nitrogen (Ruess et al. 1989). An essential link in the maintenance of a flow of nitrogen back to the swards grazed by these geese is the mineralization in the sward by cyanobacteria of the nitrogen in the faeces, returning it to the sediment where it is again available to the plants (Hik and Jefferies 1990).

Horsetails *(Equisetum* spp.) are an important, and often major, spring and summer food for several species of geese, including the Canada goose, *Branta canadensis*, the lesser snow goose in America, and the barnacle goose and others in Europe. Spring is a time when adults are still building reserves and forming eggs prior to breeding. Summer is a time when they are both moulting and replenishing reserves lost during reproduction. More importantly, summer is

Fig. 24. Snow geese, like many herbivores, maximize their access to readily assimilable nitrogen by eating the flush growth of plants that they repeatedly graze. Photo courtesy of R. L. Jefferies

the time when goslings are growing rapidly. Thomas and Prevett (1982) investigated the nutritional value of *Equisetum fluviatile* growing in the feeding grounds of the lesser snow goose at James Bay, Canada. They took samples during spring when geese were migrating into the bay, in summer when goslings were hatching, and in autumn when geese were migrating south again. They demonstrated that the succulent young emergent stems of horsetails provide a source of high protein food for the geese. Horsetails are among the first vegetation to appear after the snow and ice recede. They develop continually throughout the summer months, and are selectively grazed by the geese. They contain in excess of 20% protein with a composition of amino acids "conducive to protein anabolism" and with high levels of minerals necessary for breeding and growing geese. Their energy content (as neutral fats and soluble carbohydrates) is low and they have very little fibre. By the autumn their nitrogen content is not high, but by then the geese are seeking energy-rich food for fattening prior to migrating south.

Coleman and Boag (1987) worked with Canada geese foraging on their autumn staging grounds in the Yukon. There this species preferred to graze on the rhizomes of *Potamogeton richardsonii*, even although *Equisetum palustre* stems were freely available. *Potamogeton* rhizomes are more nutritious than the stems of *Equisetum*, and the geese would only eat the latter when the preferred rhizomes were flooded and they could not reach them.

Buchsbaum and Valiela (1987) found that breeding Canada geese at Cape Cod, Massachusetts, fed selectively on only some species of their primary food of marsh grasses, and only on some individual plants of these preferred species. This maximized their intake of protein. Goslings, with their greater need for nitrogen, were most selective and fed on high-protein plants much later into the season than did adult birds. Non-breeding birds did not show this selectivity, and in the summer all adults, but not the goslings, switched to a greater dependence on eelgrass, a better source of carbohydrate. This reflected the greater need of adults to build up energy reserves as distinct from the continuing need of the growing goslings for protein.

The growing goslings of cackling geese, *Branta canadensis minima*, in Alaska feed very selectively on arrow-grass, *Triglochin palustris* – up to five times its frequency in the feeding meadows studied [and comprising 68% of their diet (Sedinger 1984)] – despite its low abundance relative to other graminoid species. This preferred food plant contained nearly twice as much *available* protein (29%) as any other plants accessible to the goslings, and contained less cell wall material (Sedinger and Raveling 1984). The extent of this preference was further highlighted by the goslings selectively feeding on arrow-grass in mudflats where it contained more protein and less cell walls than it did in the meadows. If their ability to select just among species is considered (rather than among individuals within a species), the protein content of their diet was 37% higher in meadows, and 43% higher in mudflats, than random selection would have provided. These authors note that as a result of this very strong preference, the amount of food available to the goslings is less than might be supposed from the large extent of the sedge meadows in which they feed. There is some evidence that the supply of arrow-grass may be inadequate in years of great abundance of geese, and that the amount and distribution of this high-quality food may be an important determinant of the distribution and abundance of geese in these areas, in spite of the high overall availability of green plants.

Sedinger and Raveling (1986) further demonstrated that the concentration of nitrogen in arrow-grass (and other graminoids) peaked within a week of the end of the hatching period, in spite of a higher and prolonged peak in grazed plants. So, the nutritional quality of the diet of goslings declined as brood-rearing progressed, putting late-hatching broods at a considerable disadvantage. Sedinger (pers. comm. 1989) has recently confirmed the importance of the decline in plant protein to the growth of goslings. He has also found a seasonal decline in the growth rate of goslings of black brant geese, which nest sympatrically with the cackling geese.

Sedinger and Raveling had also calculated, however, that goslings that fed on green plants alone, and under the best foraging conditions measured, got less than the minimum level of protein needed for maximum growth rates. Even the first-hatched goslings obtained less than an adequate supply of nitrogen in their food, and goslings that hatched 1 week after peak hatching got from 20 to 30% less than those that hatched a week prior to the peak.

So, in spite of their special adaptations of feeding constantly on the best flush growth and passing it rapidly through their gut, very young geese are usually only just getting by. Most of the time their supply of protein is dangerously close to being inadequate. It is not surprising, therefore, that some goslings are found to eat invertebrates. Domestic goslings readily and commonly eat insects (M Wink pers. comm. 1989). For the rest, the evidence in the literature is sparse and conflicting. The species I have mentioned here have been studied in great detail and the consensus of those studying them is that invertebrate food is not significant in the diet of either breeding females or their young. Sedinger (1986), and Sedinger and Raveling (1984) did find insects in some wild goslings, and their captive ones readily ate insects and high-protein feed pellets. Jefferies (pers. comm. 1989) says that both adult and gosling snow geese eat insects, but they make up only a small component of the birds' diet.

Ydenberg et al. (1984) report that both brent and Canada geese are recorded as eating annelid worms, but Ydenberg (pers comm 1985) says barnacle goslings are totally herbivorous. Perhaps for those geese which feed on salt marsh halophytes the greater proportion of soluble nitrogen in these plants normally provides the necessary extra protein which allows them to survive on plants alone.

Ydenberg et al.'s observations of night feeding of barnacle geese might relate to this idea. The geese switch from feeding by day and roosting by night to feeding by night and roosting by day *only* during the few days of full moon. This behaviour persists even on nights of heavy overcast or dense fog, but they confine their visits to when the moon is above the horizon, and at around the time of high tide. The authors say it is unlikely that the geese are drawn by the availability of marine invertebrates at that time, even though this is the accepted explanation for similar behaviour by lapwings and curlews. They suggest it may be that the quality of the salt marsh plants varies in some way at this time. Could it be that changes in salinity or of times of emersion during the spring tides stress these plants to the extent that the level of soluble nitrogen in their tissues is raised sufficiently to make the monthly switch in feeding routine advantageous to the geese?

Alternatively, do birds (especially the very young) resort to eating invertebrates if the level of nitrogen in their food falls below a critical level? There is growing evidence for other herbivores as widely divergent as rodents, primates, parrots, and wallabies, that this source of nitrogen for breeding females and growing young is more important than previously believed. It needs only a very small proportion of the total intake of food of a small gosling to be animal tissues to make a huge difference in its protein budget, especially if the extra protein is tipping the balance from not quite enough to survive and grow to just enough to do so. If they eat soft-bodied annelids or insect larvae which can be reduced to nothing but a few setae 15 minutes after ingestion (Swanson and Bartonek 1970; Coleman 1974) such food could be readily overlooked, even in oesphageal samples.

The Australian black swan, *Cygnus atratus*, is probably another illustration of this. It is considered to be entirely herbivorous throughout its life (Braithewaite 1981 and pers comm 1985). Black swans are opportunistic breeders raising young whenever sufficient food is available. The major mortality is of cygnets soon after they hatch, and this in turn is determined by the amount and type of food available to them (Braithewaite 1982). Again, it is believed that their diet does not include any animal protein. Frith et al (1969) found much more animal food in the gizzards of young birds than of adult birds, but considered that this was due to accidental ingestion of invertebrates on the plants they ate. They acknowledged that their sampling would reveal only the hard parts of invertebrates eaten and would not detect soft bodied ones at all. In light of this, and of the amount of material found, it seems likely that very young cygnets may well be supplementing their vegetable diet with not insignificant amounts of animal protein. If this is so the black swan would be more akin to its Northern Hemisphere cousin, the mute white swan, *Cygnus olor*, of Europe. It too is limited in its success largely by the survival of its cygnets, and they depend for their first weeks of life on invertebrates (Braithwaite 1981). The fact that black swans take twice as long as white swans to fledge their young suggests that their cygnets may be able to get by on less animal protein than those of *C. olor*, once they are past the first crucial days of their lives.

7.1.2 European Grouse, Ptarmigan and Capercaillie

The Scottish red grouse, *Lagopus lagopus scoticus*, is probably one of the most thoroughly investigated herbivorous birds. On the treeless moors of northern Scotland where they live, heather, *Calluna vulgaris*, an evergreen dwarf shrub, predominates. The tips of the shoots of heather comprise over 90% of the birds' food. Traditionally, large areas of these moors have been managed to maintain high numbers of red grouse as a game bird by rotational burning of small patches of heather. This results in constant regeneration of new heather and maintains much larger stocks of the birds than could persist on unmanaged moors.

The general picture of the ecology of the red grouse on these moors has emerged from the long and fruitful studies of Adam Watson, Robert Moss and their colleagues at the Institute of Terrestrial Ecology in Banchory, and is given by Moss et al. (1975). Many subsequent studies have confirmed and expanded on the conclusions in this paper, and while there still seems to be no consensus as to what, ultimately, is determining the numbers of red grouse, it is clear that nitrogen plays a dominant role.

Changes in the number of young which are successfully reared each year are due to variations in the proportion of chicks which die a few days after hatching. This early mortality is due, in large part, to the quality of the eggs from which the chicks hatch. This, in turn, is determined by the quality of the hens' diet prior to laying their eggs, and the plane of nutrition of the laying hens depends on the level of nitrogen that they are able to obtain from their food. The density of birds does not influence their breeding success, but where they are living does. Grouse on "rich" moors overlying base-rich rock, such as limestone, raise, on average, more chicks per brood than do those living on "poor" moors over acidic granites; and their breeding success varies less than those on poor moors. Consequently, the density of grouse is greater and their numbers are more constant on the rich moors than on the poor moors. More species of plant grow on the rich moors and the heather there contains more nitrogen than it does on the poor moors. The rich moors also support greater numbers of hares and insectivorous pipits and golden plovers than do the poor ones (Moss at al. 1981).

The birds also breed more successfully in some years than in others. However, this was found *not* to be related to changes in the average nitrogen content of the heather in the spring. Rather, their breeding success was better when the heather was denser, and had been growing for longer before the hens finished laying. But, most importantly, it was better *the longer the concentration of nitrogen had been increasing in the tips of the heather* before the hens laid their eggs; and this before there was any physical elongation of these tips. In the spring, breeding hens also eat the growing tips of *Erica* spp. These contain less nitrogen than the heather tips, *but they begin to grow earlier.*

When a heather moor was experimentally fertilized, the grouse living there bred better and the nitrogen content of the fertilized heather increased; but the heather also *began to grow earlier in the spring.*

Moss (1977) found that in order to attain an adequate nitrogen balance red grouse have to eat large quantities of heather – far more than their energy requirements would dictate. As a result they are highly selective feeders. They eat only the growing tips of the heather shoots and are able to detect those richest in nitrogen. They thus can take in far more nitrogen than they could if they took the tips at random. Breeding birds are always selective for tips that are high in nitrogen. At other times birds are selective only if they are still growing, are in poor condition, or are living on the poorer sites (Savory 1983).

So, the picture emerges that on the best sites there is enough heather containing enough nitrogen to support successful breeding in most years. On such sites numbers of grouse are high and stable from year to year. But on poorer sites this is not so. Here, the success of the birds' breeding is susceptible to quite minor changes in the amount of nitrogen that is available to the hens before they lay their eggs. From year to year the success of their breeding and their abundance on the poor moors will vary in response to these changes.

Quite small differences in the amount of nitrogen they can incorporate into the eggs could make a large difference to the vitality – and thus survival – of their chicks. On these poorer sites only in years when there is more current growth on the heather, and when the heather starts to grow earlier, is breeding likely to be successful. Only then can hens gain access to sufficient concentrated soluble nitrogen *as it flows into the tips of the new shoots.* The key is not the total amount of nitrogen in the food plant; it is *the proportion of it which is available in a readily digestible form.* The understanding of this dependence on both the quantity and quality of the new shoots (i.e. the time during which they contain high concentrations of soluble nitrogen) also explained the "green world paradox" of birds not thriving in the presence of what at first sight appeared to be a large excess of food. They were, in fact, experiencing a relative shortage of good food.

Recently published work of the Banchory group lends support to my emphasis of the importance of the *form* in which nitrogen is available (Moss et al. 1990). Their analyses of a 7-year study of the relationship between changes in the total nitrogen of the food of the red grouse (heather), and the total and fractionated nitrogen of the birds' faeces, led them to conclude that the lack of correlation between total nitrogen in the food and the subsequent performance of the birds – something which has emerged from a number of studies – is "insufficient to rule out the possibility that digestible protein is in fact limiting".

In this connection they make some interesting observations about the influence of "winter browning" of heather plants. This is caused by cold and desiccating weather damaging and killing many of the tips of plants. Its extent varies from year to year; sometimes none is evident. But red grouse breed poorly following springs when there is much browning. They do not eat the browned shoots, only the remaining green ones. So, perhaps, the authors say, the damage might also have affected these residual green shoots. The nitrogen in them may have been altered so as to adversely affect the birds' capacity to metabolize it – perhaps it is more or less complexed with tannins, making it indigestible to the birds. I would suspect, from the evidence of feeding preferences and of breeding performance in relation to the flow of nutrients to new growth in spring, that they are right. Browning is restricting the amount of soluble nitrogen that is available in the remaining green shoots.

But, even with the best possible nutrition of their mothers, and the benefit of a higher protein yolk sac to tide them over the first few days after they hatch, chicks still need access to additional animal protein if they are to survive. Savory (1977) established from inspection of the crops of chicks taken in the field that while they mostly eat young tips of heather from the time they hatch, about 5% of their diet in the first 2 to 3 weeks of life is made up of insects. During this time they mostly foraged in areas where insects, and especially large soft-bodied Diptera, were more abundant than elsewhere. Other studies have confirmed this dependence of very young red grouse on animal protein in their diet (Lance and Mahon 1975). Savory found that young chicks also ate the spore capsules of mosses and the leaves and flowers of *Vaccinium* and

Rumex more frequently at this time, and were even more selective than the adults for high-nitrogen tips of heather shoots. After the first month the diet of the chicks was much the same as that of the adults, except that they continued for some time to take insects. All this suggests that young red grouse have a fairly pressing need to get as much nitrogen as possible in their diet.

A great deal of research has also been conducted in Scandinavia on the European subspecies, the willow grouse, *Lagopus lagopus lagopus*. In Norway, several studies on the small island of Tranøy, far to the north of the Arctic Circle (69° 09'N), have revealed the heavy dependence of willow grouse on the protein nutrition of the very young. Spidso (1980) found that while chicks ate plants from their first day of life, 60% of the dry weight of their food in the first 3 days was insects. By the time they were 3 weeks old this had dropped to only 4%. The most dominant component of their insect food was the caterpillars of geometrid moths. These varied considerably in abundance from year to year but were strongly preferred by the chicks. Most of the rest of the insect component of their diet was made up of small adult flies (Diptera) with the chicks selectively eating the largest available.

The selection of the reproductive parts and young growth of plants by these chicks reflected a further seeking after nitrogen as well as for other essential nutrients [e.g. the foliage of the blueberry, *Vaccinium myrtillus*; growing chicks are unable to produce sufficient vitamin C to fulfil their requirements (Hanssen et al. 1979) and blueberry is especially rich in ascorbic acid].

Myrberget (1981) similarly found that the diet of the chicks on Tranøy was dominated by insects in the first week of life (96%), while reproductive parts of plants accounted for 64% of all plant material they ate (83% of it in chicks less than 2 weeks old). He also found that geometrid caterpillars were by far the most important component of the insect food; so much so, in fact, that in 3 separate years (1966, 1978, and 1980) when these looper caterpillars were in outbreak numbers on the island, the growth rate of willow grouse chicks was especially high. Erikstad and Spidso (1982) found that when the proportion of these larvae in the diet of chicks fell from 80% to 20% from one breeding season to the next, the survival of chicks to the end of the summer dropped from a mean of 4.0 per brood to 0.5 per brood. Erikstad (1985) pointed out that variations in survival of chicks seems, therefore, to be most influenced by two factors. One is the quality of the egg and of the invaginated yolk derived from it; this in turn is dependent on the nutrition of the hen. The other is the quantity and quality of insect food available for the chicks in the first few days after they hatch; this will be influenced by the weather at this time because the weather affects both the availability of insects and the mobility of the chicks in their feeding. The evidence from these studies on Tranøy Island would indicate that the availability of insect food for chicks had the greater influence of the two. But the work of Steen et al. (1988), studying willow grouse in central Norway, showed that the nutrition of the hens can also have a major impact. They found that in one year when the survival of chicks was exceptionally poor, this was attributable to inadequate nitrogen nutrition of their mothers. The birch and willow twigs which stick out of the snow and make up the almost exclusive winter diet of these birds normally start to sprout new leaves in the weeks before the hens lay their eggs in late May. Sprouting twigs contain 25 to 40% more amino acids than they do before they sprout, providing a significantly increased level of protein in the diet of the hens. In years when the survival of chicks was low, however, they did not sprout until after the grouse had laid their eggs. These eggs were smaller, poorer quality ones, and produced smaller, less viable chicks than those laid in years when twigs sprouted before oviposition. The similarity of this response to that of the red grouse to changes in the nitrogen content of tips of heather as they are growing is obvious.

Steen and his colleagues noted that other studies with willow grouse indicated that the protein nutrition of hens can be a significant influence on the number of young that survive. For example, captive hens fed on a diet with 20% crude protein produced larger and more viable chicks than did those fed on 15% crude protein, and chicks hatched from field-collected eggs and kept without food survived for significantly different times from year to year.

However, even for birds that do manage to survive to the end of summer the struggle is far from over. They must continue to seek out food which will provide not simply enough energy to withstand the rigours of the long winter, but sufficient protein for both daily maintenance and for breeding in the coming spring.

Pulliainen and Iivanainen (1981), working in the far north of Finland, found that the willow grouse constantly selected for protein-rich food in the winter. Shoots of *Betula* spp. made up 77 to 98% of the diet and shoots of *Salix* spp. 9 to 22%. The protein-rich catkins of birch were preferred ahead of all foods, but their abundance was very variable from year to year. An indication of the importance of a high nitrogen diet in winter was that after one breeding season, which followed a particularly abundant crop of birch catkins, there was a great increase in the percentage of juveniles in the population. These field observations confirmed the findings of other workers as to the major importance of birch as a winter food. Also, Pulliainen and Salo (1973) had already demonstrated that birds brought back to the laboratory actively selected not only the species of plant but also the part of the plant they would eat, favouring tender new growth with the highest content of nitrogen.

It is clear, as Myrberget (1984) points out that variations from year to year in the number of juveniles determines changes in breeding numbers of the willow grouse. It is necessary, therefore, to find out which factors determine these variations in annual rates of productivity if we are to understand what is responsible for changes in their abundance. He says the phenology of the plants (and thus of insects) in the spring is a major factor because it affects the quantity and quality of the food of hens, before and during laying, and of the chicks in their first weeks of life. However, the phenology of plants in spring is determined by variations in the weather, and if the weather varies at random how can this result in cycles of abundance (see Chap. 17)?

The closely related ptarmigan, *Lagopus mutus*, exhibits an ability to discriminate similar to that of the red and willow grouse. Its food in winter and spring is not simply whatever is readily available. It shows strong selection for foods rich in nitrogen and phosphorus, and of high digestibility. By far its most preferred food, whenever available, is birch catkins. These have the highest content of digestible protein of any of the foods eaten by ptarmigan. Where these are not available other high nitrogen foods, especially reproductive tissues and new growth, are selected. This is true of the species in Finland, Iceland, Scotland, and Alaska (Moss 1968; Gardarsson and Moss 1970; Pulliainen 1970). There is also evidence of a sexual difference, females selecting more nitrogenous foods like herbs, than do males which tend more towards a diet of berries and other high energy foods (Pulliainen 1970; Gardarsson and Moss 1970).

Just how important the amount of available nitrogen in the winter/spring diet is to the success of the breeding of ptarmigan hens was demonstrated by Moss and Watson (1984). As with the red and willow grouse, the breeding success of ptarmigan is determined largely by the proportion of chicks which die within the first 2 weeks after hatching. To find out how much the chicks' chances of survival are influenced by the nutrition of their mother and the consequent quality of the eggs she lays, Moss and Watson collected eggs from various populations over several years and raised the chicks in captivity under standard conditions. They then compared the results of these rearings with measures of the breeding success and diet of hens in the field. Virtually all the chicks reared in captivity which died did so in the first 10 days after hatching, and from no observable cause. The numbers that died varied markedly between years and localities, and in parallel with variations in the breeding success of the populations from which the eggs had been taken. So, it would seem that the survival of chicks, both in the wild and in captivity, was determined in part before the eggs hatched.

Breeding success and survival of chicks in captivity were both better for populations of ptarmigan living in areas overlying basic rocks than for those living in localities with acidic granites. These better areas had more of the most preferred blueberry, *Vaccinium myrtillus*, which had a higher nitrogen content than the less preferred heather, *Calluna vulgaris*, and crowberry, *Empetrum hermaphroditum*. But all three species, which make up 96% of the birds' diet, had a higher content of nitrogen when growing over these better basic rocks than when growing over granites.

The breeding success and survival of captive chicks also varied considerably from year to year. However, the variation was not related to annual variations in the total nitrogen content of the three main food plants, but to *the number of days they had been growing before the hens had finished laying*. Not only does the total amount of nitrogen rise rapidly in growing plants, but as it is translocated into the growing tissues it is in a soluble, highly digestible form.

Twenty years earlier Watson (1965) studied the ptarmigan living on the poor granitic areas. He found that in years when the number of birds was increasing, and in the early years of a peak in their population, their breeding success was high, but in the later years of a population peak, and in years of declining numbers which followed the peak, breeding was poor. He also showed that predation of adult birds was not limiting numbers of breeding adults nor was predation of eggs and chicks of any significance in depressing the breeding production of the population. Fairly clearly, the major influence on the success of the breeding of these birds, and thus their abundance, is the quality of the food available to the hens and their newborn chicks (see also Part V).

The capercaillie, *Tetrao urogallus*, in northern Finland's brief summer feeds on a wide variety of plant species and parts, and on some animal food, but eats nothing but needles, shoots, buds, and cones of *Pinus sylvestris* in the winter (Pulliainen 1979). This provides a much lowered level of nitrogen in the diet but the birds can, and do, detect needles with the highest nitrogen content. The summer diet of the juveniles contains a significantly higher content of nitrogen than that of the young adults, and in the autumn and early winter, when the young juveniles are still gaining weight, they take longer than the adults to switch to a diet of nothing but pine.

Although plant food comprises the bulk of the nutrition of the adult throughout the year, animal food is of vital importance for the growing chicks. Like those of most other tetraonids, capercaillie chicks feed almost entirely on animal food for the first 3 weeks. Even after that the growing birds continue to take small but nutritionally significant amounts. Mostly they eat insects, especially ants, but also spiders, snails, and in one case, a frog. The juveniles eat twice as much as the adults, and eat it later into the autumn (Pulliainen 1979). It is also interesting that in this highly sexually dimorphic species (cocks are almost twice the size of hens) the juvenile males, which have reached only 69% of their adult weight by autumn, continue to eat more animal food than do the females, which by then have attained almost 90% of their adult weight (Pulliainen 1979). This also relates to the findings of Wegge (1980) for a population of capercaillie in Norway. The numbers in this population had been declining over a period of some 10 years. Brood size during the decline was less than 40% of the mean, and female chicks outnumbered male chicks by more than two to one. This deviation from a 1 : 1 sex ratio increased as broods became smaller. The most likely explanation for this, Wegge concluded, was not to do with predation or dispersal, but with food. A decline in the quality of the food available would result in male chicks, which must grow so much faster than their sisters, starving.

More recent work in Norway, using direct observation of chicks imprinted on to the human experimenter, rather than examination of the content of crops, has confirmed these findings. Spidso and Stuen (1988) observed that almost 60% of the diet of chicks in their first week was insects. Ants were most commonly eaten, but there was no question of chicks simply eating whatever they came upon. There were less insects available when the chicks were 1 to 3 weeks old than when they were 4 to 6 weeks, by which time the proportion of insects in their diet was falling sharply. By their seventh week they were eating almost 100% plant tissues, but strongly preferring berries (85% of plant material eaten) and flowers.

The authors note similar results by others working with imprinted chicks in Norway, and that the chicks of the closely related black grouse, *Tetrao tetrix*, in Scotland have been found to survive much better when insects are more abundant.

Having survived to become a full-sized bird, the chances of an individual capercaillie entering a breeding population the following spring are still very limited. Moss and Weir (1987) found that in a Scottish population, apart from the risk of being shot, a big but variable proportion of the chicks reared in the spring disappeared in late summer and autumn in most years. This is similar to findings for Norwegian populations. Apparently, many of the young must emigrate to less favourable locations where their fate has not been recorded. It would seem most likely that, like the young of other animals, these emigrants will die unless they can replace older birds that have died, or they happen upon a location where changed circumstances have increased the capacity of the environment so that it can support some capercaillie (see also Chap. 11, especially Sect. 11.3).

7.1.3 North American Grouse

The same pattern of the importance of selective feeding for high nitrogen to the success of breeding by hens, and of animal protein in the food of the very young, is found in the populations of native North American tetraonids.

Pendergast and Boag (1970) found that adult spruce grouse, *Dendragapus canadensis*, in Alberta, Canada, feed in the winter exclusively on the needles of *Pinus*, preferring these to those of either *Picea* or *Abies*. They will also eat berries of ground plants if these become uncovered by snow drift. In the spring, however, they eat new growth and fruit of deciduous plants, and arthropods. They also now seek out the few *Picea* trees growing among many pines, and feed upon the expanding buds and growing tips of these spruce. More recent work has confirmed this strong switch from the winter diet of pine needles. Naylor and Bendell (1989) found that in spring hens supplement their winter diet of 100% conifer needles with new buds and shoots, flowers and fruit of deciduous ground plants, spore capsules of mosses, and arthropods. They feed selectively for high protein prior to breeding. This change from winter to spring diet and its association with selection of plant tissues highest in available nitrogen is widespread among the tetraonids of both the Old and New World.

Hatchling spruce grouse also feed on insects, and as summer juveniles continue to eat significant quantities of them. They also eat fruit, new leaves, and flowers (especially of legumes), and fungi (including their contained insects), and do not start to eat conifer needles until late in the summer (Franceschi and Boag 1991). It is not until autumn that the level of animal protein in their diet falls to the same as that for adult birds. By early winter they are, like the adults, feeding exclusively on pine needles.

Gurchinoff and Robinson (1972) found that while eating nothing but pine needles during winter, spruce grouse nevertheless selected needles which were significantly higher in protein. They ate needles in only some selected trees

Fig. 25. A Canadian spruce grouse. An understanding of its ecology may help resolve whether breeding success or territorial behaviour determines the density of grouse populations (see Sect. 7.1.6). Photo courtesy of S.D. Mac Donald

which were significantly older than trees they did not choose. Furthermore, in most of these preferred trees they confined their feeding to the south side of the trees. The authors likened this behaviour to that reported for the capercaillie and the black grouse, *Tetrao tetrix*, in Finland. These birds feed at a preferred height in the same few pine trees year after year, and often overgraze these while plants nearby remain untouched (Seiskari 1962). Also, ruffed grouse, *Bonasa umbellus*, in Minnesota similarly select particular aspen trees. In both these cases the preferred trees were poorly growing and with fire scars or crooked stems. Moss (pers. comm. 1973, 1992) says that workers in Scandinavia and Russia have found this preference for particular trees is common among tetraonids, and that the chosen trees are often off-site or damaged in some way (ie. stressed). Other evidence (White 1984) would suggest that the needles of such trees may well have increased levels of soluble nitrogen.

Hoffmann (1961) had found the same behaviour for winter-feeding blue grouse, *Dendragapus obscurus*. They selected needles from the same few trees year after year, ignoring others nearby. Within these preferred trees they fed high up and on the south (sun) side of the crowns. All three levels of preference – tree, height, and aspect – were related to higher levels of nitrogen in the selected needles. This elevated level of nitrogen recorded in the preferred foliage is probably a similar reflection of the greater stress and rate of senescence of the chosen needles (Fennah 1965; White 1970a). They also switch from a winter diet of conifer needles to a spring and summer diet of broad-leaved vegetation. King and Bendell (1982) found that blue grouse on Vancouver Island seek out young growth and flowers and fruit, taking food higher in nutrients, especially nitrogen, than that available from random feeding. Hens eat more high-protein food than the males, especially prior to egg-laying. They also eat a small but continuous amount of invertebrates which males do not. Chicks feed almost entirely on invertebrates in their first 3 weeks, and continue to eat more than adults until they are almost fully grown. There was, nevertheless, a very high mortality of chicks in the first week after hatching, although the level of this loss varied greatly from one year to the next. King and Bendell noted that this suggests that the nutrition of both hens and chicks may more often than not be inadequate, and could result in the starvation of newly hatched chicks being a principal cause of limitation of numbers in the population.

Doerr et al. (1974) had found that in Alberta the ruffed grouse, *Bonasa umbellus*, which feeds in winter almost exclusively on the male buds of poplars and willows, selected those buds with the highest nitrogen content. Further south in Virginia, Servello and Kirkpatrick (1988) reported *B. umbellus* similarly preferentially eating leaves of herbs and expanding buds of deciduous plants in spring when their need for an adequate protein diet for breeding is greatest.

Beckerton and Middleton (1982) noted this selection and set out to test the likely influence of the level of protein in the diet of hens on their breeding success. They fed captive birds on a series of five diets, each with the same calorific value but containing from 7.6 to 20.1% dry weight of protein. They found that increasing the protein in the diet increased the duration and rate of egg laying, the number of eggs laid, and their weight. It also increased the weight of chicks at hatching, the success of hatching, and their subsequent survival. Many of the responses to increased protein in the hens' diet showed signs of reaching an asymptote at the higher levels. This would indicate that once there is an adequate level of protein in the hens' diet further increases are of little benefit. Yet 72% of the chicks that died did so within the first week, and 85% by the end of the second week. Ruffed grouse chicks feed on virtually nothing but arthropods in their first week (Pendergast and Boag 1970), so this, again, suggests that no matter how good their mother's nutrition, a high level of animal protein in their diet is essential to the young chicks' survival and adequate growth.

Long ago (and long forgotten?) Cridle (1930) pointed out that ruffed grouse need insects to successfully rear their young. He also pointed out that the correlation of peaks in their year-to-year abundance in Manitoba with outbreaks of grasshoppers probably represented maximums of the changes in survival of their young in response to changes in the general availability of insects as food for the chicks.

So it probably is with all tetraonids. The generality of this need for protein for neonates has recently been experimentally reinforced by Johnson and Boyce (1990) for chicks of the sage grouse, *Centrocercus urophasianus*, in Wyoming. They fed captive wild-caught and laboratory-hatched chicks diets of ad lib. quantities of a mix of the plants that they eat in the field with or without supplements of invertebrates (again, a mix of those they encounter in the field). Without insects all died in 4 to 10 days from hatching; with insects they all survived to 10 days. For the first 3 weeks from hatching the chicks must have insects in their diet to survive. Beyond that they could survive without them but their development was curtailed. The quantity of insects in the diet was correlated with their survival and growth.

7.1.4 Partridges and Pheasants

The same picture of a dependence on a diet of animal protein for the young emerges for the omnivorous but largely herbivorous phasianids. Chicks of the grey partridge, *Perdix perdix*, in southern England feed almost exclusively on insects for their first 3 weeks of life (Blank et al. 1967). In the same period more than 70% of them die and the number which die each year is the major influence on fluctuations in the abundance of the population. In turn, the abundance of insects in that crucial early feeding period overrides predation and variation in the quality of the eggs as the factor most influencing how many chicks die in the first 3 weeks. Potts' (1980) 10-year study of the partridge on the South Downs of Sussex confirmed this. With the advance of modern farming techniques since the 1950s, particularly the use of pesticides, the number of partridges has steadily declined. Year-to-year variations in the survival of chicks could be attributed to the influence of spring temperatures on the abundance of insects available as food for them. Green (1984) found that broods of partridge chicks foraged mostly on the edges of cereal fields where arthropods and weeds are most abundant, that their survival increased with the increased abundance of arthropods and with increased June-July temperatures. Rands (1985) confirmed experimentally that crops with a 6 m strip around the edge of the field left unsprayed with herbicides, fungicides, or insecticides had higher densities of arthropods and much larger broods of partridges (a mean of 6.38 vs 2.15 chicks per brood). The same dominance of insects in the diet of the young was found for populations of *P. perdix* established in North America (Hupp et al. 1988).

Hill (1985) demonstrated the same story for the ring-necked pheasant, *Phasianus colchicus*, in England by measuring the content of the faeces of radio tagged birds. This is another omnivorous bird, but the adult diet is dominated by seeds and vegetation. Hens do eat insects but only in small amounts (about 4% of that eaten by chicks). For the chicks, however, not only is 95% of their diet arthropods, their chance of survival depends on getting enough insects, and of the right kind, in their first 6 days after hatching. Variation in the density of arthropods in their home range explained 75% of the variation in survival of chicks, and 2-week old captive chicks preferred to eat larger, soft-bodied insects from among a range collected in the field and offered to them. So, even at this carnivorous stage of their life, selective feeding is likely to be advantageous (Whitmore et al. 1986).

Clearly, dependence on animal protein for successful breeding is widespread and general in the galliforms (Crowe 1978) and is not restricted to the herbivorous tetraonids.

7.1.5 Galliforms as Hindgut Fermenters

All these galliform birds exhibit another adaptation which enhances their capacity to extract useable nitrogen from their environment. They are hindgut fermenters with a pair of well-developed caeca containing a bacterial microflora (Gasaway 1976). As in other hindgut fermenters the caecal microorganisms break down the cellulose cell walls, allowing access to the cell contents; synthesize essential amino acids; and recycle nitrogen from the breakdown of metabolic uric acid. This recycling can account for 10% of the birds' total uptake of amino acids (Mortensen and Tindall 1978, 1981).

The uric acid is transported as urine (up to 20–30% of it) from the cloacal area to the caeca in a retrograde flow of digesta, resulting from anti-peristaltic movement (Fenna and Boag 1974a; Gasaway et al. 1975; Bjornhag and Sperber 1977; Moss and Hanssen 1980). The caeca and intestines are longer in the browsing tetraonids than they are in the variously omnivorous quails, partridges, pheasants, and turkeys. Tetraonids are thus better equipped to survive on their relatively low quality bulk food (Leopold 1953). As Pulliainen (1976) found with willow grouse, the smaller females with a relatively greater length of small intestine and caeca than the heavier males, are better able to cope with having to consume relatively more food of a given quality than are the larger birds.

In addition, the galliform birds have an adaptation like that found in lagomorphs and some rodents, of separating small particles of food in the colon and concentrating them in the caecum while coarser particles pass out with the faeces. They, also, like lagomorphs and rodents, produce two types of faeces, one containing the coarse particles and the other soft caecal material (Bjornhag and Sperber 1977).

However, unlike the mammals, the tetraonids apparently do not practise coprophagy (Moss and Hanssen 1980). It seems strange if this is true, even although they do absorb significant quantities of nutrients from the caeca (Obst and Diamond 1989). The obvious way for chicks to be inoculated with their caecal bacteria is by eating their mother's caecal droppings; they would at the same time gain a significant boost to their intake of protein. So too would the adults, especially the breeding hens. Moss et al. (1990) report that the caecal faeces of the red grouse contain more than twice as much nitrogen as do their woody faeces, and almost all in the form of bacteria or their end products. But the mounds of droppings under the night perches of red grouse contain only a few caecal faeces among many woody ones – they comprise a mere 12% of the total dry matter output. For chicks and adults alike, then, it seems strange that, with a staple food plant as low in nitrogen and high in polyphenols as *Calluna*, they should waste such a rich source of readily digested bacterial protein. Does the small amount of caecal faeces produced indicate that most of this good food *is* absorbed before it leaves the digestive system (Fenna and Boag 1974b), or do the birds indeed eat these droppings but nobody has observed this?

7.1.6 Changes in Abundance of *Lagopus* Species

In the light of all the foregoing evidence for the vital importance to the breeding success of galliform birds of nitrogen nutrition of the hens and chicks, it is interesting to look at a recent controversy in the literature about what processes are responsible for changes in the abundance of some tetraonids. It serves to unite what I have just been reporting about this importance of proteinaceous food and what I say in Section 11.2.2 about the influence of food on their territorial behaviour. It also adds to the wider focus on the generality of the inter-relationship of these two aspects of animal ecology.

The argument is between two groups of scientists, both with many years' experience of intensive study of populations of tetraonids; one in Canada and the other in Scotland. The difference in opinion revolves around which of two processes determines the annual change in numbers of breeding birds in populations of species of the genus *Lagopus*. Data are analysed from populations of three species (the red and willow grouse and the ptarmigan) studied in Alaska, Canada, Scotland, Iceland, and Norway. Discussion by the Canadian group is said to be addressed only to influences on annual changes in numbers and not to differences in the density of populations in time and space. Clearly, however, whatever influences the former must influence the latter (see Part V).

The Canadians (Bergerud et al. 1985; Bergerud 1987) argued that the annual success of breeding – the number of young fledged each year – directly influences the number of juveniles that survive the winter to breed in the following spring. Thus, they say that the proportion of the habitat which is occupied, and the size of the territories established thereon, are a *consequence*, and not the cause, of the number of young recruited to the breeding population each spring. To them territorial behaviour is "...a breeding tactic to space birds relative to environmental requisites..." but does *not* result in significant numbers of surplus birds which are doomed to an early death.

The Scottish group (Watson and Moss 1987; Moss et al. 1988) on the other hand, maintained that the density of birds each spring is determined directly by the size of the territories the birds establish. The size, and hence the number, of these territories is influenced by the nature of the habitat and by the inherent spacing behaviour of the birds which, in turn, may be influenced by their nutrition (sect. 7.1.2). The number of young fledged each year is therefore of little consequence, because usually more are produced than there are territorial vacancies to fill. So, of the many potential recruits, those which are unsuccessful in acquiring a territory are expelled and must emigrate and/or soon die.

Much manipulation of both data and words have gone into both sides of this debate, but there is still no agreement. Is "breeding success" or "territorial behaviour" the final arbiter?

As an aside, a third group of workers (Boag and Schroeder 1987) concluded from their 21-year study of the southwestern subspecies of the spruce grouse, *Dendragapus canadensis franklinii*, in the foothills of the Rocky Mountains in Alberta, that they needed to invoke *both* these hypotheses to explain *their* observations. They found that when spring numbers of spruce grouse were below the mean of the 21 years of their counts, the birds' breeding success in the previous spring seemed to influence the numbers breeding in the current year. On the other hand, when the number in the spring population was higher than this 21-year mean, there was evidence that the birds' spacing behaviour was limiting recruitment to the breeding population.

They also recorded three other things. First, when numbers had been increasing there was a high proportion of yearlings in the population, and when numbers had been declining there was a low proportion. Second, in spite of considerable fluctuations, there had been a long-term decline in the size of the population over the 21 years of the study. Third, this decline paralleled advancing maturity (characterized by the increased size and dominance of pines) of the mixed pine/spruce/poplar fire-sere forest in which the birds live. The authors attributed the decline in abundance to (unspecified) changes in the favourability of the habitat consequent upon its advancing maturity.

While I wholeheartedly support the calls by the proponents of both hypotheses for more experiments and less talk, their experiments are likely to be unproductive if they are testing the wrong hypotheses, and I believe that all three groups have missed the main point. However, the key to the resolution of their differences probably lies in the latter three observations of Boag and Schroeder. Consider the following: Ultimately it is the capacity of the habitat to support the birds – the resources it provides – which decides how numerous the birds will be in that habitat. This capacity will vary in both the short and the long term. The most commonly limiting resource in the birds' habitat is enough protein food in a form that is available to breeding females and growing young. The number of young that survive to breeding age will vary as the supply of this food varies for both. The number and size of the birds' territories will vary according to the amount of this food that is available in the habitat. That is, both breeding success *and* territorial behaviour are *consequences* of the variable, but mostly chronic, shortage of protein food. Yet the former will have the major influences on the abundance of the birds; and the latter only a secondary modifying affect.

Jenkins et al. (1967) working with the red grouse of Scotland 25 years ago, saw that this was so. They said, "Clearly our results in general show that territorial behaviour does limit breeding populations. However on most moors it is so greatly modified by high or low production of young birds that territorial behaviour appears to be less important for population control than breeding success. These differences reconcile in this species the views of both those who believe that territorial behaviour does limit numbers (e.g. Wynne-

Edwards 1962) and those who believe it is of doubtful importance (e.g. Lack 1954)".

In the specific case of Boag and Schroeder's population of spruce grouse, if there is an increase in the amount of good food, more young will survive and the population will grow (high proportion of yearlings in an expanding population); and vice versa. Whether numbers are above or below the average of their 21 annual counts means nothing, because the average means nothing in nature. It is simply the mean of the 21 annual counts which they made and is dictated by those counts, not they by it. What matters is whether the numbers are going up or down from one season to the next, and why.

If the population is expanding after a "low", because there is more food available in the habitat, there are likely to be new territories being established on previously inadequate parts of the habitat; and a shrinking of existing territories as previously adequate habitat gets better. So there will be little or no evidence of territorial behaviour excluding young breeders – little evidence of a doomed surplus. At times of a decline in numbers, however, because food is becoming scarcer, the reverse is likely – there will be fierce competition for a shrinking resource, fewer and larger territories, and more young driven out. If the habitat changes gradually over a long time (maturation of the forest) so that it provides less good food (less understorey, less broadleaf buds. And less insects?), the population will gradually decline.

Compare this story, both here and in Section 11.2.2, with that of the response of the Grants' Galapagos finches to the sudden explosion in abundance of their food (Sect. 7.3.2), or with the results of the manipulations of the chipmunks' territories by Mares and Lacher (Sect. 11.2.1), of Whitham's poplar aphids and their population of "floaters" (Sect. 11.2.3), of Riechert's spiders (Sect. 11.1.4), or with my general discussion of philopatry (Sect. 11.3). Furthermore, the parallels between this story of tetraonid birds and that of the snowshoe hares (Sect. 6.2.5.3) must surely be apparent.

So this is the hypothesis that further experiments should test:

The amount of nitrogenous food available to breeding females and young chicks is what determines the size of the population from one year to the next. Usually there will not be enough of this food for all that are trying to use it. The birds' territorial behaviour will ensure that as much as possible of what *is* available is used effectively, by restricting the number of birds that gain access to the supply (see Sects. 1.2 and 11.2.2).

It seems to me from the results of the studies I have outlined here that changes in two things in the environment of these birds emerge as crucial to an understanding of the problem. First is the quantity, quality, and timing of the flow of soluble nutrients into new growth and inflorescences of the birds' food plants. Second is the availability of suitable arthropods when the chicks hatch. All else is likely to be driven by these and they, in turn, will be driven by the weather.

At the same time, due note should be taken of the caution of Boag and Schroeder – studies need to encompass both large-enough areas and long-enough times to detect what is really happening in natural populations – and those of Bergerud et al. echoing my previously expressed concerns – it is dangerous to extrapolate from data from intensively managed artificial populations (where the supply of good food has been greatly increased and maintained at a high level) to explain what is determining the abundance of populations in the natural environment (where good food is nearly always in chronically short supply).

There is one further caveat. The territorial behaviour of the male grouse seems not to be the appropriate behaviour to measure. It may have an influence on the nutrition of the hens, but it may not. Their fitness to produce quality eggs has probably already been decided by the time they mate. It certainly cannot influence the neonates' nutrition; the home feeding ranges established by the hens after their eggs hatch will do that. As Lance (1978b) pointed out, hens with newly hatched chicks move quickly away from their nests and their subsequent movements bear no relationship to where they fed when they were incubating their eggs. Mostly they and their chicks feed outside the territory of the parents – and the cock stops defending the territory and accompanies the hen and chicks (see Sect. 11.2.2)! Possibly much more attention should be paid to the diet of the new hatchlings and how their mother's behaviour modifies this, rather than to her diet. Weak hatchlings may survive better if they have access to an abundance of rich food as soon as they hatch; the most robust chick, from the best possible egg, will not last long without it.

7.1.7 The Takahe

Another herbivore which is far removed both taxonomically and geographically from the tetraonids, yet shows the same dependence upon protein food for its successful breeding, is the flightless gallinule from New Zealand, *Notornis mantelli*. Long thought to be extinct, and still very rare, the takahe is confined to the rugged mountainous fiordland of the southwest of the South Island. In the summer these birds live in alpine tussock fields above the tree-line where they feed on tussock grass (*Chionochloa* spp.), a mountain daisy *(Celmesia petriei)*, and grass seeds when they are available. When snow forces them down into the *Nothofagus* forest in the autumn they begin to feed on the rhizomes of a fern, *Hypolepis millefolium*, which makes up 60 to 80% of their winter diet. In the grasslands they feed mostly on the tussock grasses but preferentially on the least abundant species of *Chionochloa*, and on individual plants within a species. This selection is according to the level of nutrients in the plants and especially for the nitrogen content (Mills et al. 1991). This results in a large proportion of what looks like a great excess of food being left untouched because it is not nutritious enough (Mills and Mark 1977). However, they are even more selective than this, taking only 2 to 3 cm of the basal meristematic tissues of the tussock grasses, stripping away and discarding the sheathing leaves. This meristematic tissue contains about twice as much nitrogen as the older leaves (Williams et al. 1976), and, of course, the nitrogen is

Fig. 26. The New Zealand takahe feeds selectively on protein-rich basal meristems of tussock grass, but feeds its young on insects exposed by clearing characteristic "scrapes" in the litter. Photo courtesy of J.A. Mills

in much more readily available soluble form, just as it is in the growing shoots of heather preferred by the red grouse.

In the autumn *Notornis* starts to feed on the rhizomes of *Hypolepis*, when the starch and nitrogen contents rise steeply as the fronds die off and nutrients are translocated to storage in the rhizomes for the winter. These rhizomes thus provide both a good protein source and a high level of carbohydrates for the high metabolic requirements of thermoregulation in sub-freezing temperatures (Mills et al. 1980).

Yet, winter and summer, the survival of the species is tenuous. In some parts of their distribution hens nest and lay eggs regularly, but never produce chicks. Many pairs fail to produce chicks even on better sites where their preferred foods are more abundant. Breeding success is generally correlated with the quality of the site (Kean 1956; Mills and Mark 1977). The level of nutrients in their preferred tussock grasses is low by comparison with improved pasture grasses. The birds have a rapid transit of food and leaf fragments in their faeces appear structurally unchanged. They spend more than 90% of the daytime in spring and summer foraging for food. Whenever they do manage to produce chicks these depend almost entirely on insects to survive. Adult birds scratch away moss and litter from the ground in order to expose insects for the chicks to eat. In doing so they produce characteristic "scrapes" several square metres in extent, and found only where and when the birds have young chicks (Gurr 1951).

7.1.8 The Hoatzin

The South American hoatzin, *Opisthocomus hoazin*, is another apparently obligate folivorous bird, but one which has evolved a very different strategy for maximizing the amount of available nitrogen in its food from that used by the other leaf-eating birds I have discussed (Grajal et al. 1989). More than 80% of its diet consists of green leaves; the rest is flowers and fruit. Of these leaves 90% come from only 17 species of plants, and 56% of the leaves they were observed to eat in the breeding season came from only four species, including two high-protein legumes, *Lonchocarpus cruciarubieres* and *Acacia articulata*. Within these species they feed preferentially on the buds, shoots, and new leaves. Analysis showed that the eaten portion of their food was significantly more nutritious than the uneaten part, with higher protein and lower lignin and cellulose. So, like all herbivores, they feed selectively on high nitrogen tissues.

Furthermore, the hoatzin is a foregut fermenter, and the only bird known to have evolved as such. The much enlarged crop and oesophagus have developed as the site of fermentation. The crop is still highly muscular with interior cornified epithelial ridges. So the bird is essentially a ruminant, but achieving both "chewing" and fermentation at the same site! Experimentally, an extract from a hoatzin crop digested cell wall material as efficiently as an extract from the rumen of a cow. Like other gut fermenters they have a long transit time of the food with differential retention of fine particles in the crop and oesophagus.

The young do not eat any animal matter. They are, however, like the young of many other herbivorous birds (Sects. 7.2 and 7.3), fed on regurgitate while in the nest (Grajal pers. comm. 1989). This no doubt serves to inoculate the nestlings with gut microflora, but it probably also contains enough microbial material to provide a significant protein supplement to their plant diet. As has often been pointed out, flight is incompatible with long retention of a lot of food in voluminous fermentation chambers (Morton 1978). Not surprisingly, therefore, the hoatzin is a poor flier. The sternum is much reduced to make room for the much enlarged crop and oesophagus and consequently the area for attachment of flight muscles is reduced. Life as a herbivore entails many sacrifices!

7.2 Birds Eating Nectar and Fruit

One group of birds considered (at least in more popular accounts of natural history) to be specialist and obligate herbivores are the tropical hummingbirds.

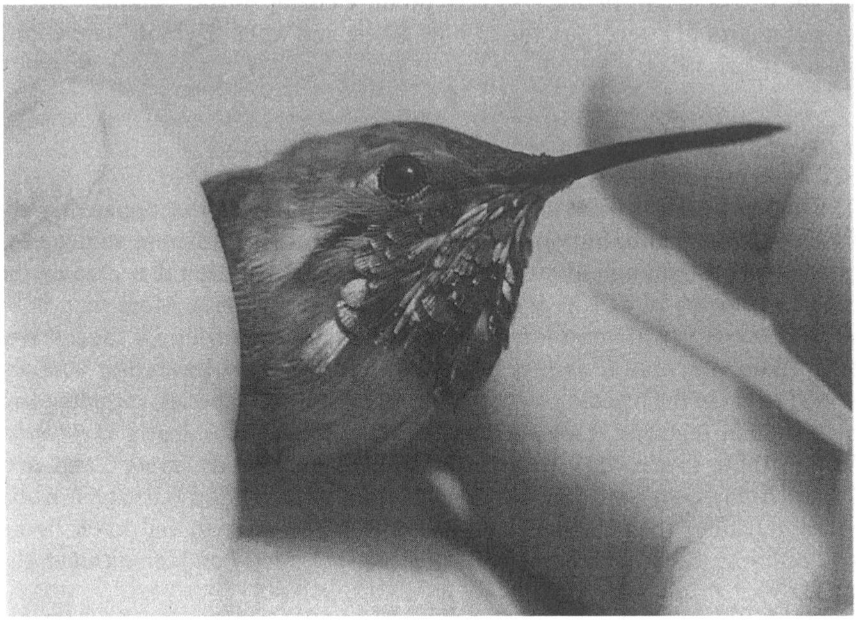

Fig. 27. Tiny hummingbirds, usually thought to subsist on a diet of nothing but nectar, must eat some insects each day, and feed their growing young on a diet of insects. Photo courtesy of D.C. Paton

Fast moving, very active, and with a high metabolic rate these tiny birds seem to need a near-continuous supply of high-energy nectar to survive. Yet it is now apparent that they are far from being true herbivores. Remsen et al. (1986) have shown that almost all tropical hummingbirds routinely eat arthropods, and conclude that they feed on arthropods on a daily basis and probably at regular intervals throughout the day. Not surprisingly, they also feed their nestlings on insects (Carpenter and Castronova 1980; Montgomerie and Redsell 1980). Carpenter and Castronova also found considerable quantities of pollen in the crops of nestling hummingbirds, although Brice et al.(1989) have shown that they appparently cannot digest significant quantities of this pollen.

However, this situation is not so for others long assumed to be exclusively dependent upon nectar as their staple food. The Australian brush-tongued lorikeets feed upon flowers, mostly of species of *Eucalyptus*, and their very specialized tongues have been cited as ideal adaptations for the collection of nectar. However, closer study of the feeding behaviour and the structure of the tongue and alimentary tract of the purple-crowned lorikeet, *Glossopsitta porphyrocephala*, and the rainbow lorikeet, *Trichoglossus haematodus*, reveals that pollen is their *staple* food (Churchill and Christensen 1970; Richardson and Wooller 1990). In particular, these birds have very short intestines and very fast passage of food through them. Digestibility of pollen by birds seems to rarely exceed 50%, so a continuous flow with quick absorption of this 50%

allows for maximum digestion of pollen. It is more advantageous than longer retention which would be unlikely to significantly increase the proportion of grains digested. These birds can readily collect enough nectar to satisfy their energy needs. In the apparent absence of insects or other sources of nitrogen in their diet, however, pollen becomes absolutely essential as the only significant source of nitrogen available to them – not just for breeding, but for day-to-day maintenance, as is the case for other known pollen-eaters (Sects. 6.4; 6.5.2).

A similar situation holds for other nectivorous birds. The various Australian honeyeaters, specialists in collecting nectar from flowers and obtaining not insignificant amounts of protein from the digestion of pollen (Wooller et al. 1988), must nevertheless supplement this food and manna (true plant exudates) with honeydew (the faeces of sap-sucking insects) and insects on a daily basis. They spend nearly equal time feeding on insects and nectar, but expend far more energy in collecting insects than they gain from eating them (Ford and Paton 1976). Also, during the breeding season they increase their consumption of insects far more than they do of carbohydrates (Pyke 1980; Paton 1982).

Many species of tropical birds are said to be partly or wholly frugivorous, but on closer examination few prove to be totally dependent on plant food. Most are only facultative frugivores, foraging for insects as an essential supplement to the protein they can obtain from fruit (Snow and Snow 1971; Foster 1978); and this is in addition to being highly selective for fruit which has the highest protein content (Morton 1973). This is especially true during the breeding season when protein becomes crucial to the successful raising of young (Levey 1988). The Australian mistletoebird, *Dicaeum hirundinaceum*, is a good example. It feeds – exclusively it was thought – upon the fruit of parasitic mistletoes (Loranthaceae) and has the specialized behaviour, internal anatomy, and physiology typical of birds adapted to a diet of fruit (Richardson and Wooller 1988). But it eats insects and feeds its nestlings nothing else for the first few days of their life. So too does the painted honeyeater, *Grantiella picta*, another Australian bird once thought to be wholly dependent upon a diet of mistletoe fruit (Reid 1986).

Foster (1978) had estimated that many tropical fruit contained enough protein to meet the requirements for the rapid growth of nestlings of frugivorous birds, but only if they ingested carbohydrates at three to five times the amount needed to meet their needs for energy. She concluded that this would account for why few, if any, frugivorous birds feed their young solely on fruit, and why insects would be a much more appropriate diet for young nestlings. She noted that nestlings which ate more insects grew faster than their counterparts eating more fruit, and birds which do feed mostly fruit to their nestlings feed them a much higher proportion of insects just after they hatch. Others, which do not feed their young on insects, must have some way of concentrating the nitrogen in the food they supply to their nestlings. Thus, the oilbird, *Streatornis caripensis*, feed their nestlings on a special regurgitated semi-liquid diet for the first 12 days after they hatch.

A recent exchange of views and hypotheses as to why so few birds can survive on a diet of nothing but fruit, still did not come to any definitive conclusions but did at least highlight that it is this shortage of nitrogen, especially for the raising of young chicks, which is the limiting factor involved. Izhaki and Safriel (1989, 1990) hypothesized that the reason there are so few wholly frugivorous birds is not because there is not enough protein in the fruit they eat, but because secondary chemicals in the fruit render much of the protein unavailable, thus "forcing" the birds to seek insects as an alternative source of nitrogen (and, incidentally, in so doing, more effectively distributing the seeds of the plant which produced the fruit). Their critics (Mack 1990; Sedinger 1990) point out, however, that it is equally, or more likely that there is, in fact, not enough nitrogen – or enough nitrogen in an appropriate form – for the birds' requirements within the limits set in all animals by times of passage and digestion, and the inevitable loss of metabolic nitrogen.

Karasov and Levey (1990) report that the fast rate of passage of food through the gut is typical of fruit-eating birds generally, but the faster the passage the less the efficiency of digestion (of sugars). They presume there is "some compensatory advantage" for this short retention time of the food, but were unclear as to what it might be. My bet is that, as with other fruit-eating herbivores, it is the same "creaming-off" strategy that the Australian lorikeets adopt with pollen. It is more profitable to take the immediately available nitrogen and "dump" the rest. Holding the food for longer and more complete digestion of the more recalcitrant nitrogen would not produce a concomitant gain. The strategy is not confined to those eating fruit. Many other herbivores, for example *Pieris* caterpillars (Sect. 3.4.1), sap-sucking insects (Sect. 3.5), geese (Sect. 7.1.1), and seed-eating birds (Sect. 7.3), do the same thing.

Another not inconsiderable way of gaining additional protein, especially for breeding birds and their young, may be through the consumption of fruit which is infested by insects (Crome 1975). For example, Drew (1988) knew that the breeding season of the Australian brown pigeon, *Macropygia phasianella*, coincides with maximum abundance of the ripe fruit of *Solanum mauritianum* in Queensland, and that at this time these birds eagerly devour this fruit. Also at this time, 100% of these fruits are infested by the fruit fly, *Dacus cacuminatus*. The female flies inoculate the fruit with bacteria when they lay their eggs and the larvae remain and feed within the developing bacterial "soup" (see Sect. 3.6). Drew analysed separate samples of clean fruit, infected fruit, and the eggs and larvae taken from the infected fruit. Pulp with insects in it contained nearly twice the amount of protein as uninfected fruit. Clearly such a diet must significantly improve a bird's chance of breeding successfully.

7.3 Birds Eating Seeds

7.3.1 Columbids

Many pigeons are partly or wholly seed eaters. As such they have access to a more concentrated source of protein than can be obtained from the flesh of fruits. Thus, Crome (1975) found that species of Queensland frugivorous pigeons which also ate seeds were less selective of high protein fruits than were those pigeons which are apparently obligate frugivores. Again, however, a diet of nothing but seeds is seldom likely to be adequate for successful breeding, so it is not surprising to find that pigeons (and other columbids) have evolved yet another way of supplementing the nitrogen that is available to their young with animal protein. They feed their nestlings on crop milk (Blockstein 1989; Shetty et al. 1990). Crop milk is unique to columbids, produced by both males and females, and comprises semi-solid cheese-like clumps of desquamate cells sloughed off the germinal epithelium of the crop. It is rich in proteins, lipids, and minerals, and is the sole food of nestlings in their first 4 days. It contains levels of essential amino acids and fatty acids equivalent to that found in mammalian milk or the yolk and albumen of eggs, and higher levels of many trace

Fig. 28. The seed-eating mourning dove feeds its chicks on crop milk – epithelial cells from the parents' crops – a protein food that enables them to grow faster than any other young vertebrate. Photo courtesy of J. Rathert, Missouri Dept. of Conservation

elements. Its dietary protein is more digestible than that of muscle or egg albumen. Chicks feeding on crop milk can grow to 22 times their birth weight in 3 weeks, a rate of growth not known to be achieved by any other vertebrate. The rate at which the birds can produce crop milk probably limits the size of the clutch that they can rear to two, but it does enable them to produce a series of broods much more rapidly than other birds.

Blockstein tested the first supposition by adding a third nestling to two groups of nests of the mourning dove, *Zenaida macroura*, that contained two chicks each. In one group he added the extra chick at the beginning of feeding when chicks are fed only crop milk, and in the second group after 4 days, when the chicks are fed seeds as well as milk. Fewer chicks survived in both groups with a third chick added than did in the two-chick control nests, but significantly fewer survived in the nests where the extra chick was added at the beginning of feeding. The chicks from this group also grew much more slowly than the controls; those from the later period only slightly so.

7.3.2 African Queleas, European Finches, and the Great Tit

Another large and well-known group of seed-eating birds is the finches, yet few of the many New and Old World finches are true herbivores. Seeds are by far the most protein-rich of all plant material (naturally enough, as the seed must nurture the new plant when it first attempts to establish and grow in a harsh environment). So seed-eaters have access to a much better source of nitrogen than do those birds which eat only foliage or fruit. There are some finches which are known to be able to subsist exclusively upon seeds (Ziswiler 1990). They have evolved very specialized digestive structures and functions which enable them to maximize their proteinase activity so as to obtain all of the small amount of nitrogen present in seeds which are 90% starch. However, Ziswiler has shown experimentally with captive finches that even this is not sufficient for their needs, especially at times of increased demand for nitrogen – moulting, egg-laying and rearing of the young. They consume, daily, double or treble the amount of food they need for their energy metabolism, and pass the unneeded excess carbohydrate out in their faeces, mainly in its original form of starch. They have adopted the same strategy as that of aphids and psyllids feeding on phloem sap (Sect. 3.5). Most finches, however, must supplement their diet of seeds, especially in the spring just prior to breeding, with growing buds and shoots, and insects. Very many feed their young exclusively on insects.

A very dramatic example of this dependence on insect protein for breeding by seed-eating sparrows and finches is the case of the African red-billed quelea, *Quelea quelea*, (Jones and Ward 1976, 1979). In the dry season these birds feed exclusively on grass seeds, from which they gain sufficient protein for normal maintenance and for synthesis of feathers following moulting, but not enough for storage; and they do not breed. In the wet season they feed also on the insects which then become very abundant. By taking advantage of this protein-rich food the queleas can accumulate sufficient protein reserves to breed. They follow the rains across thousands of kilometres and flocks of birds sometimes numbering millions make these breeding migrations. As soon as individual females have accumulated enough reserves of protein, and there are large quantities of insects present to provide food for themselves and their young, the birds stop and form large synchronized communal nests. Individual birds may breed several times during one wet season at widely separate sites, whenever they can find sufficient insects to build their protein reserves up to the high level needed for successful breeding. Once this food source is gone they cease to breed.

The adults of many passerines do not eat insects but feed them to their young. In others, even the young are not fed insects but only the "chyme" of regurgitated and partly digested seeds. Whether this chyme (or the regurgitate fed to many young birds, both herbivorous and carnivorous) contains any addition of animal protein from the parent, like the crop milk of pigeons, I do not know. However, Newton (1967) found that while the adults of the European bullfinch, *Pyrrhula pyrrhula*, continue to feed exclusively on seeds during the breeding season, they feed their young on a regurgitated mixture of seeds and invertebrates (caterpillars, slugs, snails, and spiders). This food for nestlings is stored in special pouches evident only in the breeding season beneath the lower jaw of the adult birds. The proportion of invertebrates in this special food declines markedly as the nestlings grow (from approximately 40% in the first 5 days after hatching to less than 5% after 12 days). Young birds do not eat invertebrates after leaving their parents.

Yet, even on a diet of 100% seeds, bullfinches are selective, taking seeds of only some of the many species available, and then not in proportion to their availability. Even within one species they may be selective. The seed of ash, *Fraxinus excelsior*, is especially important to bullfinches as a winter/spring food once other more preferred seeds become scarce. In Marley wood, Oxford, Newton recorded them consistently, year after year, taking virtually every seed from the same few trees while other nearby trees remained untouched. The basis of this preference has not been investigated, but it is a fairly safe bet that it is similar to the preference of capercaillie and other tetraonids for individual trees; they are seeking the highest level of available nitrogen.

Each spring, prior to breeding, bullfinches take the young swelling buds of several deciduous trees and shrubs, but prefer those of hawthorn, *Crataegus* spp. (Newton 1964). As with the ash seeds, the birds do not feed at random. They concentrate on a few trees bearing the larger (and presumably more nutritious) buds, while ignoring others nearby. Normally they do not start eating buds until late in the spring, just prior to breeding, but in years when there has been a poor crop of ash seed, and this is exhausted in late winter or early spring, they turn earlier to buds. With the clearing of their native woodlands and the growing of cultivated fruit trees in the south of England, they have come to prefer the buds of these fruit trees. Now, in years of poor ash mast, they can be a serious problem in large commercial orchards. They eat only the central part of those buds destined to become fruit, and will gorge themselves on this food. But, again, they are selective, even to the extent of preferring the buds of a particular cultivar of the one species of fruit tree. Wilson et al. (1983) found that of two varieties of pear, the buds of the variety preferred by bullfinches had three times more free amino acids than those of the less preferred variety.

Although birds maintained in captivity on a diet of preferred buds lost less weight than those fed the less preferred buds, none could survive on buds alone (Newton 1964). In the field, in years when ash seed ran out and the birds were forced to feed on buds in the winter, many died. At this time even the preferred pear buds contain less protein than the seeds, and are therefore not an adequate substitute, even though the birds gorge themselves, eating the maximum volume of food of which they are capable. However, the level of all nutrients in the buds increases suddenly just as they start to swell in the spring – the time when the birds normally start to feed intensively on them. At this time the buds become a better source of *available* nitrogen than the seeds for a breeding hen. Prior to this they are not, and will be eaten only as an inadequate substitute source of protein when there are no seeds. Again, we are seeing the same story as that of the growing tips of *Calluna* for red grouse (Sect. 7.1.2) – access to soluble nitrogen that is being translocated to growing buds and forming seeds.

For the first few days after the young hatch the hen bullfinch broods continuously and is fed by the male. During this time she eats the faeces of her nestlings; these are produced enclosed in a gelatinous sac. Thereafter she drops them away from the nest. This eating of the very young nestlings' faeces is common in passerine birds. Its value is usually thought to lie in keeping the nest clean or in providing the brooding hen with additional water or energy (Morton 1979).

Gluck (1988) monitored this behaviour in the European goldfinch, *Carduelis carduelis*, to demonstrate that the latter is the most likely explanation. The goldfinch is apparently a true herbivore. Gluck's experimental birds ate – and fed their young – a diet of nothing but *milky ripe* seeds of goatsbeard, *Tragopogon pratensis*. He found that the mean energy content of the nestlings' faeces was very high in the first 3 days after hatching, but had dropped nearly 80% by the time they were 10 days old. Clearly the digestive efficiency of very young birds is minimal, but quickly improves. The hen ceased to eat her nestlings' faeces once the energy content had fallen to approximately that of adult faeces; when, Gluck said, it becomes inefficient, from an energy point of view, for her to do so.

But is this the answer? As Gluck points out, the partly digested faeces of the young may well be held in the hen's crop and fed back to the young along with the chyme. I think this is most likely. I doubt that gaining extra energy – for either adult or nestling – is the answer. The hen remains on the nest during this period and is fed constantly by the male. Her consumption of energy would be relatively low. Similarly with the young; they are constantly kept warm by their mother's body and need do little but open their mouths, swallow and defaecate. However, they must grow at a tremendous rate. To do so they need as much protein as they can get. Eating the seeds when they are milky ripe ensures that virtually all the nitrogen in their food is available as soluble amino acids [just like the mice (Sect. 6.2.2.1) and parrots (Sect. 7.3.4) which breed only when they can eat milk-ripe grass seeds]. The nestlings will still void undigested and metabolic waste nitrogen. This behaviour is therefore most likely to be just another form of coprophagy operating to maximize the extraction of protein from the vegetable diet of the young.

The European red poll, *Carduelis flammea*, was considered to be totally dependent on seeds as food. Certainly in winter this seems to be so, as is the case with most seed-eating birds. Also, it has been reported that they can raise a brood of young on a pure seed diet. However, the 10-year study of this small passerine, carried out by Pulliainen and Peiponen (1981) in the subarctic forests of Finland, would make this seem improbable. At best it would be a rare event in a year when spruce seed was both superabundant and reached the top of its reported range of 10 to 22% dry weight content of protein. Mostly, Pulliainen and Peiponen found that the birds must rely heavily on animal matter in their diet if they are to successfully raise their young. In summer this can comprise 80% of the food of adults and 90% of that of the young. The nestlings are fed entirely on the regurgitated contents of the crops of parent birds and fledge in 9 to 13 days. The authors considered that to fledge in this time the nestlings must be fed large volumes of food containing enough insect protein to support such rapid growth.

Another example of apparent dependence upon a winter diet of seeds is that reported by Perrins (1966) for the European great tit, *Parus major*, and the seed mast years of beech, *Fagus sylvatica*. The great tit is as much an insectivore as a herbivore, feeding its young on insects, mostly leaf-eating caterpillars, but in winter it feeds heavily on the seeds of beech and other trees. There seems little doubt that a staple diet of seeds is important to the survival of these birds in winter – they will leave an area where there is little or no seed. Also, their populations fluctuate in synchrony with the availability of beech seed, peaking in each year of a mast crop. However, there are two complications to this story. First, the numbers of tits fluctuate synchronously over wide areas of Britain, Scandinavia, and Europe, *including areas where there are no beech trees*. So it cannot be access to beech seed which determines how many tits survive. Perhaps they eat the seeds of other trees which also tend to peak in abundance at the same time as the beech? The second complication not only discounts this idea, however, but reveals that the amount of seed that is available for the birds to eat in winter has little, if anything, to do with their abundance. Their numbers increase when more young survive in the spring of years when there is going to be a mast crop in the succeeding winter – *before* the good supply of seed is available. This includes that of other species of tree which seed in the same year as beech.

There must be some other factor in the environment which simultaneously improves the reproductive success of both beech trees and great tits. What is it? Perrins noted that more aphids are found on parts of trees which are flowering, so possibly young birds survive better because they get more animal food in the spring of those years in which more seed is going to be produced in winter; but this idea, he said, was untested, and so, it would appear, it remains to this day!

If confirmed, such an explanation would show the causal chain to be the mobilization of nutrients into developing seeds that improve the plants as a source of nitrogenous food for various herbivorous insects (not just aphids), and these in turn provide an increased supply of protein for carnivores (which is what young tits, and almost certainly their mothers prior to egg laying, are). However, it would leave unexplained the cause of all plants, over wide geographical areas, mobilizing nutrients to grow seeds at the same time. A possible explanation is suggested from the next example and I attempt, in Chapters 16 and 17, to elaborate upon this.

7.3.3 Darwin's Galapagos Finches

A recent and compelling demonstration of the extent to which the successful breeding of a species is limited by a shortage of food, and specifically of protein food, and the extent to which animals are constantly pushing against these limits, is that of the response of populations of Darwin's finches on the Galapagos Islands to the 1982–1983 El Niño (Gibbs and Grant 1987; Grant and Grant 1987).

Very little was known about the ecology of these birds until the Grants and their colleagues started to study some of them in the mid 1970s. By the time of the 1982–1983 El Niño – the strongest so far this century – they had already assembled an impressive knowledge of many of the species present on a number of the islands. Their monitoring of the responses of finches to the changes wrought by the oceanic warming caused by El Niño was concentrated on populations of three species living on two islands of the archipelago. On Isla Daphne the two most abundant of four species on the island, the medium ground finch (*Geospiza fortis*) and the cactus finch (*G. scandens*), had been monitored in the first 6 months of every year since 1976. Isla Daphne is only about 40 hectares and it had been possible to census the entire population of both species on the island. In the years spanning the El Niño virtually *all* nests of both species were found, and *all* birds born on the island individually banded as nestlings. On Isla Genovesa a population of the large cactus finch (*G. conirostris*), one of four species on the island, was subject to a similarly intensive study on a site occupying about 10% of the island.

The story that emerged is essentially very simple. The Galapagos Islands have a hot wet season from January to April/May and a cool dry season from June to December. The finches all breed in the wet season. The timing and amount of the few heavy falls of rain which characterize the wet season determine the amount of growth of vegetation, and hence the production of seeds and arthropods upon which the finches depend for their food. Rainfall thus determines the success of the birds' breeding. In years when little or no rain falls in the wet season the birds do not breed at all. Of the several species of

Fig. 29. An outbreak of Galapagos finches occurred during the 1983 El Niño. They bred explosively, feeding their young on caterpillars which flourished on the lush vegetation generated by the heavy rainfall. Photo courtesy of Heidi Snell

Geospiza present on the islands, some eat mostly insects and others eat mostly seeds; but this latter group all eat arthropods in the breeding season and feed them (particularly lepidopterous caterpillars) to their young (Schluter 1982).

In 1983, the El Niño year, the wet season started exceptionally early and finished exceptionally late. The total rainfall for the season was approximately *ten* times the previously recorded *maximum* rainfall for a single wet season. Plants began to grow some 2 months earlier than usual and continued much later than usual. The vegetative cover was much greater; there were many more plants, individual plants were larger and more lush, and they flowered and fruited continuously. Not surprisingly, the biomass of seed also increased dramatically – up to 11 times that in 1982 – and remained almost four times as great until the end of 1984. The proportion of this biomass which was small and soft seeds (the preferred food of the finches) rose from less than 25% of the total in 1982 to more than 80% in 1983, and persisted at about 40% for at least 2 years after this peak.

On this great flush of vegetation in 1983 there was a dramatic rise in the abundance of herbivorous arthropods, especially the lepidopterous larvae which are fed to the young of all species of *Geospiza*. These caterpillars were up to six times as abundant and present for four times as long as in the previous year.

The finches responded appropriately to this explosion of good food. They bred continuously and for an unprecedented length of time; up to twice as long as the usual 4-month breeding season. Individual females, on average, produced four times as many clutches (one female produced ten), laid five times as many eggs, and fledged nearly four times as many young (25 for one female) as in a normal good breeding season. Many more individuals bred, including those which, in a normal year, would not have done so. Some bred where none had bred before; new territories were established in previously unoccupied areas. Nests were built in places where they would not normally be built. Species bred on islands where they had not previously been recorded breeding. Many more immigrants appeared in the populations being studied (and many of the presumed deaths in these populations may have been emigrants). Many of the birds born early in the season matured and bred before the end of the season (they would not normally have bred until they were 2 years old).

In spite of the fact that much of this explosive production of young was damped by desertion of nests and death of nestlings (largely as a result of the incessant stormy weather with high winds and heavy rain), by the end of the 1983 season densities of finches were exceptionally high and young birds predominated. A veritable outbreak of finches!

While detailed measurements had been confined to the populations of only three species of finches on two islands, other species of *Geospiza* on these islands had also increased markedly in abundance. In fact, the numerical response was typical for finches throughout the archipelago. Other species of land birds responded similarly to this bonanza year. For example, the Galapagos mockingbird, *Neosomimus parvulus*, on Isla Genovesa increased dramatically in abundance, mainly due to repeated breeding in a prolonged season. Females fledged about twice as many young as the previously recorded maximum. Also, the cuckoo, *Coccyzus melacorhyphus*, which had not been seen on Genovesa from 1978 to 1982, arrived on the island in considerable numbers in 1983. One pair bred in the study area and one nestling fledged. A few remained in 1984, but by 1985 all the cuckoos had disappeared from Genovesa "... together with the bulk of the orthopterans and caterpillars on which they had been feeding".

The party was similarly soon over for the finches. On Genovesa 1984 and 1985 were both dry years. In 1984 16% of the young of *G. conirostris* which fledged in 1983 were still alive, 4% in 1985 and only 2% by the commencement of breeding in 1986. On Daphne *G. fortis* and *G. scandens* fared a little better. High mortality of both adults and juveniles had reduced the population of both species by more than 30% by January 1984, but the 1984 wet season was more or less normal, and by January 1985 their numbers were still one and a half to two times that of the pre-El Niño populations.

It is clear that in spite of the supply of preferred seeds remaining higher than usual for up to 2 years after the great rains, the birds could not sustain their numbers at the 1983 level, nor repeat the breeding frenzy of that year.

Increased production resulted from a greater *amount* of reproductive activity in both space and time, not from an increased *rate* of reproduction. More individual females were able to breed for longer. These females were not able to increase the number of eggs they laid in a clutch, nor could they speed up the time it takes to incubate and rear a batch of young. Also, because of the atrocious weather they hatched a smaller percentage of their eggs and fledged a smaller percentage of their nestlings than they did in other normally good breeding seasons. But, like the quelea, they bred whenever, wherever, and for as long as there was enough protein food for them to produce eggs and feed their young, and especially the latter. This was dictated by the quantity of arthropods available to them, not by the supply of seed. That this was so showed up clearly on both Daphne and Genovesa. Breeding declined and stopped when the supply of caterpillars finished, despite the fact that rains had not ceased and preferred seeds were still abundant. The authors expressed puzzlement as to why caterpillars disappeared when they did, suggesting perhaps that predators (other than the finches?!) or allelochemicals, produced by damaged plants, were responsible. Little appears to be known about these insects – not even the number of species or their names. I would guess that there are several species, but that only one or a few are of most importance as food for young finches. They are all likely to be univoltine Lepidoptera, adapted to breeding quickly on the flush each wet season and to surviving the dry season as pupae. In 1983 many more of *their* young would have survived on the abundant and prolonged flush of new growth, but once mature would have entered an obligate pupal diapause.

The dependence of Galapagos finches on insects for the successful breeding and raising of their young had already been established in previous studies (Millington and Grant 1983; Boag and Grant 1984). Adults commonly take insects during nesting periods and nestlings are fed little else (although at times their growth is sustained on a diet of pollen and nectar). The number of nests initiated in a dry year, as against a wet year, correlates specifically with the number of *caterpillars* (not all arthropods) available rather than with the abundance of flowers or fruit.

These same studies have revealed that the nitrogen-hunger of the birds is not confined to the breeding season. During the dry season finches commonly supplement their diet of seeds with animal protein. They scavenge broken eggs, fish scraps and faeces from the nesting sites of seabirds, eat any available carrion, the detached tails of escaping lizards, placentae of sea lions, and the blood in developing tail feathers of seabirds.

The greatest significance of these studies of Galapagos finches, however, is the way in which the exceptional event confirms the usual condition. The abundance of these birds is determined by the success of their breeding, which in turn is determined by the abundance of the insects they eat. The abundance of the insects is determined by the influence of the weather on the supply of their food: flush new growth of plants. The weather is variable such that shortage is the norm, and abundance an unpredictable rarity. The birds are adapted to make maximum use of whatever resources are available. They are *at all times* pressing hard against the limits set by the availability of nitrogenous food. They breed as fast as they can and for as long as they can under variably unfavourable and exceptionally favourable circumstances. That is why they breed explosively when there is a sudden increase of good food, but just as quickly die when the supply runs out. As Grant and Grant conclude, the finches are always breeding to the maximum, and not "hedging their bets" by breeding submaximally.

What the birds' immediate response to the exceptional event also illustrates, however, is that during "normal" times it is this shortage of food, not any other environmental factor, which is limiting their increase. [Most populations of Darwin's finches are anyhow free of predators (Price 1985).]

The Grants' Galapagos finches were responding in just the same way as do locusts (Sect. 3.4.7), kangaroos (Sect. 6.5.4), rabbits (Sect. 6.2.5.2), and pelicans (Sect. 13.4) in Australia when unpredictable rainfall produces a short-lived abundance of good food for them. All these studies of populations of animals, which live in widely varying and usually harsh environments, and which accordingly fluctuate violently in abundance, serve to reveal how other animals, which live in more stable (but often equally harsh) environments, and which do not fluctuate in abundance very much, may nevertheless be similarly limited in their abundance by their food supply (see also Part V).

7.3.4 The Australian Galah

Australia provides another illuminating example of the response of seed-eating birds to an increase in the supply of good food – but this time a permanent increase. The range and abundance of the beautiful pink and grey parrot, the galah *(Eolophus roseicapillus)*, have been greatly enhanced by the agricultural development which followed European settlement 200 years ago. Nowhere is this more apparent and dramatic, however, than in the wheat belt of Western Australia in the last 50 years (Rowley 1990). Here, the seed of wheat, *Triticum aestivum*, oats, *Avena sativa*, and barley, *Hordeum vulgare*, make up more than three-quarters of the galahs' diet; but, more importantly, the remaining one-quarter of their diet comprises the seed of several species of *Erobium*, the introduced and widespread corkscrew grass weed. On this great abundance of food the galah has multiplied and thrived, invading great tracts of land where it previously could not live. The most significant component of this ongoing bonanza is the provision of high-nitrogen food for newly hatched chicks and for breeding females. The birds breed in spring. The *Erobium* is flowering and setting seed then, about a month before the wheat does so, and coincident with the peak of hatching in the galahs' nests. The nestlings are first fed a clear liquid, dribbled to them by their parents. Their first solid food, regurgitated by their parents, is delivered on their second day. It must be soft and easily digest-

Fig. 30. Seed-eating corellas gut large insect larvae, discard the exoskeleton, and feed the contents to their nestlings. What chance of detecting this food from crop or faecal samples? Photo courtesy of Graham Chapman

ed for the first 10 days. Thereafter they are fed upon the *unripe soft milky* seeds of *Erobium*, and later upon the *milky heads* of ripening cereal grains. The females also feed extensively upon developing *Erobium* seeds when they are initiating and maturing their eggs. In years when seed production by *Erobium* is little or late many females do not lay, and many fewer nestlings are fledged. Clearly, early access to a concentrated supply of soluble nitrogen as it is imported to the forming seed is vital to the success of their breeding.

Several questions arise at this point. The feeding of regurgitate to the young of herbivorous birds is obviously widespread. What is the protein content of such regurgitate? More interestingly, what proportion of the regurgitate is animal protein – parental, as in the crop milk of pigeons, and/or arthropod?

How widespread is the consumption of arthropods by seed-eating birds? Crop samples of galahs in the Western Australian wheatfields contained insect larvae from May to November. In August to October (the southern late winter – early spring) a quarter of all crops contained larvae, reaching a peak of 13% of total crop volume in October. It seems unlikely that these insects were not making a significant contribution to the protein diet of both mother and young. Insects form an important part of the diet of at least five other species of Australian seed-eating parrots (Long 1984; Temby and Emison 1986; Smith and Moore 1991). In all, they take most insects in late winter – early spring, and feed them to their very young nestlings. Smith and Moore found that the crops of a sample of nestling corellas *Cacatua pastinator* contained about 50% insect larvae and 50% small green seeds (mostly *Erobium*). Of particular interest, however, they observed the breeding females of this corella searching for and digging up large unidentified insect larvae; gutting them, feeding the contents to their nestlings, and discarding the exoskeleton. What price crop, stomach, or faecal sampling in the face of such behaviour?

Finally, how commonly does the success of breeding of seed-eating birds depend upon access to seed at the milk-ripe stage rather than just to the quantity of seed? Furthermore, on the broader front, the parallel between galahs and feral mice (Sect. 6.2.2.1) both living in grain-growing areas, and both depending upon access to milk-ripe seeds for the success of their breeding, is surely noteworthy.

Chapter 8 Reptiles

Reptiles have metabolic rates which are an order of magnitude lower than similar-sized mammals on the same diet; their rates of feeding are some ten times lower, as is the transit time of food through their gut. They have equal or lower extraction efficiencies, so their intestines absorb nutrients ten times more slowly than do those of mammals with their constant high temperatures and the larger surface area of their longer intestines (Diamond 1987). On top of all this, young reptiles start their existence with several other disadvantages compared with young mammals. While they will normally have a small reserve of yolk to supplement their nutrition for a short period, they are not fed on milk or helped in any other way by their parents. From the first moment they are on their own and must subsist as best they can with what food they can find for themselves. Furthermore, they are usually smaller as a fraction of their final adult size, so must ingest and assimilate nitrogen at a higher rate than adults just for daily maintenance, as well as for the higher additional requirement for rapid growth. If they are herbivorous the whole situation is exacerbated. Not surprisingly, therefore, most reptiles are carnivorous, or, at best, omnivorous. Very few are truly vegetarian in that at no stage of their life do they eat animal protein. There are a few apparently truly herbivorous turtles and lizards, but all depend on symbiotic microorganisms and an array of behavioural and physiological adaptations to gain enough nitrogen from their diet. Even so, the success of their reproduction – and thus of their abundance – is still precariously balanced on the edge of a varying insufficiency of protein.

8.1 The Giant Tortoises of Aldabra Atoll

Probably among the best known examples of herbivorous reptiles are the giant tortoises of the Galapagos Islands and the Aldabra Atoll. Malcolm Coe and his colleagues from Oxford University have carried out extensive and intensive studies of *Geochelone gigantea* on Aldabra. These animals, once widespread in the Indian Ocean, were almost exterminated by excessive human predation.

Gibson and Hamilton (1983) were able to demonstrate the extent to which the food supply influences the feeding behaviour of these giant tortoises. They are highly selective feeders; grazers by preference, browsers by necessity. During the wet season they concentrate on the taxonomically diverse "tortoise turf". Their intensive and repeated grazing of this turf (they remove 90% of its annual production) maintains all species in it as new growth less than 1 cm high, maximizing the content of soluble nitrogen in their food. During the dry season, however, when the growth of the turf has ceased, they are forced to turn to available, but non-preferred foods such as fallen leaves of shrubs, sedges, and grasses. With the latter two they concentrate on the flowers, developing seeds, and new growth; again seeking out the parts with the maximum content of soluble nitrogen. Interestingly, the juveniles show distinctly different feeding preferences from the adults. They also concentrate on the turf in the wet season, but select even more strongly within this than do the adults, especially for rare herbs (less than 0.5% of all vegetation) and also for the long grasses. In addition, they seek out plants growing in holes in the rocks much more than do the adults. All these highly preferred foods of the young are scattered and typically occur in small patches.

The authors assumed that the preferential grazing of the turf confers the advantage of increased access to high nutrient and low toughness food, as does the selection of flowers, fruit, and fresh growth of the other preferred species in the diet. They also observed that the small particle size consequent on continued grazing would increase the efficiency of assimilating energy. This is aided by the long time the food spends in the gut. Food took 6 to 19 days to pass through the gut of experimental animals fed on tortoise turf, but when they were fed on the tougher *Sporobolus* sward, which they eat at the end of the dry season, it took 10 to 22 days (Hamilton and Coe 1982). However, the transit time was considerably faster in smaller tortoises. The authors presumed that the juveniles' preferences reflected the generally higher reward, compared with that of larger animals, that small animals gain from feeding on small patches; and that the vegetation in holes in rocks is generally not accessible to larger tortoises. They also confirmed earlier reports that giant tortoises feed on carrion (including tortoise), goat faeces, and (unspecified) human camp debris, but they make no comment on the significance of this behaviour.

Gibson and Hamilton (1983) concluded that the differences between the giant tortoises and other large herbivores is one of degree only; they behave like typical large grazing animals. Both their growth and reproduction are "tightly restricted by food supply". Their feeding behaviour and preferences certainly reflect just how tight this is.

Swingland and Coe (1979) showed quite clearly that once recovered from the earlier overexploitation by humans, the abundance of *G. gigantea* on Aldabra is limited by its food supply which, in turn, is limited by the weather. They discovered that when there was more rain and an extended wet season, the growth of the tortoises' principal food, the tortoise turf, was increased and prolonged. This improved the nutrition of females during late follicular development with the result that mean egg mass, clutch size, numbers of nests, and total production of hatchlings all increased significantly. The larger eggs produced larger hatchlings, which survived better during their first year than did those from smaller eggs. (But the authors found that mortality of hatchlings in their first year was still in the order of 80 to 95%.)

That more water produces more and better food, which results in greater abundance of tortoises, was reflected in the higher density of tortoises living on the eastern end of the Grande Terre sections of the atoll. There the production of turf is increased by freshwater lenses which are forced up by spring tides (Coe et al. 1979).

So, it is clear that these tortoises experience a shortage of protein, especially for reproduction. Their feeding behaviour and response to any increase of flush growth illustrates this; so too does their ready acceptance of animal protein whenever it is available. What is more, I suspect that the feeding preferences of the hatchlings reflects a seeking out of plants from which they can get more available protein, rather than anything to do with the size or distribution of the patches of their food.

The very long time food takes to pass through the gut of a tortoise would allow greater breakdown of cell walls and extraction of more high-nitrogen cell contents, but there is no record of a gut microfauna breaking down cellulose. Hamilton and Coe (1982) say that it is not clear whether cellulose breakdown is a regular feature of tortoise digestion and the question cannot be solved without further study. Nor is there any record of coprophagy. They suggest that the giant tortoises resemble the elephant in being able to support a high biomass on relatively low consumption and assimilation rates, and that they would be assisted in this by their poikilothermy. In the light of evidence from other herbivorous reptiles (Zimmerman and Tracy 1989; see also following sections) and the very slow passage of food through their gut, however, it seems likely that these animals do depend upon active fermentation of cellulose in their hindgut.

Gibson and Hamilton (1983) conclude that the dynamics of all the subpopulations of *Geochelone* on Aldabra are "driven by their resources". The amount of plant food, in turn mediated by the substrate and the rainfall, determines the success of their breeding, and thus their abundance. The tortoises, on the other hand, have no significant effect on variations in the amount of food that is available to them. However, the authors found that some subpopulations of tortoises were more hard pressed by the limits set by the supply of their food than were others. In places where their density was still low the animals were large, had many young, and grew fast; in dense subpopulations the animals were small, had few young, and grew slowly. In fact, at very high densities many juveniles did not grow or mature sexually. There was one subpopulation where numbers were still relatively low, but which had not increased in density for at least 10 years and probably for much longer. Gibson and Hamilton's data revealed that the answer to this apparent anomaly must lie in the fate of the very young hatchlings. Either all of them are killed by predators or their food supply is inadequate. Tantalizingly, they know nothing about either predation on these small tortoises, or what they eat. My bet is that they starve through lack of sufficient protein, and that predators (other than humans) are of minor significance.

8.2 The Green Turtle of the Bahamas Islands

The green turtle, *Chelonia mydas*, is a marine herbivore which feeds on extensive beds of seagrass or algae in tropical waters (Garnett et al. 1985). Bjorndal (1980) demonstrated that in the Caribbean Bahamas Islands this reptile feeds selectively and exclusively on the young leaves of the seagrass, *Thalassia testudinum*. It is able to do this because it maintains "grazing plots" by consistently recropping the same plants. When Bjorndal first released turtles into her fenced 3-hectare enclosure they systematically bit the lowest part of the *Thalassia* leaves and allowed the upper part to float away. In three months the turtles had established areas of cropped grass which they no longer enlarged but repeatedly grazed, leaving sharp boundaries with stands of tall grass which they never touched. The flush growth of *Thalassia* on these grazed plots had a higher protein and lower fibre content than did older leaves in neighbouring ungrazed stands of grass. However, it still had a high percentage of fibre and a low proportion of cell contents, so in terms of available protein it seemed a low quality forage. Again, however, the flush growth is supplying nitrogen in a more readily assimilated and digestible form.

Chelonia is also a hindgut fermenter with a well developed caecal microfauna, digesting cellulose as efficiently as ruminants (Bjorndal 1979). The vol-

Fig. 31. The green turtle enhances its intake of nitrogen by eating flush growth of grazed plants, and has a caecal microfauna. Yet it eats animal food when opportunity allows, and is a carnivore when young. Photo courtesy of A. Carr

atile fatty acids produced in the caecum constitute a significant source of energy for the turtles, but the great benefit of microbial fermentation is the breakdown of cell walls to release the easily digested protein in the cell contents (and, presumably, the provision of microbial protein, although I do not know of any evidence of digestion of caecal contents in the hindgut, or of coprophagy).

Yet, in spite of these adaptations, the turtles still have a low rate of intake, and grow slowly on a food with a demonstrated low protein digestibility. Bjorndal suspected that this might be why records from mark-recapture studies showed such slow growth of juvenile turtles.

But the young are carnivores (Wood and Wood 1981). They hatch with a reserve of yolk which fuels their frantic swimming far out to sea; and there they vanish, not to be seen again until they are 2-year-olds. Only quite recently was Carr (1984) able to explain this "mystery of the missing year". In their first year these young turtles live in the Langmuir bands found on the surface of the sea. These bands are formed by wind and currents which concentrate floating debris, including the dispersed food resources, on the surface of the sea. Here the small turtles live, eating the macro-zooplankton that also concentrate in these bands.

There is evidence, however, that *Chelonia*'s carnivory does not end there. Bjorndal (1985) reports that some populations of turtles commonly eat marine invertebrates and fish (one captured subadult had a stomach full of fish eggs). Also, on turtle farms, animals fed on fish, invertebrates, or high-protein pellets, consume 12 times as much nitrogen as wild animals eating *Thalassia*. Consequently, they grow much more rapidly and have a much higher rate of reproduction. Clearly *Chelonia* is limited by a shortage of nitrogen when feeding on *Thalassia* alone, even when this grass is available in excess. It seems most unlikely to me that animals which so readily eat animal protein in captivity – and thrive on it – would not do so in the wild whenever the opportunity arose.

8.3 The Marine and Terrestrial Iguanids of the Galapagos Islands

A number of lizards, usually only the larger species which weigh 50 to 300 g or more, are partially herbivorous, but their young are insectivorous until they reach a critical minimum body mass for herbivory (Pough 1973). Until quite recently it was thought that this was true of all lizards; that none were truly herbivorous. While, as adults, they may be totally vegetarian, breeding females, like the very young, will often seek out animal food. Except possibly for the iguanids. However, with one exception to date, this would seem to be true of them as well. Thus, the famous marine iguanas of the Galapagos Islands, *Amblyrhynchus cristatus*, feed only on algae as non-breeding adults, and mostly on just one species, *Ulva lobata*. Just how essential these algae are for the survival of such specialist herbivores as these iguanids was shown very dra-

Fig. 32. Most Galapagos marine iguanids starved to death during the 1982–1983 El Niño, while the number of finches (sitting on its head) exploded on a superabundance of food (see also Sect. 7.3.3). Photo courtesy of Fritz Pölking

matically by the advent of the 1982–1983 El Niño, and the associated rise in temperature of the sea surrounding the archipelago. Most of the iguanas' preferred species of algae disappeared and were replaced by a brown alga which they could not digest. Over 60% of them starved; and it was 2 years before the surviving females bred (Laurie and Brown 1990). The diet of the juveniles, on the other hand, is approximately only 80% algae (and mostly of one species of *Sargassum*). Crabs, molluscs, diptera, and arachnids make up 11% of their diet, and 2% of it is formed by the faeces of the adult iguanas (Nagy and Shoemaker 1984). Clearly, the young iguanas need – and get – significant quantities of animal protein. Eating the faeces of the adults would be essential for them to establish their own hindgut microfauna, but it would also provide another valuable source of high-protein microbial food.

The terrestrial iguanas of the Galapagos, *Conolophus pallidus* and *C. subcristatus*, are, however, said to be herbivorous even as hatchlings. Christian et al.'s (1984) detailed observations of the feeding of these two species on Isla

Santa Fe revealed that the adults strongly prefer to eat the leaves – and especially the flowers – of species of *Lantana* (which also happen to have the highest levels of nitrogen of all the plants available to the iguanas). In the cool season, when there is little *Lantana* to eat, they mostly feed upon the flowers and leaves of a species of *Cordia* (which is the next richest source of nitrogen after *Lantana*) and the flowers of *Opuntia*. However, of all the observations the authors were able to make of the feeding of the hatchlings, none showed them to eat anything other than flowers (mostly of *Lantana*) and a few leaves. Christian and his colleagues therefore concluded that these iguanas are totally herbivorous – that they do not make the transition from juvenile carnivory to adult herbivory. But they observed females eating iguana eggs, and juveniles eating iguana faeces; and they and earlier workers had observed the lizards commonly eating dead crabs, birds, rats, and sealions. So the evidence again suggests that both breeding females and juveniles need animal protein. Perhaps the young can get by without this supplement to their plant food providing they can get enough high-protein *Lantana* flowers, and the microbial supplement derived from being coprophagic.

8.4 The Desert Iguanid of California

Mautz and Nagy (1987) claim to have shown for the small Californian desert iguana, *Dipsosaurus dorsalis*, that the hatchlings do not need to defer herbivory until they attain adult size, because they fed hatchlings in the laboratory on a pure vegetable diet for 7 days. At the end of that time they recorded that the animals were maintaining an adequate level of digestion of their food, but they had to be force fed – the young lizards would not voluntarily eat a mixture of desert plants offered to them in a dish. Yet, in the field they found that 1–3% of the stomach contents of hatchlings consisted of arthropods, and hatchlings in the laboratory avidly ate insects. When fed a diet of nothing but insects in the laboratory they grew at a rate as fast as the fastest ever recorded for hatchlings in the field. Furthermore, Norris (1953) had already reported that these lizards eat "important amounts" of animal food. The bulk of their spring food was flowers, mostly those of the creosote bush, *Larrea*, but they also ate a variety of insects, carrion, and faeces. Apparently faecal pellets (both their own and those of mammals) are a regular component of their diet, especially in the spring. So, while it may be possible to sustain hatchlings for short periods in the laboratory on only the vegetable component of their diet, and while they may well be able to survive for short periods in the field without eating any animal food, it seems to be stretching a point to classify them as strict herbivores. Without a significant (albeit small percentage) input of animal food, and active coprophagy, it seems unlikely that this species could successfully breed and persist in the field.

Significantly, not one of these studies of iguanas mentions protein food. The emphasis of all is entirely upon energy.

8.5 The Green Iguanid of Panama

To date, the one exception to the rule that all young lizards are carnivores seems to be the green iguana, *Iguana iguana*. Adults are known to readily eat carrion and insects, and it was suspected by early workers that the very young were almost entirely insectivorous (Loftin and Tyson 1965). Troyer (1984a, b, c), however, has conclusively demonstrated that *I. iguana* can be entirely herbivorous. She set out to determine for a population of these lizards living along the shoreline of Gatun Lake, Panama, how the immature iguanas are able – on a completely vegetable diet – to satisfy their higher requirements than the adults for nutrients. She did this by closely observing the animals feeding in the field, and by conducting feeding and digestion trials in the laboratory. This latter task called forth particular devotion to the cause of science because captive animals refused to eat. She had (with the assistance of others, admittedly) to hand feed her experimental animals. Over a period of 3 weeks she fed each of 123 lizards one leaf at a time, waiting for it to swallow each one, until it had eaten its daily estimated intake (a stomach-ful!).

Troyer had already shown that green iguanas are hindgut fermenters with microbes degrading plant fibre as effectively as they do in mammalian herbivores. This characteristic they apparently share with all herbivorous lizards (Nagy 1977; Iverson 1980; Foley et al. 1992).

In the field she found that adults would eat the leaves of a number of species, but the juveniles, and especially the hatchlings, were much more selective. A small tree legume, *Lonchocarpus pentaphyllus*, formed 82 to 88% of the dry weight of the diet of the iguanas. This legume was notable for producing new leaves almost continuously throughout both the wet and dry seasons. Furthermore the animals fed selectively on its youngest leaves; those containing the highest levels of digestible protein. Troyer also found, however, that, unlike most plants, the youngest leaves also contained the highest levels of fibre. All ages of iguana ate many more of these best leaves than the proportion of them available on the trees. The juveniles ate more of them than the adults; but the hatchlings were even more selective, eating practically nothing other than the classes of leaves with the highest content of nitrogen.

In the laboratory Troyer established that the young cannot ingest more food per body mass than can adults, but they can process food much more quickly through their gut. They may be aided in this, at least in part, by maintaining a higher body temperature. Regardless of this, relative to their body mass hatchlings and juveniles can process more food per unit time, and therefore assimilate nutrients at a greater rate, than can adults. However, she found that they do not digest energy any more efficiently than adults, and they do show the same true digestibility of protein. So, even with their faster rate of passage of food through the gut, they are extracting as much of the nitrogenous cell contents from it as the adults. This is probably because the younger leaves that they eat have a higher true digestibility of protein. And the young could degrade fibre just as efficiently as the adults. The only surprise, said Troyer, was that the young did not select food with less fibre. As the young leaves with the highest protein also contained the highest fibre this would seem to be difficult. Their selection highlights the fact that maximizing *available* nitrogen in the diet is of greater importance to growing iguanas than is avoiding fibre in their food. They were able, on a strictly vegetarian diet, to obtain the higher level of nitrogen they need by selecting a diet higher in digestible proteins, "creaming off" that which is available during fast passage, and digesting it up to twice as rapidly as do the adults. We have already encountered this strategy among diverse sorts of herbivores.

Troyer (1983) found they have other adaptations which are vital to the survival of a young iguana in its initial period of independent existence. First, like other oviparous reptiles and birds, they hatch with a store of maternal yolk. This amounts to about 14% of their body mass. They metabolize 70% of this yolk in their first week, but it continues to contribute to their maintenance for up to 4 weeks after hatching. Hatchlings that started with larger yolk masses grew faster in their first 3 weeks of life. Second, they must eat both soil and the faeces of adults to obtain the bacteria and protozoa essential for their hindgut fermentation (Troyer 1984c). Hatchlings eat soil in the first few days as they burrow out of the nest. Those that she hatched and reared without access to soil grew more slowly in their first 3 weeks of life. Once on the surface they exhibit a quite specific behaviour. They seek out and stay close by adult iguanas in the canopy of the forest; and continue to do so for the next 3 weeks. After that they disperse to low vegetation far removed from the adults, but continue to associate closely with each other. Troyer did not observe the young eating the faeces of adults in the field, but she reared hatchlings with and without access to adult faeces. Those without access grew more slowly from 3 to 5 weeks of age, and did not have the characteristic protozoans in their gut – nor did wild hatchlings prior to association with adults. She also noted that the young of other herbivorous lizards, including the Galapagos iguanid *A. cristatus*, were already known to eat the faeces of adults (Troyer 1982).

So, it seems clear that it is critical -and difficult – for young iguanas to acquire sufficient protein from their vegetable diet for survival and growth. It is also obvious that they must acquire from the adults the microorganisms essential for the fermentation of their food. There the records are left tantalizingly incomplete. Do they need 3 weeks to acquire these microorganisms? Eating one faecal pellet should be sufficient. Or do they continue to be dependent on coprophagy among themselves during their continued close association with each other? Do the adults also benefit from coprophagy? The extent and known benefits of coprophagy among other herbivores with hindgut fermentation would lead one to suspect that they might.

Chapter 9 Fish

9.1 The Carnivorous Young of Fish

Most fish produce enormous numbers of pelagic larvae with non-functional mouths and digestive tracts. These larvae start to feed only after their yolk sacs have been absorbed and their alimentary systems are functional (Laurence 1974, 1977; May 1974; Houde 1975, 1978; Hussain and Higuchi 1980; Lemly and Dimmick 1982). Fish biologists call this the "critical period". It occurs when larvae become juveniles, and when most die. Seemingly, at this stage, all juvenile fish, including those of species which are strictly herbivorous as adults, are carnivorous. They eat zooplankton, and it is the relative shortage – the dilution in space – of this animal protein which is the major cause of death. Variations in this accessibility of zooplankton determine the abundance of many species of fish from year to year. This is reflected in the year-classes of many commercial species for which detailed records are available. It is also becoming apparent that this death of young pelagic larvae/juveniles limits the abundance of sedentary, territorial marine fish which have traditionally been thought of as saturating their habitat (Victor 1983; Wellington and Victor 1985). For an individual juvenile fish survival is likely only when it happens upon a patch of zooplankton; the probability of its death from starvation is very high (Houde 1975; Houde and Schekter 1978). As Jones and Hall (1974) pointed out, it is unlikely that any predator would be able to make significant inroads into a population of young fish spread thinly (and patchily) through the top 50 or so metres of hundreds of square kilometres of ocean. Any predator in such a situation would experience a relative shortage of food similar to that experienced by its prey (see also Sects. 4.1, 4.2 and Part IV).

An interesting adaptation, which highlights this dependence upon an adequate concentration of protein food for the very young, is the breeding systems of the golden perch, *Macquaria ambigua*, inhabiting the large river systems of Australia (Rowland 1983). Each summer the gonads of adult fish develop and mature, but they breed only if there is a flood. If there is no flood they resorb their gonads in the autumn. When there is a flood water spreads over hundreds of square kilometres of land. In these shallow waters there is a large "bloom" of phytoplankton and zooplankton; a rich source of protein food for young fish larvae. What the token signal is that enables adult fish to recognize and respond to a flood is not known, but I imagine that it involves the fish detecting dissolved chemicals generated by the planktonic bloom (Lake 1967; see also Sect. 4.1).

The well-known story (summarized in Andrewartha and Birch 1984) of the Peruvian anchovy fisheries and their collapse following a bad El Niño perhaps epitomizes the position for the majority of fish. The pelagic larvae of the anchovy, *Engraulis ringens*, feed on zooplankton that have fed on the phytoplankton growing in the rich concentration of nutrients in the oceanic upwellings of the Humboldt Current. Adult anchovies feed directly on the phytoplankton (so are classified as herbivores rather than microbivores). "Guano" birds (cormorants, gannets, and pelicans) feed almost exclusively on adult anchovies, and more recently human predation on these fish has become a major factor. The periodic arrival of the warm El Niño current highlights the dependence of this trophic chain on limitation from below by the supply of nitrogenous food. The warm, but barren, El Niño overrides and suppresses the cold nutrient-rich upwellings. The growth of phytoplankton is depressed or ceases altogether. The result is starvation: first for the zooplankton, then the anchovies, and finally the guano birds. Stimultaneously, the fishing industry collapses.

Not all very young fish are pelagic and dependent upon the vagaries of the supply of zooplankton for their survival, however. There are fish which care for their young for some time after they hatch, and some also provide them with a high-protein food. This is common behaviour in several genera of brooding Cichlidae. The mucus cells in their skin increase markedly at mating time. Thereafter, they produce copious amounts of mucus upon which their newly hatched young feed by picking at their parents' flanks. Goldstein (1973) showed experimentally that it was impossible to raise the fry of some of these species if they are prevented from feeding on this parental supply of rich protein food. So rich and plentiful is this mucus that some small catfish have evolved the life style of subsisting on a diet of nothing else; picking the slime from the sides of other, larger species of fish (Winemiller and Hong 1989).

Most fish continue as obligate carnivores throughout their life. Many are facultative, opportunistic omnivores; some are exclusive detritivores. Only a few are obligate herbivores, but all experience a relative shortage of protein in their diet at one stage or another in their life, and require special adaptations to overcome this. As we have seen, even the strictly herbivorous ones must start life as carnivores. The need for animal protein for the growth of omnivores is clear, including those which, like the grass carp, *Ctenopharyngodon idella*, may at times be almost solely herbivorous. They grow ten times faster when fed animal food rather than plant food (Fischer 1973; Birkeland and Neudecker 1981). They can survive for long periods when fed a diet of nothing but plants, but grow hardly at all; a sensible adaptation for a poikilotherm experiencing a shortage of protein.

A good example of the importance of animal protein for the successful growth and reproduction of fish, including those which are mostly herbivorous as adults, is the work done by Vaughan (1978) on the sea bream, *Archosargus rhomboidalis*. This is an omnivorous tropical fish inhabiting beds of seagrass along the west coast of Central America. Vaughan studied the species in Florida's Biscayne Bay where he found, from analyses of stomach contents, that it changed its diet as it grew. As pelagic larvae they eat crustacean zooplankton. Small juveniles ate benthic crustaceans on the seagrass flats. As they grew they ate more algae; at first the fine filamentous species, then the coarser ones. They next included vascular plants (mostly the seagrass *Thalassia testudinum*, but other species as well). Fish over 20 cm long ate mostly vascular plants, but molluscs and other animals like tunicates growing on the blades of the seagrass became an important part of the diet of fish bigger than this.

Vaughan fed captive fish a series of diets: as much chopped shrimp as they could eat; a tenth of this amount of shrimp; various amounts and species of algae; a mixture of shrimp and algae; and a control group fed nothing. Offered a choice of shrimp or algae, the fish always chose shrimp. Offered algae alone they would eventually eat it, but ate least of the species with the lowest content of nitrogen. Only those fed to satiation on shrimp gained weight. The fish on all other diets lost weight to an extent not significantly different from that lost by unfed fish. His results showed unequivocally that the sea bream, although largely herbivorous as an adult, needs more protein in its diet than it can obtain from algae and vascular plants for growth – and probably for survival – when it is young. It is also likely that the females will breed successfully only when they can supplement their diet with animal protein.

9.2 Fish Which Eat Detritus

Then there are the detritivores. Detritus is dead organic matter, for all practical purposes derived solely from vascular plants, and mostly from old and senescent tissues of those plants. As a food it is therefore very deficient in nitrogen. So, animals which feed exclusively on detritus face similar or greater problems to those faced by herbivores feeding on living plants. Not surprisingly, they have evolved similar adaptations to increase the proportion of nitrogen in the food they ingest, and the efficiency with which they can extract and absorb nitrogen from that food.

The extensive and detailed work of Bowen (1976, 1979, 1980, 1981, 1984, 1987) on the freshwater tilapias (Cichlidae) probably best illustrates these adaptations in detritivorous fish. There are broadly three different, but necessarily overlapping and integrating ways in which they counter the inadequacy of their food.

1. They have evolved selective feeding behaviours, seeking out the type and age of detritus, the species of microorganisms growing on the detritus, and the sites where both occur, in such a way as to maximize the nitrogen in the food they ingest.
2. They have evolved several digestive specializations such as a longer gut, higher gut acidity, or grinding mechanisms. All these specializations contribute to an increased capacity to break down the cell walls of plants and microorganisms, allowing intestinal enzymes access to the cytoplasm, and,

in the case of the very long gut, allowing time for complete digestion of the contents, including non-protein amino acids from the detritus.

3. They benefit from the presence of microorganisms living on the detritus in two ways. First, the microbes break down the detritus before the fish ingest it [and possibly as gut microbes after the food is ingested (Dabrowski 1982)]. Second, the fish digest the bodies of the microbes. It is still not agreed just how great a contribution to the protein nutrition of detritivores comes from digestion of microorganisms. They may not be the sole source of nitrogen in the diet, but it seems that without them a diet of detritus may be barely adequate for survival, and certainly not for growth and reproduction. Either way, it is clear that fish which eat nothing but detritus are not solely herbivorous.

9.3 Fish Which Eat Algae

It seems that the young of herbivorous fish are all carnivores; a diet totally lacking in animal protein is inadequate for their survival and growth. Even so, like all fish, most of the young die while they are pelagic carnivorous larvae (Doherty 1983; Roberts 1991). Should they happen to survive to settle as juveniles, however, their chances of successfully establishing and adapting to a herbivorous diet are still very slim. Thus, Victor (1986) recorded that while the number of bluehead wrasse, *Thalassoma bifasciatum,* settling on coral reefs in the San Blas Islands of Panama was determined by the number of pelagic larvae in the current above the reef, 22% of the newly settled juveniles disappeared in the first 3 days, and in 3 weeks 56% of them had gone. Sale et al. (1984) reported that 25 to 30% of the young of sedentary species of coral reef fish, and 50% of those of the more mobile planktivorous species, living on the Australian Great Barrier Reef are lost during the first 5 days after settlement. Their total mortality in the first year was 40 to 50% and 60 to 75% respectively. Again, Sale and Ferrell (1988) found that, for 17 species of reef fish, most of them died in the first 1 or 2 weeks on the reef.

Most adult herbivorous fish feed on algae, not vascular plants. Such algal-eating fish tend to dominate the fauna of tropical reefs where they graze microalgae growing on the surface of rocks or dead coral. Others feed on macroalgae, and so encounter the general problem of herbivores of coping with large quantities of indigestible structural material, including cellulose, in their food. In both cases the fish experience a relative shortage of sufficient protein to support growth and reproduction. Consequently, they exhibit a similar range of behavioural and physiological adaptations to those shown by other herbivores for increasing their capacity to extract nitrogen from their diet.

First and foremost, they are selective in what they eat. As with vascular plants, all algae are not equally nutritious; in particular, their nitrogen content varies from species to species and from time to time. Most tropical and temperate fish which feed on macroalgae selectively feed on the green and red algae which are nutritionally superior to the brown algae.

Montgomery and Gerking (1980) found that two pomacentrids, the cortez and giant blue damsel fish (*Eupomacentrus rectifraenum* and *Microspathodon dorsalis*), living in the gulf of California, eat red and green algae and ignore brown and calcareous algae. The former fish does this by choosing particular species of green and red algae from a mixture of green, red, brown, and calcareous species; the latter by feeding only from a mat composed almost exclusively of one species of red alga. For both, the result is a marked concentration of nutrients generally, but especially of nitrogen, in the food they ingest compared with that available from a random grazing of the algae in the habitat. The fish are actively selecting the most nutritious food available to them, maximizing their intake of protein.

Horn et al.(1982, 1986) measured the selective feeding and food preferences, and the consequent assimilation of energy and protein throughout the year, of two species of stichaeid fish, *Cebidichthys violaceus* and *Xiphister mucosus*, in the temperate seas of California. Both species are almost exclusively herbivorous once past 4.5 cm in length. They feed selectively on eight to ten species of red and green algae which fluctuate in abundance seasonally. In the winter they feed mostly on a single species of perennial red alga which is rich in carbohydrate, but in summer they shift to feeding on species of annual red and green algae, especially four species which have a higher protein content than the others. This gives them a diet high in protein and low in carbohydrate throughout the summer and into autumn. So neither species is "optimally foraging" in terms of maximizing their intake of energy; rather, they are maximizing their intake of protein. Only during winter is their diet low in protein, either at or below maintenance level, and certainly insufficient to allow for growth. At this time there is little choice. The only generally available food is species of algae which are low in protein.

As adults, both species continue to ingest small quantities of animal food. The authors concluded, however, that the amounts recorded would have little, if any, influence on the overall diet; that they ingest by chance the few small invertebrates that happen to be on the algae they eat. In light of the unreliability of measuring accurately just how much animal food *is* ingested (as discussed in several earlier sections), and of the quite small amounts of such protein supplement which are needed to make a great difference to the survival and growth of an animal, this conclusion should be treated with caution.

Odax cyanomelas is one of a relatively few species of fish which eat brown algae. It feeds almost exclusively on the dominant kelp *Ecklonia radiata* growing on shallow rocky reefs along the central New South Wales coast of Australia. Andrew and Jones (1990) investigated the influence this fish has on the distribution and abundance of its food plant, but in so doing revealed the presence of a behaviour already encountered in many other herbivores, and reinforced the importance of breeding females gaining access to soluble nitrogen.

Odax repeatedly graze cleared patches in the kelp forests. They do not, however, feed repeatedly on regrowth of previously grazed plants, nor do they feed here all year. They eat the basal meristems and primary laminae of each plant, decapitating and "wasting" most of the older laminae. These plants do not regrow, and soon die. All this happens in the early spring each year when new *Ecklonia* are establishing, so the fishes' feeding establishes dynamic patches consisting of a single age-class of plants, in marked contrast to the surrounding mixed-age forest. Each spring the fish eat the meristems of vigorously growing young plants, establised in the previous season. The total area of patches in any locality does not change much from year to year, but their location changes slightly – a kind of "shifting cultivation".

Clearly this is another strategy that increases access to nitrogen as it is translocated into growing tissues.

However, the story does not end there. These patches are maintained exclusively by females, and close by where they will spawn. They congregate and feed in them prior to and during the season when they will enter the nearby territories of individual males to breed. The males, meantime, do not feed in these patches, nor do they eat meristematic tissue. They browse randomly through the surrounding forest, eating only lamina tissue.

9.4 Gut Microbes and Coprophagy in Fish

Choosing to ingest food with the most protein in it is, however, not necessarily sufficient. Herbivorous fish, like detritivores, have evolved various physiological and anatomical adaptations which serve to increase the proportion of ingested protein which is available for digestion. They do this by breaking down the cellulose and other complex polysaccharides of algal cell walls, thereby making the contents accessible to action by digestive enzymes. Two fairly general adaptations of this sort which are found in herbivorous fish, are grinding mechanisms in the gut and high gastric acidity. The first is an obvious alternative to mastication of the food in the mouth, something which fish do not do. The very acid pH values of 1.4 to 4.5, which are found in the stomachs of some species of fish, are as effective at releasing cell contents from ingested algae as are the grinding actions of pharyngeal mills or gizzard-like stomachs found in other species (Lobel 1981).

There has been much debate as to whether herbivorous fish produce their own cellulase or harbour microorganisms which might do this for them. Generally the consensus has been that there is no evidence for either, certainly among marine fish. Now, however, there is clear evidence that herbivorous fish, too, have evolved the same association with gut microbes as have so many other herbivores. Rimmer and Wiebe (1987) recorded the first example of fish having guts specialized to accommodate intestinal microorganisms. Two species of herbivorous buffalo bream, *Kyphosus cornelii* and *K. sydneyanus*, which are abundant in temperate and subtropical waters of Australia, are true hindgut fermenters. They both have greatly enlarged, thin-walled, caecum-like pouches housing an abundant and diverse microflora of bacteria and protozoans which produce volatile fatty acids. Both species have a long retention time of food in the gut (a mean of 21 hours compared with only a few hours for most other herbivorous fish). Both species browse exclusively and selectively on benthic macroalgae; *K. cornelii* on red algae (92% of the diet), and *K. sydneyanus* on brown algae (94% of the diet). The cell walls of both red and brown algae contain cellulose. Animal material was virtually absent from their diets (less than 0.25%). The fish feed by biting pieces from the thalli and swallowing them whole. They do not chew them up, nor do they possess a mechanism for grinding up the food in the gut.

However, these adaptations still cannot provide the protein needed by the very young. Rimmer (1986) demonstrated for *K. cornelii* that the post-settling juveniles are still partly carnivorous. While they feed mostly on the preferred red algae, biting off pieces like the adults do, they also eat a variety of small crustaceans. These invertebrates make up some 11% of the diet of the approximately 3-cm-long juveniles when they first settle, falling to about 3% some 3 to 5 months later when the juveniles have nearly trebled their length. Although the guts of newly settled juveniles have fully developed caecal pouches they contain no microorganisms; however, they soon acquire them, apparently by ingesting fragments of adult faeces. (The adults feed in the same zone as the juveniles, often in such densities that the water in the vicinity of a school is cloudy with their faeces.) As their caecal microflora developed, and once past the early stage of very rapid growth, their need for animal food would decline.

It would seem that coprophagy is far more common and widespread among herbivorous fish than this. Robertson (1982) set out to measure the extent of coprophagy among fish living on a coral reef at Palau in the Caroline Islands. He found a fascinating inter-specific web of consumption of faeces involving over 45 species, and constituting what he called a "food trap". There was a constant flow of faeces, starting with the high-protein ones of the top carnivores and passing down (through as many as four different species) to a range of species – mostly herbivorous – with increasingly poorer diets. Little if any reached the sea bottom. It is another good example (Sect. 4.1.4) of how valuable nitrogen is recycled rather than lost to deep waters.

All species except the top carnivores were coprophagic, many extremely so. Robertson calculated that some herbivores could ingest from 3 to 25% of their own wet weight during a 2 hour feeding bout. Furthermore, these herbivores practise coprophagy repeatedly and predictably. They regularly associate with the carnivores which produce the richest faeces, seeking them out and following schools of them while they feed. Individuals of any one species often ate the faeces of more than one other species, but always of those eating a higher protein food than themselves. None ate their own faeces or those of conspecifics. Robertson found that the protein and lipid content of carnivores' faeces is equal to or greater than that of the various algae eaten by the herbivores, so these faeces are more than just a supplement or chance addition to the herbivores' diets. They comprise a regular and significant protein component without which their diets of algae would barely be adequate for growth and reproduction. This process of serial coprophagy also serves as a food trap

Fig. 33. Cascading coprophagy. Three herbivores in the foreground feed on faeces of smaller planktivores above. They eat faeces of carnivores. Over 45 species, and passage through four different species, may be involved. Few faeces reach the sea floor. Photo courtesy of D.R. Robertson

in a second way. It brings considerable additional nutrients to the reef via the many carnivores which feed at night in surrounding beds of sea grass and rest on the reef during the day. Robertson has observed similar behaviour off the Caribbean coast of Panama, so he says it is probably a widespread phenomenon. He also makes two important observations about this coprophagy. When diets of fish are determined by analysis of stomach contents, either the faeces are unrecognizable as such and their contribution will go undetected, or, because of detectable remains from the diet of the *producer* of the faeces, a wrong picture will be gained of the herbivores' diet. I would add that his research is an excellent illustration of the fact that we find only what we have first thought to look for – the hypothesis must come before we count and experiment. How widespread and important is coprophagy as a source of protein food for herbivorous fish? Or has it just gone unnoticed, or been dismissed by observers as a casual and unimportant behaviour?

To return to the matter of symbiotic microorganisms. There are other herbivorous fish which have microorganisms in their gut but without specialized structures to contain them. Fishelson et al. (1985) found what are clearly obligate symbiotic microorganisms in the gut of the herbivorous surgeon fish, *Acanthus nigrofuscus*, in the Red Sea. These microorganisms included a species of the giant *Metabacterium*, also recently discovered in the gut of six different rodents (Kunstyr et al. 1988). They were absent from the stomach and reached maximum density in the mid region of the posterior gut. There was no distinctive change in the anatomy of the gut, but the microorganisms were closely associated with the lining of the gut rather than being dispersed throughout the food. It appears that the fish may readily lose their symbionts, as individuals starved for 2 or 3 days were apparently free of them. However, the authors observed behaviour in the field that would ensure constant reinfection. The fish maintain a bolus of incompletely digested algae plus symbionts in the posterior intestine overnight when they are not feeding. When they start to feed in the morning they discharge this bolus over the turf of algae which they eat.

Most recently Clements et al. (1989) have found that *A. nigrofuscus* living on Australia's Great Barrier Reef also carries these large *Metabacteria*. The authors also found them in other herbivorous and detritivorous acanthurids, but not in planktivorous members of the family. Nor could they find them in algal-eating members of six other families of fish, although they did find a wide range of other microorganisms, including various bacteria and flagellated and ciliated protozoans, in many of the taxa they examined. So it would seem that marine herbivorous fish are much more dependent on an active gut microfauna than was previously thought.

There are also records of freshwater herbivorous fish containing microorganisms in the gut. Dabrowski (1982) reports *Tilapia* possessing bacteria which synthesize essential amino acids and are probably digested as a source of protein. Similarly, it has been reported for five species of freshwater salmonids, including the rainbow trout, *Salmo gairdneri*, that their guts contain large numbers of bacteria which may be essential when these fish feed on "material not immediately digestible" and which may represent a source of nutrients for them (Trust and Sparrow 1974; Austin and Al-Zahrani 1988). And, of course, there were Dr. Obispo's 200-year-old carp with their intestinal flora ("So rich, so wonderfully varied!") which gave them immunity to old age through their production of fatty acids. Unfortunately Huxley (1939) did not reference the source of this information!

Part III: Survival in an Inadequate Environment

Let me now return to the six strategies within which, in Section 2.6, I grouped the adaptations which herbivores have evolved to ameliorate a shortage of nitrogen in their environment. For all but strategy F little more need be said; their significance and generality is apparent from the discussions in Part II, and will be further expanded for carnivores in Part IV. For all, however, a brief recapitulation may help to clarify the overall picture before I deal in more detail with strategy F.

Chapter 10 Strategies to Counter Shortage of Nitrogen

10.1 Strategy A: Synchronize the Life Cycle with Availability of Good Food

The first of these strategies should need little further expansion. The benefits of timing the breeding of females and the growth of neonates to coincide with the peak supply of good food in the environment are pretty obvious. By now I hope it is equally obvious to the reader that "good food" is that which contains the maximum amount of nitrogen in a form which is most immediately available for absorption and digestion. So the timing will be with that of the growth and reproduction of the plants – when they are mobilizing and concentrating nutrients to new growth and developing seeds. Sometimes, however, the association is more cryptic, or the timing may be out of phase with the obvious. The chief source of nitrogen might be animal prey which, in turn, depends upon the presence of new flush growth. Alternatively, the herbivore may be a senescence feeder. Often these events in the life cycle are programmed by indirect cues: temperature; day length; or a non-food chemical.

In this periodic world the timing of flushes of good food is fairly predictable – but synchrony is seldom exact. Also, the amplitude of peaks of good food is widely and unpredictably variable. Furthermore, the patchiness of the world means that the distribution of good food may be such that it is difficult to find enough of it often or soon enough.

Few things in nature are perfect. Sometimes caterpillars hatch before new leaves flush and they all die; a river does not flood, and there is no bloom of phytoplankton so gravid fish resorb their ripe gametes; the rains fail and there is no new grass so kangaroos do not breed.

Such perturbations of the periodicity and amplitude of abundance are rarely absolute, however, nor of equal magnitude everywhere at once. This, plus the patchiness of the environment, means that while many may die and many local populations become extinct, a few widely scattered individuals will survive to breed next time. Were it not so the greater population would go extinct.

On the other hand, when there is good synchronization with a great abundance of food that is easy to find (an equally rare and unpredictable coincidence) the herbivore will become widespread and abundant. Good rains – much new grass – a plague of rabbits. El Niño rains – much new growth – many more caterpillars – explosive breeding of Galapagos finches. But only for a short time. Such events are also short-lived. And the story is no different for carnivores (see Parts IV and V).

10.2 Strategy B: Concentrate or Prolong Availability of Nitrogen in Food

Having synchronized with the presence of good food the simplest way for any herbivore to increase the level of nitrogen in its diet would be to selectively seek out and eat only those plants, or parts of plants, containing the most nitrogen. The evidence is that all herbivores do this. They quickly detect and feed upon any food plant in which the level of nitrogen has increased, no matter what might have caused that increase.

More specifically, however, they seek out just those tissues which contain nitrogen in a form that they can most quickly and efficiently absorb and digest – simple molecules which do not need the expenditure of time, energy and expensive enzymes to break them down, and which are not cloistered inside impregnable cell walls. This narrows the choice to tissues in which nitrogen is being mobilized and concentrated to sites of vegetative or reproductive growth; or away from sites of senescence or stress-induced breakdown of tissue. And it is the breeding females and fast-growing neonates for which access to this soluble nitrogen is so important. Fully grown animals, needing just to maintain themselves, have much less call on this scarce resource.

There are innumerable variations and gradations of this theme: aphids "plug in" to the petiole of an expanding leaf; red grouse select the actively growing tips of heather; rabbits and kangaroos choose new flushing grass; mice, goldfinches, parrots, and mirid bugs select milk-ripe or germinating seeds; *Helicoverpa* caterpillars select the forming seeds of beans and tomatoes in my garden. Others select pollen; some bats spit out fibre and swallow only the juice of the fruit they eat. And these preferences override the presence of various secondary chemicals or fibre.

A refinement of flush feeding is to induce the same plants to produce more new growth for longer by repeatedly grazing and cropping them. Many herbivores – bovids, geese, hares, turtles, limpets, caddisflies – have gone down this evolutionary path. Another way to prolong and concentrate the flow of good food is to induce the plant to grow a gall – a "nutrient sink".

Less obvious, and it would seem less common, are the senescence feeders. They use the soluble nitrogen that the plant exports from the breakdown of tissues. This tends to be a slow business compared with the flush of new growth. Consequently senescence feeders grow much more slowly than flush feeders. In addition some have evolved the capacity to induce the plant to speed up the breakdown process, releasing more nitrogen more quickly than it would otherwise have done; the herbivore hastens senescence. This adaptation is particularly well developed by some aphids and psyllids.

There is a third sort of herbivore, like the aphids on sycamore, or some gall formers and leaf-miners, which takes advantage of both the import and the export of nutrients.

Again, nature is not perfect, but patchy and changing. So, selection of the best food will usually be only partially successful. It will achieve better success than random feeding, but not every individual will succeed in finding the right food at the right time in the right place. Nor will any one individual necessarily find enough food soon enough to achieve its maximum breeding potential. For a hungry animal – and animals in nature are almost always hungry – there is no single perfect solution. It has little time to find and try all kinds of potential foods. Even with the best developed strategy for detecting the best food, in a sea of inadequacy an individual will often make the wrong decision. Natural selection will, therefore, settle for a workable compromise for variable conditions in a world where life will always be risky.

It follows that no matter how highly selective individuals might be for a particular resource, they do not have the capacity, singly or collectively, to exhaust that resource. Before this can happen the combination of the patchiness of the habitat plus the stochastic nature of their behaviour, will result in a relative shortage of the resource acting as an upper limit before there is an absolute shortage. The resource is present but not *available* because it is not attainable or accessible in time or space. A population of individuals will press against this level of availability. And all this is as true for carnivores as it is for herbivores (see Part IV).

10.3 Strategy C: Eat More Food More Quickly, and Digest More Efficiently

Having synchronized with the times when the plant is most nutritious, and learnt to select and eat those tissues which are richest in nitrogen, there are yet other ways to concentrate this precious resource. An obvious one is to eat faster and/or spend more time eating. Many herbivores – from mammals to cladocerans – have been shown to vary their rate of intake of food in response to changes in its quality. The paramount nutrient influencing this behaviour is nitrogen, except when the food already contains sufficient for survival and growth, or the herbivore is fully grown and has a greater need of carbohydrates for energy. And, always, it is the proportion of that nitrogen in accessible form, not the total amount in the food, which is crucial.

Mammals and birds which feed upon the flesh of fruit, or nectar and pollen, have developed this process to a fine art – some process food through their alimentary canal in a matter of minutes. So, too, do others. Geese feeding on flush growth; sap-sucking insects, especially xylem feeders; and, less obviously but achieving the same end, some leaf-eating caterpillars. For all such animals the advantage gained seems to be the same. They "cream off" that which can be immediately absorbed and discard the rest. This way they get more nitrogen from their food than they would by holding it for longer and breaking down more cell walls and complex molecules. The less nitrogen there is in bound form, the truer this will be.

However, this tactic means that all these sorts of animals take in far more carbohydrate than they can use. The volume relative to their own body weight becomes huge. Most can extract all the energy they need from the sugars already in solution. The rest, mostly starch, is discarded along with any fibrous material in their voluminous faeces. All this surplus energy has provided the wherewithal for the evolution of some adaptations not otherwise possible – like the raised temperature and long foraging flight of some bats, and migratory flights of many birds. Sap-sucking aphids, burdened with this surplus carbohydrate in the form of sugars, have overcome the potential osmotic problem of so much sugar in the gut by converting much of it to more complex oligosaccharides. A curious end point of this adaptation is found in many lerp-forming psyllids of Australia. In their alimentary tract they convert the surplus sugar to starch – and eject it as liquid faeces which quickly solidify and are used to construct the lerps beneath which the psyllids shelter.

Because of the low level of nitrogen in plants *all* herbivores pass far greater volumes of faeces than do equivalent sized carnivores. However, there are limits to the volume of food that can be passed, just as there are limits to the speed with which nutrients can be digested, especially once the readily absorbed contents of the cell have been used.

This situation is commonly encountered and best documented for adult mammals, both grazers and browsers. At times, usually in winter or during the dry season, their food has so little nitrogen in it that they lose weight, and sometimes starve to death, even though they are eating constantly and always have a full stomach. They just cannot process enough food quickly enough for basic maintenance metabolism. The examples of the quokka and the koala clearly illustrate this.

Even when the food *is* adequate to maintain a full-grown animal, however, it may not be so for breeding females and neonates. In spite of the benefits of placental feeding and mother's milk it may still not be possible to take in sufficient nitrogen for the successful production and growth of new individuals. For this reason it is unlikely that studies of the feeding efficiency of near-adult mammals (or mature larvae of insects for that matter) have much relevance to the growth and abundance of their populations. Such animals are past the stage of exponential growth, and may well subsist from day to day on a diet which would quickly have been lethal were that all they had when young; or, at best, have been inadequate had they been females trying to develop viable young.

10.4 Strategy D: Enlist the Help of Microorganisms

No vertebrate is known to be able to produce cellulase, and few invertebrates seem to have this capacity (Martin 1983). Consequently, most animals cannot digest the cellulose which comprises the bulk of the tissues of plants; nor, therefore, can they readily gain access to the contents of the cellulose-walled cells of plants. We have seen that one way to overcome this problem is to abandon all nutrients which are thus not readily available, and quickly absorb those which are. On the other hand it is not unexpected to find that herbivores have evolved a variety of associations with microorganisms which can digest cellulose. Many of these associations have been mentioned in the sections on different sorts of herbivores, but there are many more that I have not discussed. They range from the complex of different associations that are found in the alimentary tracts of all mammmalian herbivores (Langer 1988; Hume 1989) through equally diverse interactions in the gut of birds, reptiles, fish, and a range of invertebrates (Brooks 1963; Buchner 1965) including those which live on plant detritus (Kurihara and Kikkawa 1986).

There is still debate, however, no matter what the animal involved in the association, as to whether the microorganisms merely provide the herbivores with access to the plant's cell contents and the by-products of microbial metabolism, or whether the microorganisms are themselves a significant protein food for the herbivores. The former is symbiosis; the latter cryptic carnivory. I think the evidence supports the existence of both functions, but the latter is far more extensive and important to the nutrition of herbivores than is generally conceded. The cow grazing on grass is actually a carnivore eating the microbes which eat the grass.

In mammals it seems clear that microorganisms provide a large part of the nitrogenous food of ruminants (Hungate 1966; Parra 1978; Orskov and Kay 1987; Wallace and Cotta 1988). The same seems true for all forms of fermentation in the foregut. Microorganisms that grow there on the plant material are themselves digested in the small intestine. This is not necessarily true for all mammals which have hindgut fermentation, but it is for those with caecal fermentation. They reingest the microorganisms concentrated in special faecal pellets which are then digested in the intestine (Janis 1976; Hintz et al. 1978). It is generally considered, however, that those mammals in which fermentation is chiefly or solely in the colon cannot benefit from the digestion of their microfauna. That they do not practise coprophagy would support this contention. Rather it is thought that they may benefit by extracting more nutrients per unit time than can foregut fermenters (Duncan et al. 1990). Nevertheless there is clear evidence that mammalian hindgut fermenters which are not coprophagic can absorb and use microbial protein as efficiently as can the ruminants (Parra 1978). This certainly seems to be true for many birds which have caecal fermentation (Fenna and Boag 1974b; Obst and Diamond 1989) so why not for mammals (Foley and Cork 1992)? It would be much stranger to find that only this one group of the many herbivores which have evolved such associations with microorganisms was not benefiting from that association.

Similar associations with microorganisms in the gut are found across the whole taxonomic spectrum of herbivores, and all confer the same major benefit – additional protein in the diet (Conover 1968; Breznack 1982; Connor and Edgar 1982; Martin 1983).

However, the microorganisms themselves cannot grow and reproduce without access to nitrogen. They, in turn, depend for their success upon the quality of the diet of their host, and, like their host, they need to supplement the small amounts available in the plant material. Those housed in the alimentary canal do this by using the metabolic nitrogen of the herbivore, delivered to them in a variety of ways: as urea in the saliva or absorbed through the walls of the rumen; as urine recycled from the cloaca of birds to the caeca; as uric acid from the Malpighian tubules to the paunch of termites. And in so doing they benefit the herbivore, converting its unavoidably wasted nitrogen to digestible microbial protein (Hungate 1966; Breznak 1982; Mackie 1987). Finally there is growing evidence that bacteria, both those living in the guts of animals and free-living ones, can fix atmospheric nitrogen. In so doing they also make a significant contribution to the protein nutrition of many herbivores (Breznak 1982; Waterbury et al. 1983; Leschine et al. 1988; Carpenter and Romans 1991).

There is another large and diverse group of herbivores which depends on microorganisms breaking down the plant food *before* it is ingested. These herbivores range from ants and termites which maintain "gardens" of specific fungi growing on either fresh or partly decayed plant tissue, to the detritus feeders, those which simply ingest dead and variously decayed plant material along with the microorganisms growing on that material. Again, there is controversy as to whether these microorganisms are a direct source of protein for the herbivore, or merely aid it by breaking down its plant food.

Many detritivores can survive for a time on detritus alone, depending upon its intrinsic nitrogen content. However, they grow little or not at all, and cannot complete their life cycle without the additional nitrogen provided by the microorganisms which can digest the otherwise refractory cellulose. This has been demonstrated for, among others, collembolans (Booth and Anderson 1979), caddisflies (Becker 1990), chironomids (Ward and Cummins 1979), mosquitoes (Fish and Carpenter 1982), and polychaete worms (Tenore 1977; Findlay and Tenore 1982).

So it seems likely that for most detritivores the microorganisms are their major – if not their sole – source of protein food (Kurihara and Kikkawa 1986; Smith and Douglas 1987; Werner and Dindal 1987).

10.5 Strategy E: Supplement Plant Food with Animal Protein

It would seem, then, that the main benefit that a herbivore gains from its association with microorganisms, whether internally or externally, is a significant and often essential supplement of protein to its vegetable diet. But this may still not be enough, even in association with one or more of the other adaptations already outlined. Whenever and wherever we look carefully and thoroughly enough it seems we find that herbivores need to supplement their diet with the tissues of other animals. Some, both male and female, must do this to survive from day to day; many females are unable to breed without animal protein; all neonates must have it (White 1985). Fast growing young herbivores obtain animal protein either directly from their mother (placental feeding, stored egg protein, milk, yolk sacs, crop milk, skin mucus, proteinaceous eggshells, trophic eggs, eating siblings or their mother's body) or by predation upon other species of animals (with or without the help of their parents). There are, however, some apparent exceptions to this general rule. I say "apparent" because so many strict herbivores have proved not to be upon closer inspection (a strict herbivore is an animal which obtains *all* its protein from vascular plants). Some seem to have achieved true herbivory by a very fast passage of food through the gut. Others eat only plant tissue which is especially high in readily digestible nitrogen. However, even among these sorts of animals there are many which have been found to eat arthropods, usually soft-bodied immature insects. Often such food is but a small fraction of the total diet, and may be eaten only by the very young. The point is, however, that without that small but vital input the animal would have died. How many others, I wonder, rely on such a supplement which has never been recorded? Often its ingestion may be cryptic and intermittent. Once swallowed it is quickly digested so that it escapes notice in stomach or faecal samples. This would seem to be true for studies of animals as diverse as gorillas, snails, and fish, as well as for the better known examples of geese and rodents; it is true even for detritivores (Anderson 1976). And if, like the Western Australian corellas, the contents of insect larvae are ingested, and the exoskeleton discarded, there would be nothing left to notice.

All this is especially true if the investigator is not looking for it! I am reminded of a story (possibly apocryphal) from my time in the Pacific Islands about the anthropologist who could not understand why the children in the island community he was studying were so fat and healthy. From his recording of the meals they ate once weaned – mostly very starchy vegetables (their elders ate what meat was available!) – they should have been suffering from severe protein malnutrition. What the anthropologist had not recorded was the things they ate *between* meals: lots of insect larvae, marine invertebrates, and seeds. McKelvey's observations of the carnivory of Kangaroo Island wallabies is a good, and definitely not apocryphal, example. When I asked a colleague who had worked on the nutritional physiology of these animals what he thought of these observations, he simply rejected them as a ridiculous mistake. Thus can our preconceptions blind us to the obvious.

There is one group, however, for which there seems no doubt as to their credentials as true herbivores, namely larval insects which eat leaves. But not all of them; many rely upon microorganisms, eat proteinaceous egg shells, or are opportunistic predators or carrion eaters. There are many, however, especially among the Lepidoptera and Orthoptera, which show none of these adaptations, or if they do, can yet be raised for many generations on a diet of nothing but the leaves of their host plant. They do not have a particularly fast passage of food through their gut, and in many their faeces are made up of large pieces of leaf with many of the cells still intact. Certainly most feed very selectively upon flush new growth, but they do not necessarily eat reproductive tissues. What adaptation do they have which enables them to do what no other herbivore seems to be able to do? They eat enormous quantities of food, and grow as much as five times faster than mammals and birds (Reynolds 1990). Is it their recently demonstrated capacity to digest the contents of intact plant cells in their gut (Barbehenn 1992)? What have we missed? Whatever it is, it can make them remarkably successful from time to time. Many of these insects are among the most destructive pests of our crops. Yet, apart from the artificial world of human agriculture, they are, like all other herbivores, rarely able to destroy more than a tiny fraction of their food plants. They lose most of their young each generation soon after they start to feed. Some occasionally explode to huge abundance without any prior human intervention in their ecology (see Part V), but for most of the time they too remain on the knife edge of inadequacy.

10.6 Strategy F: Apportion and Concentrate the Limited Food to a Selected Few

I think most readers would agree that the various adaptations I have been discussing are consequences of the shortage of nitrogen in the food – they represent "attempts" by the animals to gain sufficient of this scarce resource to survive and reproduce. They are an *inevitable* outcome of the process of natural selection operating in an inadequate world.

However, the proposal that the evolution of territorial and social behaviours of both carnivores and herbivores, including the phenomena of philopatry and "helping" by the young, have also arisen in response to this same inadequacy may not find such ready acceptance. The further suggestion that cannibalism is a widespread and similarly driven behaviour, and that *all* these behaviours operate to *increase*, not decrease, the numbers which successfully use this scarce resource, by the simple strategy of allowing only some of the many to gain access to it, may prove even less readily acceptable.

So I have kept a discussion of examples illustrating these points until now. I feel they are better presented together rather than scattered throughout the earlier discussions of different groups of herbivores. Furthermore, an understanding of the true function of these social behaviours may be more readily apparent to the reader who has already absorbed the evidence of those earlier examples.

Chapter 11 Territorial and Social Behaviours

It is generally accepted that territorial and social dominance behaviours are of major importance in directing the course of *evolution*; they lead to competition to decide which individuals will survive and contribute to the next generation. But an explanation of the *ecological* origin and significance of these behaviours is more contentious. Most people would agree that the proximate cause is a shortage of a resource, usually, but not always, of food. Many, perhaps most, also support the concept that social and territorial behaviours function to *reduce* numbers in a population to below levels which would be attained in the absence of such behaviours, that they act in a density-dependent manner to prevent all the resources in the habitat from being used.

I would agree that these behaviours are in some way caused by a shortage of resources, but I would not agree that they decrease the number which ultimately survive in populations. On the contrary, they *increase* it. There are not enough resources for all. One way to make sure that what resources are available in the environment are used as effectively as possible is to channel them to only *some* of the individuals seeking to use them. All but a "favoured" few are denied the "privilege" of using the limited resources for the production of another generation.

This is what these behaviours do, and in so doing they *maximize* the number of individuals that the environment can support in each generation. They do not conserve resources for the next generation. Every last bit of resource that can be found by members of each generation is used to produce the maximum possible number of breeding members of the next generation. Members of the next generation must, in turn, look to their own devices to maximize *their* production of young.

So, these behaviours, in all cases, serve not to decrease but to increase the numbers that are produced with the limited resources available. They maximize the efficiency with which the individuals making up a population convert the food to which they can gain access into more individuals in succeeding generations. Those individuals which are most successful at doing this will contribute most genes to those succeeding generations. It is the shortages – the inadequacies of the environment – which drive natural selection. Competition, and the selecting-out of some individuals that goes with it, are both consequences of this inadequacy. The fact that these behaviours force decisions as to *which* individuals are successful in gaining access to a resource is immaterial in maximizing the *number* of individuals that a given amount of resource can support.

Many reject this concept as the very antithesis of what a structured society should be about. They are drawn instead to concepts of fair-sharing of resources, and nature in balance. But in nature such terms have no meaning; the human concepts and ideals they represent do not exist. For any form of life to survive and persist in nature it must be constantly and selfishly pressing at the limits set by its environment. Any traits which cause an individual *not* to do this will quickly be eliminated by natural selection. Only phenotypes which inherit traits which make them better than other phenotypes at gaining access to what resources are presently available will persist and contribute offspring to the next generation (Dawkins 1989). And if this is so, ideas like those of Wynne-Edwards (1962) and Chitty (1960) about "self-regulating mechanisms" can have no substance. Such hypotheses are anyhow unnecessary to explain the levels of abundance of animals if, indeed, all are constantly faced with a shortage of resources.

Different sorts of resources may be in short supply, and, in general, an individual will most often, and most vigorously, defend the resource most often likely to limit its capacity to reproduce. A female's reproductive success will depend most on her access to proteinaceous food; a male's on his access to females. So, quite independently of the female's defence of a feeding territory for her and/or her young, a male may defend a mating territory, or an actual or potential mate, against other males. There are innumerable variations, combinations, and overlaps of the defence of these two resources by the two sexes, but, by and large, the supply of females for males is dictated by the supply of good food for the females.

In this section I am concerned to illustrate the way in which territorial behaviour functions to increase the numbers in a population by better apportioning of the food that is accessible to the breeding females and their young. To achieve this it is not important which males are successful in mating with which females. So, I will set aside consideration of this aspect of territorial behaviour, except to note some examples which clearly distinguish it from the females' behaviour, and which show how the success of males in acquiring mates and producing young is ultimately dictated by the availability of food for the females. For those who wish to confirm this difference for themselves, an obviously distinct male territoriality is that of adult insects (Baker 1983; Fitzpatrick and Wellington 1983). Other examples (some of which I discuss here) where male territoriality is variously but clearly distinct from the sequestering of the available food resource by the female are recorded for hedge sparrows (Davies and Lundberg 1984), passerines in general (Temrin and Arak 1989), microtines (Ostfeld 1985), chipmunks (Gregory et al. 1988), squirrels (Armitage 1988), bears (Rogers 1987), lizards (Simon 1975; Waldschmidt 1983; Guyer 1988a, b), and fish (Andrew and Jones 1990). But, of course, in nature there are always exceptions, as illustrated by the example of male pied wagtails defending a supply of winter food (Davies and Houston 1981).

There are other resources which might sometimes be limiting. Nesting holes or burrows are commonly mentioned examples. Often the use of such resources is flexible. Kluyver and Tinbergen's (1954) study of titmice (*Parus* spp.) in managed broadleaf and pine plantations in The Netherlands illustrated this. Even though artificial nesting boxes were provided in excess, the number of birds breeding in the plantations (and thus the number of boxes occupied) varied between pine and broadleaf, and from year to year, according to the quality of the habitat.

However, what happens in nature when human interference is not an influence is more revealing. There, apparently less "desirable" but adequate sites are left unused when few survive; when numbers increase in response to an increase in food they are occupied. At very high densities various makeshift arrangments may be made. The digging of warrens by rabbits in the Australian desert country is a case in point. When numbers are low the few surviving rabbits occupy only the best sites which provide maximum protection from the weather and predators. When numbers increase these good warrens are extensively enlarged, and many new good ones are dug. At very high densities many rabbits will attempt to establish breeding burrows where their chances of successfully raising a litter are small, but some will succeed so long as there is an abundance of green feed (Sect. 6.2.5.2).

The kangaroo rat, *Dipodomys spectabilis*, living in the Chihuahuan desert of Arizona provides a very similar example (Jones et al. 1988). The rats construct breeding mounds which provide protection from the weather and predators, and a place to store seeds. These mounds last for decades once constructed, but may lie vacant and neglected for years. When numbers of rats increase following unpredictable summer rains, old mounds are readily rehabilitated and reoccupied.

Another example where the unused resource was not so readily apparent is that of the large cactus finch on the Galapagos Islands outlined in Section 1.2. In response to a sudden superabundance of food not only did they breed continuously, and in parts of the habitat where they had never been recorded breeding before, they changed their nesting behaviour. Previously all birds had nested in cactus bushes, but during the El Niño year more than half their nests were built in trees. What seemed an upper limit of supply – cactus bushes – was not.

There are numerous examples of individuals in expanding populations modifying their behaviour in this way to accommodate greater numbers. In so doing they reveal that an apparently limiting resource was not limiting. There is simply a gradation of suitability, and thus of preference and use. Only when there is more food do more survive and increase pressure on other, less limiting resources.

The weight of published evidence clearly shows that the territorial behaviour of females is driven first and foremost by a relative shortage of food with which they may breed and raise their young. It is only then that the great majority of females defend a territory. At other times they are tolerant of, or indifferent to, conspecifics. I could not attempt to review the immense number of these studies. Sufficient that I describe a few examples which will serve to illustrate how general this adaptation is within the animal kingdom, and how numerous are the variations on the theme produced by natural selection.

11.1 Territorial Behaviour in Carnivores

Although this book is mostly about herbivores, it is worthwhile at this point devoting a little space to carnivores. The link between their territorial behaviour and the supply of food in the habitat is more obvious and more widely recognized than it is for herbivores.

11.1.1 Birds

This link seems to be especially true for birds of prey. Newton's long-term studies of the sparrowhawk, *Accipiter nisus*, in Scotland have clearly shown the overriding importance of the supply of its prey as the ultimate factor determining its distribution and abundance (Newton and Marquis 1986). Newton et al.(1977) demonstrated that in any one area the sparrowhawks' territories are regularly spaced, the mean distance between territories being strongly correlated with the altitude and agricultural productivity of the land. These proximate factors were important through their influence on the supply of woodland songbirds which comprise 97% of the prey of sparrowhawks (Newton and Marquiss 1982).

Other studies confirm this dependence of the size and spacing of various raptors' territories (and consequently of their breeding success) on the supply of their food. Examples include kestrels and long-eared owls (Village 1981, 1982), goshawks (Kenward 1982), ferruginous hawks (Woffinden and Murphy 1978), tawny owls (Hirons 1982), and the African hawk-eagle (Hustler and Howells 1988, 1989).

In some cases, such as that of the tawny owl, the size of the territory does not vary; the number of young raised is related to the abundance of the prey available in the territory (Southern 1970). When supplies of food dwindle, first those pairs holding the poorest territories, and ultimately all pairs, fail to raise *any* young. In very lean years many pairs may not breed at all.

There is a further refinement of the business of adjusting the number of individuals gaining access to the available food which is very common among birds of prey. The oldest/largest nestling kills or expels its lesser brethren, or the parents ignore or reject the weakest young. This phenomenon is not confined to raptors, however, and asynchronous hatching may help adjust this behaviour between good and lean years for many sorts of birds (Magrath 1989).

A good illustration of both extremes of the continuum of response to the changing abundance of food is the study by Wallace and Temple (1988) of the carrion-feeding Andean condors, *Vultur gryphus*, of Peru. These large, long-lived birds breed continuously at high elevations in the Andes where there is a fairly constant and adequate supply of carrion. By contrast, in the deserts of the coast and foothills food is limited, except during and immediately after an El Niño. Here, condors defend their nesting sites and maintain pair bonds in the drought years between El Niños, but they do not breed. During an El Niño the food supply of the coastal condors decreases because changes in ocean currents prevent so many dead marine animals being washed up on the beaches, but the food supply of the foothills birds increases because of the great increase in the number of deaths among free ranging livestock attributable to the huge increase in rainfall during an El Niño. (In 1982–1983 there was as much as 4.26 m of rain in 9 months in an area which had averaged 9 cm per annum in the previous 5 years.) Pairs which have not bred for years promptly initiate breeding in response to this sudden bounty of good food, while none of the coastal pairs breeds. Because of the vagaries of the El Niño cycle, however, the desert dwelling condors seldom produce sufficient young to maintain their populations, which are replenished by immigration of some of the "surplus" production of the highland birds.

Similar relationships exist between the size of territories and the supply of food for a variety of other carnivorous and omnivorous birds. Temrin and Arak (1989) illustrated for European passerines the relationship of female territoriality to the supply of food. They also showed the distinction between this and territoriality of males as a means of sequestering females, and that it is the amount of food available to the females which drives the whole complex "polyterritorial" system.

Davies and Lundberg (1984) working on the omnivorous hedgesparrow, *Prunella modularis*, found that where food patches were dense female ranges were smaller. When they experimentally provided extra food on some territories, female ranges became smaller, and males' access to females increased! The territories of males overlap the feeding ranges of females. Males may remain unpaired, share access to one female, have sole access to one female, or, ultimately, gain sole access to more than one female. The smaller the home range of females, the easier it is for males to control more than one female; the larger the home range, the more difficult is control. So, the gradient of access from none to more than one for the males follows decreasing size of the females' home ranges which, in turn, decrease as food supply increases. The more food there is in the environment, the more hedgesparrows there are successfully breeding; but when food is scarce, only the few most dominant females (and males) will gain access to it and breed.

11.1.2 Lizards

There are many examples of exactly the same relationship of food to territoriality among other sorts of animals, but these are mostly not so well known. Guyer's (1988a, b) work with the small anoline lizard *Norops humilis* in Costa Rica is a good illustration. He demonstrated that by attracting many more insects to pieces of rotting meat he could greatly increase the supply of food for the lizards, and double the density of their population. This created, as he aptly put it, a "packing problem" for these animals. They did not resolve the problem by decreasing the size of home ranges of individuals, but by a complex series of overlappings of these ranges. The greatly increased number of juveniles in the population clumped themselves around the sources of the supplementary food. They also increased the extent to which their ranges overlapped; not only with those of other juveniles but with those of adult males. Adult males increased the overlap with those of other males, of adult females, and of the juveniles. Adult females increased their overlap with both males and juveniles, but *not with other females*. Instead, extra females occupied areas which had previously been unoccupied.

So the pattern is clear. Food is limiting the abundance of these animals, and the available supplies are channelled via home ranges to those most able to successfully defend them. "Surplus" young disperse and are doomed unless they can find other adequate and unoccupied areas. Any increase in the amount of food available is immediately converted to more offspring. Females produced more young more quickly and more of these survived; also extra females were able to breed where previously there was not sufficient food for them to do so. Furthermore, they tolerated more of these young (and more immigrating males) on their territories, but each female maintains her minimum home range separate from those of other females. The males responded to the increase in their food by growing faster and bigger, channelling the additional resource into an increased ability to gain access to females.

11.1.3 Insects

As I have mentioned, most attention to territoriality in insects has been directed towards mating behaviour. There are abundant examples of males defending mating territories, potential mates, or inseminated females against other males. There are examples of males and females protecting young or eggs from predation (Baker 1983; Fitzpatrick and Wellington 1983). However, as with other animals, it is to the breeding female and the fast-growing young that we should look for examples of territorial defence of limited supplies of food. Jermy (1985) draws attention to some examples of female beetles, butterflies, and flies marking the place where they have laid their eggs to deter others of their kind laying more eggs at the same site. He sees this, however, as behaviour which prevents competition for food – which keeps the population *below* the level that the available food could support. Two other such examples of females marking the feeding territory of their young – that of the predatory wasp (*Asolcus basilis*) laying her eggs in the eggs of the green vegetable bug, *Nezara viridula*, and of the *Dacus* spp. fruit flies laying their eggs in fruit, are elegantly described (and correctly explained!) on pages 46–47 of Andrewartha and Birch (1984). The females' behaviour sequesters the available food to only as many of those seeking to use it as can do so effectively. In the case of *Asolcus*, in the rare event of a second egg being laid in a *Nezara* egg, the first larva to hatch kills the second. If it did not do so neither would survive on the amount of food in the one *Nezara* egg. It is not difficult to see the similarity of this behaviour to that of larger nestling birds killing smaller siblings when food is in short supply.

More "conventional" forms of territorial behaviour have been demonstrated for the larvae of aquatic insects. Hart (1985a, b, 1987a, b) records territoriality in 29 species spanning seven orders of insects. Twelve of these feed on invertebrate prey, ten are filter feeders, and seven graze microalgae. One example which clearly demonstrates how the maintenance of a territory operates to sequester the available food is his study of a black fly, *Simulium piperi*, (Hart 1986). The larvae of these flies are almost entirely filter feeders. They are found in patches in a stream, clumped where the speed and turbulence of the current concentrates the flow of food particles in the water. Within these patches they are regularly spaced and exhibit strong aggressive behaviour associated with the defence of filterable food in the current. Hart presented three kinds of evidence which indicated that this aggressive behaviour is associated with the defence of the filterable food in the current:

1. A larva nearly always acts aggressively to its upstream neighbour – the only neighbour which can seriously influence its supply of food.
2. Following the displacement of its upstream neighbour a larva's short-term ingestion rate increases significantly.
3. When Hart experimentally increased the filterable particles available in the current the frequency of territorial interactions declined.

11.1.4 Spiders

Perhaps, for many, spiders would seem to be even less likely than insects to exhibit territorial behaviour in defence of food. *Agelenopsis aperta* is such a spider. It is a ground dwelling species building a non-sticky sheet web with a funnel in which to shelter. The spiders capture prey hitting the web and "choose" sites according to their temperature and the availability of prey. They defend an area larger than the area of their web. Good sites may be in short supply in many years, but spiders that occupy excellent sites can achieve 13 times the reproductive success of those in sites lacking good features. Riechert (1981) studied populations of this spider at two sites: a "stringent" desert grassland in New Mexico, and a "more favourable" riparian site in Arizona. The better site provided three times more prey and a higher temperature which allowed 2.5 hours more hunting time per day than did the desert site. The average density of adult spiders on the poor grassland site was three times less than on the riparian site, and the mean size of territories was six times larger. Riechert demonstrated that the size of these territories is fixed and inflexible for each population (at least in one generation). When she established spiders from each population in experimental enclosures with isolated but equal-sized territories, and then removed the barriers and allowed the spiders in each enclosure to interact, they adjusted the spacing of their territories to nearly that of their respective field habitats. This spacing held whether the spiders were satiated or had been deprived of food for up to 4 weeks, and came about as the result of aggressive encounters between neighbours with 90% of disputes being won by heavier individuals.

While spiders maintained their fixed territories in these experiments, the size of the web they maintained varied with the availability of food. Satiated spiders maintained smaller webs than deprived ones. In fact the majority of satiated spiders in her experiments let their webs deteriorate and merely came to the entrance of their funnel each day to accept the insects she offered to them in forceps! She found that satiated spiders in the field also had smaller webs.

Riechert found that there were large numbers of "floaters" in the field which quickly replaced the owners of territories which she removed. These floaters live in cracks and crevices within occupied territories and were frequently observed trying to catch prey on the web of a territory holder. In all cases the intruder was challenged by the owner and withdrew immediately or became involved in a "dispute". By repeatedly catching, weighing and releasing spiders which she could identify as either owners of territories or floaters, Riechert was able to show that over a season owners gained weight while floaters lost weight. Individuals which remained as floaters for any length of time were unlikely to survive, let alone produce offspring.

The author believed that her results demonstrate that territorial behaviour stabilizes populations of this spider *below* the maximum carrying capacity of its habitat and that it cannot adjust the size of its (genetically fixed?) territory in response to changes in the abundance of its supply of prey. If this were so, however, it seems unlikely that the territories where there was more food would be *six times* smaller than those at the site where prey was less abundant, and their occupants *13 times* more productive. Furthermore, her results showed that in years when food is most scarce, only those occupying the best territories contribute to the next generation. On the other hand, in years when food is abundant, as many spiders occupying the poorest sites survived and reproduced as did those occupying the best ones. (It would also be interesting to know how many floaters manage to survive *and* reproduce in the very best years.) Surely the behaviour has evolved under selection pressure of the worst years when only spiders most capable of holding the best territories would contribute genes to the next generation? At the other end of the spectrum they seem to be well adapted to quickly absorb any increase in the availability of food in the good years and convert it to more spiders.

11.2 Territorial Behaviour in Herbivores

The preceding few examples illustrate some of the ways in which territorial behaviours are driven by the supply of accessible (but not necessarily all) food in the environment, by the number of prey that the animals can catch. What is more, they highlight that it is food with which to produce the next generation that is so crucial. But these are all examples of carnivorous animals. What is the picture like when we look at the herbivores? It is the same. They have various adaptations which enable them to use a limited supply of food more effectively, and (where the investigator has looked) it is found that it is a specific relative shortage of available nitrogen which is the limiting resource. Either there is not enough nitrogen in a form they can use in their plant food, or they cannot obtain enough animal or microbial protein to supplement that plant food.

Some of the following examples are about the same animals that I have discussed in earlier chapters. They complement and supplement those accounts at the same time as they illustrate the link between food and territorial behaviour.

11.2.1 Mammals

Mares and Lacher (1987) pointed out that the relationship between the size of a territory and the abundance of resources has been noted since the 1860s, and that in general there is an inverse relationship between the size of the home range of small mammals and the abundance of their resources. The studies by Mares et al. (1982) of the eastern North American chipmunk, *Tamius striatus*, elegantly demonstrate this. In an earlier study (Mares et al. 1976) they had established that the addition of supplemental food to an area both increased the number of animals in that area and decreased the size of their home ranges. After they stopped adding food resident chipmunks forced immigrants to emigrate. So the authors carried out three manipulative experiments to separate the influence of added food from that of increased immigration on the size of home ranges:

1. They first removed half the resident animals from the experimental area, leaving the remaining occupied home ranges spread evenly over the whole area. At the same time they removed animals from surrounding areas to prevent immigration. A month later they systematically added food to the centre of activity of each experimental chipmunk's home range. With half the home ranges vacant and no new chipmunks entering the experimental area, there was no change in the size of the remaining occupied home ranges. However, as soon as supplementary food was added, the resident animals reduced the size of their home ranges by an average of 70%.
2. They added food at random throughout the experimental grid, but did not remove any animals. The response was the same as in 1976; the size of the home ranges decreased and the population density increased.
3. They again removed half the resident animals, but this time taking all from one half of the grid, and again prevented immigration from the surrounding areas. No food was added. The remaining residents did not alter the size of their home ranges.

So, the observed decrease in the size of home ranges of these chipmunks is a direct response to supplemental food. The increased density of animals (when immigration is possible) is *not* the cause of the decrease in the size of home ranges but a *consequence* of the augmented food supply (as is the decreased size of the home ranges). It made no difference whether the food was available centrally or scattered in the home range, or whether the home ranges were spread or clumped. However, they further established that this reduction in the size of the home range occurred *no matter what the original size of the home range* (Mares and Lacher 1987). They also discovered that the larger home ranges have more trees than smaller home ranges. These trees supply the chipmunk's staple food – seed mast – so they concluded that something other than the supply of simple calories (i.e. energy) is influencing the size of the home ranges. "Perhaps", they say, "information on potential nitrogen production...will be more useful". I hope they are looking!

Sullivan et al. (1983) reported a similar reduction in the size of the home range in response to supplemental feeding in the western North American chipmunk, *Eutamius townsendii*, and noted that such experiments by other workers had achieved the same response with deer mice (*Peromyscus* spp.) and voles (*Microtus* spp.) as had their own studies with the Douglas squirrel, *Tamiasciurus douglasii*, in British Columbia (Sullivan and Sullivan 1982).

Ostfeld (1985) succinctly reviewed the relationship of the availability of food to territoriality in microtines. He says that because microtines eat low quality food they should be limited in their distribution and abundance by food. The reproductive success of females should therefore depend upon their ability to acquire food and convert it into weaned offspring. The reproductive success of males, on the other hand, should be limited by access to females. Therefore, territoriality in females should be food-based, and in males it should be female-based. He further proposed that females should be territorial only when they eat food which is patchily and sparsely distributed and poorly renewable, because such food is more readily defendable and the cost of allowing intruders to exploit it is greater. In species which eat food that is abundant, evenly distributed, and quickly renewable, females should not be territorial.

Laboratory studies, field studies, and supplemental feeding experiments with microtines support his conclusions. For example Ims (1987) carried out controlled field experiments in southern Norway with populations of the vole *Clethrionomys rufocanus*. He used animals which he had raised in the laboratory, and released enough of them to exceed their maximum recorded field density, saturating the habitat on a small island from which he had previously removed all wild voles. The home ranges of reproductive females that had access to additional food became considerably smaller, and these same females were more dominant towards strangers. He concluded that the spatial organization of breeding females is determined by the distribution and abundance of their food.

Similarly, Taitt and Krebs (1981) established, by the provision of supplemental food, that the home ranges of *Microtus townsendii* became significantly smaller in response to extra food, while Jones (1990) demonstrated that the size of the territories of female *Microtus pennsylvanicus* is a function of the availability of food. Smal and Fairly (1982) found that the home ranges of both wood mice, *Apodemus sylvaticus*, (an omnivore) and bank voles, *Clethrionomys glareolus*, in Ireland increased and decreased in response to the availability of food.

Rogers (1987) studied black bears, *Ursus americanus*, in northern Minnesota. They are largely herbivorous, but opportunistically omnivorous/carnivorous and generally solitary animals. These traits are good adaptations to their primary food of small, widely scattered items – fruits, nuts, colonial insects – all high in nutrients and low in cellulose. The supply varies from year to year and while they are good scavengers, and cannibalism is not uncommon when food is short, they are not adapted to browse foliage when fruit and nuts are scarce. The success of their reproduction and the survival of their cubs and yearlings is influenced mainly by variations in abundance of their primary food. Many young starve to death in lean years. The females are territorial and defend extensive feeding areas, but the males range much more widely. Daughters stay with their mothers, but sons disperse widely before they reach sexual maturity. Females holding exclusive feeding areas gain more weight, and when food is generally more abundant their territories are smaller and overlap more. As Rogers concluded, territoriality "...seems mainly to influence *which bears, rather than how many bears*, obtain easiest access to food" (my emphasis).

Human settlement has provided a new and significant supplementary source of good food for these bears: garbage dumps. Around these clumped supplies the normally solitary animals gather, with a marked diminution or total absence of agonistic behaviour. They become "...integrated into a hierarchy, as has been reported for grizzly and brown bears at dumps and salmon streams... thus, social organisation as a whole was governed by the distribution and abundance of food ...". This change to hierarchical behaviour also ensures that more of the sudden bonanza of food gets converted to more bears than would be possible if the animals remained scattered in their territories (see Sect. 11.4).

11.2.2 Birds

Gass et al. (1976) studied migrating rufous hummingbirds, *Selasphorus rufus*, in north west California. Migrants move in a series of flights interspersed with short stays in mountain meadows where they may hold feeding territories for a week or more. There these birds adjust the size of their feeding territories so that the supply of food (nectar from flowers) is kept at a level which approximates their metabolic requirements. Ewald and Carpenter (1978) studied individuals of another species of hummingbird, *Calypte anna*, defending their non-breeding territories. These territories are concerned only with the defence of a source of energy (flower nectar) for migratory flight, so the authors could easily manipulate the amount of potential energy available to the birds by using feeding bottles of artificial nectar. When there was plenty of food, birds vigorously defended highly exclusive territories, expending a great deal of energy in chasing intruders. As the supply of food declined so did the exclusiveness of the territories, and the owners used less expensive forms of defence and spent less time on their territories.

In Sweden, resident nuthatches, *Sitta europaea*, occupy all-purpose territories all year round. A pair defends a territory and in autumn and winter store and feed extensively on beech mast when it is available. Enoksson (1990 and Enoksson and Nilsson 1983) demonstrated that the size of the territories in autumn was closely correlated with the size of the beech mast crop of that year. However, the birds significantly reduced the size of their territories when supplied with a supplement of sunflower seeds in the winter. This represents a significant adjustment to the availability of food for maintenance during the non-breeding season. The winter territories are smaller in parts of the forest which have a preponderance of beech. In the breeding season, however, while territories are smaller than in winter, they are about the same size in beech as in oak-dominated forest. But breeding success is higher in the oak forest. The birds' major food at this time is arthropods, mostly lepidopterous caterpillars on new growth, and these are more abundant on oak than on beech trees. This highlights another aspect of the relationship of food to territoriality. The *kind* of food which is in critically short supply differs. To survive a cold winter – in contrast to the breeding season – a good supply of energy is more important than is a supply of protein, provided always that you start the winter with a good reserve of body protein built up in the spring and summer.

Andreev (1988) demonstrated a strong negative correlation between the size of the territories of the willow grouse, *Lagopus lagopus lagopus*, in Siberia and the biomass of willows, the birds' principal spring diet. The average size of these territories increased some eight times northward to the limit of the distribution of both willows and grouse. Territories on the southern tundra were two to three times larger than those in nearby valleys. Both trends reflected a decrease in the quantity and quality of food available in the area. Furthermore, the content of nitrogen in a diet of willow declines throughout the winter. The extent of this decline, combined with the availability of high quality foods in the spring, was the major factor determining the breeding success of the birds.

The hens of the willow grouse in northern Norway establish home feeding ranges for their broods of chicks (Erikstad 1985). These sometimes overlap but mainly only in lower-quality habitats where they also tend to be larger. The size of the home ranges increased as the chicks grew older, but at different rates and to different ultimate sizes for different broods. Broods generally preferred to feed in areas especially rich in insects. In years when insects were scarce few chicks survived and ranges tended to be larger. Chicks from broods with smaller home ranges grew faster and survived better than did those in the largest home ranges. Andersen et al. (1986) found the same behaviour for willow grouse hens in central Norway, and again showed that this behaviour is related to the provision of food for the chicks.

Red grouse, *L. l. scoticus*, are monogamous, and form pairs for long periods. This contrasts with most tetraonids (including willow grouse) which are polygamous; hens nest and rear chicks independently of cocks. Male red grouse usually outnumber females, whereas in subspecies feeding on willow and birch – a more nutritious source of food than heather – the sex ratio is near 1:1. Red grouse cocks establish exclusive territories during autumn, and both cock and hen feed almost entirely within their territory in the winter and spring. Heather, *Calluna vulgaris*, is their main food, and its quantity and quality will have a major influence on the breeding success of the hen. The better her nutrition the better quality her eggs, and the greater probability that her chicks will survive. Lance (1978a) demonstrated that the size of a pair's territory was inversely related to nitrogen in the tips of the heather shoots on that territory, and that hen grouse reared more young on the territories where the heather was richer in nitrogen. Moss et al. (1988) have confirmed and extended Lance's findings. They demonstrated that cocks with hens had bigger territories and richer feeding sites than unmated cocks, and that the size of a cock's territory was inversely related to the level of nitrogen in his food.

Of more significance to the success of the birds, however, is the fact that hens actively choose between males, and thus between their territories. This means that some males are left unpaired for the breeding season and do not contribute to that year's brood. All females will breed, but some will be more successful than others, depending on their choice of territory. A hen will move from one cock to another before finally choosing which territory she will select. The authors are not clear whether it is the quality of the male per se or his territory which she uses to make her choice. They even suggest that some males may *choose* to remain celibate! Whatever the proximate stimulus to her final choice, the ecological function of that choice is clear.

Hens move about more and take longer to make their choice where the food is poorer. Where the quality of the heather had been enriched by the application of fertilizer the ratio of hens to cocks increased. The nitrogen content of the food on different territories clearly influences a hen's choice of a mate. She in fact chooses to settle where she is best able to gain access to high-protein food, and so maximize her chances of producing high quality eggs, and thus maximally viable chicks. In so doing she is concomitantly choosing for perpetuation of the genes of a male which has been best at selecting a site which will provide her with this essential level of nutrition. It is *this* which is important about a male's choice of territory – gaining him access to a female – not that he chooses one that is big enough to provide just the minimum food requirements for a pair.

However, the story does not end there. From the moment they hatch the chicks must find their own food. They start with a small reserve of nourishment from the invaginated yolk sac (Erikstad 1985), but must quickly gain access to a supply of high-protein food. For their first few days of life this means animal protein – largely insects – until they can subsist on a purely vegetable diet. Like the willow grouse, immediately after hatching the hens leave their home territory and establish a home feeding range for their brood of chicks (Lance 1978b). This has been interpreted as demonstrating that the function of the territory of the adults is not to provide food for the chicks. Clearly it is not. It is to provide food for the breeding hen. The nutritional needs of the chicks are quite different – they must have still more protein – and so a new behaviour is needed.

The same strategies are found operating in the closely related but omnivorous phasianids (pheasants, quail and partridges). Hill (1985) found that chicks of the common pheasant, *Phasianus colchicus*, in southern England, which depend for the first 2 weeks of their life on insects for food, survived better in home ranges which contained a greater biomass of arthropods. These richer ranges were smaller in area, and the chicks spent less time ranging over them than did chicks in larger home ranges with less animal food. He also reported similar findings for populations of *P. colchicus* in the USA, and for the partridges *Perdix perdix* and *Alectoris rufa* in England.

11.2.3 Insects

Craig et al. (1988) found in northern Arizona that females of the gall-forming sawfly, *Euura lasiolepis*, exhibit a strategy essentially the same as that discussed for carnivorous insects, whereby the limited supply of available food is sequestered to the offspring of those females that first lay their eggs in the places which will provide the best food for the larvae. These females lay their eggs in the petioles of leaves as they first unfold on the growing shoots of the arroyo willow, *Salix lasiolepis*. They show a strong preference for the petioles of the leaves on the fastest growing shoots. The chances of the larvae forming galls and surviving is greater on such shoots than on slower growing ones. Up to 100% of available sites may be galled on these fast-growing shoots, even though they comprise but a small proportion of all shoots growing on the willows to which the sawflies have access. Through a series of experiments the authors demonstrated that a female sawfly will avoid laying an egg where another female has already oviposited; it is the presence of chemicals generated by the wound tissues of the plant which deters her. Females responded to artificial scars made to simulate an oviposition scar, or to an aqueous solution of finely ground scar tissue painted on to undamaged nodes. They did not respond (as do *Asolcus* females) to a pheromone generated by another female, or to any tactile or visual cues. Both experimental results and field observations showed that the oviposition scars of *Euura* females were more uniformly distributed than would have been the case if the sawflies had been ovipositing on the best quality sites, ignoring previous ovipositions. So, this behaviour reserves the small number of top-quality sites for a few. The remainder are forced to lay their eggs where the probability of their young surviving is, while not zero, very much less than it would have been if they had secured one of the good sites.

There are several other examples of herbivorous insects being deterred from laying eggs on a supply of food already occupied by the egg or larva of a conspecific. Some involve pheromones, others the release of plant substances; some are generated by the activity of the adults, others by activity of the larvae. All serve the same function of more effectively using the limited supply of good food by rationing it to only some of those seeking to eat it.

However, as with all things in nature, no system is perfect. Sometimes more than one egg will be laid where there is enough food for only one. Then all would perish and the resource would have been wasted, unless there is a "fall-back" strategy to prevent this from happening. There is such a strategy, as the example of the carnivorous *Asolcus* showed; the strongest larva kills the weaker (and usually younger) larva. This is the strategy which operates in the case of the strictly herbivorous codling moth, *Cydia pomonella* (Geier 1963). The female lays her egg on the outside of an apple. The larva burrows to the core where it feeds as a first and a second instar on the soluble and highly digestible nitrogen in the developing seeds. When populations are high, many eggs may be laid on one apple and several larvae may bore into a single core. A large apple might support two or three of them, but in small apples the seeds provide only enough food for one larva to survive to pupation. Then the first and biggest will eat the others, just as happens in the nests of many raptors.

Perhaps the nicest story illustrating the influence of a limited supply of food on the territorial behaviour of a herbivorous insect is that of Whitham's (1979) account of actual fighting between first instars of the aphid *Pemphigus betae* for a place where they can survive and reproduce. These aphids form galls on the leaves of the narrow leaf cottonwood, *Populus angustifolia*, in Utah. Stem mothers hatch from overwintering eggs in close synchrony with budbreak of the trees in spring. Over a 3 day period nearly all of these black, 0.6-mm-long, highly mobile first instars migrate from the base of the tree to

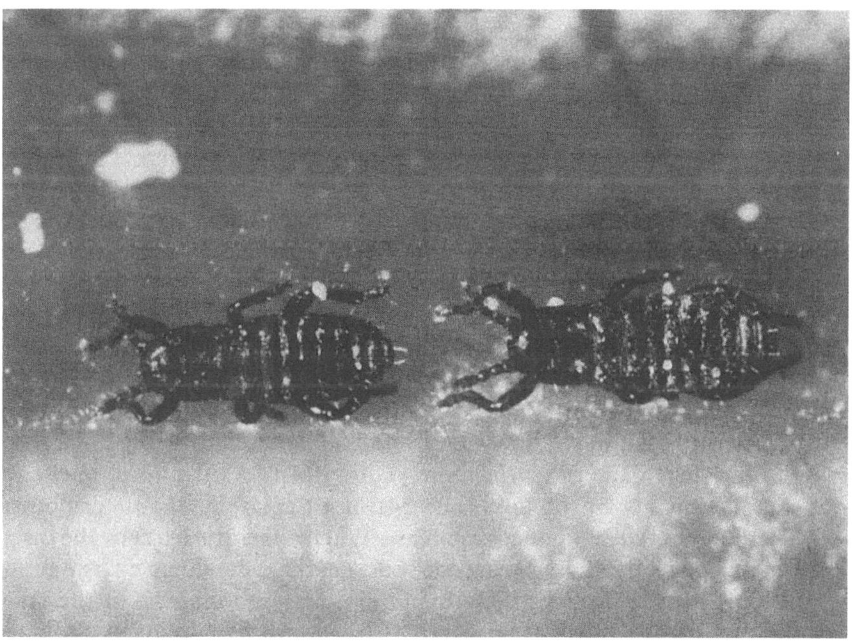

Fig. 34. Newly-hatched 0.6-mm-long poplar aphids. These "stem mothers" fight each other, often for days and to the death, over possession of the prime site to form their galls at the base of a leaf. Photo courtesy of T.G. Whitham

developing leaves where they attempt to initiate galls. By probing the expanding leaf tissues each stem mother causes a small depression to form. If she is successful, within 3 days she will be enveloped in a lineal, hollow, pocket-shaped gall on the petiole of the leaf. There she may produce up to 300 parthenogenetic offspring, but her chance of dying is very high at the crucial initial probing stage. On some resistant trees no first-instar stem mothers survive from among the many thousands which attempt to feed.

The preferred sites for initiation of galls are the bases of the largest leaves. These leaves provide the highest quality food, are least prone to fall prematurely, and stem mothers settling on them produce the highest number of young. Such leaves are relatively rare, so only a few of the many stem mothers preferentially seeking to settle on them will be able to do so. The result is intensive territorial interactions, with two stem mothers engaging in kicking and shoving contests which Whitham has recorded lasting as long as 2 days and sometimes causing the death of one or both aphids. The aphids "move like boxers in a ring, continually sparring to determine the winner". In doing so they move up and down the mid-rib of the leaf, vying for the superior basal position. Whitham found that the largest stem mother usually won such contests, successfully defending a territory about 3 mm long at the base of the leaf. Removal of the stem mother holding the basal territory can result in the challenger moving down to occupy the better site. Removing the challenger resulted in the dominant aphid expanding its basal territory. Normally the loser either departs to find another leaf or settles on the less favourable site distal to the winner.

Whitham's (1986) later experiments showed quite clearly that:

1. Solitary stem mothers produced larger galls than did those in the presence of a competitor. Alone, they are free to probe and feed over a greater length of the leaf, but competing aphids spent nearly all their time at the boundary between two territories engaged in kicking and shoving contests instead of feeding.
2. Removing one or other contestant resulted in the residual aphid incorporating the extra leaf area in a single large territory equal to a territory in the absence of a competitor.
3. The largest stem mothers claimed the (best) solitary sites, the bigger of two fighting aphids won the better basal position of a shared site, and a resident aphid will be ousted by a larger challenger.
4. The largest stem mothers occupying solitary sites produced more offspring than did those sharing a site, and the larger aphid occupying the basal position of a shared site produces more young than does the distal one.

It was the quality of the food, however, not the size of the stem mother, which determined her reproductive success. When either the basal or the distal member of a pair of aphids sharing a leaf died early in development, the survivor not only produced more young than did other stem mothers of the same size, but produced as many as her larger sisters which had occupied a solitary site from the beginning.

Finally, Whitham (1978, 1980) demonstrated that the supply of food which is adequate for a settling first instar to survive and reproduce is very limited

(Sect. 3.3.5). Less than 1.6% of all available leaves are good enough to ensure 100% survival. Smaller leaves, on which 80% of the aphids attempting to settle fail, make up 32% of all available leaves. On even the superior leaves there is a steep gradient of suitability of the tissues of each leaf as food. Of all galls 84% are positioned within the bottom 1 cm of the leaf blade, and solitary stem mothers in galls 6 to 9 mm from the petioles produced 65% more progeny than solitary aphids 18 mm or more from the leaf base. Success goes to those establishing nearest to where all nutrients translocated in and out of the leaf must pass.

So, again, we see territorial behaviour restricting access to a limited supply of food so that a favoured few in a population produce more young. This means that more of the food is converted to more individuals in the next generation than would have been the case in the absence of this sequestering behaviour. At the same time the capacity is there to immediately absorb any increase in the abundance of food, making sure any such increase is used to good effect. There is a large "floater population" of first instar stem mothers. Most of these are the classic "doomed surplus" which search in vain for a suitable site. Their mortality rate was 76% (compared with 28% for residents) so some floaters find a site just sufficient to survive and produce a few young. There is a gradation of success with all accessible food being used; and always a "reserve" available to quickly take up any increase in the number or quality of feeding sites. By removing stem mothers before they had produced a gall, Whitham found that floaters quickly occupy any vacated site. He also observed that resident stem mothers were constantly being challenged by new arrivals. Most challengers departed after a brief skirmish, but a stronger aphid could oust the resident to the inferior distal position, or altogether from the leaf. There is constant pressure on the limits set by the chronically short supply of available food.

Hart's studies of territorial behaviour of aquatic insects, which I mentioned when discussing carnivores, included some insects which are classified as herbivores but feed on algae and detritus. Both foods are a much richer source of nitrogen than are the tissues of plants. The caddis fly, *Leucotrichia pictipes*, is one such insect. Its larvae feed mainly on diatoms. In the final instar a larva spins a silken case which is fixed to the substrate. It never fully leaves this case but stretches out to feed and defend a territory from conspecifics and other grazers such as mayfly larvae and snails. The territory comprises a more or less elliptical area around the case, and the upper surface of the case. When Hart scraped some of the algae from within a territory, or when he shaded some territories to decrease the rate of growth of the diatoms, the larvae increased the area of their territories. Also, while larger larvae had bigger territories, it was not the size (i.e. the maximum "reach" from the case) which decided the area defended, because even the large larvae increased their reach when food was reduced.

It seems that these larvae are living in a habitat where the supply of algae for various grazing animals is in chronic short supply. While the larvae are very evenly spaced on a particular piece of substrate by their territorial behaviour, they are patchily distributed on the "macro" scale of the river bed, concentrating where the food supply is richer. The boundaries of their territories are clearly defined by the marked change in abundance and composition of species of the algae within and without a territory. Hart found that the average density of food available within a defended territory was four times greater than in an undefended area. If he removed a larva from its case other grazers quickly cleaned out the algae on its territory. By being able to monopolize an area to themselves the caddis larvae prevented other grazers gaining access to preferred microhabitats. They, in effect, create "resource oases" of dense patches of algae.

It is also possible that they actively maintain "algal gardens" of higher quality species as well as sequestering the quantity of algae that is available. Hart (1985b) found that they apparently weed out filaments of blue green algae from their territories, preventing these filaments from overgrowing their preferred (and much more nutritious) but slow-growing diatoms.

11.2.4 Fish

Parma victoriae is a small herbivorous fish living among the rocky reefs in the shallow waters along the southern coast of Australia. It feeds on a number of different species of small red algae which grow in turf-like layers on boulders. Both males and females defend separate territories all the year round. In the summer breeding season females make brief sortees into the territories of males to lay their eggs.

Norman and Jones (1984) studied the factors influencing the territorial behaviour of these fish. They confined their experiments and observations to the winter to avoid any possible influence of breeding activities on their measurements. They found that whether they increased or decreased the total amount of algae in the winter territories of these fish this had no effect on the size of territories. However, when they removed fish from neighbouring territories the remaining residents rapidly expanded their territories to take over the vacant areas of reef. Furthermore, when their food was manipulated or their territories expanded, the fish did not change the time they spent defending their territories, foraging, or feeding. Norman and Jones therefore concluded that food could not be a limiting resource for these fish.

But why, they asked, do these fish defend an apparently unlimited supply of food, and why do they expand their territories when their neighbours are removed if they already have sufficient food on smaller territories? Of a number of possible hypotheses to explain their observations they preferred the one which suggested that an increase in the size of its territory might allow a fish to feed more selectively on more preferred species of algae. The answer, they thought, may lie not in terms of an absolute shortage of food, but in a relative shortage of preferred (and, I would suggest, therefore more nutritious) species of algae.

Jones and Norman (1986) confirmed their choice of hypothesis. They measured the volume of different species of alga in the field and in the stomachs of fish, and discovered that certain species of alga were being selected and others being excluded from the diet of these fish. They then offered the fish a "cafeteria" of equal volumes of a range of preferred and non-preferred algae on trays. The fish ate these in the same order of preference as indicated from the comparison of abundance in the field and in stomach contents. Clearly, the fish do have preferences for certain algae which could be ranked in order of decreasing preference.

All this was done without reference to the size of territories. When they repeated their experiments with fish from large and small territories they found that the size of territory had no influence on the preference ranking in "cafeteria" feeding experiments. However, the fish from small territories ate more of the preferred species than did the fish from large territories. This, the authors said, was because individuals from large territories tended to be less oriented towards feeding trays (could it be, rather, that the fish on the small territories were hungry for a diet of these preferred species?).

By contrast with these feeding trials, the natural diet of fish from large territories contained more of the highly preferred species, while the diet of the fish from the small territories contained more of the least preferred ones. When Jones and Norman measured the abundance of algae present on the two sizes of territory they found some of the less abundant but high-ranking species in greater quantities per unit area on large territories and in significantly lower quantities on small territories. It seems that fish with large territories fed in a highly selective manner while those in the small ones ate algae in proportion to their abundance in the territory.

They then looked at what happened after some fish expanded their small territories when their neighbours were removed. These fish did not eat a greater volume in their new expanded territory. Instead, the single most preferred species of alga present in the territory now dominated their diets. They ate more of all the higher ranking species and less of the lower ranking ones. Their diets in their expanded territories were now the same as those of fish naturally holding large territories.

Jones and Norman concluded that an animal's food should not be assumed to be homogeneous. What *appears* to be an unlimited resource may not be, and for fish that are confined to a small territory good quality food is in short supply. To survive they are forced to eat food of low quality. But, the authors said, they know nothing of the influence of the different diets on the survival and reproductive capacity of individual fish, or of the relative quality of those diets. I hope they are trying to find out.

Meantime, it would seem clear that these fish are behaving in the same way as Whitham's aphids (Sect. 11.2.3). The best territories for the fish – those containing enough high-quality algae – are the equivalent of the best site for the aphid on the base of the largest poplar leaves. The most successful individuals (in terms of contribution to the next generation) will, in both cases, be those which can gain and hold the best territories. Those less successful must make do with what is left, and may or may not get enough food to survive and reproduce. However, these less favoured individuals are ever-ready to take better resources should they become available to be used. So, the population as a whole is primed to track any increase in abundance of good food and incorporate it in more of its own kind. Every bit of accessible food is used and, by channelling it through only some of all those attempting to use it, as little as possible is wasted.

11.2.5 Molluscs

Another group of aquatic animals, many of which feed exclusively on algae, are the molluscs. Here too we can find individuals defending their territory from conspecifics and other herbivores, while a "pool" of less successful individuals eke out an existence on less preferred sites, waiting their chance to move into any vacant or less-than-well defended territory. And for such apparently "simple" animals they can show remarkably complex behaviour in defending their territories.

Stimson's (1970, 1973) study of the owl limpet, *Lottia gigantea*, is a case in point. *Lottia* is a common mollusc along the coast of California and Mexico where it lives on smooth rock surfaces in the upper half of the tidal zone. For a limpet *Lottia* is quite large; individuals can grow to 8 cm in length. Each limpet lives in a clearly visible territory of approximately 900 cm^2. Within this territory it maintains and feeds upon a dense mat of algae (primarily filamentous blue-green algae). This mat grows only within a *Lottia* territory. Outside the territory the rock is stripped bare. Stimson, by simply marking and observing the limpets, established all this and demonstrated that the limpets graze only within their own territories; and that the size of a territory is proportional to the size of the incumbent. When he removed limpets from their territories the algal mat disappeared in about 2 weeks. When he established limpets on new sites on bare rock an algal mat was visible around each limpet within 3 weeks. So, clearly, the film of algae is maintained within the area of a territory by the activity of each limpet. Counting the number of other grazers (principally five species of another limpet genus, *Acmaea*) inside and outside territories revealed no more than about three *Acmaea* within any territory; but 15 to 20 in an equivalent area outside any territory. Removing *Lottia* resulted in an increase in the number of *Acmaea* on the empty territory within 2 to 3 weeks; the reverse happened when he placed limpets on new sites.

Stimson then decided to see how all this came about by closely observing the behaviour of individual limpets. He placed an *Acmaea* limpet immediately in front of a grazing *Lottia* and watched to see what happened: "After touching the *Acmaea*, the *Lottia* stopped, then, after about half a minute, the *Lottia* retracted its shell about one centimetre, lowered the forward edge, and suddenly thrust its shell and moved its foot forward a distance of 2 to 3 cm, striking and pushing the *Acmaea*. This shove was much faster than the usual grazing movements of *Lottia* and often the shove dislodged the *Acmaea* so that it was washed away by the next wave". He observed that this shoving was maintained until either the intruder did get washed away, or was pushed a few centimetres beyond the edge of the territory. One *Lottia* pursued an *Acmaea* for 0.5 m across its territory before dislodging it. Stimson repeated these staged encounters with *Acmaea* and with other species of grazers, including other *Lottia*. The reaction was always the same. But when placed outside their own territory *Lottia* did not respond to others in this way.

He then tested the response of resident *Lottia* to two species of predatory snails, *Thais emarginata* and *Acanthina spirata*: "Sometimes the *Lottia* responded as they did to *Acmaea*, but much more quickly. At other times the *Lottia* responded quite differently. Instead of lowering its shell, the *Lottia* raised it and then brought it quickly down on the anterior edge of the foot of the *Thais* or *Acanthina*, causing the snail to retract its foot, lose its grip on the rock, and be washed away". However, when confronted by herbivorous snails which also extend the foot beyond the shell when grazing, the *Lottia* did not do this to them, but merely shoved them away in the same way as it did for *Acmaea*. Large potential competitors for space, like anemones and barnacles, were also dislodged by shoving. Small ones were simply rasped off as the *Lottia* grazed.

Finally, Stimson tested their response to foreign objects held in front of them (including his finger!); they did not respond. So, a limpet is able to distinguish between and act differently to simple obstructions and animals on its territory which, on the one hand, would eat its food or occupy its space, and, on the other hand, would attack it.

Having demonstrated this active maintenance and defence of a supply of preferred food within a territory, Stimson went on to demonstrate that the amount of food available influenced the size of the territory which a limpet would defend. The thickness of the algal film varied, and where it was thickest there were more *Lottia* living in smaller territories grouped more closely together. When he removed the film from some territories the residents increased the size of their territories by 25% in 2 weeks. He also found that the rate of growth of the limpets was greatest where the film of algae was thickest. In places where it was very sparse they did not grow at all.

Smaller limpets occupied less favourable sites, usually the shells of mussels growing near the smooth areas of rock preferred by large *Lottia*, but quickly colonized these better places when Stimson removed the residents. Clearly, once again, the supply of good food is far from abundant and the population is pressing hard against the limit of that supply.

11.3 Surplus Young, Dispersal, and Philopatry

So it seems that, within a habitat, holders of territories, by driving off other potentially breeding animals, sequester the limited supply of food to themselves and/or their offspring. But, paradoxically, by denying access to many, they enable more to survive.

But this perpetuates the situation. In every generation there are more individuals seeking to use the available food than it can support. There is a constant production of young which must be expelled from the territory. These young must seek their own future. They may disperse or remain nearby. The former behaviour means potential habitats are found, and all adequate ones are occupied; the latter means that vacancies in existing territories are quickly filled. All these behaviours combine to maintain pressure on available resources. In terms of the food which is accessible to be used by the animals, their environment is effectively, if not apparently, saturated; so most newly produced young soon die. They are, indeed, for most of the time, Errington's (1946) "...harassed and battered lot congregating about the fringes of areas dominated by [those] already in residence,... a biological surplus, largely doomed through one medium or another".

Furthermore, the supply of food is never constant. Whenever it increases (e.g. a mast seed year, a flush of grass growing after good rains, or an "outbreak" of scarce prey) many more will survive, packing more densely into already occupied habitats and occupying previously inadequate and vacant ones. When the supply declines many fewer will survive. For the population there is constant flux, for the individual a constant walking of the tightrope of adapting to the moment.

In an environment in which food is more or less regularly renewed at a fairly stable level and where, therefore, all available habitats have long since been occupied, a best bet may be to stay close to home. To subsist, but not breed, within or near established territories might increase the probability of eventually breeding successfully, either at home or away. Whenever and wherever this is so, such behaviour would persist and, over evolutionary time, spawn innumerable variants. As Armitage (1988) points out, one way for a female to increase the probability of her daughters living long enough to reproduce is to retain them in their natal area. This, as a form of sociality, is a trait which continues a female's reproductive investment beyond immediate raising of her young, and increases the probability of her offspring reproducing.

Morse (1980) makes the point that territories and social dominance hierarchies can probably be thought of as opposite ends of a spectrum along which there are many situations combining aspects of both. So, too, this tendency for young animals to stay at home – this philopatry – may profitably be thought of as one point on this spectrum; a point only marginally removed from simple territoriality involving immediate expulsion of all young that are capable of independent existence. It is a more behaviourally complex way of retaining "surplus" young during the time when there is enough food in the habitat to support new young. This increases both the chances that any vacancies in the home territory are quickly filled, and that eventual dispersal will successfully locate and use any adequate habitats not already occupied. For the individual, to stay at home means a greater chance to survive and eventually breed than does leaving. It has nothing to do with supposed altruism towards parents or siblings.

The risks of dispersal, regardless of the chance of breeding elsewhere, are high. The widespread taxonomic distribution of philopatry would support this view (Waser 1988). The ecological factors that promote philopatry decrease the odds of successful dispersal; and they are involved in increasing its advantages and decreasing its costs. This concept of the inadequacy of the environment applying selection pressure against those that leave the parental territory too soon is encapsulated in the concept of "habitat saturation". However, as I pointed out at the beginning (Sect. 1.2), this concept is true only in terms of what is accessible, attainable, and *useable* by the animals *at any particular moment*, not the human perception of under- or unoccupied habitat. An environment is saturated when opportunities for a new individual to breed in it are near zero and the odds of surviving dispersal are correspondingly low. Then, philopatric young and parents which tolerate them are at a selective advantage (Jones et al. 1988). The other side of the coin is that, in the long term, the availablity of food will change, often dramatically. Then it is the capacity to disperse which sees to it that all such fluctuations are absorbed. So natural selection is constantly "balancing" staying against dispersing.

In its broadest sense philopatry itself exhibits a spectrum of complexity and variability. Examples range from individuals remaining near or on the natal territory for only short periods, through various types of "helpers" remaining for one or more breeding seasons, to complex societies where some offspring may never leave the group. In all, the "motivation" of those individuals that stay is an increased chance of ultimately breeding successfully. Where individuals stay at home after reaching sexual maturity, it is most often only females which do so; young males depart. Females must maximize access to protein in order to successfully pass on their genes in surviving offspring; males must seek females to pass on their genes. Seen from this perspective, the ecological derivation of the evolution of philopatric behaviour is a relative shortage of food for females to produce viable young – the inadequacy of the environment. Any individual which exhibits a behaviour that increases the number of young which it can raise with this limited resource will be at a selective advantage.

But the literature is vast. Whole books have been written on a single facet (Brown 1987). Here a few examples of animals previously discussed in sections 6.2.1 and 7.1.3 must suffice to illustrate my point.

Rusch and Reeder (1978) working with the red squirrel, *Tamiasciurus hudsonicus*, in Alberta found that they were most abundant (and with the smallest territories) in spruce stands, less so in pine stands, and least in aspen woods. Yet, when the season's young dispersed in the autumn, most of them dispersed to the aspen stands. By early winter virtually 100% of squirrels in these stands were juveniles compared with 50 to 70% in the pine stands and only 15 to 20% in spruce stands. The "surplus" young are driven out to habitats which mostly will not have enough food to support a squirrel through the winter. Few will survive. A similar system is found with *T. douglasii* (Sullivan and Sullivan 1982); juveniles disperse to a suboptimal habitat.

The gray squirrels, *Sciurus carolinensis*, unlike the red squirrels, do not hold territories but have strong social dominance hierachies and stable overlapping home ranges (Thompson 1978); yet their abundance is similarly restrained by the availability of food. Gorman and Roth (1989) studied the influence of habitat and food on a population of gray squirrels in Delaware over a period which spanned a year with a large mast crop followed by a year with a very poor crop. The number of squirrels dropped sharply in the poor year. The major components of this decline were the loss of juveniles produced in the good year, and the curtailment of breeding by surviving females; they produced fewer young and fewer of these survived past weaning. As with the red squirrels, when food is scarce juveniles and subordinates are relegated to suboptimal sites where most of them die, and none breeds successfully; even the dominants holding the best sites experience a marked decline in breeding success.

Zwickel (1980) and his colleagues working with the blue grouse, *Dendragapus obscurus fuliginosus*, on Vancouver Island, British Columbia, showed in a series of removal experiments that there is a large surplus of juveniles present on the breeding range of these birds. These juveniles, both male and female, are non-breeding and there were twice as many or more of them present in spring than would have been needed to replace adults which had died since the previous breeding season. Following removal of virtually all adult and yearling grouse from two large experimental areas, both populations were completely replaced by the next spring, mostly by yearlings. While the non-breeding young on a breeding range weighed less than breeding birds, they had the same potential of both weight and breeding capacity given the opportunity. The factor setting them apart was the failure to acquire a space where they could settle.

This interpretation of the processes driving philopatry has now been elegantly demonstrated in a just-published story of the 32-year recovery of the last population of the Seychelles warbler, *Acrocephalus sechellensis* (Komdeur 1992).

There were only 26 birds left on one island when careful management to restore their habitat began in 1968. A spectacular recovery followed. By 1973 all of the island was occupied by territories, and from 1982 onwards the population stabilized at approximately 300 birds.

In 1973 helping by young birds was first observed in a few rich territories. By 1982 it was widely observed all over the island. From then on all offspring that failed to become breeders remained as helpers in their natal territory – but they constantly foraged into neighbouring territories searching for openings in the breeding population.

Komdeur found that while predators were evenly distributed over the island, and nesting sites were abundant, the supply of food (insects) was variable. He was able to classify territories as high, medium, or low quality on the basis of the supply of insects available to the birds in them. Reproductive success and survival were much greater in the high quality territories. Only 10% of recorded vacancies occurred in them, compared with 30% in the medium, and 60% in the low quality ones.

So, the options open to young birds were either to fill a vacancy in a low quality territory where the probability of immediate breeding was higher, or remain as a non-breeding helper in a territory of high quality.

Komdeur calculated that for a given quality of territory a yearling bird breeding immediately had a greater lifetime reproductive potential than one that stayed. However, in a high quality territory, a bird which stayed for several years and then bred in a good quality territory would still produce more offspring in its lifetime than one which bred immediately in a medium quality territory. Birds in low quality territories, even though starting to breed as soon as they were mature, had little chance of rearing *any* young.

These expectations were reflected in the observed behaviour of the birds. Yearlings born in high quality territories were more likely to remain as helpers than those born in medium or low quality ones. Compared with individuals born in medium or low quality territories, they were more likely to wait for a vacancy in a high quality territory than disperse and breed in a poorer territory.

Experimental demonstration of all this came when Komdeur transferred birds to two unoccupied islands. At first, juveniles left their natal territory at a mean age of 4 months compared with 1 year on the source island, and bred at a mean age of 8 months (4 years on the fully stocked island). *None* of them stayed as helpers. On both islands, once all high quality territories were occupied, young birds born in them began to stay as helpers *even though there was still abundant space left in low quality areas to establish territories.*

Komdeur concluded that his data demonstrate that both habitat saturation and territory quality are involved in the evolution of cooperative breeding. But it is the quality of the territory – the amount of food available for successful breeding – which determines when the habitat is saturated. It is saturated when it will support no more animals. The number and size of territories, and the breeding success and number of animals occupying them, are all consequences of this carrying capacity of the habitat.

Finally, his experiment demonstrated, as did that with the feral donkeys in Australia (Sect. 6.1.1), but by the reverse process, how populations will grow exponentially through increased production and survival of young, until stopped by the exhaustion of the supply of food for breeding females and for growth of their neonates.

If we now move further along Morse's spectrum to look at more complex and more permanent parent/offspring/sibling associations we find that food, or a lack of it, is still the pivotal factor driving these associations.

In his review of the social organization of herbivorous ground-dwelling squirrels (*Marmota* spp.) Armitage (1988) reports, for a number of different species, that either natural or manipulated changes in the abundance of their food did not influence their social structure. They still maintain their matrilineally based groups, occupying systems of burrows that are essential to their survival. These burrows are, however, aggregated on the better quality sites; those with more food. On these better sites there are larger social groups (those with three or more adult females, each defending her own system of burrows, and having a female offspring or sibling share her territory with her). In addition, on those better sites – and on those where supplemental food was added – females bred earlier, had more successful pregnancies, and larger litters. Their offspring were heavier, grew faster, reached sexual maturity more rapidly, and dispersed earlier. And when the amount of food available to them was artificially decreased or increased, the size of a defended territory decreased or increased accordingly.

Fig. 35. Female yellow-bellied marmots live in social groups ("matrilines") wherein numbers vary with the availability of food. Only the dominant female of a group breeds. Subordinates and juveniles function in the group like "helpers" of some territorial birds. Photo courtesy of Dennis W. Johns

In natural situations the number of matrilines (a resident female either alone or with associated kin) can change. Immigrants may successfully establish a new matriline in the presence of residents in larger habitats when conditions are better; or a resident may be squeezed out by expansion of a more aggressive neighbouring matriline.

Daughters are highly philopatric while males usually disperse before they reach sexual maturity. Adult males may be monogamous, defending one matriline, or polygamous defending a harem of more than one matriline. Their status is determined by how scattered or clumped the females are and this, in turn, is determined by the amount of food available to the females. This is a situation essentially the same as that of the polyterritoriality of European passerines (Temrin and Arak 1989).

Within a matriline the mother is dominant over her daughter; where sisters share a territory, one is dominant. The subordinate females fail to produce litters. However, a subordinate female quickly recruits to any vacancy following the death of the breeding female, whether it is her mother or a neighbour. So they are functioning in the same way as "helpers" in birds. They are "surplus" young which, by staying at home when there is enough food to support them, increase their individual probability of eventually reproducing more than they would by dispersing immediately, with a high probability of not finding a habitat with enough food.

11.4 Social Structures and Dominance Hierarchies

Moving yet further along Morse's spectrum we come to more complex social behaviours. These also are widespread but, apart from the special case of the social insects, are most highly developed in mammals. Here, a group of adult and immature animals of varying ages live together, "sharing" the same physical resources; but all in the group are subservient to a ruthlessly maintained dominance hierarchy. All resources are *not* shared equally. On the contrary, the physically strongest and behaviourally most aggressive animals (of both sexes) get the "lion's share" of the available food. Individuals at the bottom of the "pecking order" may or may not, from time to time, get enough nourishment to survive. But rarely, if ever, will they breed. Only those which get the most and best food (and benefit in many other ways at the expense of their subordinates) produce young. The available resources are efficiently concentrated for the well-being of a favoured few. Only they contribute their genes to the next generation. Again, however, in ecological terms, the behaviour sees to it that as many individuals as possible – more than in the absence of the social structure – are produced in each generation.

A socially structured group may live in a geographically defined and defended territory, or it may range more or less widely over an area which is frequented by other groups of the same species. As numbers in the group increase, newly maturing young and those of lowest rank (the two are usually synonymous) will sooner or later be driven away from the group. If they can find sufficient resources they will survive, but most will soon die. They are the "doomed surplus".

There are, however, many examples in which the young remain for varying periods of time with their parents in the group, although they seldom breed. Either way most young produced will be more than the existing resources can support, so they must sooner or later run the gauntlet of being selected against. Any which inherited a trait that increases their chance of staying alive until a vacancy arises in the group, or until more food becomes available, will have an increased chance of escaping selection and eventually reproducing. Because the amount of food in the environment is never static, and because dominant animals grow old and die, there are always some opportunities for new individuals to breed. Being able to remain with the greater resources and security afforded by the group must be such a trait. Nor does it matter whether to do so you help feed or protect the next generation, keep a superior's fur clean, or guard the group against predators and hostile conspecifics. It is better than moving away to an almost certain and lonely death in a strange and inhospitable land. Also, like parental care, allowing their young to stay and consume food increases the parents' genetic fitness more than would immediately driving them away. A side benefit of this individually motivated behaviour is that the population retains a "reserve" of individuals with which to maintain pressure on the limits set by the environment. It retains the capacity to quickly absorb any increase in the supply of food, converting it to more individuals in the next generation.

A symposium with the goal to "explore relationships between ecology and the origin and maintenance of social behaviour" (Slobodchikoff 1988) seemed to produce a consensus that the basis of the evolution of social behaviour is the *effective* use of limited resources (food, space, nest sites, etc.) to increase the number of an *individual's* offspring that survive and reproduce. Sociality is, they say, an ecological rather than a phylogenetic or anti-predator device. In spite of this, papers presented to the symposium included statements like "...the dynamics of social behaviour can be viewed as an interaction between resources and individuals seeking to exploit those resources. Through cooperation, the group is able to collectively exploit or defend resources that each animal individually would not be able to exploit or defend" (Slobodchikoff and Schulz 1988). "Most [ecological] hypotheses...suggest that groups originate in order to exploit resources that cannot be exploited as efficiently by solitary individuals. They imply that the presence of conspecifics generates a direct and nett benefit for an individual joining or interacting within a group" (Slobodchikoff and Shields 1988).

They seem to miss the point. Social groupings do not give all individuals joining a group *equal* access to resources that would not otherwise be available to them. Quite the opposite. All forms of dominance – all forms of social grouping – result in an *uneven* distribution of limited resources. By sequestering resources to themselves the dominant few limit or deny access to them by others; many get none and die.

It is perhaps more likely that social grouping originated because of this and because subordinate adults by staying (with or without some trade-off "help" to the dominants in exchange) may garner enough crumbs to live a little longer than they otherwise would have done, and so increase their probability of eventually breeding. (The same is true of the young of the dominant animals once the limit of the territory or group's capacity to support them is reached.) Natural selection, by eliminating more of the subdominants and young which disperse than of those that stay, would have made *inevitable* the evolution of the great variety of groupings of dominant and subdominant individuals which we observe today as social hierarchies of differing complexity.

While it might be more profitable to examine the evolution of social structure from this point of view, my aim here is to illustrate the ecology of these behaviours; how, in today's environment, they influence the distribution and abundance of animals which have evolved these behaviours. They channel what food is available at the moment to only some of the individuals seeking to gain access to it. This increases the number of individuals in the population which survive and breed. More genes are passed to the next generation than would otherwise be the case. How many favoured ones there are in each generation will vary as the amount of available food varies. All this is accommodated by varying either or both the number tolerated in a territory or group, or the size and number of territories or groups in a locality.

There are, nevertheless, upper and lower possible limits to such processes. There must be a practical limit to the area that an animal can defend and still leave time for it to feed and breed. Thus when food is scarce the size of a territory may increase, but not necessarily in proportion to the food it contains.

Dill et al. (1981) found this to be so for juvenile salmon, *Onchorhyncus kisutch.* They defend territories in the streams where they spend the first 2 years of their lives. Those unsuccessful in obtaining a good territory are forced further downstream, and frequently out of the stream altogether. The territories are smaller when the food in them is more abundant and the pressure from intruders is high. When food is scarce, however, feeding territories, although larger, were smaller than expected. Fish holding territories where food is less abundant must not only defend a larger territory than those on good territories, but they obtain proportionally less food for their increased effort.

Alternatively, as is commonly found with predatory birds, the size of the territory does not increase, but fewer young are produced; or breeding ceases altogether. Very commonly all semblance of territorial and social behaviour is abandoned when food becomes very scarce, individuals simply foraging where they can. As Morse (1980) points out, the fact that some animals change from territorial to hierarchical behaviour when resources deteriorate suggests that hierarchies may be more efficient than territories at such times. This also seems to be true at the other extreme, when food becomes atypically abundant, as Rogers and others have found for bears (Sect. 11.2.1).

Territorial and social behaviours do have built into them this flexibility to enable animals to quickly take advantage of any amelioration of the usual chronic shortage of food in their environment. However, because the behaviours have evolved in response to this continuing inadequacy, there are limits to the short term flexibility of the response to any increased abundance in food by most animals inheriting those behaviours. They have evolved and are "programmed" to delineate and defend a territory, or to maintain dominance in a group, at or above a level which evolutionary "experience" has shown to be an essential minimum.

Occasionally in nature food becomes *so* abundant that defending a territory or dominating others is unnecessary. There is plenty for all. Yet if such abundance is so great or so infrequent as to be outside the evolutionary "expectation" of a species, much of it may go unused. The behaviours evolved to survive the usual will prevail. Morse (1978) found an example of this during an outbreak of spruce budworm in northern Maine. For the small passerines that feed on budworm larvae, this outbreak represented a superabundance of food. Yet, while the number of birds increased, there were not anywhere near enough of them to use all the available food. They maintained their territorial behaviour, chasing and attacking each other, and removed no more than about 6% of the caterpillars.

Similar constraints apply to social hierarchies. Variations in the abundance of food beyond the range experienced over evolutionary time in a world of shortages may result in apparently non-adaptive behavioural responses. Extreme examples, where human manipulation has created situations quite beyond anything ever experienced in nature, perhaps best illustrate this point; examples such as many of the problems associated with maintaining farm or laboratory animals. Most dramatic, however, have been experiments where rats and mice have been confined with ad lib food and removal of wastes (e.g. Southwick 1955). Breeding continues in such an "ideal" environment, but the young cannot disperse. Numbers continue an unrelenting increase and space in which to live becomes the limiting resource. Genetically fixed and normally adaptive social spacing and dominance behaviours become increasingly pathological, and finally all semblance of social structure breaks down. This is truly an exception that proves the rule of limitation of abundance by a relative shortage of food.

We can predict, with fair certitude, that if the supply of good food in nature continued indefinitely at a high level, natural selection would act against the conservatives fighting to maintain the status quo of territorial rights. Soon only those able to adapt, by whatever means, to ways of converting the additional food into more offspring would be left. That is until physical space became limiting – so far an unlikely event in the natural world.

Chapter 12 Cannibalism

Early attitudes to cannibalism were that it was an aberrant and abhorrent form of human behaviour which was occasionally seen in animals under extreme stress, such as being artificially crowded in a laboratory, but not in nature. Many biologists today still maintain that the killing – let alone eating – of members of one's own species is a rare event, that "intra-specific conflict" is largely ritualized and rarely causes serious injury or death.

Cannibalism is, however, a widespread and normal phenomenon in populations of animals from protozoans to mammals, including many species otherwise considered to be strictly herbivorous (Fox 1975; Polis 1981). Even among ecologists who recognize this, however, most see cannibalism as a behaviour operating as a self-regulating mechanism to control a population below the maximum size it would attain if the animals comprising it consumed all the food that was accessible to them "...in the same manner as territory and social dominance [maintains] the size of the population below the carrying capacity" (Polis 1980).

On the contrary, I would argue that cannibalism, like other forms of social behaviour, has evolved under the selection pressure of a shortage of protein, and does the *reverse*. It ensures that a population is at all times using its food supply as efficiently as possible, using as much of the nitrogen in the environment as possible, to *maximize* the number in the population that survive and reproduce. Cannibalism enables a more effective use of the limited food in the environment by making more of it available than otherwise would be the case, and to only a "favoured few" which survive and reproduce at the expense of the "doomed remainder".

Nor is cannibalism "important in the maintenance of equilibrium densities" (Polis 1980). There are no equilibrium densities in nature, only the maximum number that can survive on the resources available in the habitat at the time. (Witness the various examples discussed in this book of violent increase and decrease following sudden and unexpected changes in the amount of food available.) It is perhaps better to consider cannibalism as part of the broader concept of animals using "conspecific tissues and secretions as sources of nutrition" (Joyner and Gould 1987). Cannibalism is then only part of a suite of behaviours evolved to increase the number of animals in a population which can obtain sufficient of the scarce supply of nitrogen that is available in the environment.

Cannibalism takes many forms – mate eats mate, parent eats offspring, offspring eats parent, sib eats sib – and most cannibals are female and herbivorous. I have already referred to examples of this behaviour in lepidopterous larvae (Sect. 3.4.5), locusts and other orthopterans (Sect. 3.4.7), termites (Sect. 3.7.2), and snails (Sect. 5.6).

I will exclude from this discussion two other forms: the (usually) male cannibalism sometimes associated with killing in defence of territory or social status; and infanticide of the progeny of ousted rivals. The former is a consequence of those fights, although it does further highlight the chronic shortage of protein in the world – none can be wasted. The same is true of the scavenging of the bodies of conspecifics which have died from any cause. This is another common phenomenon among herbivores, for example, locusts (Sect. 3.4.7) and Aldabra tortoises (Sect. 8.1), the latter so graphically illustrated in one of David Attenborough's television programmes.

The killing by a male of the offspring of another male is now fairly well documented for what it is – the replacement of the departed or killed male's genes with those of the victor. This infanticide is widespread among vertebrates, both carnivores and herbivores, and some invertebrates. There is even some evidence that the behaviour may still persist in human populations (Daly and Wilson 1988). Occasionally the same behaviour (for the same apparent genetic advantage) is found in females (Hoogland 1985; Spradbery 1991). Often those killed are not eaten; the behaviour has nothing to do with food.

Cannibalism usually occurs at two points in the life of an animal, both where access to an increased supply of nitrogen is vital: the production of young by females; and the initial survival and growth of neonates. At both points the concern is to sequester as much as possible of the limited amount of nitrogen accessible in the environment, and successfully convert it into the maximum possible number of surviving offspring in the next generation. All else flows from this.

The advantage gained is not that the population is reduced when food is short so that all do not starve. Rather it is that, even though *without* cannibalism some *might survive* such shortages, *with* cannibalism many more *will survive*. To understand the difference, and how natural selection has moulded this behaviour, one must understand the usual inadequacy of the environment – most of the time there is a chronic shortage of adequate food interspersed with occasional (and usually unpredictable) bursts of abundance. With these points in mind we can look in more detail at these two critical points in an animal's life history where cannibalism is most likely to prove adaptive.

12.1 Cannibalism by Females Producing Young

A breeding female must obtain enough protein, over and above that needed for normal maintenance, to form her new young, and she must do this in an environment in which the supply of available nitrogen is all too scarce. Cannibalism is one way in which this supply might be enhanced, so it is not surprising that cannibalism by breeding females is widespread and common. It ranges from the well-known propensity of female spiders and mantids to eat their mates during or immediately after copulation, through (if we take the broader view of Joyner and Gould) the eating or direct absorption of proteinaceous spermatophores transferred during copulation, to the resorption of embryos or eating of live young when food becomes scarce.

The cannibalism of living young ranges from a female eating her own still-dependent offspring to the active predation of free-living conspecifics (not necessarily her own) when there is a shortage of food. The latter is probably the most studied form of cannibalism and the one which commonly leads to the conclusion that cannibalism is a density-dependent self-regulating process (Fox 1975; Polis 1981, 1988). But is it? It is found in populations of carnivores and herbivores. In both the members of the population face a chronic but fluctuating shortage of protein food. With carnivores this shortage has been readily identified; with herbivores it has not. Herbivores are seen to live in the midst of apparent abundance – a great quantity of plant food which they never eat. However, for most of the time the *quality* of this plant food is inadequate. Its content of available nitrogen is insufficient to meet the protein demands of breeding females. This situation is strictly analogous to that which carnivores face when individual prey are so thinly spread in time and space that breeding females cannot catch enough, often enough, to provide their minimum requirements of protein for reproduction.

From his studies of carnivorous desert scorpions, Polis (1980, 1981, 1988; Polis and McCormick 1987) identified two strategies by which cannibalism could circumvent periods when food is in short supply. The first, the "lifeboat" strategy, is obvious. Big scorpions eat little scorpions as their supply of normal prey (insects) declines. Thus a small population of cannibals could survive a period when food is scarce where a similar population of non-cannibals would have died out.

The second, the "grazer" strategy, is not so obvious. By eating their smaller fellows, large scorpions gain more than the one-for-one benefit of eater over eaten in the lifeboat. Small scorpions capture prey which, while they may be abundant, are too small for large scorpions to catch. So, by eating the small scorpions, large scorpions are able to "graze" a source of food otherwise inaccessible to them – a more effective use of a resource. More of the limited amount of protein in the environment becomes available to the population than would have been the case without cannibalism, and it is converted, via the smaller scorpions, into fewer, but successfully breeding, scorpions. Without cannibalism the available protein would have remained in many smaller scorpions, most or all of which may have died of starvation before reaching reproductive age.

Another well documented example of the benefit of the grazer strategy is recounted by Gabriel and Lampert (1985). The copepod crustaceans (*Cyclops* spp.) which they studied feed exclusively on algae when young, but become carnivorous when adult. When suitable prey are in short supply these carnivorous adults cannibalize their algal-feeding young. Gabriel (1985) and Gabriel and Lampert (1985), using mathematical models based on physiological data and field and laboratory observations taken from the literature, showed how such cannibalism can be advantageous to *Cyclops*. Whenever the abundance of their zooplankton prey falls below a level where the growth of its population cannot compensate for losses to the predatory *Cyclops* – even when adequate supplies of its own food, the phytoplankton, again become available – the *Cyclops* would eliminate its prey, and thus itself. But adult *Cyclops*, by eating their own young, gain access to some of the primary production of the algae, with the result that the populations of both predator and prey persist. The small disadvantage of cannibalism continuing when food is abundant is more than compensated for by its advantage during times when the prey is in short supply. This study is another good example of cannibalism acting as a strategy to use more of the food in the environment than would otherwise have been the case, and to use it more effectively by distributing it to a few successful individuals rather than many unsuccessful ones. Daan et al. (1988) also came to this conclusion after finding much the same story for another primary herbivorous copepod, *Temora longicornis*. Cannibalism of small nauplii by adult copepods became significant only when there was little algal food available. Cannibalism does not, as Polis (1981) would have it, *decrease* population size and increase *average* access to general resources.

The lifeboat strategy is common, and it may well be that the grazer strategy is equally so. Fox (1975) and Polis (1981) give examples including squids, newts, insects, and fish where the young eat markedly different food from the adults and the adults frequently prey upon younger conspecifics.

However, it is when the same strategy is found operating in a population of herbivores that the "bottom line" is truly revealed. It is not the quantity of food which is in short supply, it is the protein that it contains. Any process which makes more of this scarce resource available for conversion into successfully breeding individuals in a population will have great adaptive value.

Such an example is provided by Wolcott and Wolcott's (1984) study of the response of the Bermudan land crab *Gecarcinus lateralis* to supplementation of its natural diet with a high-nitrogen food. They set up two groups of individually caged adult crabs freshly emerged from winter dormancy, and prior to the breeding season. The control group were fed ad lib. with freshly collected plants which they normally eat in the field. The experimental group received the same diet plus ad lib. raw soybean (a naturally high-nitrogen plant material acceptable to the crabs). Crabs fed with soybean supplement consumed more nitrogen and grew more than the non-supplemented group. The Wolcotts calculated that to accumulate the amounts of nitrogen they observed in the crabs on the soybean supplement, crabs fed only on grasses would have had to consume grass at six to 24 times the maximum rate possible under the constraints of the capacity and clearance rates of their guts.

After 43 days the individual crabs in both groups were each provided with three juvenile *G. lateralis* in addition to their diet. The crabs fed only grasses consumed more juveniles than did those given the supplement of soybean; especially in the first 8 hours after the juveniles were added to the containers.

The Wolcotts concluded that *G. lateralis* shares with other herbivores the problem of acquiring sufficient nitrogen from food which is low in nitrogen. These strategies include selective feeding on plant tissues with the highest content of nitrogen, scavenging (especially on carrion, mammalian faeces, and human refuse), predation on both vertebrates and invertebrates, and cannibalism of their young. They also pointed out that this cannibalism of the young by *G. lateralis* adults fulfils Polis' criteria for a "grazer system". The adults, experiencing a shortage of nitrogenous food, consume some of their own young which as pelagic larvae eat plankton – a source of food not otherwise available to adults. This allows a more effective use of the limited food resource, enabling more breeding individuals in the population to survive and reproduce than would have been the case without cannibalism. It is *not* a mechanism to reduce the size of the population below that which the available supply of food could support.

These authors commented, however, that the adults being limited by the quality rather than the quantity of their food is "a novel aspect of the land crab system". I hope that what I say throughout this book will dispel this notion. It is not in the least novel. *All* herbivores face this relative shortage of nitrogenous food. As Joyner and Gould (1987) point out "Although food quantity is obviously important, it is increasingly clear that food quality is a significant factor in the occurrence of cannibalism".

Furthermore, they quote examples of various herbivores, ranging from termites and honeybees to armyworms, where this lack of quality is, in fact, a lack of protein.

12.2 Cannibalism by Growing Young

In the examples discussed so far the major advantage of cannibalism accrues to breeding females, increasing the number of them which are able to contribute more and stronger young to the next generation. But strategies which have evolved to enable females to produce fitter young are only variably successful, and rarely fully so. Neonates will, therefore, more often than not, be less than maximally viable; yet they must somehow continue to grow at a rapid and initially exponential rate. To do this they must have access to a continuing supply of high-nitrogen food. One way of achieving this is to have continuing access to the protein food already stored by their mother, either in her own body or in that of their siblings. Various forms of cannibalism by neonates gives them this access and at the same time continues the more effective use of the limited supply by concentrating it in fewer favoured individuals.

There are many well-documented cases of cannibalism by neonates (Polis 1981; Joyner and Gould 1987) ranging from the devouring of the body of the mother by some arachnids, through the suckling of milk by mammals, or the eating of large, infertile "trophic" or "nurse" eggs, to the active predation of one sib by another. Usually it is the biggest and strongest (normally the first born or hatched) which devours the weaker sib. As Polis (1980) expressed it, "siblings functionally serve as packages of live meat [or a transitory food cache] for their kin". Even more weird and wonderful (to those unaccustomed to the vagaries of natural selection) are specializations of sibling cannibalism such as the in utero cannibalism in some sharks and salamanders, often by individuals equipped with enlarged jaws and teeth not possessed by their less fortunate fellows. There are also variations on the theme concerned with passing on the maternal store of protein without which many herbivores cannot survive on a diet of nothing but plant tissue. Eating of eggshells and the faeces of the mother are two widespread examples. The effect of all these different forms of cannibalism by the young is to increase the probability that those practising it will survive their early exponential growth at the expense of those that do not. An extension, if you like, of the whole process of obtaining the maximum possible amount of protein from the environment and converting it as effectively as possible into viable individuals in the next generation.

That the limiting resource is nitrogen is most clearly brought out in examples of cannibalism by the young of otherwise herbivorous animals. One such is that illustrated by the Baurs' study of newly hatched terrestrial snails eating their unhatched siblings (Sect. 5.6). A more specialized form of this behaviour is seen in the evolution of nurse or trophic eggs. It is an adaptation found in several taxa, but is especially common and well studied in prosobranch gastropods (Rivest 1983; Baur 1992). These snails are scavengers or predators, not herbivores, but this simply illustrates again that the advantages of adaptations that improve the protein nutrition of the young are not confined to herbivores. The snails' eggs, which are laid in a capsule, are of two kinds: normal ones which develop embryos, and infertile ones which do not. The latter outnumber the former, often by several thousand, and are the nurse eggs. Rivest experimented with the egg capsules of a predacious intertidal snail, *Searlesia dira*, collected from sites along the coast of Oregon, British Columbia, and Alaska. At the trochophore stage of their development the embryos feed on the nurse eggs for 1 to 3 weeks. One embryo may ingest anything from 10 to 50 eggs. The more it eats the larger it is when it hatches from the capsule as a fully formed snail, and the better its chance of surviving. Rivest kept groups of hatchlings of different sizes and supplied them with abundant food for 36 days after hatching. Of those less than 2 mm long at the start of the experiment, 70% died, but 77% of hatchlings larger than this survived, and they grew more than twice as fast as the smaller ones.

This form of behaviour is not, however, confined to eating unhatched eggs. The otherwise strictly herbivorous caterpillars of many species of moths will eat other larvae, and readily devour conspecific pupae within moments of ecdysis (Sect. 3.4.5). Al-Zubaidi and Capinera (1983) produced a nice experimental demonstration of the influence that the level of nitrogen in the diet has on cannibalism by such caterpillars. They fed generalist-feeding larvae of the noctuid moth *Spodoptera exigua* on the foliage of sugar beet plants which had been fertilized with different levels of nitrogen. As the larvae reached the third instar the authors presented them with pupae of their own species, one at a time, along with foliage in excess. They found that cannibalism of pupae increased as the level of nitrogen in the foliage decreased; and, within different levels of nitrogen in the foliage, as the amount of foliage fed to the larvae was decreased. Females emerging from larvae which had eaten foliage containing increased levels of nitrogen laid more eggs. But females reared from cannibalistic larvae within each of these treatments laid more eggs than those reared from larvae fed only on beet foliage.

However, caterpillars will eat other caterpillars as well as pupae, and the behaviour is not an aberration induced in laboratory cultures. Semlitsch and West (1988) observed in the field, and demonstrated in the laboratory, that large caterpillars of the noctuid *Litoprosopus futilis* eat small conspecifics but not large ones, and small ones do not eat large or small ones. All sizes will readily eat dead conspecifics of any size. Their results sprang from observing cannibalism in an outbreak population of this species in South Carolina. Of particular interest was their demonstration that cannibalism by these larvae was neither an experimental artifact nor a simple function of crowding (the usual interpretation of cannibalism in laboratory cultures). Large and small caterpillars both survived well (95 and 89% respectively) when confined with others of the same size, but when mixed together at similar densities nearly all small caterpillars were eaten, while the larger ones mostly escaped unscathed (11% survival of the small and 95% survival of the large). What is more, the *lowest* densities in their experimental containers, where caterpillars segregated by size remained unharmed, were some three times higher than the average *highest* density in the field, where the authors repeatedly observed cannibalism of the small by the large at all densities. Finally, for reasons they could not discover, the size of larvae in the field was strongly bimodal; the mean head-width of large caterpillars was about 50% greater than the mean of the small ones. It would seem that this is another example of the grazer effect.

I think the oft-quoted statement of Kirkpatrick (1957) that some plant-feeding insects are cannibals "even in the presence of an abundance of food" should be modified to read, "in the presence of an *apparent* abundance of food".

12.3 Cannibalism, Warfare, and Protein

That the adaptive value of cannibalism is to improve access to the limited supply of nitrogen in the environment is perhaps best illustrated by an example of what may be thought of as cannibalism's ultimate evolution – organized warfare between conspecific populations. Mabelis (1979) proposed that warfare and cannibalism in European wood ants, *Formica polyctena*, is the result of a lack of protein food. The growth of a new (usually sexual) brood of larvae in the spring puts a very high demand for protein food on a colony at a time when prey are scarce. He demonstrated that dead worker-ants brought to the nest are fed to the larvae. Thus, preying on workers of neighbouring nests

Fig. 36. Fighting European wood ants. In spring, before insect prey are available, well-organized, structured wars between neighbouring conspecific nests provide scarce protein, in the form of the bodies of worker ants, to feed the larvae of the sexual brood. Photo courtesy of Albin Hunia

would enable the spring brood to be raised before, and until, other prey became available. But the neighbours are confronted with the same problem. Warfare is the inevitable outcome.

Driessen et al. (1984) set out to test this hypothesis by making a detailed study of one nest, and its relationship with neighbouring conspecific colonies, over a period of 80 days in spring in Holland. They were able to make quantitative estimates of the colony's demand for protein, the supply of protein available in its foraging area, and the amount of protein taken into the colony.

As large final instar larvae are the major consumers of protein, an estimate of their number gave a fair representation of the demand of the colony for protein. Counting the number of empty cocoons carried out of the colony each day, and extrapolating back by the pupation period, gave a measure of the number of these large larvae. The availability of lepidopterous larvae and bibionid midges (which together with workers from other nests provided 86% of the protein diet of the colony during the study), and of other minor prey, was estimated with pitfall and frass traps and by direct observation and counting. Workers from other nests were abundantly available at all times. The rate at which protein was being taken into the colony was approximated from the dry weight of all items taken in by workers that entered the nest during the first 6 minutes of every daylight hour.

Approximately 1 month after the ants in this colony first began to be active in the spring, and 1 week after significant foraging away from the nest was observed, war broke out. Fighting continued at varying intensities for 1 month, mostly on one battlefield with workers from the nearest neighbouring colony (the colony killed 75000 enemy workers during this battle) but included another separate battle involving two other colonies. During this time more corpses of wood ants were taken to the nest than were all other items of prey. The intensity of fighting declined quite quickly over the last few days of this period when bibionid prey first began to be brought to the nest. Caterpillars did not become available in quantity until about a week before warfare ceased, and bibionids became very abundant a few days later. Over the entire 80 days of observation the proportion of total prey taken was evenly divided between corpses of conspecifics and the two major types of prey. But in the 3 weeks before these other prey became available conspecifics comprised 83% of the intake; in the 3 weeks following, only 17%. A small number of ant corpses continued to be brought to the nest after the battles stopped, but apart from casualties from a 4-day battle 5 weeks later, these were regarded as "natural mortality in the field in time of peace" (Driessen et al. 1984).

Daily counts of cocoons carried from the nest showed that demand by the colony for protein had increased dramatically just at the time that warfare started, and continued at a high level until later in the summer when it gradually declined. Warfare and the accompanying cannibalism therefore represent the more effective use of limited nitrogenous food by these ants. When normal prey is abundant it is converted to workers which can be recycled in times of shortage of protein. Battles thus represent an exchange of food stored as live individuals. Even when a nest loses more workers than it gains (a nett loss of protein) warfare is still adaptive. Fighting workers belong to the oldest generation with the shortest life expectancy. These are "exchanged" for food with which to raise a sexual generation in the spring. The nest could not do this without cannibalism because there is no other source of protein available at the time. Even the rarer summer and autumn wars are advantageous because they convert old workers which are less likely to survive the winter and therefore would not be available in the next season. However, in all cases the nett recovery of protein from cannibalism is low. It costs more than one individual to raise another. So, only in the absence of alternative prey – when there is a shortage of nitrogen in the habitat – can cannibalism be adaptive.

The same sort of response as that evinced by these ant wars is illustrated by the cannibalism of conspecifics by colonial wasps. Cannibalism appears to be quite common among temperate species, and may be either intra- or intercolonial within a particular species of wasp, as illustrated by two species of *Polistes* from Japan (Kasuya et al. 1980). In autumn the number of workers of these paper wasps declines, as does the proportion of workers to developing reproductives. These declines, combined with the decreased abundance of lepidopterous larvae which constitute the normal prey of these wasps, means that food for new, still-growing reproductives becomes scarce; there is less of it and fewer to collect it. It is then that the remaining workers, faced with unsuccessful foraging trips for caterpillars, start cannibalizing the larvae and pupae from either their own nest or those of nearby conspecifics. They are recycling the shrinking supply of protein, concentrating it into the production of a new generation of fewer but successful males and females.

Human cannibalism and its associated warfare between groups of conspecifics has, as a regular practice, been most commonly found in populations living in places in the world where protein is in short supply. It is quite possible, therefore, that even although it is usually disguised by religious or other cultural explanations or justifications, cannibalism serves a similar function in these human populations. Dornstreich and Morren's (1974) studies in Papua New Guinea would support this suggestion. They found that small groups of people practising pay-back warfare and cannibalism were living where the supply of protein food was marginally and chronically inadequate. Cannibalism bridged the gap, providing 5 to 10% of each population's protein needs, and so keeping the people from suffering an otherwise inevitable protein insufficiency.

But to return to herbivores. Whitham (1987) in discussing the evolution of cannibalism among herbivores in relation to the defences of their food plants, asks, "with herbivores killing herbivores. . .who really benefits from cannibalism?" He then extended the view implicit in his question – that cannibalism reduces the numbers that survive to reproduce within a population – by supposing that cannibalism must therefore *benefit the plant* by reducing the number of herbivores feeding on the plant. I find it hard to imagine how a behaviour could have evolved which causes members of a population, by killing some of their own kind, to convert *less* of their available food into future generations. Obviously the reverse must be true. Cannibalism, by more effectively redistributing a limited amount of available food among individuals, ensures that more small herbivores grow to be big herbivores than would have done so had there been no cannibalism. The end result *must* be *greater* damage to the food plant.

Finally, Joyner and Gould (1987) ask "... is feeding on other members of the same species more or less energetically efficient than feeding on other sources of nutrients?" This is not the question to ask. Cannibalism is not to do with the efficient use of energy. It is the supply of *protein* that is critical. Individuals that feed on their own kind must have a greater chance of surviving and reproducing than those which do not, else natural selection would have eliminated such behaviour which destroys members of one's own species. As I have tried to illustrate in the preceding discussion, a careful look at recorded cases of cannibalism (and other forms of feeding on the tissues of conspecifics) reveals that *all* are directed towards *increasing* the number of young that survive to reproduce, either by directly concentrating the available supply of nitrogenous food in fewer young individuals, or indirectly through the nitrogen nutrition of the mother. Cannibalism is another evolved behaviour which increases the effective use of the limited amount of nitrogen available in the environment of animals, especially, but not exclusively, of herbivorous ones.

Part IV: Predators in an Inadequate Environment

Having devoted most of the discussion so far to herbivores and their relationships with their food plants, it is now time to look at those *usually* credited with responsibility for limiting the abundance of the herbivores – their predators.

I would think there is general agreement among ecologists that the abundance of top predators is limited by the accessibility of their food. Some might argue that they are kept below the carrying capacity of their environment by parasites and diseases – in effect that they are *not* top predators. But this is just moving the problem on a notch. What then controls the abundance of viruses, bacteria, flukes, nematodes, ticks, and the like? They are also predators. Let us leave these on one side for the moment, and consider the more "conventional" vertebrate and invertebrate predators.

The food of carnivores (the bodies of other animals) is, unlike that of herbivores, a concentrated source of readily digestible nitrogen. Food and nitrogen are one and the same; if there is enough food there is enough nitrogen. So it seems unlikely that they too would be limited by a shortage of nitrogen. However, their food, again unlike that of herbivores, exists in discrete "packages" which are mostly very mobile and usually fairly thinly, and patchily, spread in the environment. Also their prey are equipped with a wide variety of behaviours for avoiding predators, so are difficult or impossible to catch. They might, therefore, be restricted for much of the time by a relative shortage of food. While there may be a great deal of prey present in their environment, much of it may be inaccessible or unattainable, and therefore there is not enough nitrogen available for reproduction and growth. Intensive behavioural studies have found (and modern television programmes have shown us all) that for predacious mammals and birds this is more often true than not. While living in a habitat which seems (to us) to be teeming with enough prey to feed many hundreds of times their number, they can take only the very young, the old, the sick, the injured, or the simply unlucky or momentarily incautious. Even then their best efforts to run down and kill a prey often fail. Frequently they must subsist by scavenging or catching invertebrate prey. Usually their attempts to breed and raise young fail, or are of only limited success.

Those of us who have spent time watching invertebrate predators such as insect parasitoids, carabid beetles, or hunting spiders, have observed the same story. In action these sorts of animals are equally as spectacular and ferocious as any attacking lion – and equally as often unsuccessful.

I have seen newly hatched New Zealand katipo spiders, *Latrodectus katipo*, distressed and repelled by the secretion from the cornicles of aphids they tried to capture, and spider-hunting wasps caught and eaten by the spider. I have watched various species of hunting parasitoid wasps unsuccessfully try to sting small caterpillars or aphids. Often they are flung away by the wild thrashings of the prey or repelled by secretions or vomit (large larvae and adult aphids are virtually immune from attack because of such tactics). Often, upon encountering potential prey on a plant, a female wasp will go into an attacking frenzy, thrusting repeatedly but aimlessly in all directions with her ovipositor, more often missing than hitting the prey. I have watched ichneumonids attempting to sting moth pupae in forest litter. They were repelled by and unable to sting the thrashing prepupae in their cells, and their ovipositors could not penetrate the hard brown cuticle of the mature pupae. It was only during the short time immediately after ecdysis when the pupae are soft and green that they were able to lay an egg in their prey. Yet as often as not these hunting wasps fail to find the pupae in the first place, even when they are (to our perception) close at hand.

Nor do "sit and wait" predators, like orb-web spiders or praying mantids, fare any better. They depend upon the capture of individuals which move essentially at random with respect to the predator. Praying mantids miss as often as they hit, and not every fly that hits a web sticks long enough to be subdued. Unless prey happen to be especially abundant, catches tend to be few and far between, often too few to sustain a growing juvenile or a female developing her eggs. As we shall see in the examples which follow, these sit and wait predators are equally as inefficient as the active hunters (and as the herbivores!) at obtaining the food they need to reproduce and grow. Neither of them is an optimal forager. Both rely upon chance encounters. At best they can improve upon the randomness of the distribution of their prey by behaviours which tend to maximize their encounters with them.

Predators, then, are confronted with an environment every bit as harsh and inadequate as that of the herbivores. Their populations, too, are constantly pressing hard against the limit set by the accessibility or attainability of food for the individuals making up that population. While living in a habitat which *seems* to contain an abundance of food they are, in fact, saturating that habitat in terms of the food that they are able to use. As many animals as the environment can support at any given time are living there. As a corollary, they cannot be "controlling" their prey, or can do so only exceptionally or for short periods.

The points I have made so far are not, I think, matters of serious contention and I need not dwell on them, or take time to expand them with further examples; at least not for vertebrate predators. However, many ecologists, and especially those attempting the biological control of insect pests, would not agree that invertebrate predators are inefficient. So I will give some examples which demonstrate just how inefficient most of these animals are at locating and capturing their prey and, as a consequence, how little effect they have upon the density of their prey.

There is a further point which is not commonly realized for either vertebrate or invertebrate carnivores, and therefore needs some further elaboration, namely, the process by which this shortage of food operates. It is not the success or failure of the very visible adult predators which decides how numerous a population will be. Most adult carnivores in their prime will get enough food to maintain themselves but, like herbivores, precious few of their kind ever reach maturity. Most starve at an early age, and females of breeding age are hard pressed to get enough good food to produce viable offspring in the first place.

Finally, as with the herbivores, it is not any shortage of energy which limits this production and growth of young, but the availability of an adequate supply of protein.

Chapter 13 Vertebrates

13.1 Lions, Lynx, and Feral Cats

A review by Van Orsdol et al. (1985) of studies of the African lion, *Panthera leo*, living in ten separate habitats, illustrated very well the way in which carnivores too are under the same sort of constraints as are the herbivores. The density of lions in any area depends upon the amount of prey in that area at the time of year when prey is least abundant. Prides have more lions in them and occupy smaller home ranges in places where there are more prey to eat in the leanest season. The amount of this food which is available influences their numbers by way of the number of young which live for the first 12 months. Cubs burdened with rapid growth, getting only poor nutrition from their underfed mothers, and later unable to get a fair share of the diminished number of smaller prey that the pride can catch, are the first to succumb to nutritional stress.

Recently, careful observation of the hunting behaviour of lions has revealed that this comes about because, in common with other carnivores, they are inefficient predators (Stander 1992). There is no absolute shortage of prey, but because of their incapacity to catch most of it, they suffer a chronic relative shortage of food. This is alleviated only briefly in the wet season when ungulate herds are concentrated and producing young.

While lions do hunt large animals, they eat whatever they can catch, and kill the easiest prey. So, the greater bulk of their food comprises vulnerable small species, and neonates of larger ones. However, even when hunting such easily captured prey half of their attempts at capture fail, and several separate studies have shown that their average rate of successful capture varies from 1 in 3 attempts to only 1 in 7.

In Canada the staple food of another cat, *Lynx canadensis*, is the snowshoe hare, *Lepus americanus*, (Brand and Keith 1979; see also Sect. 6.2.5.3). When numbers of the hare decline so do those of the lynx, and the contribution of hares to the lynx diet falls from 90% to 35%. Lynx then eat mice, voles, squirrels, and a miscellany of other small animals and birds. In spite of this shift in diet their total intake of biomass falls drastically, as does their body condition, and their reproduction. Females take longer to reach sexual maturity, fewer of them breed, and those that do have smaller litters. Also, the number of kittens surviving from spring to the beginning of winter falls precipitously. Postpartum mortality of kittens, reaching as much as 95%, was the predominant cause of lowered recruitment to populations at these times. The kittens

Fig. 37. Lions, renowend as supreme hunters, are, in fact, inefficient predators. Half of all attempts to catch their easiest prey fail, and most cubs die of malnutrition before they are 12 months old. Photo courtesy of P. E. Stander

died when they were still completely dependent upon their mothers for food, and malnutrition and outright starvation were the major causes of their deaths.

The domestic cat, *Felis catus*, is similarly limited as a feral animal by the amount of food for breeding females and their new young (Jones and Coman 1982). In locations scattered around the world this species has been found to exhibit the same reproductive potential, and with it the capacity to rapidly increase in response to an increase of food. However, it persists at very different densities set by the level of attainability of the food in different habitats. In farmland, forest, and national parks of Australia and New Zealand, where they eat mostly feral rabbits and mice, native birds, lizards, and insects, cats are found at densities varying from two or three per square kilometre to fewer than one. In a Californian refuge for waterfowl they reached nearly eight, and in the Portsmouth docklands of England, where there is an abundance of food from human refuse, *200* per square kilometre. On subantarctic islands, like Macquarie Island, they can reach densities as high as eight or nine per square kilometre, feeding largely upon feral rabbits or the indigenous petrels which nest in the ground. When first introduced to Marion Island where there are many of these burrow-inhabiting mutton birds (easy prey for cats), they were recorded at densities of 16 per square kilometre, and increasing at annual rates of 23%. In contrast, on Campbell Island, where most birds have been eliminated and there are no alternative prey, cats are too scarce for a meaningful density to be gauged. In their 4-year study of cats in Australian mallee scrub Jones and Coman found that the animals, feeding mostly on rabbits, were maintaining a fairly stable population from year to year. They considered that they were at the carrying capacity of the habitat. Each summer the population reached its maximum density, the population including adults, adolescents, and juveniles, but each winter their numbers dropped by more than a third and only adults were left. Most of each year's crop of kittens had vanished well before winter, the nutritional stress of lactating females and newly weaned kittens taking its toll.

13.2 Coyotes, Wolves, and Foxes

The coyote, *Canis latrans*, is another predator of snowshoe hares in Canada (Todd and Keith 1983). Like the lynx, its numbers also vary in synchrony with those of the hare, and declines in its abundance are clearly caused by a shortage of hares depressing the breeding capacity of their females and the survival of their young. Like the lynx, the coyotes subsist at times when hares are scarce, on other mostly smaller prey and on carrion. Again, however, this switch does not compensate for the loss of abundant good food, and they lose condition. Fewer females breed, those that do breed produce smaller litters, and the proportion of juveniles in the population drops by some 60%.

The coyote's larger cousin, the wolf, *Canis lupus*, is similarly limited by the influence of the supply of food on its breeding success (Jordan et al. 1967; Fuller 1989). Their major prey are ungulates like the moose, *Alces alces*, and various species of deer, depending upon their location. They kill mostly the very young, in both summer and winter, but also take adults, with packs running down old and debilitated individuals. They hunt smaller mammals and invertebrates only when their preferred prey is not available. Their numbers in North America are directly related to the biomass of ungulates; in years when ungulates produce many young, many more wolf cubs survive. This survival of pups is the major factor in the growth of wolf populations, the annual rate of change in a population directly reflects the proportion of pups present in the population. Again, the critical factor for survival of these pups is the attainability of food during parturition and rearing.

Angerbjorn et al. (1991) conducted an experiment in Swedish Lapland to test the influence of food on the reproduction of another canid, the arctic fox, *Alopex lagopus*. It is a small opportunistic omnivore with a circumpolar distribution. Its numbers vary greatly but correlate well with the numbers of voles in the area, as does the number of surviving cubs produced each year. In late winter and early spring, over 5 years, Angerbjorn and his colleagues placed fresh meat near selected dens to supplement the supply of prey that the foxes could catch, leaving the occupiers of control dens to fend for themselves.

The response was strong and consistent. Each year more of the supplemented dens were occupied, and they produced many more cubs at weaning than did the control dens. None of these years had a peak in the number of either voles or foxes, so clearly the response was one of animals experiencing a shortage of good food. More females attempted to breed and more were successful when they were able to get adequate food immediately before and at the beginning of the breeding season. But this was still not enough. Of the 65 cubs that the authors tagged during the study only three survived until the following spring. Without an assured supply of food after they had been weaned few of these still-growing cubs would have escaped starvation.

The red fox, *Vulpes vulpes*, in the north of Sweden has also been shown to respond in the same way to changes in the number of voles, and to the provision of supplemental food (Lindstrom 1989). Clearly, it too is limited by the availability of food for successful breeding and rearing of its young. Further south, however, its numbers are more stable and many individuals do not breed. It is believed that this is because the foxes' social and territorial behaviour prevents them from increasing and using all the available food. But this is the wrong interpretation. The success of their breeding and the survival of their young in the south are still limited by the supply of food, albeit at a more stable and predictable level. Young females stay within the territory of their parents but do not breed. Young males leave their natal territory and soon die. The foxes' territorial and social behaviours operate to channel the available food to successful breeding by a few dominant individuals, while maintaining a "surplus" of young females ready to breed whenever more food becomes available (Chap. 11).

13.3 Stoats, Mice, and Seed Mast

The stoat, *Mustela erminea*, is a small but ferocious predator of other small mammals. Its favoured prey are rabbits and hares, but it has a fairly catholic taste, including birds, and even lizards and insects if pushed. Like its principal prey it has been widely distributed around the world. King (1983) conducted a very thorough study of stoats in beech (*Nothofagus* spp.) forests of the fiordland of the South Island of New Zealand. There are, of course, no rabbits in these forests, but she was able to demonstrate that the number of stoats increased markedly following peaks in abundance of the feral house mouse, *Mus musculus*. The summer peaks in the number of stoats in these years were two to five times higher than in other years, and consisted mostly of juveniles. The number of these juveniles caught was significantly correlated with both the density of mice in spring and the frequency with which mice were found in the guts of stoats in the same summer.

Stoats produce one litter each spring, but have evolved the strategy of delayed implantation. At each ovulation the mean number of ova shed is fairly constant, providing for the most young that a female could rear when food is abundant. The young are born only 28 days after implantation, but mating takes place 8 to 10 months before birth. The ecological tactic is then (via unknown environmental cues at the time of gestation and lactation) to fail to implant some blastocysts, or to resorb some or all of the implanted embryos. If food is very scarce many females may not produce a litter at all; those that do, produce only small litters, and many of their young die as nestlings. But when there are very many mice in the spring the response by stoats is marked; most fertilized eggs implant, and large numbers of juveniles appear in the population early in the following summer. Mice not only provide a ready source of easily caught food for pregnant and lactating females, they are ideal prey for young stoats before and soon after they leave the nest. So, the level of nutrition of both females and young limits how many viable young are produced and survive, while at the same time the capacity of the population to swiftly respond to any unexpected nutritional bonanza is anticipated. The population is constantly pressing hard against the limit set by the accessibility of good food for individual stoats.

This, however, was not the whole story. King also demonstrated a long suspected "three-way link" in these New Zealand beech forests between seedfall of the beech trees and changes in the populations of mice and stoats. Species of *Nothofagus* set seed only irregularly and patchily, but occasionally all trees over wide areas produce large crops of viable seed at the same time. The peaks in abundance of mice were found to follow hard on the heels of such bumper crops, and were associated with improved recruitment of young in those years. The subsequent declines were associated with failure of recruitment and disappearance of aging adults.

However, King pointed out that beech seed, while the immediately most plausible source of improved nutrition for the mice, may not be so. It has not been established that the mice do eat the beech seed. Furthermore, although the seed always falls at the same time of year no matter what its abundance, peaks of mice have been observed at various times, and sometimes at levels much lower than expected from the size of the beech crop.

King (1985) found some evidence that predators might at least have a dampening influence on the number of mice similar to that found for predators of rabbits by Newsome et al. (1989) (Sect. 16.2). As in that and so many other examples, however, it is unlikely that the stoats would have any significant influence on the mice once they have access to an abundant supply of good food. I suspect that if it is not the highly nutritious new seed to which the mice are responding (Sect. 6.2.2.1), then it is to an associated flush of invertebrates generated by the same pattern of weather responsible for the seed mast (Sect.7.3.2). Whatever the final answer might be, however, this association does nicely highlight the way in which the "domino effect" of improved breeding with improved nutrition flows from herbivore to carnivore.

13.4 Pelicans, Puffins, and Other Sea Birds

Most biologists are reasonably familiar with the ecology of terrestrial birds of prey – hawks, eagles, kestrels, owls – and of the close and consistent link between the abundance of their prey, their breeding success, and the survival of their young. When food is very scarce many of them do not even attempt to breed. At other times the number of young they raise each year will vary in proportion to the abundance of prey (Sect. 11.1.1).

The relationship of the accessibility of food to successful breeding and abundance is also well documented, but possibly not so widely recognized, for sea birds. There is probably no more clear demonstration of this link – and one which must now be abundantly clear to all who look – than the influence of the El Niño Southern Oscillation phenomenon. The suppression of the upwelling of nutrient-rich water by this warm current has a "flow-on" effect which moves rapidly up through the food chain, ending in widespread starvation for a variety of sea birds (Schreiber and Schreiber 1984; Hays 1986; Boersma 1987; Glynn 1988).

Unfortunately, however, this spectacular and widespread death of grown birds tends to deflect attention from the more usual situation. In most years there is no indication of a shortage of food. Rarely are starving birds seen – unless one looks closely at what is happening at nesting times. Then, even in "good" years, starvation, or at best debilitating malnutrition, is the lot of many young birds because their parents are unable to gather enough food fast enough to fuel their rapid growth. Furthermore, the influence of food is two-pronged. The adequacy of the breeding females' diet determines the protein content of the eggs that they are maturing. This, in turn, has a major influence upon the viability of the newly hatched chicks (Nisbet 1978; Galbraith 1988), exacerbating the impact of a lack of food on their chance of fledging.

Anderson et al. (1982) provided a succinct picture of this relationship of food and breeding success for a carnivorous bird with their study of the brown pelican, *Pelicanus occidentalis californicus*, in the southern Californian Bight. These birds eat practically nothing but anchovies, *Engraulis mordax*, and the influence of El Niño on changes in the supply of this food is strong. Consequently, as the numbers of this pelagic fish change in the bay, so do those of the pelican. This is because the number of young the pelicans are able to raise depends upon how much fish they can feed to their young nestlings. Typically when food gets scarcer many more nestlings starve, many nests are deserted, and many fewer chicks are fledged. In especially poor times whole nesting colonies may abandon their nests.

This relationship between anchovy and pelican is a very clear example of the operation of a relative shortage of food. Usually there are plenty of anchovy in the bay and adult birds, ranging freely over large areas, can readily catch enough fish to maintain themselves in good health. There are more than enough fish in the bay to feed all the young that these birds could possibly produce in any one year, but the fish are not uniformly distributed. Not surprisingly, therefore, the pelicans' nesting sites tend to be located near areas where the anchovies are more abundant. In spite of this, unless the fish are very concentrated and remain within the foraging range of nesting pelicans, many chicks will not get enough to eat and, ultimately, will starve. So the year to year abundance of the pelicans is set specifically by the amount of food that is available for their growing young. In very bad years – when El Niño suppresses the upwelling of nutrient-rich waters and the number of anchovies drops precipitously because *their* food supply has failed – even adult birds will starve.

Populations of other species of pelicans have been recorded responding in the same way, but even more violently. The African subspecies of the great white pelican, *Pelicanus onocrotalus roseus*, breeds on many of the large lakes of the great rift valley so long as these contain an assured and large stock of *Tilapia* fish (Brown and Urban 1969). They feed their newly hatched chicks on a liquid regurgitate for their first 14 days, but thereafter must have a continuous and abundant supply of fish for them. Even so more than 40% of nestlings usually starve in the first month. For much of the time relatively small populations of pelicans persist where there is permanent water with enough fish to support some breeding. However, quite unpredictably, infrequent heavy rains may fill the usually dry lake beds. Then there is a huge bloom of algae and invertebrates on the rotting vegetation in the water, the *Tilapia* breed in millions, and equally quickly so do the pelicans, until the water starts to evaporate and become increasingly saline. The *Tilapia* then die, and so do the pelicans. First the young, as the adults abandon the whole colony, leaving the chicks to starve, but then many adults too as the supply of fish continues to shrink. Ultimately there is an absolute shortage of food for the much enlarged population.

In 1989 the dry salt flat that is Lake Eyre South, 650 km north of Adelaide in South Australia, filled with water for only the third time this century. Pelicans, *Pelicanus conspicillatus*, migrating from as far away as Queensland and Papua New Guinea, were soon nesting on the three small islands in the lake, feeding upon the enormous numbers of small fish in the water (Lewis 1990). By the middle of 1990 national park rangers estimated that there were more than 200000 birds on the islands' 7 ha, and more than 50% of them were young chicks. This is about 80% of Australia's estimated total population of pelicans!

By then, however, the lake was already drying up. Previously 2-m-deep water was now only ankle-deep, and was eight times as salty as the sea, killing the fish. The islands had a salt-encrusted band of rotting fish all around them, and were carpeted with thousands of decaying bodies of mostly unfledged chicks, and unhatched eggs. Fledged young that were still alive were mostly too weak to fly. Many adults and early-bred young had already departed, and birds were still deserting nests (it was estimated that birds would have faced a round trip of more than 300 km to bring food to their young on the islands).

So, we see for three species of pelican, each living in quite different parts of the world and eating different species of prey in either fresh or sea water, that their numbers change in exactly the same way in response to unpredictable and violent changes in the availabilty of their food. And the process which brings about these changes is the same in each case – a change in the amount of food for the production of young. Because these events are so large and spectacular this process is clear and obvious. It is also very obvious that these birds have the innate capacity to quickly take advantage of any increase in

Fig. 38. Australian pelicans breed to outbreak numbers eating fish multiplying in rarely-filled Lake Eyre, then die in thousands when the food disappears as the lake dries up (see also Chap. 16). Photo courtesy of Tony Lewis

abundance of their food. Quite simply, at such times more of the young they produce get enough to eat and survive. It is equally apparent that once the food runs out the situation is reversed and their numbers quickly decrease again. At all times the population is pressing hard against the limit set by the amount of food that is *accessible*, and can be used *at the time*.

There are similar stories for many other birds which live where they are dependent upon an unpredictably patchy and elusive supply of prey in the sea. The North Sea puffin, *Fratercula arctica*, is one of these. Harris (1978) experimented with two populations in Scotland, one in the western Outer Hebrides and the other in the east on the Isle of May in the Firth of Forth. The west coast population had declined over the previous 30 years but had started to increase again, while the east coast population had been increasing for the past 10 years. Chicks in the west, unlike those in the east, were thin and constantly begging, and fledged at a lower proportion of the mean adult weight than did those in the east.

In both populations Harris supplemented the diet of some chicks with 50 g fish, took some chicks to the laboratory and fed them ad lib., and removed one parent from each of a third group. In the west the chicks on the supplement put on significantly more weight than did control ones, and those in the laboratory put on even more weight. Those with only one parent died or developed very slowly to fledge much later than did control birds. His companion study in the east produced similar but less dramatic results. Clearly the puffins were having difficulty supplying their young with enough food, even in the obviously much better-off east coast population. All the signs were that the young would do much better with access to more food.

Later work in Norway (Barrett et al. 1987) has shown how sensitive the breeding success of puffins is to changes in the quantity and quality of the food they are able to provide for their chicks. Using a combination of data from studies of commercial fisheries, and their own measurements of the food the adults brought to their nestlings, these authors showed that as good food became harder to find the birds brought less and poorer quality prey to their young, and more chicks starved before fledging. Ultimately, a decline in available food for the chicks caused the adults to abandon them.

Other North Sea birds, like the shag, *Phalacrocorax aristotelis*, (Aebischer 1986), the arctic tern, *Sterna paradisaea*, (Monaghan et al. 1989), and the kittiwake, *Rissa tridactyla*, (Harris and Wanless 1990), experience the same capricious changes of fortune. When food is short their breeding success declines, mostly because more chicks starve, but also because some birds fail to breed, and others abandon their eggs or nestlings. When stocks of fish are very low most birds fail to breed at all and their numbers crash.

At the other end of the world Thomas et al. (1983) found the same situation for the most southerly ranging albatross, the light mantled sooty albatross, *Phoebetria palpebrata*. One of its breeding sites is on South Georgia. There the newly hatched chicks are fed frequently for the first few days upon a mostly liquid regurgitate; thereafter they are fed on prey caught at sea. This entails long excursions away from the nest by the parents, sometimes up to a week, during which time the nestlings do not eat. Yet they increase their weight nearly six times in their first 20 days; 19 times in 80 days, by which time they weigh 120% of the mean weight of adult birds. Not surprisingly, many starve, and by 6 months fewer than half of them are still alive. Further north of their main feeding grounds on Marion Island, fewer birds breed and with less success. They produce lighter chicks which take longer to fledge, and only 30% of them survive. This is because the parents must make much longer trips to collect the best food. So, the nestlings are not only fed less frequently, but on less-preferred species of prey.

Such rigours of survival are not confined to the cold polar waters, however. In the Galapagos Islands where El Niño periodically wreaks havoc with the supply of planktonic fish and crustacea, Audubon's shearwater, *Puffinus lherminieri*, suffers a similar fate (Harris 1969). When the supply of planktonic food starts to decline the birds bring smaller feeds to their nestlings less frequently. Some eggs are deserted and no new ones are laid. Eventually nearly all birds desert the colony, leaving the chicks to starve. Harris made an observation about these starving nestlings which is important in a later context

(Sect. 14.5). They all had eye infections and were heavily infested with lice (Mallophaga) and fleas (Siphonaptera).

These responses of carnivorous birds are unusually clearcut and unambiguous because, in each case, the birds are living in an especially harsh environment where changes are large and dramatic, and both cause and effect are readily observable. Other factors which could conceivably be responsible for their changes in abundance, in particular predators, are either absent or clearly seen to be ineffectual. Their particular value, however, lies in the fact that the huge fluctuations of their numbers represent merely the two extreme tails of the normal distribution of what is happening all the time in populations of *all* animals, carnivore or herbivore – the interplay between food and successful breeding. Most animals live in more stable and apparently less harsh environments where changes are much smaller and less dramatic, and where cause and effect are often cryptic and subtle; so the link is not found. The norm of failure to breed and/or death of most young is not observed, or is discounted from consideration. Then other consequential factors, like predation, which impinge upon the individuals in a population, loom large as the apparent arbiters of the success of that population (Part V).

Chapter 14 Invertebrates

14.1 Triclad Worms

These free-living Turbellaria include a large group of predators (e.g. species of *Planaria*, *Dugesia*, *Polycelis*) of detritus-feeding gastropods, crustaceans, oligochaetes, and insects. Reynoldson's (1966, 1977, 1983) extensive studies of these triclads in Wales have shown that their distribution and abundance are determined by that of their detritivore prey. These organisms, in turn, are limited by the chemical richness of the water influencing the success of microorganisms in the sediment which they eat. The chemistry of the water is the product of the geology of the drainage basin of the lake or stream concerned. In any body of water both the abundance of populations of individual species of triclads and the diversity of species directly reflect this "upward" flow of nutrients through the food chain.

Once again, this limitation of the abundance of predators by their food operates via the fecundity of females and the survival of their young. *Their* predators and parasites have but a negligible effect upon their numbers. This means that, in nature, triclads are usually at the carrying capacity of their environment because they are experiencing a relative shortage of food. There is more food than they ever use in the habitat, but it is so thinly scattered that they cannot find it fast enough or often enough. Their females produce few or no eggs, and their young, once their reserve of yolk is exhausted and they must hunt for themselves, mostly soon starve. Experiments have shown that both the fecundity of the females and the survival of the young can be dramatically increased by the provision of extra food.

The extent to which these carnivores are pressing against the limits set by the amount of food which is accessible to them, and just how dramatically they can respond as soon as it becomes more abundant, Reynoldson recorded for *Dugesia polychroa*. For 5 years he monitored a newly established population of this triclad in a recently constructed pond. Water snails (the favoured prey of *D. polychroa*) had already been introduced to the pond and were very numerous. The colonizing population of *Dugesia* was therefore well below the carrying capacity of its food supply. In the breeding season following its introduction to the pond the numbers of *Dugesia* increased by 800%. This explosively increasing population was characterized by unusually large individuals with 50 to 60 times greater fecundity than individuals found in established populations, and a much higher survival of the young. By only the fourth generation in the new pond, however, the growth of the population had stopped, and fecundity and survival had dropped to levels found in old established populations. Food was once again limiting. Reynoldson discussed this rapid increase in response to abundant food as an apparent "K" strategist behaving as an "r" strategist. But as I discuss in Chapter 15, a K strategist, is really only an r strategist held in check by its inadequate environment.

14.2 Spiders and Scorpions

I referred briefly in Section 1.4.1 to the lonely struggle of young orb-web spiderlings to catch enough prey to survive. It is the death of such neonates which is principally responsible for the limitation of the number of spiders. These young carnivores live for up to 2 weeks after they hatch on their reserves of yolk from the egg. Only after they have moulted to the second instar do they spin their first web and attempt to catch their own food. Then they are confronted by a relative shortage of food which will soon prove fatal for most of them, whether or not others of their own kind are seeking prey in the same locality. Because these and other sorts of young web spinning spiders capture only a small fraction of the insects which contact their webs, doubling the number of spiders present would not make any difference to the shortage of food for the individual (Wise 1975, 1983a).

Smith and Mommsen (1984) observed such young spiderlings spin and dismantle several webs without apparently catching any prey, and wondered whether they might be subsisting on microscopic organic matter caught in their webs. So they did an experiment with second instar *Araneus diadematus*, a common garden orb-web spider in British Columbia. They allowed them to build their first webs in individual closed cages and randomly assigned them to three groups. They dusted birch, *Betula papyrifera*, pollen on the webs of the first group, spores of the mould *Cladosporium herbarum* on those of the second group, and starved the third group as a control. Those spiderlings getting the pollen doubled their life expectancy, and spun more webs more frequently than did those in the other two groups. Other spiderlings reared to third instars on aphids and then treated the same way responded in the same way.

These experiments demonstrate that, in nature, these young carnivores are doing the reverse of what young herbivores do – supplementing a relatively scarce supply of animal food with an abundant supply of plant protein! The birch pollen is not an adequate food on its own (it is deficient in the amino acid tyrosine without which the spiderlings cannot form a new cuticle and therefore cannot moult to the next instar). However, eating it significantly increases a spiderling's otherwise very slim chance of surviving between captures of the odd small insect.

For those few which do survive to later instars, and even as adults (Anderson 1974), the problem continues. For both web-spinning and free-ranging hunting spiders the amount of food they catch is rarely, if ever, sufficient to enable them to reach the reproductive potential that they do when fed plenty of prey in the laboratory. Nor is this due entirely to the scattered and patchy distribution of their prey. It is also a function of their inefficiency as predators; they capture only a small fraction of the insects which contact their webs or which they encounter while hunting. Although their rate of success will vary with the size and type of prey, it seldom exceeds 60% of attempts, is more often around 20%, and can be less than 2% (Wise 1975; Morse 1979; Olive 1980). There is a great deal of evidence of the generality of this sort of relative shortage impinging upon the success of populations of spiders. On the other hand there is a lack of any clear evidence that the spiders have any significant effect upon the density of their prey, or that there is any meaningful competitive interaction between closely related species of spiders (Wise 1983b).

Morse's (1979) study of the crab spider, *Misumena calycina*, in Maryland provided a clear illustration of how the inefficiency of hunting spiders, like that of the web spinners, contributes to the relative shortage of food that they too experience. These spiders wait in flowers and regularly attack all insects which visit them, attacking up to 20 times an hour. Their most important prey (in terms of food gained per capture) are bumble bees, *Bombus terricola*. The next most important, and most frequently taken, are the much smaller syrphid flies, *Toxomerus marginatus*. The spiders that Morse observed managed to catch only 1.6% of the former and 39% of the latter, yielding them an average catch of a little over one prey per day – slightly less than one syrphid per day and one bumble bee every third day. Not only are they missing most of their potential food, they are unlikely to have much impact upon the abundance of their prey. But worse than this, they are being far less efficient than they could be. Morse calculated that a spider which, instead of attacking all comers, concentrated its efforts on just catching bumble bees would have gained over 7% more food for its effort. Furthermore, had it confined its hunting to those times when prey was most accessible, it could have got 20% more food. However, his spiders showed no tendency to do either – they were clearly not foraging optimally.

Another large group of generalist invertebrate predators are the scorpions. The work of Polis and his associates with several species of desert scorpions in California, but especially with the large *Paruroctonus mesaensis*, has produced abundant evidence that these nocturnal carnivores, too, are limited by the restrictions that the availability of prey places on the survival of their young, and of the reproductive performance of their females (Polis and Farley 1979, 1980; Polis and McCormick 1987). These, and scorpions in other parts of the world, experience a period of high early mortality in the weeks immediately after birth. Usually only about 40%, and often 20% or less, survive for long once they have absorbed their endogenous food reserves and must hunt for themselves (Smith 1966). At sites where prey are more plentiful more survive, they eat more frequently, grow faster and to a larger size, and they are more abundant. At any one locality their numbers change from season to season as the accessibility of their prey changes. The scarcer prey become the more they cannibalize their own (see Sect. 12.1) and prey upon other species of scorpions. When captive pregnant females are fed ad lib. they grow faster and larger, and develop larger and more advanced embryos than do equivalent females caught in the field.

14.3 Ground Beetles, Tiger Beetles, and Ant-lions

In The Netherlands Den Boer and his associates have clearly established that several species of predatory ground beetles (Carabidae) are limited in their abundance by the quantity and quality of their food. Careful field studies and extensive laboratory experiments have shown that, as with other animals, this shortage operates both via the nutrition of the females influencing the number of eggs they can produce (Van Dijk 1983, 1986), and via the nutrition of the larvae influencing how many of them survive (Nelemans 1987; Nelemans et al. 1989). Of these two processes, recruitment and survival of the young that hatch from the eggs far outweighs changes in the number of eggs produced as the major influence on their abundance (Den Boer 1986b). There is, at all times, an apparent abundance of prey available in the field for these beetles, but they experience a relative shortage of food because they never find much of the prey that is present in their habitat, nor are they capable of catching and devouring

Fig. 39. A carabid beetle, hunting by random search with attack stimulated only by encounter, is inefficient because it finds but a small fraction of the potential prey in its habitat. Photo courtesy of J. v. Osch

all that they do encounter (Den Boer 1986a; Van Dijk and Den Boer 1992). Other studies have shown similar responses by carabids to changes in the attainability of their prey (Lenski 1984). Unlike vertebrates these insects, and invertebrates generally, do not have the neurological complexity to develop true hunting behaviours. Theirs is a system of attack stimulated and reinforced by encounter with a potential prey, but soon decaying to random search when no prey is encountered (Mols 1987). Theirs is an effective strategy for finding some of the food, which is thinly and patchily dispersed in a heterogeneous environment, but not for finding more than a small fraction of it. There is little doubt that most sorts of carnivorous insects are similarly limited by a relative shortage of food for successful breeding and growth of immatures in habitats which appear (to us) to contain an abundance of prey.

Pearson and Knisley (1985) demonstrated that this was so for a Californian assemblage of species of tiger beetles belonging to the genus *Cicindela*. In the laboratory they showed that females fed ad lib. developed and produced many more young than did those fed about half that amount of prey. These, in turn, were more fecund than beetles given half as little again.

In the field the feeding rates of females that Pearson and Knisley measured were up to five times less than those of females given ad lib. prey in the laboratory. Only in the most favourable habitats, where prey were especially plentiful, did they approach this level of success. Mostly there were just not enough prey available for them to eat all the food that they could. When the metabolic rate of these free-living beetles was compared with that of the sedentary lab-fed ones (some 85% greater for the former) the beetles in the field were clearly not achieving anywhere near their reproductive potential because they experience a permanent relative shortage of food.

That this shortage was linked to the number of potential prey in their environment was further demonstrated by the year-to-year variations in the rate at which females caught their prey. This varied by more than 280% over the 7 years of the authors' study, and was significantly correlated with the number of prey that they recorded in the beetles' habitats each year.

When they fed the larvae in the laboratory, the less food they gave them the longer they took to complete their development, and the smaller the resultant pupae and adults. In the field the larvae do not actively chase and catch their prey like the adults, but sit and wait at the mouth of a burrow for prey to pass within striking distance. They managed to catch only 40 to 60% of those that did. This meant that they were catching much less prey than that which allowed the greatest rate of growth of larvae in the laboratory.

In this study the most obvious way in which a lack of food was depressing the numbers of these predators was via its influence upon the fecundity of the females. Unfortunately, the authors did not record its effect upon the survival of the newly hatched young in the field, and in the laboratory they did not subject larvae to different feeding regimes until they had raised them to the second instar on an abundance of food.

The larvae of ant-lions (Neuroptera) dig pits in dry sandy soil and buried, all but for their long poison-loaded mandibles, wait at the bottom for a passing insect – usually an ant – to fall in. Their survival and growth is totally dependent upon how often this happens and how successful they are at overpowering those that do fall in, not by how many other larvae have dug pits nearby.

Griffith (1980) found all this to be true for *Morter obscurus*, a species common during the dry season on the campus of the University of Sierra Leone. He showed that the density and biomass of prey in several local habitats was such that all larvae experienced a chronic relative shortage of food, and that this was the critical factor determining their survival and growth; so much so, in fact, that it was doubtful if any of the generation of larvae that he studied survived to pupate. This shortage of food was a product of both the number of potential prey (and hence the frequency with which one would fall in a pit) and the size of these individuals. Of those prey which could fall in a pit, there was a limited range of size which any individual of each of the three instars of *M. obscurus* could handle or use effectively: too large and they could not be overpowered; too small and either they could not be caught or the expenditure of energy and time to do so exceeded the nutritive return obtained from eating them.

This situation was most acute for the smallest larvae – the newly hatched first instars. They were the least efficient at using the prey that entered their pits, being unable to handle those greater than 2 mm in length or shorter than 1 mm. As the few surviving larvae grew, the increased diameter of their pits and the speed with which they could consume their prey improved their chances of getting enough to eat.

Paradoxically, once they were large final instar larvae they became almost as vulnerable to a relative shortage of prey as they had been as neonates. They were then unable to process the smaller prey, and the larger individuals, which they could economically process, were so infrequent that often not enough of them fell in their pits to provide the minimum supply of food necessary for survival.

14.4 Praying Mantids

The mantids are another well-known group of "sit and wait" predators of other arthropods in which a relative shortage of food restricts the reproductive output of females and the survival of their young. They too rely upon sufficient prey passing within reach sufficiently often, and on their ability to catch sufficient of those that do. Field and laboratory studies of the introduced Chinese mantid, *Tenodera sinensis*, in Delaware (Eisenberg et al. 1981) clearly established that a lack of available prey operates to limit their abundance in three ways. First, it reduces the survival of the nymphs, especially the first instars when they are newly emerged from their egg cases in the spring. At this time potential prey are scarce and 90% or more die before reaching the second instar. Second, those nymphs which get less food take longer to reach maturity so that, as adult females, they have less time to feed, and so mature and lay fewer eggs before they are killed by autumn frosts. Third, nymphs which do survive in spite of not getting enough to eat will develop to smaller adults which will produce fewer eggs (even if, as adults, they obtain abundant food).

Of these three effects the death of neonates is, without doubt, by far the largest and most important. Even quite a big increase in fecundity of females from one generation to the next will be of no significance if the limited availability of prey in the habitat of the young nymphs does not change. More young will be born, but about the same small number will get enough food to survive.

That this is so was beautifully (if inadvertently!) illustrated by a field experiment that this group of researchers carried out (Hurd et al. 1978). They placed unhatched egg cases of *T. sinensis* in six 15×15 m plots of mixed grasses in a field which did not have a resident population of mantids because it had been mown each year. These plots were separated from each other by wide mown strips and sticky barriers, and were randomly selected to provide two plots with three times as many newly hatched nymphs (about 18000 each) as another two (about 6000 each), and two control plots without any nymphs. They sampled all plots repeatedly and thoroughly for both mantids and their potential arthropod prey, ending with a very exhaustive "extinction sampling" of mantids just before they were due to moult to adults which could fly away.

Their samples revealed that while all plots had the same biomass and size range of potential prey throughout the 2 months that the experiment lasted, over 90% of the nymphs had disappeared from each of the plots in the first 2 weeks, and by the time of the extinction sampling the nymphs left in all the plots totalled a mere 1% of the original number hatched. *But there was no significant difference between the numbers left in each of the four plots.* Nymphs at the highest initial density had simply reached the final density more quickly.

When they later repeated this experiment (Hurd and Eisenberg 1984a) they obtained the same result; more than 90% of the nymphs died soon after they hatched, and there was no difference in the final numbers in any of their plots. In view of these results they then did a laboratory experiment to look at the effect of density on the survival of nymphs. They placed freshly emerged nymphs in cages at a series of deliberately unrealistically high densities, and provided them with unlimited live prey and water. At the same time they set up a similar series of caged nymphs with water but no food to see if cannibalism increased with density in the absence of food. Only at the ridiculously high densities of 50 and 100 nymphs in a 29×30×40 cm box did crowding lower the survival of well fed first instars. Furthermore, cannibalism among unfed nymphs, while higher than when prey were provided, was not related to their density. Nor was there any correlation between density and survival of starved nymphs; 95% died in the first 2 weeks, and none survived to the second instar.

In this second field experiment Hurd and Eisenberg had also measured the dispersal of mantids from their plots. Not surprisingly, they found that more mantids emigrated from the plots in which most mantids had hatched, but nearly all of the dispersing individuals were in their second instar. Most had died before they could disperse.

What they *did* find, however, was that the final densities in the experimental plots differed markedly between years and fields. This supported their general observation that in nature the densities of this mantid are highly variable between sites and from year to year, and that the number of egg cases present in spring is not a good predictor of the subsequent number of mantids in the population. It also suggests that the number of mantids that survive in each generation depends on the changing carrying capacity of their habitat – on the amount of food that is available to them.

From the results of the same experiment the authors tried to measure what influence the addition of so many mantids had had on the abundance of potential prey in the plots (Hurd and Eisenberg 1984b). They found only highly variable, often conflicting, and inconclusive relationships which gave no indication that the mantids had made any significant impact upon the numbers of other arthropods in the plots. Looking at the simple numbers of animals present this does not seem surprising. The herbivores ranged from 50 to 180 per square metre, and other carnivores from 5 to 60, throughout the experiment. Mantid nymphs surviving to the second instar numbered only 1.3 per square metre in the plot with the highest density, and by the time all were past the second instar they ranged from 0.12 to 0.05 per square metre. It seems highly unlikely that so few very small mantids could have made any measurable impact upon such a mass of prey, especially when they would never have encountered many of them, and many that they did encounter would have been too big for them to handle.

The authors attempted to explain their results in terms of density-dependent and density-independent mechanisms and "regulation" of the population to within the carrying capacity of the habitat. However, the density of mantids has no influence upon how many mantids the environment can support. The simplest interpretation of their results is that even in the plots receiving the highest number of hatchlings (27 nymphs per square metre) the presence of mantids made no measurable difference to the number of their prey in those plots. They are inefficient predators; most potential prey were simply not attainable. Furthermore, this relative shortage of food was most acute for the neonates. The habitat provided only enough available food to support to maturity between a half and one mantid per square metre (only 0.05 to 0.12 of a mantid in the second experiment). This "carrying capacity" could not have been influenced in any meaningful way by the number of young nymphs born in the habitat. Where there were more nymphs to start with more died more quickly. Nor is dispersal "regulating" the population according to the varying carrying capacity from season to season. The animals simply disperse when they fail to get enough to eat. As the authors point out, it is the individual's inability to locate sufficient prey, not the absolute number of prey in the habitat, which is limiting their numbers. They suffer a relative shortage of food and, as a consequence, have little significant influence on the changes in abundance of their prey.

Fig. 40. Mantids are voracious predators, but most die young because they cannot catch enough prey. Given excess prey they *maximize*, not *optimize*, how much they catch. They have evolved to grab what they can get in an inadequate environment! Photo courtesy of L.E. Hurd

It seems, then, that like Morse's crab spiders, these predators are unlikely to be optimal foragers. That this is so was demonstrated by Hurd and Rathet (1986). To investigate the functional response of first instar nymphs to varying levels of prey, they fed groups of 20 newly hatched nymphs, individually housed in small vials, on live *Drosophila* flies. These groups were fed at rates ranging from one fly every 3 days, through quantities that they would never encounter in nature, to unlimited and constant access to flies. Under these conditions the young mantids' mean rate of feeding conformed to a conventional type II Holling response; it platformed off at two flies a day (about double what they would get in the wild). However, when the authors eliminated from their analyses all but those mantids which had fed and survived to the second instar, the picture was quite different. These successful nymphs increased their rate of feeding with each increase in the quantity of prey they were given, eating more than they needed to survive, and more than that which maximized their growth and size at the time of moulting to the second instar. (In fact, those with constant and unlimited access to live prey were *smaller* than those eating two flies a day. Presumably in this unnatural situation they devoted so much time and effort to catching flies that, in the process, they either ate less food or burnt up more than they ate.) Obviously these neonates did not optimize their rate of feeding when confronted with more food than they needed. They have evolved to catch and eat everything they can in a world where they are lucky to encounter prey often enough to survive. Under such conditions it seems unlikely that they would exhibit any choice for an optimum size or type of prey. In their inadequate environment they must simply grab whatever they can get.

Hurd (1991) reanalysed the data from this experiment and came to this same conclusion. Beyond the peak of efficiency of growth at about 0.7 flies per mantid per day, a subsequent increase in biomass eaten *reduced* the growth of these first instars by 30%. Clearly, because these animals evolved in an environment where they never experienced an excess of prey, natural selection could not operate against those getting too much to eat – it could leave only those which *maximized*, not optimized, their rate of predation.

14.5 Parasitoids, Parasites, and Diseases

Wasps and flies which prey upon other insects and invertebrates comprise a large group of carnivorous insects. Females lay their eggs in or on their prey and the young feed, usually internally, on the victim's body while it is still alive, and usually while it continues to feed and grow. For this reason they are often not thought of as predators. Yet they eventually kill their prey, so they are not conventional parasites. They are called parasitoids.

As with other carnivores, a relative shortage of prey can prevent female parasitoids from finding enough food for all the offspring that they are capable of producing. Many of these animals have been shown to be just as inept as other predators at finding their prey, and in overpowering those that they do encounter. But there would seem to be little chance of the survival of the young being much influenced by a shortage of food. Once hatched in or on the prey, a larva is confronted with an abundance of high quality protein food, and is usually equipped with mechanisms to prevent wasteful sharing of this resource with its conspecifics (Sect. 11.1.3).

Yet it seems that this is not necessarily the end of the story. House (1969) demonstrated how important a balanced protein diet is for such carnivores when he was able to rear *Agria housei*, a fly which parasitizes the larvae of the spruce budworm, *Choristoneura fumiferana*, in Canada, on a completely synthetic and aseptic diet. Given a choice, *A. housei* larvae always selected the diet which was nutritiously superior – the one that contained the optimum proportion of amino acids – and grew fastest and biggest on it. Not surprisingly, the absence or very low concentration of any one of the essential amino acids killed the larvae or stopped their growth in the first instar. More interestingly, larvae could detect quite small changes in the proportions of one or more amino acids in the diet, and these apparently quite minor manipulations of the food could mean the difference between survival or death for all first instars that fed on them.

So, the vigour and survival of a larval parasitoid is just as dependent as that of any other animal on the quality of its food. Moreover, in nature, the quality of the body in which it finds itself is far from constant. In particular the range and proportion of amino acids in the host's body will vary between species, and between individuals of one species, according to their age, sex, and physiological condition. Free-living predators like spiders have been shown to select different species of prey in proportions which tend to maximize the intake of essential amino acids in their diet (Greenstone 1979). But no such choice is available to a parasitoid larva hatching inside its prey; it has nowhere else to go. It was therefore to be expected that the success of parasitoids would be found to be influenced by the size (King 1989) and the level and balance of amino acids of their preys' bodies (House and Barlow 1961; Thompson 1976), and, ultimately, by the quality of the food that their prey eat (Greenblatt and Barbosa 1981; Greenblatt et al. 1982; Fox et al. 1990).

Just how sensitive the developing parasitoid is to the condition of its prey, how much this is influenced by the latter's diet, and the dominant role of nitrogen in these interactions, was clearly demonstrated by Bloem and Duffy (1990). In California they raised larvae of two common noctuid pests, *Helicoverpa zea* and *Spodoptera exigua*, on artificial diets containing different forms and quantities of protein. When the larvae were moulting from second to third instar, they exposed batches of them from each diet to attack by females of the ichneumonid wasp, *Hyposoter exiguae*, and then reared them until they pupated. On diets containing inadequate protein fewer, smaller noctuid larvae survived, and the parasitoids in them were more severely stressed; unless a host larva reached a minimum size the parasitoid larva feeding in its body was unable to complete its development and pupate.

This "tritrophic" flow of the nutritional quality of the food plant through the herbivore to its parasitoids is now becoming widely recognized. Recent work in India has demonstrated this very clearly for small wasps that develop in the eggs of pentatomid bugs (Senrayan and Annadurai 1991). One such bug, *Coridius obscurus*, feeds preferentially, grows more quickly, lives longer as an adult, and its females lay more eggs when it feeds upon the nutritionally most superior (in terms of content of amino and fatty acids) of three species of cucurbits. The parasitoid *Anastatus ramakrishnae*, which lays its eggs in the eggs of *C. obscurus*, grows faster, survives better, and produces more females, when growing in the eggs laid by the bugs which have fed on the more nutritious plants. In addition, fewer *A. ramakrishnae* emerge from the eggs of bugs which have fed upon the poorer plants, and more of those that do are males.

There is another group of carnivorous invertebrates – the fleas, ticks, lice, bedbugs, mosquitoes, etc. – which prey upon vertebrates. These animals are classed as external parasites rather than predators because individuals do not kill their prey, but merely eat part of its body, usually some small amount of its blood. So, once such a predator has found an individual prey it has far more food than it can use. But for most of these sorts of predators the risk of dying before they can find a prey is very high, especially when they are very young, and even when plenty of prey may be close by in their habitat. This starvation in the midst of plenty was elegantly demonstrated by Milne (1950) when he studied the ecology of the sheep tick, *Ixodes ricinus*, in the Scottish Highlands. Browning (1963) provided a particularly succinct account of this work, but the essence of the story here, is that the ticks must feed on the blood of a bird or mammal three times in order to complete their life cycle. After each meal, and before they moult to the next stage, they must return to the damp grass sward where they first hatched from the eggs. Each time they must climb to the top of a blade of grass and wait for a warm-blooded creature to pass close enough for them to grasp it and feed on its blood. While they wait they desiccate, so if they are not soon successful they must return to the humid sward to resorb water. After several such journeys they die of starvation. Even in a paddock where there are many sheep and other sorts of mammals and birds, any one of which could provide a more than adequate meal for many ticks, most ticks die before a prey comes close enough to be grasped during the relatively short time that they are at the top of a blade of grass. Milne's work showed that of all the ticks which hatched in a field, only 6% of them would get three meals. The rest starved because of a relative shortage of food in the midst of an absolute abundance.

Moreover, even when a tick has attached to a host it may not necessarily obtain an adequate meal. The species, sex, age, diet, and physiological condition of the animal a tick happens to find will materially influence the quality of its blood as food for that tick. Gladney et al. (1973) illustrated this point with their work on the cattle tick, *Boophilus annulatus*, in Mexico, especially the major effect that the hosts' nutrition may have on the success of the ticks. They kept groups of Holstein steers on either high- or low-protein diets for 7 weeks, and then infested each animal with approximately 20000 newly hatched larval ticks. By this time those steers fed high protein were fat, sleek, and gaining weight; those on low protein were thin, lethargic, and losing weight. In all their trials, at best only about half the ticks reached adulthood, and these were on the malnourished beasts. On the well-fed steers as few as 1% survived, and never more than about 15%. Three times more female ticks matured on the weakened animals, and they were heavier than those on the vigorous animals. Then when steers were switched from high- to low-protein diet, and vice versa, more ticks matured on the previously well-fed cattle losing weight on the poor diet than did on the malnourished ones now rapidly gaining weight on the high-protein diet. Finally, by preventing the steers from removing ticks by grooming or rubbing, the authors confirmed that the difference in the survival of ticks was caused by physiological differences in the steers, not by their behaviour. O'Kelly and Seifert (1969) obtained almost identical results with *Boophilus microplus* on Shorthorn-Hereford cross steers in Australia.

It seems that the same is also true for other sorts of parasites of cattle. Lice, like ticks, are usually noticed first on poor, unthrifty animals. Ely and Harvey (1969) found that steers in Kansas fed on nothing but silage gained less weight and supported significantly more of the louse *Haematopinus eurysternus* than did steers given protein supplement with the silage. Fewest lice survived on steers fed the highest protein diet.

So, too, it would seem to be for other external parasites of vertebrates. Finding a meal is a chancy business and, once found, it may still be barely adequate. The tsetse flies (*Glossina* spp.) of Africa provide another good example. The larvae of these flies develop to maturity in the bodies of their mothers, and so depend entirely for their nutrition on the diet of the females. The females feed on the blood of various ungulates, but must search out a new prey for each of the many meals they need to nourish a series of larvae to maturity. Unlike the ticks, which must wait where they fell from their previous host for the next one to pass, these flies can actively hunt for their prey. Yet, like the ticks, they are faced with a situation where, although food is in great excess when found, it may be widely scattered and difficult to find. If it is too scattered many flies may die before they can find a meal, or, at best, produce only a few of the many larvae that they have the potential to produce. That this is so was very dramatically demonstrated by a field experiment carried out by Potts and Jackson (1953) in Tanzania. They employed hunters to shoot most of the game in an area of 1554 km^2 with the specific aim of exterminating three species of *Glossina* from the area. After 5 years the number of mammals had been reduced by over 8500, but there were still sufficient of them left to provide many times more than enough blood for all the flies that had been there at the beginning of the experiment. However, they were now so few and so scattered that any one tsetse fly had but a small chance of finding a meal. So small a chance, in fact, that at the end of the 5 years two of the species of *Glossina* were successfully eliminated, and the third, if not exterminated, was reduced to very small numbers.

However, to find a sufficient number of meals is not necessarily sufficient. The survival and vigour of the young depends upon the quality, as well as the quantity of food that the flies eat. Furthermore, as Kabayo et al. (1986) have shown, this is chiefly related to the content of amino acids in the blood. In their Texas laboratory they fed female *Glossina palpalis palpalis* with batches of pig and cow blood which contained different amounts of amino acids. The flies did better on the pigs' blood, which has a higher content of amino acids, than on cows' blood. But within this difference between species of host, on the blood which contained the least amino acids the flies' survival dropped by as much as 40%, and the number of mature larvae that they produced by as much as 90%.

In Section 13.4 I noted that Harris (1969) had recorded that puffin chicks which subsequently died of starvation had become heavily infested with lice and ticks before there were any overt signs that the chicks were starving. This sort of event is one manifestation of a widespread phenomenon. Animals which are malnourished or otherwise physiologically debilitated, and which are senescent or growing rapidly, are more prone to suffer heavy infestations of external and internal parasites than are healthy mature animals. This has long been understood by those who work with domestic animals, and is encapsulated in the aphorism "sick dogs have fleas". In studies of natural populations, on the other hand, the significance of such events is often not understood; so they go unrecorded, are mentioned only in passing, or the parasites are thought to have caused or contributed to the animals' condition.

For example, a population of koalas, *Phascolarctos cinereus*, in Victoria had heavily defoliated their locally preferred food plant, *Eucalyptus ovata* (Martin 1985b). As a result the animals were severely malnourished. By comparison with a nearby population in which the koalas were not short of food, they weighed less, were anaemic, the juveniles grew more slowly, and the females were less fertile. The females also had an average of five to seven *Ixodes tasmani* ticks in their ears, while in the well-fed population the 70 koalas that Martin examined produced only two ticks. It was clear from his investigation that this heavy load of ticks was not the cause of the koalas' condition, but he speculated that they might be exacerbating it. He made no comment as to why, in the first place, the ticks should be so much more abundant on the debilitated than the healthy animals (see also Sect. 6.5.1).

The explanation is straightforward enough once the cause and the effect have been reassessed. It is well known that animals which are starved, or deprived of adequate protein in their diet, soon start to break down and burn their own body tissues. They go into negative nitrogen balance with the result that there is a marked elevation and change in the proportion of essential and non-essential amino acids in their blood. Other forms of great metabolic demand for nitrogen, like gestation and growth of new feathers after moulting, have a similar effect. This process has long been documented in domestic and laboratory animals (Widdowson and Whitehead 1966; Leibholz 1970) and reported in a wide range of animals in nature: oysters (Riley 1980), caterpillars (Bosquet 1977), scorpions (Sinha and Kanungo 1967), ground squirrels (Bintz and Strand 1983), and ducks (Harris 1970). In effect, these physiological disruptions convert much more of the nitrogen in the body of an animal to a form which is more readily available to, and more quickly assimilated by, those which feed upon its blood or other body fluids. The examples I have just quoted of the responses of parasitoids, ticks, and tsetse flies to such physiological changes of their hosts and the level of nitrogen in their food illustrate this relationship. Furthermore, the realization of the generality of this relationship suggests an explanation for observations like that of Nash et al. (1966). They found that female tsetse flies which had been fed on pregnant goats lived longer and produced more and heavier pupae than did those fed on goats which were not pregnant. Nor is this phenomenon confined to arthropod predators, or predators of vertebrates. Slater (1988) found that laboratory mice which had been fed on an inadequate protein diet suffered much higher levels of infestation with nematodes, and the success of nematode parasites of a variety of insects is strongly influenced by the protein content of their hosts' bodies, and the food that they eat (Kaya and Moon 1980).

The third group of predators in this section are those which we call diseases – simpler forms of life like protozoa, bacteria, and viruses. They can multiply only in the bodies of their prey, and either do not kill their host or do so only once they have reproduced and there is a high probability that some of their progeny will find a new host. It has been fairly clearly established that the success of these organisms, like that of other parasites, is intimately linked with the physiological condition of their hosts, and that anything which stresses a host animal is likely to make it a more favourable environment for the disease organisms (Neuvonen et al. 1990). In particular, animals which are malnourished or starving are particularly prone to disease, and it is the quality and quantity of the protein component of the diet which has the most profound effect (Tanada 1965; Gross and Newberne 1980; Crompton and Hall 1981).

Much evidence for this relationship accrued for viral diseases of insects in the 1950s and 1960s, when there were great hopes that viral diseases could be used as an alternative to insecticides for the control of insect pests, especially lepidopterous defoliators. It was also a time when attempts to maintain populations of these insects in the laboratory were frequently frustrated by a viral disease killing most of them. Transmission of infective particles between crowded larvae, and/or poor laboratory hygiene, were thought to be the causes.

What is more, they are often still blamed today. Lindroth et al. (1991) at the University of Wisconsin were investigating the influence of dietary deficiencies upon the performance of fourth-instar larvae of the gypsy moth, *Lymantria dispar*, when they obtained an unexpected and unexplained result. All the larvae fed a vitamin-deficient diet died in the fifth instar from a polyhedrosis virus. No other test animals were affected. They assumed that the virus must have somehow slipped past their rigorous quarantine and sterilization procedures. A much more likely explanation would seem to be that the extreme deficiency of vitamins triggered some biochemical change in the larvae which, in turn, provided the resource needed for the virus to grow and multiply.

Thirty years ago Jacques (1962), at Cornell University, demonstrated that this is the more likely explanation. He found that when he took caterpillars of the cabbage looper, *Trichoplusia ni*, from greenhouse populations apparently free of viral disease and crowded them in the laboratory, many of them died from nuclear polyhedrosis. From this point, with a logical progression of elegantly simple experiments, he showed that the stress of crowding was largely a result of the caterpillars getting less to eat because of the constant jostling interrupting their feeding, and that it was this that was responsible for them succumbing to the virus. There was also an apparent additive effect of virus and shortage of food. Total mortality of his experimental larvae was greatest among those which were starved after he had fed them with polyhedra. The results of all his experiments suggested that the cause of the deaths from nuclear polyhedrosis was a change in the caterpillars' metabolism which allowed a previous infection with virus, which normally would not have been fatal, to develop sufficiently to kill the stressed host. Simple transmission of the disease between crowded larvae was not the cause. He confirmed this by raising groups of individually contained *T. ni* caterpillars on a shaker and a vibrator to approximate the stress resulting from continual jostling in crowded containers. They grew and developed more slowly, and had up to a four-fold increase in the incidence of nuclear polyhedrosis compared with larvae reared at the same time in stationary racks. In nature these non-gregarious animals would never have experienced the degree of crowding and associated jostling encountered in these experiments, but they might frequently have experienced a similar or greater shortage of food, or a marked fluctuation in its availabiity. Also they would have been carrying latent virus (Smirnoff 1962).

That this is the sort of thing which happens in nature was reported by Wellington (1962). A cold snap in British Columbia resulted in temporary starvation of the first instars of a population of the forest tent caterpillar, *Malacosoma pluviale*. This activated an outbreak of nuclear polyhedrosis in these caterpillars over a large area. The only sensible explanation of how so many animals could have simultaneously developed the disease over such a large area was that starvation-induced metabolic changes had allowed the development of viruses which were already present, but latent, in many, if not all, of the insects' bodies.

Today it is known that viral diseases may pass for an indefinite number of generations as provirus integrated into the chromosomes of the host without inducing symptoms of disease in the carriers. What has not been established is what the physiological change is which triggers the growth of such latent virus when the host is stressed. For such simple forms of life it seems unlikely that it would be the simple release of nitrogen in soluble form in the blood and other body fluids. I would posit that it is much more likely to be some change in the availability of the basic resource essential for the replication of viruses – the chromosomal nucleic acid of the host's cell nuclei. But this is still a matter of availability of nitrogen!

Other diseases, too, may be strongly influenced by the availability of nitrogen in the required form in their potential hosts. Garber (1954, 1956) showed that bacterial pathogens of both plants and animals, while they will attack any host, depend upon an adequate supply of nitrogen in the tissues of the host for their growth and multiplication as virulent pathogens. What is more, a single amino acid can make the difference between virulence and avirulence, and while it may be present in the potential host, it must also be *available* to the pathogen in sufficient concentration at the site of infection. He demonstrated this with a mutant of the bacterium *Erwinia aroideae* which will not grow without arginine. It would not grow on blocks of host tissue (e.g. carrot or radish) placed on a culture growing in minimal agar. These blocks contained enough arginine to support the growth of the mutant, but not enough at the surface of the blocks to allow initiation of growth. However, when Garber placed drops of homogenized host tissue on the same culture the bacteria grew around these drops. He obtained a similar result with purine-requiring mutant bacteria. They were non-virulent to mice lacking this amino acid, but virulent in mice with purine added to their diet.

Hirano (1964) obtained a similar result with larvae of the rice stem borer, *Chilo suppressalis*. When fed on a synthetic diet with a high-protein content, these insects had a much increased level of amino acids in their bodies and were much more susceptible to bacterial disease than those fed on a low-protein diet.

Also, when Bauer and Nordin (1988) fed larvae of the eastern spruce budworm, *Choristoneura fumiferana*, on diets containing either 2.5 or 4.5% nitrogen, many fewer of those on the latter died from infection with the microsporidian *Nosema fumiferanae*. Both diets were adequate for growth, but larvae ate more of the 2.5% diet, grew more slowly, and achieved a lighter final weight on it.

At the beginning of this chapter I put to one side a consideration of the concept that parasites and diseases may be limiting the "top" carnivores. Pursuing this line entails explaining what then limits *these* predators, and risks entering the never-ending "big fleas have little fleas..." debate. But, of course, there is no need to do so. It is clear that all these organisms in turn depend upon, and are limited by, *their* access to food of sufficient quantity and quality for their successful reproduction and growth. This is why most predators will be able to limit the abundance of their prey only in special circumstances or for short periods. In Part V I will try to explain in more detail why this might be.

Part V: The Alleviation of an Inadequate Environment: Outbreaks

Chapter 15 What Is an Outbreak?

15.1 Some Examples

In the Australian bush you can sometimes find a *Eucalyptus* tree with one or more of its lower branches heavily infested with the scale *Eriococcus coriaceus*, while the remainder of its crown bears none of them. There is no obvious reason why this should be so. The scales' predators and parasites have free access to all parts of the crown. In fact nearly all of them will be found on those parts of the tree where their prey are living. Yet the scales remain abundant and flourish, forming a dense covering on the bark of the branch and its twigs, and clustering well out along the mid-ribs of its leaves. Such populations may persist for varying lengths of time and usually only die out when the branch on which they live dies. Occasionally whole small trees are infected and may become sickly or even die, but usually they recover and the scales disappear.

This sort of small, localized and very dense population is not all that uncommon. Nor is it confined to *Eucalyptus* or scale insects. One which puzzled me for a long time was a kurrajong tree, *Brachychiton populneum*, in the grounds of the University of Adelaide. Its leaves became heavily infested with a free-living psyllid, *Protyora sterculiae*, so much so that the concrete path beneath its crown was coated with a dark patch of sticky honeydew. The insects eventually became so abundant that nearly all the leaves on which they had been feeding were prematurely shed. Many psyllids remained, however, now feeding densely packed along the bare twigs. But this was on only one side of the tree. The boundary between the two extremes was sharp; only a few centimetres separated totally defoliated and heavily infested twigs from those bearing a full quota of healthy leaves free of psyllids. There was, again, no impediment to the free movement of predators, which were quite common. Nor was the attack confined to the sunny north face of the crown, as I had come to expect with other species of psyllids attacking *Eucalyptus*. What is more, it was winter time. Not until I happened to walk past this tree after dark one evening did I realize what was happening. All the defoliated twigs, bearing large numbers of psyllids, were within the arc of illumination from a large electric light which had been installed that autumn, and was left burning all night to light the path.

There is a billabong (oxbow lake) of the Murrambidgee River at Wagga Wagga in New South Wales which I used to pass each day on my way to work. Usually it was dry and grassed, as its bottom is high above the normal river level, but occasionally the river would flood sufficiently to overflow its banks and fill the billabong to the rim. In a few days the river would return to its normal level leaving the water in the billabong perched many metres above. This meant that the roots of the river red gums, *Eucalyptus camaldulensis*, growing on the usually dry banks of the billabong, remained flooded for months while the water slowly seeped and evaporated away. If this happened immediately before or during a dry summer, the foliage of these trees became heavily infested with the psyllid *Cardiaspina retator*, while those trees nearby and up and down the river bank remained unaffected. Once the water had gone the psyllids also disappeared, and this always happened before there had been any serious defoliation of the trees. Predators and parasites were everywhere present both before and after such events.

A similar thing happened on the Torrens River in South Australia. There, the red gums were growing on the banks of a lake formed behind a weir across the river. During a very dry summer they were attacked by *C. retator* when the lake was drained, after it had been kept full for much longer than usual. Trees further away from the bank, and those up and downstream of the lake, were not affected.

At other times I have seen red gums, growing over many square kilometres in the Mount Lofty ranges of South Australia, so heavily attacked by this and other species of psyllids that the trees appeared to be dying. And in the southeast of the State I have witnessed, on several occasions, *Eucalyptus fasciculosa* trees over large tracts of land heavily attacked by *C. densitexta*. One spring, after several seasons of particularly heavy attack, I flew over hundreds of hectares where every tree had lost all its leaves. Yet, in both localities, trees were again fully leafed within 12 months, and showed no sign of injury or ill-health.

What is the point of these anecdotes? They are all examples of the same phenomenon – they were all outbreaks. Most people would have no difficulty applying this name to the latter examples. The animals suddenly became enormously abundant and widespread where previously they had been rare, or at least not common enough to be noticed or to cause any damage. Now they were visibly destroying all, or a large proportion of their food. A sudden and dramatic departure from normal. On the other hand, few people would describe an increase in numbers on one tree among many as an outbreak – even less so on only one small part of that tree. Yet, albeit on a much smaller scale, the same thing had happened. Also, apart from differences of scale of both time and space, all these events are no different from the explosions in number of flatworms, finches, snowshoe hares, mice, pelicans, etc. that I have talked about in earlier sections.

But outbreaks are not something apart. They are simply one extreme of a continuum of constant change in the number of animals in response to constant change in their patchy and variable environments. Nor, I suspect, are the magnitudes of change in numbers that they represent any different from the quite wide fluctuations in abundance of many animals which, to our perception, are rare, scattered, and stable in numbers. Mostly this is because the changes of the latter are of quite small numbers (in terms of numbers per unit area or volume) so go unnoticed, even though the *rate* of change is the same.

Fig. 41 A, B. Two sorts of outbreaks. **A** The kurrajong branch on the *left* was defoliated by pysllids now feeding on the bare twigs. It was lit all night by a street lamp. The one on the *right*, free of psyllids, was on the shaded side. **B** Pink gum trees (in the *foreground*) completely defoliated over hundreds of hectares by psyllids. Other species of gums were not attacked

A doubling of a population is of little consequence if this is represented by four adults surviving in one or two generations instead of an average of two. It is more apparent if 100, instead of 50, neonates survive. However, if 100 or 1000 neonates survive where before only two had done so, and if that level of recruitment persists for more than one generation, then the impact will be sudden and dramatic.

15.2 What Causes Outbreaks?

Is there, then, a common pattern or are they each in their own way unique? In all cases the pattern of change is the same – a sudden increase to a peak, then a crash to a trough. Both peak and trough may be of varying amplitude and duration, and occur periodically or intermittently. In all cases something changed so that more females bred successfully and more of their young survived, so there were many animals where before there had been few. Then something changed again so that few females bred and few young survived. Once more there were few animals where recently there had been many. In most cases it was not obvious what had changed, or whether the rise and subsequent fall in numbers had a common cause.

A major problem when trying to find the cause of outbreaks is that we do not know when a peak or a trough has been reached until *after* the event: a peak is apparent only after the numbers have started to fall; a trough after they have started to rise. By then it is too late. If we are to find the seeds of the cause of each event, we must look for that cause *before* the event. What, then, do we measure in order to detect the cause of an event when we don't know when (or if) it is going to happen? Unfortunately, funding and other priorities usually result in research efforts being mounted after the event when, presumably, whatever was responsible for the dramatic increase or decrease in numbers has ceased to operate, and so can no longer be measured. Such post hoc searching incorporates a double danger. Not only will the true cause not be found, but some other factor, trivial when the true cause is operating, may seem to be the significant cause.

So, we must start by asking the right questions. We know that all organisms have the potential fecundity which would enable them to increase exponentially, therefore the question should not be "what causes an organism to increase to outbreak numbers?" but, rather, "what stops an organism from increasing to outbreak numbers?" In other words, "what is it about the environment which prevents an organism's potential fecundity from being realized?"

The *potential* fecundity of females in a population is a function of the historical probability that any one neonate would survive and reproduce; "wastage" is necessary if that particular genotype is to persist. It is of evolutionary, rather than ecological, significance and has nothing to do with the number of individuals that the environment of the moment can support.

The *actual* fecundity of a female is her potential fecundity modified by the amount and quality of resources she can acquire from her immediate environment prior to breeding. It is mostly less than her potential, may differ over time and from that of her conspecifics, and is of direct ecological significance.

The *realization* of her actual fecundity is a function of the capacity of the environment at the time to support the young that she does produce, *irrespective of how many she produces.* As the history of the growth of the human population shows, numbers in a population will increase up to the level that the available resources will support, *not* as a function of either the size of the existing population, or of the fecundity of the females in that population. The fecundity of the females becomes ecologically significant only when the environment can support more young than some females can produce. Then those populations with the most fecund females will increase the fastest.

Most actual fecundities that have been measured in nature are still much more than sufficient to produce exponential growth, so it seems probable that the key to answering the question lies in the *realization* of those actual fecundities – in the capacity of the environment to support neonates – and not in the number produced.

With the examples of psyllids and scale insects that I have just given, and with most other herbivores which increase to outbreak levels of abundance on the foliage of plants, there is no visible change in the amount, form, and availability of food, at least not until numbers become so great that they begin to destroy the plants faster than they can grow.

So, the usual conclusion is that something which previously killed most of the animals before they could use their food must have ceased to operate. Predators are the most likely candidate, especially when nothing else in the environment seems to have changed; they were the only thing found to be killing any significant number of the small populations that existed before and after the outbreak, and during the outbreak they remained plentiful and ubiquitous, but clearly ineffectual.

Yet, in all the examples I gave, but especially in the case of just one branch among many nearby and often touching ones being affected, such an explanation seems fairly credulous. Is it likely that predators could virtually eliminate their prey from most of the tree while remaining plentiful but ineffective in the one place where their prey are clearly flourishing? I think it is more likely that something quite independent of the activity of the predators has happened to one part of a tree and that this has enabled many herbivores feeding on that part to survive where previously only few could do so.

Fortunately, there are recorded outbreaks where the true cause is obvious and unambiguous; both what happened and how it happened are clear. The examples of the pelicans in Section 13.4 and of *Cactoblastis* in Section 3.4.6 are cases in point. The supply of fish for the pelicans was seen to increase, the cause of that increase was apparent, and there were no other factors like predators, previously holding the numbers of pelicans in check, which had ceased to operate. Soon there were a great many pelicans where before there had been few. When the supply of fish disappeared, many pelicans died, and soon there were again only a few pelicans where before there had been many.

No predators were attacking *Cactoblastis*, but until Dodd deliberately increased the level of nitrogen in the plants which were not destroyed when he first introduced the moths, only a few caterpillars could survive on those plants. There were few caterpillars and many plants. Once nitrogen in the plants was increased beyond a minimum threshold very many caterpillars survived. They quickly became so abundant that they destroyed the plants. Then many caterpillars died before they could complete their development.

The only difference between these two examples is that the activity of the pelicans had no influence over the increase or decrease in the supply of their food, but the activity of the *Cactoblastis* caterpillars made a significant contribution to the decline of theirs – they overrode their "prey".

15.3 The Paradox of Enrichment and "r" and "K" Strategists

When animals have been detected overriding and eventually destroying a suddenly much increased or enriched food supply in this way, it has sometimes been referred to as "the paradox of enrichment" (Rosenzweig 1971). An explosive increase in survival leads to the eventual death of many, often many more than were there originally, and even to the extinction of local populations. But there is no paradox. There is simply a sudden, infrequent and unpredictably larger-than-usual increase in the abundance of food. Animals which use that resource have evolved, under the selection pressure of a usually chronic shortage of the resource, to get as much of it as possible, and to be "programmed" to quickly take up any increase in its availability. Such increases in availability, whether they be of prey for a carnivore, or plants for a herbivore, are usually quite small and often only seasonal. The corresponding changes in abundance of the predator or herbivore are small and short-lived as well, and so go unnoticed. Furthermore, where the food that is changing in abundance is a constituent of plants, not the plants themselves, there is little or no observable damage to the plants.

However, when there is a *great* increase in the resource there is a consequent great increase in the number of animals using it. As long as the resource continues to be abundant the animals will continue to press hard on it, "knowing" (in the evolutionary sense) no other way. Sooner or later, however, the amount of food available to them will decrease, either, as in the case of the pelicans, because whatever caused it to become so abundant in the first place ceases to operate, or, as in the case of *Cactoblastis*, because the animal "catches up", overrides the resource, and uses it all up. The end point is the same whichever it is. The population of the animal will crash behind the disappearing supply of its food.

The same sequence of events has been recorded for other examples that I discuss in this book: outbreaks of spruce budworms which kill whole forests of balsam fir in North America when age and weather combine to make the trees a better source of food; numbers of rabbits increasing explosively in inland Australia after rain, and dying off in equally spectacular numbers when the green feed dries up; and the Galapagos finches breeding beyond all expectations when suddenly provided with an abundance of food to feed to their young.

In this context there is no such thing as a "K" strategist – only "r" strategists held in check by their inadequate environment. The short generation time and huge fecundity of females of invertebrate "r" strategists are well known. In general, the number of young that they are capable of producing (their potential fecundity) is in direct proportion to the risk of the offspring not surviving. Most are "wasted", but the wide and repeated dispersal of very many ensures that what resources exist are found and used, and any increase in the availability of those resources is quickly taken up by a few more of the many dispersers. For the "K" strategist this risk is less and so, consequently, is the potential fecundity of individual females. However, the capacity to quickly produce young in response to any increase of resource is only superficially reduced.

It is not true, therefore, as Weins (1977) and others have proposed, that many birds and mammals with fecundities which are low by comparison with most invertebrates, do not have the capacity to rapidly track changes in the abundance of their food. On the contrary, the more ephemeral and unpredictable the food supply, the more finely "tuned" are individual females to producing additional young at short notice. Their fecundity is still great, but numerous devices have evolved to prevent the costly raising of young which have little prospect of surviving. Many are referred to in this book: a return to oestrus only after the loss of young, delayed implantation of fertilized eggs, resorption or abortion of foetuses, abandonment or eating of the young (human infanticide is a variant of this) – all serve this function. A step further on there are social structures which restrict the number produced, and delay the expelling of young. All these devices operate to limit the number of young actually released into the environment to those with a reasonable chance of surviving (and in the process increase the number which do survive: Chaps. 11 and 12). At the same time they retain the females' capacity to quickly and fully express their latent fecundity to take up any unexpected increase in food. When this happens even the most apparently conservative "K" strategist has the potential to swamp its environment with its own kind – to produce an outbreak. Darwin's classic example of a pair of elephants having 19 million living descendants after 750 years would constitute an outbreak. Only the time scale would separate such an event from a plague of mice in a wheatfield, or an outbreak of caterpillars in a forest.

It may often be that all the females and all the young that they can produce in a generation are unable to use all of a large and unexpected increase in food. Should such increased abundance persist for more than one generation, females may continue to realize their maximum fecundity without using all the food. At such times there would be *no* doomed surplus; nearly all would survive. However, sooner or later the numbers would "catch up", and once more production and survival of the young would fall. Witness, for example, the inverse situation when feral donkeys in Australia were culled (Sect.6.1.1), but the process is the same.

In nature, however, superabundance is invariably short-lived, and the end comes much more quickly.

Chapter 16 The Interactions of Food, Prey, and Predators in Outbreaks

16.1 Bacteria and Protozoa

Jensen and Ball (1970) experimented with about as simple an interaction between an organism, its food, and its predator, as it is possible to get – populations of bacteria and their protozoan predators maintained in culture flasks of water. They obtained the organisms from natural populations in a Michigan river, and added nothing but a sterile nutrient solution to the cultures. The bacteria grew on the nutrient solution and the protozoa ate the bacteria. By adding the same amount of this nutrient solution to flasks once every 7 days, they generated weekly peaks of great abundance of bacteria followed by peaks of protozoa. After each peak the numbers of both organisms remained at very low levels until the next injection of food. This manipulation had generated a classic feedback alternation of a cyclic prey and its predator. However, when they added exactly the same amount of nutrient solution each week, but in equal daily amounts, these regular peaks and troughs of bacteria and protozoa failed to appear. The numbers of both organisms continued to fluctuate, but instead at much lower amplitudes, and at random to each other and in time. A clear demonstration of the dependence of both organisms on their food supply.

Apart from this demonstration of how periodic injections of food can induce cyclicity in a population (Chap. 17), this study produced a result within each of the cycles that were generated, which mirrors what has been documented for hundreds of outbreaks of all kinds of herbivores and their predators throughout the world. The herbivores suddenly begin to increase exponentially, seemingly bursting away from their predators to reach a peak of abundance before declining more or less steeply and to varying levels of scarcity relative to their original density. The numbers of their predators also increase rapidly to a peak and then decline. However, the predators' cycle is always a step behind that of the herbivores. In spite of the lag, however, this pattern is more often than not interpreted as evidence that the predators have "caught up" with their prey, regained control over their numbers, and forced them back in balance with a previous equilibrium.

16.2 Rabbits, Foxes, Cats and Dingoes

A study by Newsome et al. (1989) of the influence of feral foxes, *Vulpes vulpes*, and cats, *Felis catus*, as predators of the European rabbit, *Oryctolagus cuniculus*, in pastoral and woodland areas of New South Wales in Australia clearly illustrated what really happens in these sorts of interactions in nature. Their experiments involved counting the numbers of rabbits and predators on blocks of land ranging from 50 to 180 km^2 while removing the predators from some of these blocks by intensive and repeated shooting. They were able to demonstrate that in years when the rainfall was average and there was some green feed each season, the predators could keep the rabbits at significantly lower levels than they would have achieved had the foxes and cats not been there. However, while this effect that they demonstrated is real, and of possible importance to graziers running cattle on the land or to conservationists preventing soil degradation from overgrazing, it is relatively trivial in the long-term ecology of the rabbit living in this environment. The predators have minimal influence over the major rises and falls in abundance of rabbits generated by the flush and dying-off of grass which follow above-average rains and droughts. They merely "tag along behind", breeding greatly on the increase in their food and dying-off when the rabbits die (just as the pelicans do behind the supply of their fish food, and see Sects. 16.3 and 16.5).

The extent of this dependence of the success of the predator on that of their prey was highlighted by two events which Newsome and his colleagues recorded. Two years before they started their experiments there had been a great eruption of rabbits which collapsed to near extinction as their food disappeared during a drought. The number of foxes and cats had increased rapidly in response. Then, as the drought took its toll, the newly independent young predators thrived on the huge and readily caught food resource of starving rabbits. But, as the rabbits disappeared their numbers, too, soon declined. In the meantime, however, the survival of so many young predators as the rabbit population collapsed due to the drought "presented an *apparent* predator-prey cycle" (*their* emphasis!). The underlying cause was fluctuations in the abundance of food for the rabbits, just as Gibb (1981) had found for rabbit populations in New Zealand.

Then, after good rains had again generated a big increase in the number of rabbits, and while they were still removing predators from their experimental blocks, another drought hit. All rabbit populations declined steeply to very similar levels regardless of whether their predators had been removed. But, the "crash" began and was steepest in the densest population where the predators had first been removed (and where, consequently, food was scarcer), and while Newsome and his colleages were still removing them.

Clearly, when rabbits are unable to achieve their full reproductive potential because the supply of good food is limited, their predators can "dampen" fluctuations of their numbers, even to the extent of preventing them from increasing in response to small increases in their food. But, equally clearly, when there is sufficient of the right sort of food to allow female rabbits to achieve their full fecundity, and for most or all of their young to survive and breed, the predators can no longer keep the number of rabbits down.

So, if the environment was more equable – if there were no spasmodic heavy rains or droughts – there would be no outbreaks followed by massive dying-off of rabbits. An ecologist studying the interactions of these animals in such an environment would find (and could demonstrate by removing the predators!) that the number of rabbits was being kept in check by the predators – unless a deeper look revealed the limitation being imposed by the lack of food for the rabbits.

Other Australian studies have shown the same relationship for dingoes, *Canis familiaris dingo*, as predators of the rabbit (Corbett and Newsome 1987) and for foxes, cats, and several species of carnivorous birds preying upon feral house mice, *Mus musculus* (Sinclair et al. 1990). A more recent study in Chile (Jaksic et al. 1992) has similarly shown the inability of predatory mammals and birds to either prevent irruptions or cause the crash of populations of their small mammal prey. At best they may prolong and extend the trough which follows the crash, and delay a later return to high densities. Like their invertebrate counterparts, they are not optimum foragers (Corbet and Newsome 1987).

So it is also for other animals, herbivore or carnivore, vertebrate or invertebrate. An outbreak is simply the animals' response to an infrequent and unpredictable increase in the availability of food which is adequate for the production and growth of the young. It is *not* an escape from control by predators, although frequently this will *appear* to be so.

16.3 The Varying Response of Predators to Changes in Prey

The impact of a predator depends upon, and will vary according to, the rate at which its numbers increase relative to those of its prey. The *rate* at which a population of animals increases, however, rarely, if ever, equals the *capacity* of the individuals comprising that population to increase their numbers. The former, as measured by ecologists, is the expression of the latter modified by the pressures of the environment at and preceding the time of measurement. The capacity of an individual to increase is the product of its innate speed of development, fecundity, and length of life, plus its behavioural and physiological attributes which determine its powers of dispersal, its efficiency at finding and processing food, and its ability to avoid becoming food for others: in short, everything that improves the probability of it surviving in its environment. The rate of increase of a population can be at a maximum only if all these attributes of the individuals making up that population are able to be fully expressed, if each individual's *absolute* capacity to increase is realized. In an inadequate environment this rarely happens, and when it does it is likely to be short-lived. The old analogies of "biotic potential" and "environmental resistance" still provide a simple but effective encapsulation of these concepts.

For most of the time, therefore, animals are operating at well below their potential; what the individuals achieve is an *effective* capacity to increase. In the case of herbivores it seems that, in general, their individuals have an absolute capacity which far exceeds that of their predators, but this is usually masked because they are constrained, by a shortage of food, from fully expressing it. So, they are not able to increase faster than their predators. Then, in a patchy and stochastic world, a classic negative density-dependent predation can develop. As the predators' numbers increase faster than those of their prey, they will further depress the numbers of prey until prey become so scattered and rare that fewer predators can find enough food to survive and breed. So their numbers decrease, again allowing a resurgence in the number of the prey, and so on. Over time, in an environment which does not change much (most undisturbed environments in the time scale of human experience), an appearance of "balance" will emerge, the activities of the predators seemingly keeping both their numbers and those of their prey oscillating around a long-term mean. However, it is the relative shortage of prey – albeit sometimes partly generated by the actions of the predators – which is responsible for the oscillations. At such times the exceptions which prove the rule occur. The predators are, for varying periods of usually several generations, depressing the numbers of their prey. If this depression is severe enough, or lasts for long enough, many local populations of prey will be exterminated. Under these conditions, ultimately the only hope of avoiding extinction is chance escape from detection in a patchy environment.

If, however, at any time, the environment changes so that the effective capacity of the prey to increase exceeds that of its predator, such oscillations will disappear. Instead there will be an escalation in abundance of both prey and predator, the prey steadily outstripping the predator in proportion to the extent to which its capacity to increase exceeds that of the predator. This will continue so long as the resources of the environment are sufficient for the prey to maintain this "edge". Usually it does not last long, like the regular seasonal burst of growth of so many herbivores which feed on the short-lived flush of new plant tissue each spring, or, more spectacularly, as an outbreak of rabbits, mice, kangaroos, pelicans, or Galapagos finches collapses as the food runs out.

If the food did not run out, some other resource – ultimately space – would do so, and curb the continued increase of the prey. But until that happened, and the effective capacity of the prey again fell below that of its predator, the predator could never "catch up" with the prey. In the real world, however, the food soon runs out and is in short supply for most of the time. Only on the (usually) rare occasions when this inadequacy is ameliorated, and an outbreak happens, is the true cause of the limitation of the numbers of both prey and predator revealed.

16.4 A Natural Experiment: Guano-Algae-Limpets-Oystercatchers

In Section 5.2 I outlined a study which revealed how an injection of nitrogen into a natural system flows up through the food chain, increasing the reproductive success of plants, herbivores, and carnivores alike. Guano had enriched the sea around permanent colonies of seabirds in South Africa and caused algae to grow in great profusion. Limpets which eat the algae grew faster and bigger and produced many more offspring, and oystercatchers, which feed small limpets to their young, congregated and bred on these shores. However, on most of the coast there is no such enrichment. There the algae and limpets are slow-growing and reproduce but little, and there are few or no oystercatchers.

Yet, the oystercatchers eat so many of the small limpets that grow on the enriched sites that there are *fewer* limpets there than on the poor sites. So it might seem that the oystercatchers, by gathering where there is a huge "over-production" of young limpets, and eating most of them, are "regulating" the abundance of their prey. There are fewer individuals than there are on the poor sites, so that they are prevented from eating most of the algae. Meantime, on the poor sites, the oystercatchers eat hardly any limpets, which apparently can then quickly eat nearly all the young algal sporlings before they can grow into dense mats. We are back with Hairston et al. (Sect. 2.1)! Predators control herbivores which therefore cannot control the plants, except in the absence of the predators.

Why, then, are there not many oystercatchers eating these more plentiful limpets on the non-enriched shores? There is nothing preventing them from doing so. And why, when limpets were experimentally excluded from sections of these poor sites, did the algae not grow in profusion, yet did so (in spite of the limpets) when guano was experimentally introduced to these same shores?

If the system is considered from the point of view of the reproductive success of these three organisms – of how many *genes* rather than individual phenotypes make it to the next generation (Dawkins 1989) – then it is clear that trying to explain what is happening in terms of control by predators is viewing the world from the wrong end, of putting the cart before the horse.

In most places where these three organisms live they are all struggling to survive and reproduce. Few succeed, except in the few places where the water is enriched with nitrogen. In these places all three achieve much greater biomass and *produce many more offspring*. In the case of the algae and the limpets, they achieve this *in spite* of supporting very much greater biomass of their respective predators. If the oystercatchers are not the "top" predators in this system, it is a safe bet that their increased reproductive success will, in turn, be supporting greater numbers of their predators.

It is obvious that the oystercatchers are eating many more limpets (and a greater biomass of limpets) at the rich sites. The importance of predation at these sites is like that of competition (Sect. 1.4). It decides *which* limpet genes make it to the next generation, not *how many*; the amount of food decides that. Predation by oystercatchers provides strong selection against limpets that grow slowly (they cannot prise the large ones off the rocks), that do not settle out of reach in cracks and crevices, or that do not happen to become camouflaged by algae growing on their shells (Bosman et al. 1989). In the process they strongly affect the *structure* of the limpet population – there are many larger and more fecund limpets at the rich sites and practically none at the poor ones. However, the former have a greater biomass and produce *ten times more gametes* per unit of shore than do the latter.

At the same time, on all their extra food, the oystercatchers reach a new higher asymptote of abundance, where, eventually, a relative shortage of food again stops them from taking all the food in their environment. Much remains inaccessible or unattainable, and therefore not available for their use.

Finally, it is important to see the overall picture. The enriched sites are not closed systems but microcosms within the total environment of the three organisms. At these places there are outbreaks of algae, limpets, and oystercatchers which will remain as long as the guano continues to be injected into the system. It is the same but more pronounced response of a herbivore to enrichment of its food by guano that we saw for red deer (Sect.6.1.2) and geese (Sect.7.1.1).

The system is also directly analogous to that of the scale insects concentrated on the senescing eucalypt branch, or the psyllids confined to the continuously illuminated leaves of the kurrajong tree, or red gums growing around a flooded billabong, that I discussed at the beginning of this chapter. In all cases the herbivores are concentrating where their food is best, and their predators do the same. Close by, within easy reach in the same habitat, there are few or no herbivores – or predators.

But, unlike these earlier examples, the supply of good food persists. So why is there not an ever-increasing biomass of these three organisms leading to an eventual and absolute shortage of space? Precisely because the enriched sites are not closed systems. Many algal spores will settle on the poor sites, but fail to survive and grow, limpets or no limpets. Very many young limpets will disperse to nearby parts of the habitat where there are few or no predators to eat them. However, most will starve. Similarly for young oystercatchers. They will disperse to where there are few prey to eat. Most will die, a few may eke out a bare existence.

The whole system is driven by the amount of nitrogenous food that can be used for the production and early survival of young. That mostly there is only enough of this available to support few or no animals in most of the habitat is not readily apparent, except where there is an exceptional injection of the kind exemplified in this study.

There is abundant evidence throughout the literature of this dependence of predators on their prey – that their numbers increase only behind those of the prey, and usually only override already declining populations. This is equally true for human hunting, a consciously directed, and therefore normally far more efficient and concentrated form of predation than that of any other animal. There are many records of hunting returns that show this. Yet, ironically, and all too frequently, while variations in these returns are presented as evidence that the population of the hunted animal is indeed fluctuating, the causes of its fluctuations are being imputed to other much less effective predators.

16.5 A Thought Experiment: Hot Spots in a Box of Wadding

Some years ago, in trying to explain these ideas in a letter to a sceptical Herman Klomp, I wrote the following. It seemed to work for him, so I repeat it here.

We have a hypothetical insect which feeds by chewing up cellulose wadding impregnated with a solution containing all essential nutrients. The environment or habitat is a big flat box, and the animals move around very little, spending their time feeding on the cellulose which is palatable in that it stimulates them to keep feeding. We have carefully controlled the concentration and distribution of the nutrient solution in the wadding. There is an abundance of fibre as a carrier, carbohydrates for energy, and all the necessary minerals and trace elements. However, over most of the area the concentration of amino acids in the nutrient is such that although the animals eat the *maximum* volume of cellulose that they can, feeding continuously, they do not get *quite* enough nitrogen to maintain body growth and metabolism. Under these conditions a newly born animal will very soon weaken and die from malnutrition.

However, at a few randomly selected spots in the box we place a much higher concentration of nitrogenous nutrients. Newly hatched young are deposited at random on the surface of the wadding. *If* one of them happens to land on one of these "hot spots" it will get enough nitrogen and survive (we can adjust things so that there is sufficient to get it to the third instar. It is then big enough to "process" the surrounding low-quality wadding fast enough to get the nitrogen necessary to grow to maturity).

We are also able to adjust this system in such a way that there are just enough hot spots for enough young ones to survive (by chance over many generations) to maintain the population at "equilibrium" around a mean of replacement level.

Having got this system stabilized and working, we now modify it. We increase the concentration of nutrients in the wadding generally (or increase the number of hot spots) to the extent that nearly all the young larvae get enough to eat and survive. The result? An explosion in numbers! If we now provide all the multitude of young produced by this large population with the original set-up (a few randomly located hot spots), all but a very few will perish, and we are back to the original position. A relative shortage of food was replaced by an abundance of food, which in turn was replaced by a relative shortage.

Now, instead of cellulose wadding we have pine needles, leaf sap, bark phloem, or grass, which is palatable but inadequate. But here and there there is a leaf, branch, whole tree, or group of trees, which is damaged, senescing, or growing rapidly – which is in some way a richer (and adequate) source of nitrogen. (Or consider the ocean around the nesting sites of the South African guano birds in the previous section – an example I did not know about when I wrote this!) Then we have the equivalent of the box of cellulose with the hot spots. If all the trees senesce or become stressed (by drought, waterlogging, or whatever), or there is a flush of new grass, so that the level of nitrogen is raised in *all* tissues that the young insects eat, then we will have an outbreak. A relative shortage of food has been replaced by an abundance of food. Once stress is removed, or plants adjust to a permanent change in their environment, or the flush of grass dies off, there is once again a relative shortage and most of the young will again perish (see also Chap. 18).

Unfortunately, in nature, life is not as simple as this. There will be continual variation and change in response to the constant interaction of many factors in the environment. Also, a further complication arises because there are other animals which use our animal as food – predators.

Suppose we had set things up with our original box of cellulose so that there were so few animals, so thinly scattered, that if a predator happened to arrive in the box it would die of starvation before it could find enough of them to eat. If we now move to position two (abundant food and many animals surviving), and if a predator then finds the box, two things can happen.

1. The predator can multiply *faster* than our animal. The population of the predator will rapidly increase, reducing the number of its prey. This will go on until there are so few prey (in spite of there being plenty for them to eat!) that the numbers of predators will decrease, and then the prey will start to increase again. This is a truly density-dependent cyclic set-up which, left alone, would "stabilize" around a mean population of each animal.
2. The predator *cannot* multiply as fast as our animal. Its population will still increase. We can assume *all* its offspring survive, but for every one it produces our animal produces ten (or two, or 1.3, no matter); *if* the prey produce more surviving offspring than the predator, its population will increase faster than that of the predator *as long as there is enough food for all of the young prey animals*. Obviously this would soon come to an end because the box would quickly be packed full of prey and predators – an absolute shortage of space. Or it could stop if we again returned the situation to stage one where there was a relative shortage of nutrient in the wadding. If we did this before space became limiting, what would we expect to happen? There would be a sudden and great drop in the number of our prey animal in the next generation, and the number of the following generation of the predator would override the numbers of the prey. To an observer unaware of our manipulation of the amount of food available, it would appear as if the predator had "caught up" with or "controlled" the prey population in the presence of an abundance of food for the prey (lots of cellulose which they will "happily" eat all day, and die of starvation. In the real world this happens with very small larvae and their bodies are never found. Only the few that get enough to grow bigger are seen feeding).

Again, life is not as simple as this in the real world. Because there is constant change in the environment, making more or less food available to the prey (and thus to the predator), and because some predators in turn have predators, and some have alternative prey, the end result is usually a jumble of constant change and interaction. The impression gained from all this over a sufficient number of generations is one of "stability", "balance", of populations "regulated" around a "mean". But it seems to me that because *all* animals produce sufficient offspring to literally pack the available space tight with their own kind, and because this does not happen, and because predators and prey alike can only increase to this level if their food resource is abundant, then it is likely that for most of them, for most of the time, there is not enough food for all. The species of animals that have survived have done so because they are made up of individuals which have evolved the capacity to produce sufficient offspring to ensure that enough of them find the limited amount of food available to keep the species in existence. *All animals are struggling to survive in a passively hostile environment.*

Chapter 17 Cyclic Outbreaks

In the circumpolar lands of the Northern Hemisphere the numbers of voles, snowshoe hares, and other vertebrate herbivores like the tetraonid birds and their predators (e.g. raptorial birds, foxes, lynx, mustelids) undergo changes in abundance involving order-of-magnitude differences between their peaks and troughs. The timing of these fluctuations is clearly cyclic, and peaks of abundance of a particular species are frequently synchronous over wide areas of its distribution. Although all species do not necessarily show the same periodicity, often quite different species with different ecologies (e.g. voles and tetraonids), as well as more closely related ones (like different species of microtines), cycle in synchrony in the same locality. In the more southerly parts of their distribution, however, none of these species is cyclic, nor do their numbers exhibit the wild amplitudes of abundance that they show in the polar regions (see also Sects. 6.2.3, 6.2.4, 6.2.5.3, 7.1.6, and 13.2).

There has been much debate over many years and many hypotheses proposed to explain these cycles but, as yet, there is little agreement, let alone any unanimity, about what might cause such large and regular changes. So, entering this debate from afar, I feel a little like Daniel in the lions' den without the benefit of divine intervention! Yet, certain generalities and patterns emerge from the literature which point to some fairly obvious and parsimonious conclusions. I think these might help to decide which of the many hypotheses about cyclic outbreaks are most likely to reward further testing.

Jenson and Ball (1970) had pointed out that their results with cultures of bacteria and protozoa closely mimicked the observed changes in abundance of animals in North America. Their weekly additions of food for the "herbivore" produced a pattern of cycling densities of both prey and predator akin to that seen in populations in the Arctic. When they provided the prey with exactly the same amount of food, but more nearly continuously over the same period, the two populations behaved like those in more southerly habitats – they did not cycle. They concluded that their results suggest a possible relationship between large fluctuations in population density and fluctuations in food supply which could be brought about by an extreme climate.

I would agree. Their results and much other evidence suggest that the only difference between these polar outbreaks and those of all sorts of animals all over the world is that they are cyclic. In all other respects they would appear to be the same. The similarity extends beyond the pattern of increase – peak – crash to include the restriction of outbreaks of a species to the harsher extremes of its environment, the synchronization of outbreaks of one species over large areas, and simultaneous outbreaks of quite different species in the one place. Animals like the feral rabbit and house mouse in arid outback Australia repeatedly – but irregularly – reach huge numbers in this harsh environment, but do not do so in their milder native habitats. Outbreaks of many insect defoliators, like the spruce budworms in North America, frequently, but irregularly, occur at the same time in widely separated localities. In Australia, again, native kangaroos explode to great abundance at the same time and in the same place as the introduced rabbits, as do different species of native and feral rodents in the desert. Also, something that is rarely recorded, but which I have observed on several occasions: when a herbivorous insect reaches outbreak levels on its host plant, other sorts of insects which eat the foliage of the same plants also become much more common, but their increase is usually overshadowed by the enormous numbers of the outbreak species (White 1974; Markin 1982).

Furthermore, all the evidence assembled in this book indicates that the changes in abundance of herbivores (whether or not their numbers ever become so great as to be classified as an outbreak) are driven by the supply of nitrogenous food for breeding females and their growing young. It also suggests that the success of carnivores is, in turn, driven by the supply of herbivores – the food for *their* females and young. The most probable, and parsimonious, conclusion which I can derive from all these similarities is that the cyclic changes in abundance of polar herbivores and their predators are similarly driven.

This hypothesis is, of course, far from new. There have been many variants of it proposed over many years. Most, however, have concentrated on the supply of staple food in the winter rather than on the nutrition of the breeding females and their growing neonates. And, apart from some general references to (usually unspecified) "quality" of plants, their emphasis is on the *quantity* of food. Furthermore, it is usually on the exhaustion of food by animals which have somehow been able to build up and overtake the supply.

This inevitably directs attention towards predators. As with the example of the snowshoe hares in North America, most workers recognize that food must be a major factor, but cannot eliminate the possibility that predators too play a primary role, not only in the generation of these cycles but in synchronizing those of different species (Angelstam et al. 1984, 1985; Hornfeldt et al. 1986; Summers 1986; Hansson 1987; Henttonen et al. 1987; Andreev 1988; Bergerud and Gratson 1988; Brittas 1988; Sutherland 1988; Marcstrom et al. 1990; Keith and Cary 1991).

That the numbers of animals like the insectivorous shrews (*Sorex* spp.) also fluctuate in close synchrony with those of the voles further fuels the belief in the efficacy of predators. Clearly, there must be a common extrinsic factor driving all these cycles. Their food is different, however, so predators are *assumed* to be that factor (Hansson 1984; Henttonen 1985). But shrews, too, are predators and their prey feed on the subarctic plants. The pupae of Lepidoptera are one of the major components of this food. The larvae of these insects are defoliators of trees like birch (*Betula* spp.) in Fennoscandia, and are themselves strongly cyclic. They have a 9- to 10-year periodicity compared with the 3- to 5-year cycles of the microtines. Nevertheless, the declines of their outbreaks coincide with lows of voles (and shrews), as do increases in their numbers with peaks of these small mammals (Neuvonen 1988).

The reason that the emphasis comes to be placed on predators in all these studies is, I believe, because the first step in the sequence of outbreaks usually goes undetected, or is overlooked. The factor which allows more animals to breed and survive – which allows them to suddenly approach their absolute capacity to increase – is neither an increase in the *quantity* of plants in their habitat nor any lessening of the pressure of predation. It is an increase in the amount of protein that is available to them as soluble nitrogen mobilized by the plants for the development of fruit. Then the plants become a source of food which is adequate to support successful breeding and survival of young. Once that supply is exhausted, their numbers will collapse. With this hypothesis there is no need to invoke predators. Clearly, there will be many different, and often very marked, effects of predators, including shifts in attack by generalist predators from a declining preferred prey to other less preferred ones. However, all these effects will be secondary to the major influence of food. *All* the herbivores and *all* the predators are responding to the same thing – changes in their food.

One early worker who very clearly stated this primacy of food as the driving force was Lauckhart (1957; see also Sect. 6.1). While in his review of earlier work on these cycles of northern animals his emphasis was on winter feeding, he presented an analysis which I precis here:

Plants in northern latitudes are of poor quality, and the "pressure of reproduction" tends to keep populations of herbivores which eat these plants "teetering on the brink of malnutrition". Disease, lowered reproduction, and reduced survival of young are all symptoms of a declining cycle and are all associated with malnutrition. Any small drop in nutritive quality of the food would therefore exacerbate this malnutrition. If such a change in quality was regular, it could cause cycles, even in the absence of predators. The reproductive cycles of plants in the north depart from the annual cropping of more temperate climes (where animals apparently do not cycle) and become progressively more pronounced (and animal cycles increase in magnitude) the further north the plants are growing. A heavy seed crop exhausts stored nutrients from a plant, and in northern climates it takes several years to accumulate sufficient nutrients to produce another crop of seed. So, immediately after fruiting, the quality of a herbivore's food would suddenly drop below the nutritional threshold for successful reproduction and growth of young, and would persist at a deficient level until the plants had replenished their store of nutrients. There would be a concomitant crash of the herbivore, its numbers remaining low until food was again sufficient to allow for greater success in breeding. Many subarctic trees and shrubs may bear fruit only once every 10 years or more. Most common seed cycles, however, are approximately 3- or 4-yearly, very similar to the short-term cycles of many arctic animals. On the other hand, interactions of seed cycles of different lengths could cause much longer cycles. And these interactions could be further modified by changing patterns of the weather. The timing of peaks of cycles would also be affected by the type of food eaten; seed-eaters would tend to be a year later than those eating foliage.

At the same time, others (Svardson 1957; Kalela 1962) were making similar proposals to explain cycles of birds and mammals and their predators in northern Fennoscandia. Kalela's colleagues and students have continued his investigations of voles and lemmings, and produced substantial evidence in support of the concept. They used 40-year records of cycles of five species of microtines, and 10-year records of cycles of reproduction of 15 species of plants in subarctic Finnish Lapland (Laine and Henttonen 1983). Their analyses show that, in general, the numbers of microtines in Fennoscandia cycle in synchrony. There are some phase differences in the course of the increase and peak years, but all reach the extreme low simultaneously. The main decline – the crash – is usually in winter, but characteristically continues during the following summer so that minimum densities are not found until the autumn after the winter crash, or even in the following (second) spring. The size and condition of the microtines also varies synchronously; animals are largest in the peaks, smallest during the declines. Furthermore, during and after a crash many females do not breed or are late starting. However, *the decline actually starts when breeding stops in the peak summer* – the cause of the event is well before the event.

The main predators (mustelids and a viral disease) are uncommon when numbers of microtines are increasing, but become very common in the peak years, and remain high into the decline, prolonging and deepening the eventual trough in the number of microtines. It could be argued that the predators were responsible for the synchronized decline, but this would leave unexplained the synchronous changes in the size and condition of the microtines.

The authors' explanation that the quantity and quality of food available for the rodents was the prime cause of the start of the declines provided a simpler explanation of cycles of both numbers and condition of the microtines.

As Lauckhart had predicted, Laine and Henttonen found that in the short arctic growing season all of the plants monitored, except for a few annuals, need several summers in which to accumulate sufficient reserves to be able to flower and set fruit. They found that reproduction of many of these plants which are eaten by the microtines was also cyclic and synchronous; not as precisely as the microtines, but nevertheless showing significant synchrony with them. Because of variations in the weather there was considerable variation in the rate at which plants accumulated their reserves, and the success of reproduction achieved with those reserves. However, this variable accretion culminates with the setting of fruit all in the one year. Numbers of microtines increase, with some variation in timing, peaking on the supply of seed. Once this is exhausted their numbers crash.

I would suggest that the flush of mobilized soluble nutrients associated first with the initiation of flowering, and then with the filling and ripening of seed, would have provided the major impetus for the peak years. It is this which would provide the level of protein necessary for successful breeding. But, once achieved, fruiting quickly exhausts the supply of nutrients and the process must start anew. Then, all the plants would suddenly and simultaneously become a totally inadequate source of nitrogen for breeding microtines and their growing young. Breeding would cease, young would die, and there would be a simultaneous collapse of all populations. [The parallels of this sequence with that of annual breeding of mice in irrigated fields (Sect. 6.2.2.1), of galahs in Western Australia (Sect. 7.3.4), and of many flush-feeding insects on annual new growth (Chap. 3) must be clearly evident.]

Finally, these authors pointed out that changes in abundance of the microtines with changes in latitude are linked directly with changes in the supply of food. Further south in Finland the periodicity of accumulation of reserves and flowering of plants shortens, and eventually disappears, and so does the cyclicity and synchrony of changes in the abundance of microtines.

In other words, in the south there is never enough good food for long enough to let them achieve their absolute capacity to increase for more than a short time. Then negative density dependent cycles of "regulation" by predators may be generated. But their explosive "escape" from their predators in the north, when there is a lot of food all at once, shows that, like the rabbits in Australia, given full expression of their capacity to increase, their predators are powerless to do anything about their numbers.

It would seem, then, that the picture for voles in Finland is the same as that for rabbits in Australia. Both quickly explode to enormous numbers in response to a pulse of food which will support widespread breeding and raising of young. Both die off in equally spectacular fashion once the supply of that food dries up. In habitats where there are no marked pulses of food, neither animal undergoes any great changes in abundance. The predators of both have little influence on either the increase or the decrease of their prey, merely increasing behind them. Predator and prey are each controlled by their food. However, the predators of each do override and depress the declining numbers of their prey to a lower level than they might otherwise have reached. There is just one difference – in northern Finland the pulses of food are regular and predictable, in arid Australia they are irregular and unpredictable.

These patterns of similarity are not confined to outbreaks of vertebrates. Insects, too, undergo cyclical and often synchronous outbreaks in these northern latitudes, while further south their numbers remain low and non-cyclic. They also exhibit all the other features common to outbreaks in all parts of the world (Tenow 1972; Baltensweiler 1985; Miller and Epstein 1986; Hanski 1987). Most authors recognize that food is important, but look to physical exhaustion of the supply. In the absence of any conclusive evidence, or of any other discernable cause, the action of conventional predators and parasitoids is often *assumed* to be responsible. The common appearance of outbreaks of viral diseases in collapsing populations has led to the proposal that these predators are responsible for the generation and synchronization of the collapse – presumably of an insect which would otherwise continue to increase until it had destroyed its food supply (Myers 1988; Shepherd et al. 1988). These explanations deny the evidence which reveals that outbreaks of predators, including diseases, are effects and not causes. Viruses, too, will explode in abundance if suddenly given access to an increased supply of "food" – the chemicals necessary for their replication (Sect. 14.5).

Chapter 18 The Influence of Weather on the Generation of Outbreaks

If the proximate cause of the beginning and end of outbreaks of animals is an increase and subsequent decrease in the supply of their food, what then drives these changes in the availability of the food? That there is synchrony of peaks in abundance over wide areas – not only of the one species, but of often quite dissimilar ones – suggests that a common extrinsic agent is responsible. That the same animals which reach outbreak numbers in harsh environments do not do so in gentler ones suggests that in these latter places the causative agent is either not operating or is much less variable in its action. The evidence from outbreaks which are not cyclic suggests that there are still other places where it acts only spasmodically.

Need it be the same factor? Not necessarily, although to date it would seem that all *widespread* outbreaks of animals, herbivores and carnivores, are generated by the influence of the weather upon the plants which the herbivores eat, causing a pulse of accessible nitrogen to flow up through the food chain.

Very small and local outbreaks are exceptions where the increase in food may not be due to weather. The examples with which I opened Chapter 15 illustrate this:

An individual branch of a *Eucalyptus* tree is attacked by scale insects because it senesces at an accelerated rate when it is overshadowed or outgrown by new growth on the tree. This hastens and enriches the outflow of nutrients in the phloem along the midribs of the leaves and down the twigs where the scales cluster and feed. Whole trees may similarly senesce more quickly if their roots are damaged or they happen to be growing on a particularly inhospitable micro-site.

The leaves on the brightly lit side of the kurrajong tree were photosynthesizing for longer, so produced more nutrients. However, they also senesced at a much faster rate than did the leaves on the rest of the tree, so were exporting those nutrients more quickly, to the benefit of psyllids otherwise unlikely to have found more than an occasional leaf producing enough food to sustain them.

I have previously discussed factors other than the weather, ranging from a deficiency of nutrients in the soil to irradiation, which may stress plants so as to similarly hasten the senescence of part or all of their tissues (White 1984).

The trees growing on the banks of the billabong had their roots flooded for prolonged periods at an inappropriate time. River red gums have evolved considerable tolerance to having their roots under water for long periods in the winter, but quickly deteriorate if they are flooded after growth has resumed in the spring. As any gardener knows, no plant can long tolerate the damage to its feeding roots which result from periods of alternating flood and drought. Like all forms of stress, death of these roots induces accelerated senescence and death of the aerial parts of the plant to restore the ratio of shoot to root. The consequently enriched supply of soluble nitrogen exported from those aerial parts provides the increased supply of good food for animals eating them.

On the broader scale, however, the massive outbreaks on many trees over large areas of land were associated with changes in the pattern of rainfall imposing similar stresses on the root systems of many trees at the one time. At this level the weather is directly responsible for the increased supply of food. In the billabong examples its influence is less direct because the flow of the rivers had been modified by human intervention.

I have discussed this relationship of outbreaks of herbivorous insects to the weather before (White 1969, 1974, 1976, 1986; see also Sects. 3.1 and 3.4.4). Since writing those papers I have found many more examples in the literature which suggest that this same weather–food link is the cause of widespread outbreaks of insects which feed upon trees. Their pattern is always the same as that of other outbreaks: a sudden explosion to a peak, often with extensive defoliation of the trees, followed by a "crash" in which predators (including diseases) usually play a prominent part. Frequently one species of insect erupts over wide areas at the same time, or more than one species of defoliator may erupt in synchrony.

18.1 Hot Spots Again: Outbreak Centres and Boundaries

Of particular significance are two linked phenomena which are common to most of these outbreaks of forest insects: trees growing immediately adjacent to heavily attacked ones, yet remaining immune to attack by the outbreak insect; and what are variously referred to as "hot spots", "outbreak centres", or "refuges". These latter are places where numbers first start to increase at the beginning of an outbreak, and where, when few or no insects can be found in most parts of the forest, a few may be found there. They are directly analogous to the hot spots discussed in Section 16.5.

In the first case quite sharp boundaries often separate "immune" and "susceptible" trees, which seem otherwise identical, and the distance between them may be remarkably short. Sometimes these boundaries are associated with an obvious physical boundary: a sharp change in soil type or differences in age of the trees. At other times it is more subtle: a change in drainage, slope, or aspect; or is not at all apparent.

Outbreak centres, on the other hand, seldom have precise boundaries, nor is any sort of physical barrier necessarily identifiable. Instead their limits spread as the outbreak increases, and the animals "invade" more and more land. But this is a false impression (Watt et al. 1989; Liebhold and Elkington 1990). The animals are not migrating from refuges into places where they were previously not present. They are present in all parts of their distribution, albeit in very small numbers and widely scattered. What is happening is that the places where many can survive, where before only few could do so, are increasing. This is happening because, as time passes, more plants are becoming a better source of food for the insects as more plants become stressed. Because of often slight and subtle differences in the quality of the site, however, it takes longer for roots to become stressed in one place than in another. This could be because there is a gradual shallowing of the soil in only some places. It could be because the rate at which moisture is lost or retained in the soil changes the further up a steep slope, or the further down a wet gully, the trees are growing. Typically, these centres of initially higher numbers are found at harsh sites where the food plants are likely to have their roots either physically constricted or killed by excessive wet or dry conditions (White 1974; Witter et al. 1975; Mason 1981; Stoszek et al. 1981; Larsson and Tenow 1984; Hanski and Parviainen 1985; Leather and Barbour 1987; Wallner 1987). If these trees then experience a period where the rainfall departs markedly from the norm, the stress that this change in the weather imposes upon those trees will vary in both space and time. This will be dictated by how long it takes for the soil in the different places where the trees are growing to become sufficiently wet or dry to have a significantly adverse effect upon the roots.

When, as in the first case, there is a stationary and persistent boundary beyond which trees are not attacked, this is likely to be because the "immune" trees are growing where the site is such that the imposed change in the weather has not been sufficient to create conditions in the soil which are stressful enough. A sharp change between two soils of different moisture-holding capacity could be enough (Stoszek et al. 1981; Watt 1986; Klimetzek 1990), or it may be that the soil is uniform throughout, but shallow. Then, as the trees grow, their root to shoot ratio becomes more and more unbalanced to the point where a sudden change in rainfall will seriously stress older larger trees, but not affect smaller younger ones growing nearby on the same soil. Nor would similar patterns of rainfall in the past have affected these older trees when they were younger. (Also, in such a situation, trees could grow to the stage where they become suitable for an outbreak to develop on them *without* the imposition of additional stress from the weather.)

Two related phenomena are trees being attacked by outbreak numbers when growing at the extremes of their natural range or when planted where they do not grow naturally. In the latter case the trees may thrive initially in such plantations, but become increasingly susceptible to stress from changing patterns of weather as they grow on a site to which they are not adapted. Furthermore, they may frequently be attacked by insects for which they are not a natural host (White 1974; Baltensweiler 1985; Watt 1986). In the former situation, however, the trees are, by definition, growing where they are at the limit of their tolerance to their environmental conditions, be this to temperature (altitude, latitude) or moisture. They are, therefore, most prone to be stressed by any change in that environment brought about by variation in the weather; so outbreaks are likely to be more frequent and severe in such places.

In some cases an outbreak "wave" may be generated. For example, an incipient drought would first affect trees where the boundary of their distribution is determined by their tolerance to low rainfall. As the drought intensified the outbreak would "move" to include wetter areas while becoming worse where it first started. Alternatively, it could be the result of *increased* rainfall first affecting poorly drained and flood-prone sites (Witter et al. 1975).

18.2 Spruce Budworm Outbreaks Revisited

The recorded history of both the eastern and western spruce budworms, *Choristoneura fumiferana,* and *C. occidentalis*, in Canada epitomizes what I have been saying (Blais 1983; Harris et al. 1985). Trees growing on stressful or potentially stressful sites are first and most heavily attacked during outbreaks (Hix et al. 1987) and outbreaks are associated with changes in the pattern of weather which exacerbate the stressfulness of those sites, and thus the favourability of the trees as food for young budworm larvae (Wellington 1952; White 1974; Thomson et al. 1984; Kemp and Moody 1984; Kemp et al. 1985; Shore and Alfaro 1986; Campbell 1989. And see sect. 3.4.4). The advent of such weather therefore produces a patchwork of gradually spreading areas of high numbers of insects at the beginning of an outbreak – and the false impression of invasion of uninfected areas from infested ones (Hardey et al. 1983).

Of particular interest, however, has been the response of the eastern species to changes wrought since the arrival of European settlers in North America (Blais 1983). *C. fumiferana*'s preferred food plant is the balsam fir, *Abies balsamea*, and it is most likely to reach outbreak numbers, and these outbreaks to be most severe, in climax forests which are predominantly mature and overmature balsam fir. Repeated defoliation during outbreaks eventually kills most of these senescing trees. Over the past 200 years the harvesting of forests, and their protection from fire, has increased the proportion and distribution of balsam fir. The result has been more frequent, more widespread, and longer lasting outbreaks.

However, it was the effects of the widespread use of synthetic insecticides against the great 1948–1958 budworm outbreaks in the forests of eastern Canada which revealed the interaction of the condition of the food plant with changes in weather patterns in generating these outbreaks. Aerial spraying of these forests with DDT did not necessarily stop the outbreaks, but it did prevent many millions of trees from being killed. Some outbreaks continued for more than 30 years, necessitating continued spraying to keep the trees alive. In other places, however, outbreaks stopped after 1958, even though the trees remained alive. The pattern of weather had changed, relieving the stress of drought, and allowing predators, parasites, and diseases to once again play a conspicuous part in the ecology of the budworm (Blais and Martineau 1960). Yet outbreaks started again after only 12 years – a shorter interval than ever before recorded.

In the first instance the trees which would have died without this intervention were now, obviously, sufficiently senescent to support outbreak numbers of budworm without any further stress. In the second instance, too, the trees would have died if they had not been sprayed during the time of stressful weather. But once that stress had passed, either or both their age and the inherent stressfulness of their site were necessary, but no longer sufficient conditions to support an outbreak. As they continued to age, however, they would have come ever closer to being old enough to be susceptible, and to the point where lesser and lesser deviations of weather would make their site stressful enough to trigger an outbreak. This seems to have happened.

In these and all similar situations, innumerable combinations and variations of changing patterns of weather interacting with variable sites are possible. The important point, however, is that in all there is a boundary, moving or stationary, between high and low numbers of a herbivore in an apparently uniform habitat. No physical hinderance can be found to the free movement of the herbivore (nor, note, to its predators) across this boundary. In this respect, these boundaries are no different from those found within the crown of a single tree; nor, I suggest, is the reason for the boundaries. The answer in each case is that where there is an improvement in the quality of the normally inadequate food of the herbivore, there the herbivore will be more abundant. It has nothing to do with the action of predators; they, too, will aggregate and increase where there is more food for them. Nor has it anything to do with the direct action of the physical environment upon the animals. It is the result of the action of the physical environment upon the plant. Again, consider the analogy of the box of wadding (Sect. 16.5).

A nice exception to this last point which proves the rule is that of the effect of very cold weather on outbreaks of the geometrid moth *Epirrita autumnata* in northern Fennoscandia (Tenow and Holmgren 1987). Extreme cold kills the eggs of this moth so that in very high exposed sites, and in valleys where the cold air pools, birch, *Betula pubescens tortuosa*, trees are not defoliated during an outbreak. Obviously this protection from attack has nothing to do with either the beginning or ending of the outbreaks.

As I pointed out in Chapter 15, post hoc investigations of an outbreak will not necessarily reveal what was happening. Without the additional imposed effect of the weather, or whatever the "trigger" was, there will be no difference. The differences between the sites are necessary, but not on their own sufficient to generate differences in survival. Once the outbreak is over there will no longer be any significant difference between the number, fecundity, survival, and speed of development of individuals in those places which recently held either many or few animals.

Unfortunately, these sorts of indirect relationships are not obvious, not readily apparent and, therefore, for many, not readily acceptable. It is also difficult to measure them or to manipulate them experimentally. In all of them, however, the pattern of peak and crash, first of the herbivore, then of its predators, is the same as it is in obviously weather-driven outbreaks. The difference is that in the latter we can *see* the physical change in the *amount* of food caused by the weather (green grass after rain, many fish in new flood waters). But there is no visible cause when the weather changes the chemistry of plants, or parts of plants, without any change in the number, size, or outward appearance of those plants. Then, until we have cause to look for these hidden changes, we will continue to seek elsewhere for the explanation of the response of the animals to them.

18.3 Patchy Environments and Metapopulations

The patchy and staggered responses of these defoliators and their predators to changes wrought by the interactions of the weather, the physical site, and the plants are not peculiar to insects attacking forests. They are but one example (and a good illustration) of what has long been understood and explained (Andrewartha and Birch 1954; Den Boer 1968), and is now discussed under the concept of "metapopulations" (Hanski 1991; Van Djik and Den Boer 1992). The environments of all organisms are patchy in both space and time and, as a consequence, the distribution and abundance of all organisms are forever changing. Studying any organism for more than one generation will reveal this. The dynamic and (for individual local populations) ephemeral nature of this constant change in the availability of adequate resources is perhaps best seen in the sea (Houde and Schekter 1978; see also Sect. 4.1.1) where survival depends on the number, volume, density, persistence, and movement of three-dimensional patches of food. Perhaps another thought experiment will help here.

We have an imaginary insect which completes its life cycle each generation on the foliage of trees which grow sparsely and randomly scattered in a savannah-like environment. Each tree supports an independent breeding population for an indefinite number of generations. Dispersal is by fertilized winged females. Most never find another tree and perish. Some start new populations on vacant trees. Others enter existing populations on another tree. The species is limited in its distribution by the weather: to the north by high temperatures, to the south by cold, to the west by lack of moisture, and to the east by the wet. It is therefore most abundant in the centre of its distribution where, for most of the time, a relatively dense population occupies every available tree. Towards each of the boundaries of its distribution the numbers in any one tree are usually lower, and some trees may often not have any insects living on them. Life is most risky, and extinction the norm, at the extremes. But seasonal, annual and long-term changes in the weather cause constant change over the whole of the insect's distribution. In addition, populations on individual trees may flourish or go extinct as the result of many other direct or indirect factors impinging on the insects or their food plants.

To visualize the total interplay between major and minor overall and local changes that may occur in the distribution and abundance of this insect, we model the entire area of its distribution as a large surface with a light bulb fixed at the position of each tree and connected to a separate rheostat. The presence and density of the population on each tree is represented by the brightness of the appropriate bulb (light out represents extinct; very bright represents outbreak population). From the result of field counts of many generations, each light is set to represent the state of its population, and the whole is photographed with a time-lapse camera. When this has been done for many generations and the resulting film is run at speed, a dynamic picture of the changing fortunes of the species emerges – one which would not be readily apparent in the short term. Some populations light up, glow briefly, and then disappear for varying times – some forever. Others persist, but wax and wane in intensity. Waves of light (or darkness) might sweep north or south, east or west across the area; small or large sections light up brightly, become very faint, or darken altogether. Light might be reduced to a few widely scattered and dim points, which then gradually grow and expand, merging with neighbours. Rarely, the entire area would be a blaze of light. The whole would present a pulsing, surging, constantly changing, yet persisting entity – the metapopulation comprised of hundreds of local populations.

18.4 The Role of Viral Diseases

At this point it is perhaps worthwhile harking back to Myers (1988) and Shepherd et al. (1988) (Chap. 17) who both concluded that cyclic outbreaks of forest Lepidoptera were caused by viral diseases. Both studies revealed the same picture. A remarkable degree of synchronous periodicity of outbreaks over wide areas, early increases in numbers tending not to be uniform or closely synchronized, but all populations crashing simultaneously and quickly. Early in the decline of an outbreak mortality is mostly among late instar larvae, while in the ensuing endemic populations it is mostly among the very young. Declines start with a decline in the reproductive capacity of individuals *in the peak population*. Conventional predators and parasitoids increase *after* the increase of their prey, but often override them in the decline. All these similarities between different outbreaks led the authors of both studies to the conclusion that while food and weather could clearly be important influences, the picture seemed to be one of populations continuing to build up until stopped by an epidemic of virus.

As I discussed earlier, however (Sect. 14.5), there is good evidence that outbreaks of viruses are effects of an increase in the supply of their "food" rather than causes of the end of outbreaks of the defoliators (by way of analogy nobody, I suspect, would suggest that, in Australia, the build-up of rabbits was the *cause* of the collapse of the outbreak of green grass, or the build-up of pelicans that of the fish they eat).

The sort of variable responses to changes in a heterogeneous environment that I have outlined above can equally well explain the uneven course of the build-up of numbers in different localities. A drop in the quality of the food of animals, on the other hand, is much more likely to be sudden, widespread, and simultaneous. The exhaustion of nutrients in plants following a mast year is a case in point; so, too, would be the response of trees to the ending of a period of stressful weather. While, for example, the widespread onset of a drought would be felt gradually and differentially across a mosaic of sites and age classes of trees, renewed wet weather would tend to remove this stress simultaneously from all sites. This would initiate the end of the supply of good food for the growing larvae; they would starve, heralding the end of the outbreak. At the same time, the stress of malnutrition of the larvae would trigger the multiplication of their latent virus, exacerbating and speeding the end. The fact that viruses have the capacity to multiply so much more rapidly than their victims would make their impact much more profound and sudden than that of other predators. In the subsequent low populations most mortality is again found in the early larvae which mostly starve on what is again a generally inadequate food, and the reproductive capacity of those that do survive to adulthood will again be low.

In the light of the similarity of the pattern of these insect outbreaks to those of so many other sorts of animals in which disease has not been implicated in their simultaneous collapse, this seems a much more parsimonious explanation. The contagious spread of a virus in a relatively short time to geographically separated populations of different densities which have been increasing for different times, seems a much less likely explanation than does the triggering of a ubiquitous but latent virus. It also provides an explanation for the variable beginnings of these outbreaks, which the virus hypothesis does not.

18.5 The Link to Climatic Oscillations

Ironically, it is the unequivocal evidence of the way in which the El Niño Southern Oscillation causes the *collapse* of populations which finally ties all the other evidence together. The suppression of the supply of nutrients in the oceanic upwellings generates a chain reaction of starvation, ending with many sea birds and mammals. We now know, however, that this same pattern of weather is simultaneously causing other populations to expand on a flush of food. In the Galapagos Islands, while the marine iguanids starve, heavy and prolonged rain enables the finches to breed as never before recorded. In Australia the pelicans, rabbits, kangaroos, locusts and their predators breed and then die in their millions in response to the exceptional rains and prolonged droughts now known to be driven by the Southern Oscillation. Truly "an ill wind that bloweth no man to good".

Moreover, there is growing evidence that the influence of the weather on changes in the distribution and abundance of organisms is even more pervasive and general than previously imagined. We are learning that the effects of the Southern Oscillation extend worldwide, but that the severity and timing of both the warm (El Niño) and cold (La Niña) phases differ in different places

(Vines and Tomlinson 1985; Bradley et al. 1987a). As a result, the Southern Oscillation influences events which are not synchronized or obviously linked to each other or the weather.

More detailed and accurate measurements of the weather, ever-lengthening and continuous meteorological records, and the accumulation of documented changes in the distribution and abundance of more and more organisms are providing a growing data base. From this I would predict that it will be possible in the not-too-distant future both to derive better correlations between these events, and to understand the causes of those correlations; for example, the number of outbreaks, declines, diebacks, and changes in agriculture that followed the (now abundantly obvious) sharp discontinuity in the weather in the 1940s (Folland et al. 1984; Pittock and Nix 1986; Bradley et al. 1987b; Whysall et al. 1987; Friis-Christensen and Lassen 1991).

To further advocate that changes in emissions from the Sun are responsible for the changes in the weather, and thus for changes in the biological world – and especially that the sunspot cycle is the cause of cyclic outbreaks – is to risk being labelled a crank. Over the years none of the numerous apparent correlations have stood up to statistical testing of their significance. Yet the logic is inescapable. All life on Earth is driven by energy from the Sun, so it might be expected that changes in the amount of that energy which is delivered to the surface of the Earth would generate changes in the success of various life forms on the Earth. It would also be expected that it is most likely to do this via changes in the weather. Probably the reason most attempts to establish these sorts of correlations in the past have failed is because the expectation was of a *direct* relationship between Sun cycles and biological ones. As our understanding of both organisms and weather advances it becomes increasingly obvious that this is seldom the case. There are many lags and buffers of varying duration and intensity between cause and effect. But there are some which appear to be quite direct, like the analysis by Vines (1977) of the correlation of the sunspot cycle with major droughts, yields of non-irrigated crops in Australia and Canada, and forest fires in Canada.*

Now there is gathering evidence that changes in the emission of energy from the Sun *are* having a direct effect upon the Earth's climate and weather. One such piece of evidence is the demonstration by Van Loon and Labitzke (1988) that when sunspot activity is at a maximum, the alternating east- and west-blowing stratospheric winds (the Quasi Biennial Oscillation) are in their west phase. The influence of these westerly winds on the atmospheric pressure and flow of air at sea level causes colder winters in eastern USA and milder weather in the Mediterranean. All this is reversed when the Sun is least active. However, the evidence is, of necessity, based on measurements made over only 20 years, and data from more years will need to accrue before the correlation can be substantiated (Kerr 1990). In the meantime, evidence is emerging of a link between

* *Note added in proof:* And of direct relevance to what I say here and in Sect. 6.2.5.1 are the just-published findings of Sinclair et al. (1993) that outbreaks of snowshoe hares in Canada are strongly correlated with the sunspot cycle.

the Quasi Biennial Oscillation and the El Niño Southern Oscillation (Angell 1992). The pieces are gradually fitting together.

Another strong piece of evidence is the association Friis-Christensen and Lassen (1991) have discovered which suggests that long-term variations in the Earth's temperature are closely associated with variations in the length of the solar cycle. This cycle had been lengthening since before the turn of the century, and the Northern Hemisphere had been warming in close synchrony with it. In the 1940s, however, the cycle began to shorten, and the temperatures to fall. In the 1970s both the cycle's length and the temperatures again started to climb in unison, and had continued to do so up to the time of their report. If *this* proves to be a correlation of true cause and effect, it would indicate that variation in solar activity is the single most important factor influencing changes in the weather on Earth.

18.6 Major Outbreaks Which Are Independent of the Weather

There are some organisms which are in a perpetual state of outbreak. Again, the proximate cause is an increase in the supply of their food, but it is a permanent increase and is not driven by the weather. These outbreaks are artifacts of human activity, and the organisms concerned are pests of modern agricultural and urban societies. All are enormously more abundant than in their natural state only because human activity of one sort or another has greatly increased their access to good food.

Let us take a species of wild plant. Breed from selected individuals to produce less variable, larger, more nutritious plants bearing more and larger fruit. Fertilize and water these plants to ensure that most of them survive, that their nutrient content is further enhanced, and their growth prolonged and increased. Grow them in dense and pure aggregations over large areas of land, and in places where they never grew naturally. Finally, process and/or store their seeds and other parts in large concentrations, as food reserves, buildings, and manufactured goods.

What will happen to organisms which evolved to feed on the different stages of such a plants, from the flush of its spring growth to its dry seeds or dead wood? Consider a market garden of cabbages for a *Pieris* butterfly accustomed to hunting for wild brassicas, or a silo of wheat for a weevil used to subsisting on grass seeds. They have physiologies and behaviours adapted to survival in an environment where their food is scattered, patchy, ephemeral, and/or of marginal nutritive value. They will respond as all other organisms do when a relative shortage of food is replaced with such abundance, they will breed and increase as long as the supply lasts – or until we kill them first!

Similarly, let us select, breed, feed, and multiply animals, and process and store parts of their bodies – meat, milk, hides, wool, and bone – then organisms evolved to subsist on limited access to these foods will similarly explode to pest status. Imagine what a warehouse full of cowhides, or a wall-to-wall carpet, represents to an insect evolved to persist on odd scraps of wool and skin on carcasses in the bush.

All this should cause those attempting the biological control of any pest to pause and think. I used to pose the following problem to students: In the year 2100 there are extensive plantations of "native" *Opuntia* cacti grown in Australia for the production of industrial ethanol fuel. The plants are fertilized and watered to speed their growth and maximize production. The major threat to this vital industry is continuous attack by the *Cactoblastis* moth. You are a consultant entomologist hired to find a way of controlling this pest. What would you recommend?

Those who would look for an explanation of the distribution and abundance of animals in nature from studies of pests should also pause and think. The observed responses of such animals (and their predators) cannot be used to interpret their ecology in nature, at least not until there is a clear understanding of what human manipulations have done to the original evolved relationships.

Similar caution is needed where the habitat of an otherwise wild animal is modified or manipulated to maintain its numbers for human use. The controlled burning of heather to increase the number of red grouse on the Scottish moors is a good example. Without it there would be much less good food for grouse, and many fewer grouse to shoot.

This is even more true for domesticated animals. Not only are they more abundant, they are quite different animals (and as individuals far better off!), living in an entirely different environment from that of their wild ancestors, or present-day relatives. Theirs is not an evolved relationship with humans, nor is it a mutualistic or parasitic one, it is simply an extension of human predation.

For all such animals the basic limitation with which they evolved has been removed – their environment is no longer so inadequate. In the case of pests of intensively cultivated plants, and of domesticated animals, their environment becomes (at least locally) fully adequate; for the managed wildlife it is less inadequate.

From the global viewpoint the conclusion must surely be the same. All kinds of outbreaks – all cases where a previously rare animal becomes abundant – have the same basic cause. All are the result of the inevitable evolved response of replicating life forms to increased access to nitrogenous food that can be used for successful reproduction and growth – to a lessening of the inadequacy of the environment.

References

Abrahamson WG, McCrea KD (1986) Nutrient and bioamass allocation in *Solidago altissima*: effect of two stem gallmakers, fertilization, and ramet isolation. Oecologia 68:174–180

Abrahamson WG, Weis AE (1987) Nutritional ecology of arthropod gallmakers. In: Slansky F Jr, Rodriguez JG (eds) Nutritional ecology of insects, mites, spiders, and related invertebrates. Wiley, New York, pp 235–258

Aebischer NJ (1986) Retrospective investigation of an ecological disaster in the shag, *Phalocrocorax aristotelis*: a general method based on long-term marking. J Anim Ecol 55:613–629

Agren GI (1985) Limits to plant production. J Theor Biol 113:89–92

Albon SD, Guinness FE, Cluttonbrock TH (1983) The influence of climatic variation on the birth weights of red deer calves. J Zool (Lond) 200:295–297

Albon SD, Clutton-Brock TH, Guinness FE (1987) Early development and population dynamics in red deer II Density-independent effects and cohort variation. J Anim Ecol 56:69–81

Al-Dabbagh KY, Block W (1981) Population ecology of a terrestrial isopod in two Breckland grass heaths. J Anim Ecol 50:61–77

Al-Zubaidi FS, Capinera JL (1983) Application of different nitrogen levels to the host plant and cannibalistic behaviour of beet army-worm *Spodoptera exigua* (Hubner) (Lepidoptera, Noctuidae). Environ Entomol 12:1687–1689

Al-Zubaidi FS, Capinera JL (1984) Utilization of food and nitrogen by the beet army-worm, *Spodoptera exigua* (Hubner) (Lepidoptera: Noctuidae), in relation to food type and dietary nitrogen levels. Environ Entomol 13:1604–1608

Andersen PC, Mizell RF (1987) Physiological effects of galls induced by *Phylloxera notabilis* (Homoptera Phylloxeridae) on pecan foliage. Environ Entomol 16:264–268

Andersen PC, Brodbeck BV, Mizell RF (1989) Metabolism of amino acids, organic acids and sugars extracted from the xylem fluid of four host plants by adult *Homalodisca coagulata*. Entomol Exp Appl 50:149–159

Andersen PC, Brodbeck BV, Mizell RF (1992) Feeding by the leafhopper, *Homalodisca coagulata* in relation to xylem fluid chemistry and tension. J Insect Physiol 38:611–622

Andersen R, Pedersen HC, Steen JB (1986) Annual variation in movements of subalpine hatched willow ptarmigan *Lagopus lagopus lagopus* broods in central Norway. Ornis Scand 17:180–182

Anderson DW, Gress F, Mais KF (1982) Brown pelicans – influence of food supply on reproduction. Oikos 39:23–31

Anderson JF (1974) Responses to starvation in the spiders *Lycosa lenta* Hentz and *Filistata hibernalis* (Hentz). Ecology 55:576–585

Anderson NH (1976) Carnivory by an aquatic detritivore, *Clistoronia magnifica* (Trichoptera: Limnephilidae). Ecology 57:1081–1085

Andow DA (1984) Microsite of the green rice leafhopper, *Nephotettix cincticeps* (Homoptera: Cicadellidae), on rice: plant nitrogen and leafhopper density. Res Popul Ecol 26:313–329

Andreev A (1988) The ten year cycle of the willow grouse of lower Kolyma. Oecologia 76:261–267

Andrew NL, Jones GP (1990) Patch formation by herbivorous fish in a temperate Australian kelp forest. Oecologia 85:57–68
Andrewartha HG (1970) Introduction to the study of animal populations, 2nd edn. Methuen, London
Andrewartha HG, Birch LC (1954) The distribution and abundance of animals. University of Chicago Press, Chicago
Andrewartha HG, Birch LC (1984) The ecological web. More about the distribution and abundance of animals. University of Chicago Press, Chicago
Andrewartha HG, Browning TO (1961) An analysis of the idea of "resources" in animal ecology. J Theor Biol 1:83–97
Angell JK (1992) Evidence of a relation between El Niño and QBO, and for an El Niño in 1991–92. Geophys Res Lett 19:285–288
Angelstam P, Lindstrom E, Widen P (1984) Role of predation in short-term population fluctuations of some birds and mammals in Fennoscandia. Oecologia 62:199–208
Angelstam P, Lindstrom E, Widen P (1985) Synchronous short-term population fluctuations of some birds and mammals in Fennoscandia – occurrence and distribution. Holarct Ecol 8:285–298
Angerbjorn A, Arvidson B, Noren E, Stromgren L (1991) The effect of winter food on reproduction in the arctic fox, *Alopex lagopus*: a field experiment. J Anim Ecol 60:705–714
Armitage KB (1988) Resources and social organization of ground-dwelling squirrels. In: Slobodchikoff CN (ed) Ecology of social behaviour. Academic Press, San Diego, pp 131–155
Atsatt PR, Ingram T (1983) Adaptation to oak and other fibrous, phenolic-rich foliage by a small mammal, *Neotoma fuscipes*. Oecologia 60:135–142
Auclair JL (1964) Recent advances in the feeding and nutrition of aphids. Can Entomol 96:241–249
Auclair JL, Maltais JB (1950) Studies on the resistance of plants to aphids by the method of paper partition chromatography. Can Entomol 82:175–176
Auerbach M, Simberloff D (1985) Responses of leaf miners to atypical leaf production patterns. Ecol Entomol 9:361–367
Auerbach M, Simberloff D (1989) Oviposition site preference and larval mortality in a leaf-mining moth. Ecol Entomol 14:131–140
Austin B, Al-Zahrani AMJ (1988) The effect of antimicrobial compounds on the gastrointestinal microflora of rainbow trout, *Salmo gairdneri* Richardson. J Fish Biol 33:1–14
Ayres MP, Suomela J, MacLean SF Jr (1987) Growth performance of *Epirrita autumnata* (Lepidoptera: Geometridae) on mountain birch: trees, broods, and tree×brood interactions. Oecologia 74:450–457
Bages M, Sloane L (1981) Effects of dietary protein and starch levels on growth and survival of *Penaeus monodon* post larvae. Aquaculture 25:117–128
Baker LM, Laidlaw RA, Smith GA (1970) Wood breakdown and nitrogen utilization by *Anobium punctatum* DeG. feeding on Scots pine sapwood. Holzforschung 24:45–54
Baker RR (1983) Insect territoriality. Annu Rev Entomol 28:65–89
Balch RE (1952) Studies on the balsam woolly aphid *Adelges piceae* (Ratz.) and its effects on balsam fir *Abies balsamea* (L.) Mill. Can Dept Agric Publ 867, Ottawa, pp 76
Baltensweiler W (1985) On the extent and the mechanisms of the outbreaks of the larch budmoth (*Zeiraphera diniana* Gn., Lepidoptera, Tortricidae) and its impact on the subalpine larch – cembran pine forest ecosystem. Proc 3rd IUFRO Worksh, Berlin, pp 215–219
Barbehenn RV (1992) Digestion of uncrushed leaf tissues by leaf-snipping larval Lepidoptera. Oecologia 89:229–235
Barbour DA (1988) The pine looper in Britain and Europe. In: Berryman AA (ed) Dynamics of forest insect populations. patterns, causes, implications. Plenum, New York, pp 291–308
Barrett RT, Ankernilssen T, Rikardsen F, Valde K, Rov N, Vader W (1987) The food, growth and fledgling success of Norwegian puffin chicks *Fratercula arctica* in 1980–1983. Ornis Scand 18:73–83

Bath DL, Williams CW, Torell DT (1956) The use of the oesophageal fistula for the determination of consumption and digestibility of pasture forage by sheep. J Anim Sci 15:1166–1171
Batzli GO (1986) Nutritional ecology of the Californian vole: effects of food quality on reproduction. Ecology 67:406–412
Batzli GO, Lesieutre C (1991) The influence of high quality food on habitat use by arctic microtine rodents. Oikos 60:299–306
Bauer LS, Nordin GL (1988) Nutritional physiology of the eastern spruce budworm, *Choristoneura fumiferana*, infected with *Nosema fumiferanae*, and interactions with dietary nitrogen. Oecologia 77:44–50
Baur B (1987) Effect of early feeding experience and age on the cannibalistic propensity of the land snail *Arianta arbustorum*. Can J Zool 65:3068–3070
Baur B (1988a) Egg-species recognition in cannibalistic hatchlings of the land snails *Arianta arbustorum* and *Helix pomatia*. Experientia 44:276–277
Baur B (1988b) Age specific food preferences in hatchlings of *Helix pomatia* (L.). Snail Farm Res 2:14–19
Baur B (1990) Possible benefits of egg cannibalism in the land snail *Arianta arbustorum* (L). Funct Ecol 4:679–684
Baur B (1992) Cannibalism in gastropods. In: Elgar MA, Crespi RJ (eds) Cannibalism, ecology and evolution among diverse taxa. Oxford University Press, Oxford, pp 102–127
Baur B, Baur A (1986) Proximate factors influencing egg cannibalism in the land snail *Arianta arbustorum* (Pulmonata Helicidae). Oecologia 70:283–287
Bazely DR, Jefferies RL (1989) Lesser snow geese and the nitrogen economy of a grazed salt marsh. J Ecol 77:24–34
Bazely DR, Ewins PJ, McCleery RH (1991) Possible effects of local enrichment by gulls on feeding-site selection by wintering barnacle geese *Branta leucopsis*. Ibis 133:111–114
Beatley JC (1969) Dependence of desert rodents on winter annuals and precipitation. Ecology 50:721–724
Beck SD (1956) The European corn borer, *Pyrausta nubilalis* (Hubn.), and its principal host plant. II The influence of nutritional factors on larval establishment and development on the corn plant. Ann Entomol Soc Am 49:582–588
Becker G (1990) Comparison of the dietary composition of epilithic trichopteran species in a 1st order stream. Arch Hydrobiol 120:13–40
Beckerton PR, Middleton ALA (1982) Effects of dietary protein levels on ruffed grouse reproduction. J Wildl Manage 46:569–579
Beeson M (1987) The origins of bark-stripping by blue monkeys (*Cercopithecus mitis*): implications for management. Zool J Linn Soc 91:265–291
Beeson M (1989) Seasonal dietary stress in a forest monkey (*Cercopithecus mitis*). Oecologia 78:565–570
Belanger J, Bergeron JM (1987) Can alkaloids and tannins influence food choice of meadow voles (*Microtus pennsylvanicus*)? Can J Zool 65:375–378
Bentley BL (1984) Nitrogen fixation in termites: fate of newly fixed nitrogen. J Insect Physiol 30:653–655
Berger PJ, Sanders EH, Gardner PD, Negus NC (1977) Phenolic plant compounds functioning as reproductive inhibitors in *Microtus montanus*. Science 195:575–577
Berger PJ, Negus NC, Sanders EH, Gardner PD (1981) Chemical triggering of reproduction in *Microtus montanus*. Science 214:69–70
Bergeron JM, Jodoin L (1987) Defining "high quality" food resources of herbivores: the case for meadow voles (*Microtus pennsylvanicus*). Oecologia 71:510–517
Bergeron JM, Jodoin L (1989) Patterns of resource use, food quality, and health status of voles (*Microtus pennsylvanicus*) trapped from fluctuating populations. Oecologia 79:306–314
Bergerud AT (1987) Reply to Watson and Moss (1987). Can J Zool 65:1048–1050
Bergerud AT, Gratson MW (eds) (1988) Adaptive strategies and population ecology of northern grouse. University of Minnesota Press, Minneapolis

Bergerud AT, Mossop DH, Myrberget S (1985) A critique of the mechanisms of annual changes in ptarmigan numbers. Can J Zool 63:2240–2248

Bernays EA, Barbehenn R (1987) Nutritional ecology of grass foliage-chewing insects. In: Slansky F, Rodriguez JG (eds) Nutritional ecology of insects, mites, spiders, and related invertebrates. Wiley, New York, pp 147–176

Bernays EA, Lewis AC (1986) The effect of wilting on palatability of plants to *Schistocerca gregaria*, the desert locust. Oecologia 70:132–135

Bidwell RGS (1983) Carbon nutrition of plants: photosynthesis and respiration. In: Steward FC, Bidwell RGS (eds) Plant physiology a treatise vol VII. Energy and carbon metabolism. Academic Press, New York, pp 287–457

Bintz GL, Strand CE (1983) Nitrogen catabolism during starvation and starvation with water deprivation in Richardson's ground squirrels. J Comp Physiol B 149:565–572

Bintz GL, Torgerson GE (1981) The metabolism of [^{14}C]urea by control and starved Richardson's ground squirrels. Comp Biochem Physiol 69A:551–555

Birkeland C, Neudecker S (1981) Foraging behaviour of two Caribbean chaetodontids; *Chaetodon capistratus* and *C. aculeatus*. Copeia 1981:169–178

Bjorndal KA (1979) Cellulose digestion and volatile fatty acid production in the green turtle, *Chelonia mydas*. Comp Biochem Physiol 63A:127–133

Bjorndal KA (1980) Nutrition and grazing behaviour of the green turtle *Chelonia mydas*. Mar Biol 56:147–154

Bjorndal KA (1985) Nutritional ecology of sea turtles. Copeia 1985:736–751

Bjornhag G, Sjoblom L (1977) Demonstration of coprophagy in some rodents. Swed J Agric Res 7:105–113

Bjornhag G, Sperber I (1977) Transport of various food components through the digestive tract of turkeys, geese and guinea fowl. Swed J Agric Res 7:57–66

Blais JR (1983) Trends in the frequency, extent, and severity of spruce budworm outbreaks in eastern Canada. Can J For Res 13:539–547

Blais JR, Martineau R (1960) A recent spruce budworm outbreak in the lower St Lawrence and Gaspe with reference to aerial spraying operations. For Chron 36:209–224

Blank TH, Southwood TRE, Cross DJ (1967) The ecology of the partridge. I Outline of population processes with particular reference to chick mortality and nest density. J Anim Ecol 36:549–556

Bletchly JD (1969) Seasonal differences in nitrogen content of scots pine (*Pinus sylvestris*) sapwood and their effects on the development of the larvae of the common furniture beetle (*Anobium punctatum* DeG.). J Inst Wood Sci 4:43–47

Blockstein DE (1989) Crop milk and clutch size in mourning doves. Wilson Bull 101:11–25

Bloem KA, Duffey SS (1990) Effect of protein type and quantity on growth and development of larval *Heliothis zea* and *Spodoptera exigua* and the endoparasitoid *Hyposoter exiguae*. Entomol Exp Appl 54:141–148

Boag DA, Schroeder MA (1987) Population fluctuations in spruce grouse: what determines their numbers in spring? Can J Zool 65:2434–2435

Boag PT, Grant PR (1984) Darwin's finches (*Geospiza*) on Isla Daphne Major, Galapagos; breeding and feeding ecology in a climatically variable environment. Ecol Monogr 54:463–489

Boersma PD (1987) El Niño behind penguin deaths? Nature 327:96

Bomford M (1987a) Food and reproduction of wild house mice. I. Diet and breeding seasons in various habitats on irrigated cereal farms in New South Wales. Aust Wildl Res 14:183–196

Bomford M (1987b) Food and reproduction of wild house mice. III. Experiments on the breeding performance of caged house mice fed rice-based diets. Aust Wildl Res 14:207–218

Bomford M, Redhead T (1987) A field experiment to examine the effect of food quality and population density on reproduction of wild house mice. Oikos 48:304–311

Boonstra R, Boag PT (1987) A test of the Chitty hypothesis: inheritance of life history traits in meadow voles *Microtus pennsylvanicus*. Evolution 41:929–947

Boonstra R, Krebs CJ, Kanter M (1990) Arctic ground squirrel predation on collared lemmings. Can J Zool 68:757–760

Booth RG, Anderson JM (1979) The influence of fungal food quality on the growth and fecundity of *Folsomia candida* (Collembola: Isotomidae). Oecologia 38:317–324

Bosman AL, Hockey PAR (1986) Seabird guano as a determinant of rocky intertidal community structure. Mar Ecol Prog Ser 32:247–257

Bosman AL, Hockey PAR (1988a) Life history patterns of populations of the limpet *Patella granularis*: the dominant roles of food supply and mortality rate. Oecologia 75:412–419

Bosman AL, Hockey PAR (1988b) The influence of primary production rate on the population dynamics of *Patella granularis*, an intertidal limpet. Mar Ecol 9:181–198

Bosman AL, Hockey PAR, Underhill LG (1989) Oystercatcher predation and limpet mortality: the importance of refuges in enhancing the reproductive output of prey populations. Veliger 32:120–129

Bosquet G (1977) Haemolymph modifications during starvation in *Philosamia cynthia walkeri* (Farber). II. Amino acids and peptides. Comp Biochem Physiol A 58:377–382

Bottger GT (1951) Sugar and protein in the corn plant as related to nutrition of the European corn borer. J Econ Entomol 44:40–44

Boutton TW, Cameron GN, Smith BN (1978) Insect herbivory on C_3 and C_4 grasses. Oecologia 36:21–32

Boutton TW, Smith BN, Harrison AT (1980) Carbon isotope ratios and crop analysis of *Arphia* (Orthoptera: Acrididae) species in southeastern Wyoming grassland. Oecologia 45:299–306

Boutton TW, Tieszen LL, Imbamba SK (1988) Seasonal changes in the nutrient content of East African grassland vegetation. Afr J Ecol 26:103–116

Bowen SH (1976) Mechanisms for digestion of detrital bacteria by the cichlid fish *Sarotherodon mossambicus* (Peters). Nature 260:137–138

Bowen SH (1979) A nutritional constraint in detritivory by fishes: the stunted population of *Sarotherodon mossambicus* in Lake Sibaya, South Africa. Ecol Monogr 49:17–32

Bowen SH (1980) Detrital nonprotein amino acids are the key to rapid growth of *Tilapia* in Lake Valencia, Venezuela. Science 207:1216–1218

Bowen SH (1981) Digestion and assimilation of periphytic detrital aggregate by *Tilapia mossambica*. Trans Am Fish Soc 110:239–245

Bowen SH (1984) Differential habitat utilization by sexes of *Sarotherodon mossambicus* in Lake Valencia, Venezuela: significance of fitness. J Fish Biol 24:115–121

Bowen SH (1987) Composition and nutritional value of detritus. In: Moriarty DJW, Pullins RSV (eds) Detritus and microbial ecology in aquaculture. ICLARM Conf Proc 14, Int Centre Living Aquat Res Manage, Manila, Philippines, pp 192–216

Boyd CE, Goodyear CP (1971) Nutritive quality of food in ecological systems. Arch Hydrobiol 69:256–270

Bradley RS, Diaz HF, Kiladis GN, Eischeid JK (1987a) ENSO signal in continental temperature and precipitation records. Nature 327:497–501

Bradley RS, Diaz HF Eischeid JK, Jones PD, Kelly PM, Goodess CM (1987b) Precipitation fluctuations over Northern Hemisphere land areas since the mid-19th century. Science 237:171–175

Braithwaite LW (1981) Ecological studies of the black swan II. Colour and plumage changes, growth rates, sexual maturation and timing and frequency of breeding in captivity. Aust Wildl Res 8:121–133

Braithwaite LW (1982) Ecological studies of the black swan IV. The timing and success of breeding on two nearby lakes on the Southern Tablelands of New South Wales. Aust Wildl Res 9:261–275

Braithwaite LW (1983) Studies on the arboreal marsupial fauna of eucalypt forests being harvested for woodpulp at Eden, NSW. I. The species and distribution of animals. Aust Wildl Res 10:219–229

Braithwaite LW (1984) The identification of conservation areas for possums and gliders within the Eden pulpwood concession district. In: Smith A, Hume I (eds) Possums and gliders. Aust Mamm Soc, Sydney, pp 501–508

Braithwaite LW, Dudzinki ML, Turner J (1983) Studies on the arboreal marsupial fauna of eucalypt forests being harvested for woodpulp at Eden, NSW. II. Relationships between the fauna density, richness and diversity, and measured variables of the habitat. Aust Wildl Res 10:231–247

Braithwaite LW, Turner J, Kelly J (1984) Studies on the arboreal marsupial fauna of eucalypt forests being harvested for woodpulp at Eden, NSW. III. Relationships between faunal densities, eucalypt occurrence and foliage nutrients, and soil parent materials. Aust Wildl Res 11:41:48

Brand CJ, Keith LB (1979) Lynx demography during a snowshoe hare decline in Alberta. J Wildl Manage 43:827–849

Brewer JW, Capinera JL, Deshon RE Jr, Walmsley ML (1985) Influence of foliar nitrogen levels on survival, development and reproduction of western spruce budworm *Choristoneura occidentalis* (Lepidoptera: Tortricidae). Can Entomol 117:23–32

Brewer JW, Bishop JN, Skuhravy V (1987a) Levels of foliar chemicals in insect-induced galls (Dipt, Cecidomyiidae). J Appl Entomol 104:504–509

Brewer JW, O'Neill KM, Deshon RE Jr (1987b) Effects of artificially altered foliar nitrogen levels on development and survival of young instars of western spruce budworm, *Choristoneura occidentalis* Freeman. J Appl Entomol 104:121–130

Breznak JA (1982) Intestinal microbiota of termites and other xylophagous insects. Annu Rev Microbiol 36:323–343

Breznak JA (1984) Biochemical aspects of symbiosis between termites and their intestinal microbiota. In: Anderson JM, Rayner ADM, Walton DWH (eds) Invertebrate – microbial interactions. Cambridge University Press, Cambridge, pp 187–203

Brice AT, Dahl KH, Grau CR (1989) Pollen digestibility by hummingbirds and psittacines. Condor 91:681–688

Brittas R (1988) Nutrition and reproduction of the willow grouse *Lagopus lagopus* in central Sweden. Ornis Scand 19:49–57

Brodbeck BV, Mizell RF, French WJ, Andersen PC, Aldrich JH (1990) Amino acids as determinants of host preference for the xylem feeding leafhopper, *Homalodisca coagulata* (Homoptera: Cicadellidae). Oecologia 83:338–345

Brodbeck BV, Mizell RF, Andersen PC (1993) Physiological and behavioural adaptations of three species of leafhoppers in response to the dilute nutrient content of xylem fluid. J Insect Physiol 39:73–82

Brody MS, Lawlor LR (1984) Adaptive variation in offspring size in the terrestrial isopod, *Armadillidium vulgare*. Oecologia 61:55–59

Brooker MG, Ridpath MG (1980) The diet of the wedge-tailed eagle, *Aquila audax*, in Western Australia. Aust Wildl Res 7:433–452

Brooks MA (1963) Symbiosis and aposymbiosis in arthropods. Symbiotic associations. Symp Soc Gen Microbiol 13:200–231

Brooks MH, Stark RW, Campbell RW (eds) (1978) The Douglas fir tussock moth: a synthesis. USDA For Serv Tech Bull 1585, USDA, Washington, DC

Brough CN, Dixon AFG, Kindlmann P (1990) Patterns of growth and fat content of somatic and gonadal tissues of virginoparae of the vetch aphid, *Megoura viciae*. Entomol Exp Appl 56:269–275

Brown JL (1987) Helping and communal breeding in birds: ecology and evolution. Princeton University Press, Princeton

Brown LH, Urban EK (1969) The breeding biology of the great white pelican *Pelecanus onocrotalus roseus* at Lake Shala, Ethiopia. Ibis 111:199–237

Browning TO (1963) Animal populations. Hutchinson, London

Bryant JP, Kuropat PJ (1980) Selection of winter forage by subarctic browsing vertebrates: the role of plant chemistry. Annu Rev Ecol Syst 11:261–285

Buchner P (1965) Endosymbiosis of animals with plant microorganisms. Interscience, London

Buchsbaum R, Valiela I (1987) Variability in the chemistry of estuarine plants and its effect on feeding by Canada geese. Oecologia 73:146–153

Buchsbaum R, Wilson J, Valiela I (1986) Digestibility of plant constituents by Canada geese and Atlantic brant. Ecology 67:386–393
Butler NM, Suttle CA, Neill WE (1989) Discrimination by freshwater zooplankton between single algal cells differing in nutritional status. Oecologia 78:368–372
Byron ER, Folt CL, Goldman CR (1984) Copepod and cladoceran success in an oligotrophic lake. J Plankton Res 6:45–65
Cable DR, Shumway RP (1966) Crude protein in rumen contents and in foliage. J Range Manage 19:124–128
Calow P (1973) The food of *Ancylus fluviatilis* a littoral stone-dwelling herbivore. Oecologia 13:113–133
Calow P (1974) Evidence for bacterial feeding in *Planorbis contortus* Linn (Gastropoda: Pulmonata). Proc Malacol Soc Lond 41:145–156
Calvert JJ (1985) Food selection by western gorillas (*Gorilla gorilla gorilla*) in relation to food chemistry. Oecologia 65:236–246
Campbell IM (1989) Does climate affect host-plant quality? Annual variation in the quality of balsam fir as food for spruce budworm. Oecologia 81:341–344
Campbell RW (1976) Comparative analysis of numerically stable and violently fluctuating gypsy moth populations. Environ Entomol 5:1218–1224
Cannon WN, Ortega CA (1966) Studies of *Ostrinia nubilalis* larvae (Lepidoptera: Pyraustidae) on corn plants supplied with various amounts of nitrogen and phosphorus. I. Survival. Ann Entomol Soc Am 59:631–638
Carefoot TH (1984) Nutrition and growth of *Ligia pallasii*. Symp Zool Soc Lond 53:455–467
Cargill SM, Jefferies RL (1984) The effects of grazing by lesser snow geese on the vegetation of a sub-arctic salt marsh. J Appl Ecol 21:669–686
Carpenter EJ, Culliney JL (1975) Nitogen fixation in marine shipworms. Science 187: 551–552
Carpenter EJ, Romans K (1991) Major role of the cyanobacterium *Trichodesmium* in nutrient cycling in the north Atlantic Ocean. Science 254:1356–1358
Carpenter FL, Castronova JL (1980) Maternal diet selection in *Calypte anna*. Am Midl Nat 103:175–179
Carr A (1984) Mystery of the missing year. Sciences 24:44–49
Carr A (1987) New perspectives on the pelagic stage of sea turtle development. Conserv Biol 1:103–121
Caswell H, Reed FC (1976) Plant – herbivore interactions. The indigestibility of C_4 bundle sheath cells by grasshoppers. Oecologia 26:151–156
Caswell H, Reed FC, Stephenson SN, Werner PA (1973) Photosynthetic pathways and selective herbivory: a hypothesis. Am Nat 107:465–480
Cates RG, Henderson CB, Redak RA (1987) Responses of the western spruce budworm to varying levels of nitrogen and terpenes. Oecologia 73:312–316
Catling HD (1971) The bionomics of the South African citrus psylla, *Trioza erytreae* (Del Guercio) (Homoptera: Psyllidae) 5. The influence of host plant quality. J Entomol Soc South Afr 34:381–391
Caughley G (1966) Mortality patterns in mammals. Ecology 47:906–918
Caughley G (1987) Ecological relationships. In: Caughley G, Shepherd N, Short J (eds) Kangaroos, their ecology and management in the sheep rangelands of Australia. Cambridge University Press, Cambridge, pp 159–187
Caughley G, Grigg GC, Smith L (1985) The effect of drought on kangaroo populations. J Wildl Manage 49:679–685
Checkley DM (1985) Nitrogen limitation of zooplankton production and its effect on the marine nitrogen cycle. Arch Hydrobiol Beih 21:103–113
Chew FS (1981) Coexistence and local extinction in two pierid butterflies. Am Nat 118:655–672
Chilcott MJ, Hume ID (1984a) Digestion of *Eucalyptus andrewsii* foliage by the common ringtail possum, *Pseudocheirus peregrinus*. Aust J Zool 32:605–613

Chilcott MJ, Hume ID (1984b) Nitrogen and urea metabolism and nitrogen requirements of the common ringtail possum, *Pseudocheirus peregrinus*, fed *Eucalyptus andrewsii* foliage. Aust J Zool 32:615–622
Chilcott MJ, Hume ID (1985) Coprophagy and selective retention of fluid digesta: their role in the nutrition of the common ringtail possum, *Pseudocheirus peregrinus*. Aust J Zool 33:1–15
Chitty D (1960) Population processes in the vole and their relevance to general theory. Can J Zool 38:99–113
Choquenot D (1991) Density-dependent growth, body condition, and demography in feral donkeys: testing the food hypothesis. Ecology 72:805–813
Christian KA, Tracy CR, Porter WP (1984) Diet, digestion, and food preferences of Galapagos land iguanas. Herpetologica 40:205–212
Churchill DM, Christensen P (1970) Observations on pollen harvesting by brush-tongued lorikeets. Aust J Zool 18:427–437
Clancy KM, Wagner MR, Tinus RW (1988) Variations in host foliage nutrient concentrations in relation to western spruce budworm herbivory. Can J For Res 18:530–539
Claridge MF, Wilson MR (1978) Seasonal changes and alteration of food plant preference in some mesophyll-feeding leafhoppers. Oecologia 37:247–255
Clark DA (1980) Age- and sex-dependent foraging strategies of a small mammalian omnivore. J Anim Ecol 49:549–563
Clark LR, Dallwitz MJ (1974) On the relative abundance of some Australian psyllidae that coexist on *Eucalyptus blakelyi*. Aust J Zool 22:387–415
Clark LR, Dallwitz MJ (1975) The life system of *Cardiaspina albitextura* (Psyllidae) 1950–74. Aust J Zool 23:523–561
Clements KD, Sutton DC, Choat JH (1989) Occurrence and characteristics of unusual protistan symbionts from surgeon fishes (Acanthuridae) of the Great Barrier Reef, Australia. Mar Biol 102:403–412
Clutton-Brock TH, Albon SD, Guinness FE (1986) Great expectations: dominance, breeding success and offspring ratios in red deer. Anim Behav 34:460–471
Clutton-Brock TH, Major M, Albon SD, Guinness FE (1987) Early development and population dynamics in red deer I. Density-dependent effects on juvenile survival. J Anim Ecol 56:53–67
Cockburn A (1981a) Population regulation and dispersion of the smoky mouse, *Pseudomys fumeus* I. Dietary determinants of microhabitat preferences. Aust J Ecol 6:231–254
Cockburn A (1981b) Population regulation of the smoky mouse, *Pseudomys fumeus* II. Spring decline, breeding success and habitat heterogeneity. Aust J Ecol 6:255–266
Cockburn A (1988) Social behaviour in fluctuating populations. Croom Helm, London
Cockburn A, Lidicker WZ (1983) Microhabitat heterogeneity and population ecology of an herbivorous rodent, *Microtus californicus*. Oecologia 59:167–177
Cockfield SD, Potter DA, Houtz RL (1987) Chlorosis and reduced CO_2 assimilation of *Euonymus fortunei* infested with euonymus scale (Homoptera: Diaspididae). Environ Entomol 16:1314–1318
Coe MJ, Bourn DM, Swingland IR (1979) The biomass, production and carrying capacity of giant tortoises on Aldabra. Philos Trans R Soc Lond B 286:163–176
Cole JJ, Findlay S, Pace ML (1988) Bacterial production in fresh and saltwater ecosystems: a cross-system overview. Mar Ecol Prog Ser 43:1–10
Coleman JD (1974) Breakdown rates of foods ingested by starlings. J Wildl Manage 38:910–912
Coleman TS, Boag DA (1987) Foraging characteristics of Canada geese on the Nisutlin river delta, Yukon. Can J Zool 65:2358–2361
Collins NM (1983) The utilization of nitrogen resources by termites (Isoptera). Symp Br Ecol Soc 22:381–412
Connor MS, Edgar RK (1982) Selective grazing by the mud snail *Ilyanassa obsoleta*. Oecologia 53:271–275
Conover RJ (1968) Zooplankton – life in a nutritionally dilute environment. Am Zool 8:107–118

Corbett LK, Newsome AE (1987) The feeding ecology of the dingo. III Dietary relationships with widely fluctuating prey populations in arid Australia: an hypothesis of alternation of predation. Oecologia 74:215–227
Cords M (1986) Interspecific and intraspecific variations in diet of two guenons, *Cercopithecus ascanius* and *C. mitis*. J Anim Ecol 55:811–827
Cork SJ (1986) Foliage of *Eucalyptus punctata* and the maintenance nitrogen requirements of koalas, *Phascolarctos cinereus*. Aust J Zool 34:17–24
Cork SJ, Hume ID (1983) Microbial digestion in the koala (*Phascolarctos cinereus*, Marsupiala) an arboreal folivore. J Comp Physiol B 152:131–136
Cork SJ, Kenagy GJ (1989) Nutritional value of hypogeous fungus for a forest-dwelling ground squirrel. Ecology 70:577–586
Cork SJ, Pahl L (1984) The possible influence of nutritional factors on diet and habitat selection by the ringtail possum. In: Smith A, Hume ID (eds) Possums and gliders. Aust Mamm Soc, Sydney, pp 269–276
Cork SJ, Warner ACI (1983) The passage of digesta markers through the gut of a folivorous marsupial, the koala *Phascolarctos cinereus*. J Comp Physiol B 152:43–51
Cork SJ, Hume ID, Dawson TJ (1983) Digestion and metabolism of a natural foliar diet (*Eucalyptus punctata*) by an arboreal marsupial, the koala (*Phascolarctos cinereus*). J Comp Physiol B 153:181–190
Courtice AC, Drew RAI (1984) Bacterial regulation of abundance in tropical fruit flies (Diptera: Tephritidae). Aust Zool 21:251–268
Courtney SP, Duggan AE (1983) The population biology of the orange tip butterfly, *Anthocharis cardamines*, in Britain. Ecol Entomol 8:271–281
Cowgill UM, Williams DM, Esquivel JB (1984) Effects of maternal nutrition on fat content and longevity of neonates of *Daphnia magna*. J Crust Biol 4:173–190
Cowles TJ, Olson RJ, Chisholm SW (1988) Food selection by copepods: discrimination on the basis of food quality. Mar Biol 100:41–49
Craig TP, Itami JK, Price PW (1988) Plant wound compounds from oviposition scars used in host discrimination by a stem-galling sawfly. J Insect Behav 1:343–356
Cranford JA, Johnson EO (1989) Effects of coprophagy and diet quality on two microtine rodents (*Microtus pennsylvanicus* and *Microtus pinetorum*). J Mammal 70:494–502
Crawford CS (1988) Nutrition and habitat selection in desert detritivores. J Arid Environ 14:111–122
Crawley MJ (1983) Herbivory. The dynamics of animal – plant interactions. Blackwell, Oxford
Crawley MJ (1989) Insect herbivores and plant population dynamics. Annu Rev Entomol 34:531–564
Crawley MJ, Pattrasudhi R (1988) Interspecific competition between insect herbivores: asymmetric competition between cinnabar moth and the ragwort seed-head fly. Ecol Entomol 13:243–249
Cridle N (1930) Some natural factors governing the fluctuations of grouse in Manitoba. Can Field Nat 44:77–80
Crome FHJ (1975) The ecology of fruit pigeons in tropical Queensland. Aust Wildl Res 2:155–185
Crompton DWT, Hall A (1981) Parasitic infection and host nutrition. Parasitology 82:31–48
Crowe TM (1978) Limitation of population in the helmeted guineafowl. S Afr J Wildl Res 8:117–126
Daan R, Gonzalez SR, Klein-Breteler WCM (1988) Cannibalism in omnivorous calanoid copepods. Mar Ecol Prog Ser 47:45–54
Dabrowski K (1979) Feeding requirements of fish with particular attention to common carp. A review. Pol Arch Hydrobiol 26:135–158
Dabrowski K (1982) *Tilapia* in lakes and aquaculture – ecological and nutritional approach. Acta hydrochim hydrobiol 10:265–271
Daly M, Wilson M (1988) Evolutionary social psychology and family homicide. Science 242:519–524

Danthanarayana W (1983) Population ecology of the light brown apple moth, *Epiphyas postvittana* (Lepidoptera: Tortricidae). J Anim Ecol 52:1–33

Danthanarayana W, Kathiravetpillai A (1969) Studies on the ecology and cause of outbreaks of *Ectropis bhurmitra* Wkr. (Geometridae), the twig caterpillar of tea in Ceylon. J Appl Ecol 6:311–322

Daser U, Brandl R (1992) Microbial gut flora of eight species of tephritids. Biol J Linn Soc 45:155–166

Davies AG, Bennett EL, Waterman PG (1988) Food selection by two Southeast Asian colobine monkeys (*Presbytis rubicunda* and *Presbytis melalophos*) in relation to plant chemistry. Biol J Linn Soc 34:33–56

Davies NB, Houston AI (1981) Owners and satellites: the economics of territory defence in the pied wagtail, *Motacilla alba*. J Anim Ecol 50:157–180

Davies NB, Lundberg A (1984) Food distribution and variable mating systems in the dunnock, *Prunella modularis*. J Anim Ecol 53:895–912

Dawkins R (1986) The blind watchmaker. Longmans, Essex

Dawkins R (1989) The selfish gene, 2nd edn. Oxford University Press, Oxford

DeClerck RA, Shorthouse JD (1985) Tissue preference and damage by *Fenusa pusilla* and *Messa nana* (Hymenoptera: Tenthredinidae), leafmining sawflies on white birch (*Betula papyrifera*). Can Entomol 117:351–362

Deevey ES Jr (1947) Life tables for natural populations of animals. Q Rev Biol 22:283–314

Degabriele R (1981) A relative shortage of nitrogenous food in the ecology of the koala (*Phascolarctos cinereus*). Aust J Ecol 6:139–141

Degabriele R (1983) Nitrogen and the koala (*Phascolarctos cinereus*): some indirect evidence. Aust J Ecol 8:75–76

Degabriele R (1989) Wasting disease in the koala, *Phascolarctos cinereus*. Perspect Biol Med 32:414–420

Degabriele R, Dawson TJ (1979) Metabolism and heat balance in an arboreal marsupial, the koala (*Phascolarctos cinereus*). J Comp Physiol 134:293–301

Delwiche CC (1970) The nitrogen cycle. Sci Am 223:136–146

Dempster JP (1967) The control of *Pieris rapae* with DDT. I. The natural mortality of the young stages of *Pieris*. J App Ecol 4:485–500

Dempster JP (1971) The population ecology of the cinnabar moth, *Tyria jacobaeae* L. (Lepidoptera, Arctiidae). Oecologia 7:26–67

Dempster JP (1982) The ecology of the cinnabar moth, *Tyria jacobaeae* L. (Lepidoptera: Arctiidae). Adv Ecol Res 12:1–36

Dempster JP, Lakhani KH (1979) A population model for cinnabar moth and its food plant, ragwort. J Anim Ecol 48:143–163

Dempster JP, Pollard E (1981) Fluctuations in resource availability and insect populations. Oecologia 50:412–416

Den Boer PJ (1968) Spreading the risk and stabilization of animal numbers. Acta Biotheor 18:165–194

Den Boer PJ (1985) Exclusion, competition or coexistence – a question of testing the right hypotheses. Z Zool Syst Evolutionsforsch 23:259–274

Den Boer PJ (1986a) Facts, hypotheses and models on the part played by food in the dynamics of carabid populations. In: Rep 5th Meet Eur Carabidologists, Poland: Warsaw, Agricultural University Press, Warsaw. pp 81–96

Den Boer PJ (1986b) Population dynamics of two carabid beetles at a Dutch heathland. The significance of density-related egg production. In: Den Boer PJ, Luff ML, Mossakowski D, Weber F (eds) Carabid beetles: their adaptations and dynamics. Fischer, Stuttgart, pp 361–370

Den Boer PJ (1986c) The present status of the competitive exclusion principle. TREE 1:25–28

Den Boer PJ (1987) Density dependence and the stabilization of animal numbers 2. The pine looper. Neth J Zool 37:220–237

Denno RF, Roderick GK (1990) Population biology of planthoppers. Annu Rev Entomol 35:489–520

Diamond JM (1987) Adaptations of intestinal nutrient absorption in mammals. S Afr J Sci 83:590–594
Dierenfeld ES, Hintz HF, Robertson JB, van Soest PJ, Oftedal OT (1982) Utilization of bamboo by the giant panda. J Nutr 112:636–641
Dill LM, Ydenberg RC, Fraser AHG (1981) Food abundance and territory size in juvenile coho salmon (*Oncorhynchus kisutch*). Can J Zool 59:1801–1809
Dillwith JW, Berberet RC, Bergman DK, Neese PA, Edwards RM, McNew RW (1991) Plant biochemistry and aphid populations. Studies on the spotted alfalfa aphid, *Therioaphis maculata*. Arch Inst B 17:235–251
Dixon AFG (1963) Reproductive activity of the sycamore aphid, *Drepanosiphum platanoides* (Schr.) (Hemiptera: Aphididae). J Anim Ecol 32:33–48
Dixon AFG (1966) The effect of population density and nutritive status of the host on the summer reproductive activity of the sycamore aphid, *Drepanosiphum platanoides* (Schr.). J Anim Ecol 35:105–112
Dixon AFG (1970) Quality and availability of food for a sycamore aphid population. Symp Br Ecol Soc 10:271–287
Dixon AFG (1976) Factors determining the distribution of sycamore aphids on sycamore leaves during summer. Ecol Entomol 1:275–278
Doane CC, McManus ML (eds) (1981) The gypsy moth: Research towards intergrated pest management. USDA For Ser Tech Bull 1584, USDA, Washington, DC
Dobson FS, Kjelgaard JD (1985) The influence of food resources on population dynamics in Columbian ground squirrels (*Spermophilus columbianus*). Can J Zool 63:2095–2104
Dodd AP (1940) The biological campaign against prickly pear. Commwlth Prickly Pear Board, Brisbane
Doerr PD, Keith LB, Rusch DH, Fischer CA (1974) Characteristics of winter feeding aggregations of ruffed grouse in Alberta. J Wildl Manage 38:601–615
Doherty PJ (1983) Tropical territorial damselfishes: is density limited by aggression or recruitment? Ecology 64:176–190
Dolbeer RA, Clark WR (1975) Population ecology of snowshoe hares in the cental Rocky Mountains. J Wildl Manage 39:535–549
Dornstreich MD, Morren GEB (1974) Does New Guinea cannibalism have nutritional value? Hum Ecol 2:1–12
Dorschner KW, Ryan JD, Johnson RC, Eikenbary RD (1987) Modification of host nitrogen levels by the greenbug (Homoptera: Aphididae): its role in resistance of winter wheat to aphids. Environ Entomol 16:1007–1011
Douglas AE (1988) Experimental studies on the mycetome symbiosis in the leafhopper *Euscelis incisus*. J Insect Physiol 34:1043–1053
Douglas AE (1989) Mycetocyte symbiosis in insects. Biol Rev 64:409–434
Drew RAI (1988) Amino acid increases in fruit infested by fruit flies of the family Tephritidae. Zool J Linn Soc 93:107–112
Drew RAI, Lloyd AC (1987) Relationship of fruit flies (Diptera: Tephritidae) and their bacteria to host plants. Ann Entomol Soc Am 80:629–636
Drew RAI, Lloyd AC (1991) Bacteria in the life cycle of tephritid fruit flies. In: Barbosa P, Krischik VA, Jones CG (eds) Microbial mediation of plant – herbivore interactions. Wiley, New York, pp 441–466
Drew RAI, Courtice AC, Teakle DS (1983) Bacteria as a natural source of food for adult fruit flies (Diptera: Tephritidae). Oecologia 60:279–284
Driessen GJJ, Van Raalte ATh, De Bruyn GJ (1984) Cannibalism in the red wood ant, *Formica polyctena* (Hymenoptera: Formicidae). Oecologia 63:13–22
Ducklow HW (1983) Production and fate of bacteria in the oceans. BioScience 33:494–501
Duncan P, Foose TJ, Gordon IJ, Gakahu CG, Lloyd M (1990) Comparative nutrient extraction from forage by grazing bovids and equids: a test of the nutritional model of equid/bovid competition and coexistence. Oecologia 84:411–418
Durzan DJ (1968) Nitrogen metabolism of *Picea glauca* III. Diurnal changes of amino acids, amides, protein, and chlorophyll in leaves of expanding buds. Can J Bot 46:939–937

Durzan DJ, Lopushanski SM (1968) Free and bound amino acids of spruce budworm larvae feeding on balsam fir and red and white spruce. J Insect Physiol 14:1485–1497
du Toit JT, Bryant JP, Frisby K (1990) Regrowth and palatability of *Acacia* shoots following pruning by African savanna browsers. Ecology 71:149–154
Ebbinge BK, Canters A, Drent RH (1975) Foraging routines and estimated daily intake of barnacle geese wintering in the northern Netherlands. Wildfowl 26:5–19
Edney EB, Allen W, McFarlane J (1974) Predation by terrestrial isopods. Ecology 55:428–433
Ehrlich PR, Birch LC (1967)The "balance of nature" and population control. Am Nat 101:97–107
Ehrlich PR, White RR, Singer MC, McKechnie SW, Gilbert LE (1975) Checkerspot butterflies: a historical perspective. Science 188:221–228
Eisenberg RM (1966) The regulation of density in a natural population of the pond snail *Lymnaea elodes*. Ecology 47:889–906
Eisenberg RM (1970) The role of food in the regulation of the pond snail *Lymnaea elodes*. Ecology 51:680–684
Eisenberg RM, Hurd LE, Bartley JA (1981) Ecological consequences of food limitation for adult mantids (*Tenodera ardifolia sinensis* Saussure). Am Midl Nat 106:209–218
Elliott HJ, Bashford R, Palzer C (1980) Biology of *Stathmorrhopa aphotista* Turner (Lepidoptera: Geometridae), a defoliator of *Eucalyptus* spp. in southern Tasmania. Aust For 43:81–86
Ely DG, Harvey TL (1969) Relation of ration to short-nosed cattle louse infections. J Econ Entomol 62:341–344
Enoksson B (1990) Autumn territories and population regulation in the nuthatch *Sitta europaea*: an experimental study. J Anim Ecol 59:1047–1062
Enoksson B, Nilsson SG (1983) Territory size and population density in relation to food supply in the nuthatch *Sitta europaea* (Aves). J Anim Ecol 52:927–936
Eppley RW, Peterson RJ (1979) Particulate organic matter flux and planktonic new production in the deep ocean. Nature 282:677–680
Eriksson K (1970) Ecology of the eruption and wintering of Fennoscandian redpolls (*Carduelis flammea*). Ann Zool Fenn 7:273–282
Erikstad KE (1985) Growth and survival of willow grouse (*Lagopus lagopus lagopus*) chicks in relation to home range size, brood movements and habitat selection. Ornis Scand 16:181–190
Erikstad KE, Spidso TK (1982) The influence of weather on food intake, insect prey selection and feeding behaviour in willow grouse chicks in northern Norway. Ornis Scand 13:176–182
Errington PL (1946) Predation and vertebrate populations. Q Rev Biol 21:144–177
Eshelman BD, Jenkins SH (1989) Food selection by Belding's ground squirrel in relation to plant nutritional features. J Mamm 70:846–852
Estrada A, Coates-Estrada R (1986) Use of leaf resources by howling monkeys (*Alouatta palliata*) and leaf-cutting ants (*Atta cephalotes*) in the tropical rain forest of Los Turtles, Mexico. Am J Primatol 10:51–66
Evans AC (1938) Physiological relationships between insects and their host plants. I. The effect of the chemical composition of the plant on reproduction and production of winged forms in *Brevicoryne brassicae* L. (Aphididae). Ann Appl Biol 25:558–572
Ewald PW, Carpenter FL (1978) Territorial responses to energy manipulations in the Anna humming bird. Oecologia 31:277–292
Ezueh MI (1981) The biolgical basis of resistance in cowpea to the cowpea moth, *Cydia ptychora* (Lepidoptera: Olethreutidae). Ann Appl Biol 99:313–321
Faeth SH (1985) Host leaf selection by leaf miners: interactions among three trophic levels. Ecology 66:870–875
Fair PH, Fortner AR, Millikin MR, Sick LV (1980) Effect of dietary fibre on growth, assimilation and cellulase activity of the prawn (*Macrobrachium rosenbergii*). Proc World Maricult Soc 11:359–381

Farrell GS, New TR (1980) Some aspects of the biology of the eucalypt-mining sawfly *Phylacteophaga froggatti* Riek (Hymenoptera:Pergidae). Aust J Zool 28:83–90

Farrow RA (1982) Population dynamics of the Australian plague locust *Chortoicetes terminifera* (Walker) in central western New South Wales. II Factors influencing natality and survival. Aust J Zool 30:199–222

Feeny P (1970) Seasonal changes in oak leaf tannins and nutrients as a cause of spring feeding by winter moth caterpillars. Ecology 51:565–581

Feeny P (1976) Plant apparency and chemical defense. In: Wallace JW, Mansell RL (eds) Recent advances in phytochemistry. vol 10. Biochemical interactions between plants and insects. Plenum, New York, pp 1–40

Fenna L, Boag DA (1974a) Filling and emptying of the galliform caecum. Can J Zool 52:537–540

Fenna L, Boag DA (1974b) Adaptive significance of the caeca in Japanese quail and spruce grouse (Galliformes). Can J Zool 52:1577–1584

Fennah RG (1953) Some aspects of the food-problem of Homoptera in the humid area of the neotropical region. Trans 9th Int Congr Entomol 2, The Congress, Amsterdam, pp 260–265

Fennah RG (1965) The influence of environmental stress on the cacao tree in predetermining the feeding sites of cacao thrips, *Selenothrips rubrocinctus* (Giard), on leaves and pods. Bull Entomol Res 56:333–349

Ferro DN, Sluss RR, Bogyo TP (1975) Factors contributing to the biotic potential of the codling moth, *Laspeyresia pomonella* (L.), in Washington. Environ Entomol 4:385–391

Findlay S, Tenore K (1982) Nitrogen source for a detritivore: detritus substrate versus associated microbes. Science 218:371–373

Finke MD, Scriber JM (1988) Influence on larval growth of the eastern black swallowtail butterfly *Papilio polyxenes* (Lepidoptera, Papilionidae) of seasonal changes in nutritional parameters of Umbelliferae species. Am Midl Nat 119:45–62

Fischer Z (1973) Physiology and bioenergetics of grass carp. Pol Arch Hydrobiol 20:529–557

Fish D, Carpenter SR (1982) Leaflitter and larval mosquito dynamics in tree-hole ecosystems. Ecology 63:283–288

Fishelson L, Montgomery WL, Myrberg AA Jr (1985) A unique symbiosis in the gut of tropical herbivorous surgeonfish (Acanthuridae: Teleostei) from the Red Sea. Science 229:49–51

Fisher DB, Wright JP, Mittler TE (1984) Osmoregulation by the aphid *Myzus persicae*: a physiological role for honeydew oligosaccharides. J Insect Physiol 30:387–393

Fisher M (1987) The effect of previously infested spruce needles on the growth of the green spruce aphid, *Elatobium abietinum*, and the effect of the aphid on the amino acid balance of the host plant. Ann Appl Biol 111:33–41

Fisher RC (1941) Studies on the biology of the death-watch beetle (*Xestobium rufovillosum* deGeer). IV. The effect of type and extent of fungal decay in timber upon the rate of development of the insect. Ann Appl Biol 28:244–260

Fitt GP, O'Brien RW (1985) Bacteria associated with four species of *Dacus* (Diptera: Tephritidae) and their role in the nutrition of the larvae. Oecologia 67:447–454

Fitzpatrick SM, Wellington WG (1983) Insect territoriality. Can J Zool 61:471–538

Fleming TH, Hooper ET, Wilson DE (1972) Three Central American bat communities: structure, reproductive cycles, and movement patterns. Ecology 53:555–569

Foley WJ, Cork SJ (1992) Use of fibrous diets by small herbivores: how far can the rules be bent? TREE 7:159–162

Foley WJ, Bouskila A, Shkolnik A, Choshniak I (1992) Microbial digestion in the herbivorous lizard *Uromastyx aegyptius* (Agamidae). J Zool Lond 226:387–398

Folland CK, Parker DE, Kates FE (1984) Worldwide marine temperature fluctuations 1856–1981. Nature 310:670–673

Foltz JL, Knight FB, Allen DC (1972) Numerical analysis of population fluctuations of the jack pine budworm. Ann Entomol Soc Am 65:83–89

Forbes RB (1966) Studies of the biology of Minnesota chipmunks. Am Midl Nat 76:290–308
Forcella F (1982) Why twig-girdling beetles girdle twigs. Naturwissenschaften 69:398–399
Ford HA, Paton DC (1976) The value of insects and nectar to honeyeaters. Emu 76:83–84
Forrest JMS (1971) The growth of *Aphis fabae* as an indicator of the nutritional advantage of galling to the apple aphid *Dysaphis devecta*. Entomol Exp Appl 14:477–483
Foster MS (1978) Total frugivory in tropical passerines: a reappraisal. Trop Ecol 19:131–154
Fox JF, Bryant JP (1984) Instability of the snowshoe hare and woody plant interaction. Oecologia 63:128–135
Fox LR (1975) Cannibalism in natural populations. Annu Rev Ecol Syst 6:87–106
Fox LR, Letourneau DK, Eisenbach J, Van Nouhuys S (1990) Parasitism rates and sex ratios of a parasitoid wasp: effects of herbivore and plant quality. Oecologia 83:414–419
France RL (1985) Relationship of crayfish (*Orconectes virilis*) growth to population abundance and system productivity in small oligotrophic lakes in the experimental lakes area, north-western Ontario. Can J Fish 42:1096–1102
Franceschi PF, Boag DA (1991) Summer foraging in spruce grouse: implications for galliform food habits. Can J Zool 69:1708–1711
Frankie GW, Morgan DL(1984) Role of the host plant and parasites in regulating insect herbivore abundance, with an emphasis on gall-forming insects. In: Price PW, Slobodchikoff CN, Gaud WS (eds) A new ecology: novel approaches to interactive systems. Wiley, New York, pp 102–140
Frase BA, Armitage KB (1989) Yellow-bellied marmots are generalist herbivores. Ethol Ecol Evol 1:353–366
Frazao E (1991) Insectivory in free-ranging bearded saki (*Chiropotes satanas chiropotes*). Primates 32:243–246
Fretwell SD (1987) Food chain dynamics: the central theory of ecology? Oikos 50:291–301
Friis-Christensen E, Lassen K (1991) Length of the solar cycle: an indication of solar activity closely associated with climate. Science 254:698–700
Frith HJ, Braithwaite LW, McKean JL (1969) Waterfowl in an inland swamp in New South Wales. II. Food. CSIRO Wildl Res 14:17–64
Fuhrman JA, Sleeter TD, Carlson CA, Proctor LM (1989) Dominance of bacterial biomass in the Sargasso Sea and its ecological implications. Mar Ecol Prog Ser 57:207–217
Fuller TK (1989) Population dynamics of wolves in north-central Minnesota. Wildl Monogr 105:1–41
Gabriel W (1985) Overcoming food limitation by cannibalism: a model study on cyclops. Arch Hydrobiol Beih 21:373–381
Gabriel W, Lampert W (1985) Can cannibalism be advantageous in cyclopoids? A mathematical model. Verh Int Verein Limnol 22:3164–3168
Galbraith H (1988) Effects of egg size and competition on the size, quality and survival of lapwing *Vanellus vanellus* chicks. J Zool 214:383–398
Gales RP (1982) Age-related and sex-related differences in diet selection by *Rattus rattus* on Stewart Island, New Zealand. N Z J Zool 9:463–466
Galloway M, Boonstra R (1989) Response of the eastern chipmunk, *Tamias striatus*, to sex ratio manipulations. Oikos 55:3–10
Ganzhorn JU (1992) Leaf chemistry and the biomass of folivorous primates in tropical forests. Test of a hypothesis. Oecologia 91:540–547
Garber ED (1954) The role of nutrition in the host-parasite relationship. Proc Natl Acad Sci USA 40:1112–1116
Garber ED (1956) A nutrition-inhibition hypothesis of pathogenicity. Am Nat 90:183–194
Gardarsson A, Moss R (1970) Selection of food by Icelandic ptarmigan in relation to its availability and nutritive value. Symp Br Ecol Soc 10:47–71
Garnett ST, Price IR, Scott FJ (1985) The diet of the green turtle, *Chelonia mydas* (L.), in Torres Strait. Aust Wildl Res 12:103–112
Gasaway WC (1976) Volatile fatty acids and metabolizable energy derived from cecal fermentation in the willow ptarmigan. Comp Biochem Physiol A 53:115–121

Gasaway WC, Holleman DF, White RG (1975) Flow of digesta in the intestine and caecum of the cock ptarmigan. Condor 77:467–474

Gass CL, Ange HR, Centa J (1976) Regulation of food supply by feeding territoriality in the rufous hummingbird. Can J Zool 54:2046–2054

Gauthier G, Bedard J (1990) The role of phenolic compounds and nutrients in determining food preference in greater snow geese. Oecologia 84:553–558

Geier PW (1963) The life history of the codling moth *Cydia pomonella* (L) Lepidoptera: Tortricidae, in the Australian Capital Territory. Aust J Zool 11:323–367

Giani A (1991) Implications of phytoplankton chemical composition for zooplankton production: experimental evidence. Oecologia 87:409–416

Gibb JA (1981) What determines the number of small herbivorous mammals. N Z J Ecol 4:73–77

Gibbs HL, Grant PR (1987) Ecological consequences of an exceptionally strong El Niño event on Darwin's finches. Ecology 68:1735–1746

Gibson CWD, Hamilton J (1983) Feeding ecology and seasonal movements of giant tortoises on Aldabra atoll. Oecologia 56:84–92

Gilbert F, Owen J (1990) Size, shape, competition, and community structure in hoverflies (Diptera: Syrphidae). J Anim Ecol 59:21–35

Gilbertson RL (1984) Relationships between insects and wood-rotting basidiomycetes. In: Wheeler Q, Blackwell M (eds) Fungus – insect relations. Columbia University Press, New York, pp 130–165

Gladney WJ, Graham OH, Trevino JL, Ernst SE (1973) *Boophilus annulatus*: effect of host nutrition on development of female ticks. J Med Entomol 10:123–130

Gluck E (1988) Why do parent birds swallow the feces of their nestlings? Experientia 44:537–539

Glynn PW (1988) El Niño-Southern Oscillation 1982–1983: nearshore population, community, and ecosystem responses. Annu Rev Ecol Syst 19:309–346

Goldingay RL (1986) Feeding behaviour of the yellow-bellied glider *Petaurus australis* (Marsupiala: Petauridae), at Bombala, New South Wales. Aust Mammal 9:17–25

Goldingay RL (1987) Sap feeding by the marsupial *Petaurus australis*: an enigmatic behaviour? Oecologia 73:154–158

Goldingay RL, Kavanagh RP (1990) Socioecology of the yellow-bellied glider, *Petaurus australis*, at Waratah Creek, NSW. Aust J Zool 38:327–341

Goldstein RJ (1973) Cichlids of the world. TFH, Neptune City

Gordon G, Brown AS, Pulsford T (1988) A koala (*Phascolarctos cinereus* Goldfuss) population crash during drought and heatwave conditions in south-western Queensland. Aust J Ecol 13:44–61

Gorham WT (1990) Uptake of dilute nutrients by marine invertebrates. In: Mellinger J (ed) Animal nutrition and transport processes. 1. Nutrition in wild and domestic animals. Karger, Basel, pp 84–95

Gorman OT, Roth RR (1989) Consequences of temporally and spacially variable food supply for an unexploited gray squirrel (*Sciurus carolinensis*) population. Am Midl Nat 121:41–60

Gould F, Holtzman G, Rabb RL, Smith M (1980) Genetic variation in predatory and cannibalistic tendencies of *Heliothis virescens* strains. Ann Entomol Soc Am 73:243–250

Gould SJ (1989) Wonderful life. The Burgess shale and the nature of history. Penguin, London

Grajal A, Strahl SD, Parra R, Dominguez MA, Neher A (1989) Foregut fermentation in the hoatzin, a neotropical leaf-eating bird. Science 245:1236–1238

Grant PR, Grant BR (1987) The extraordinary El Niño event of 1982–83. Effects on Darwin's finches on Isla Genovesa, Galapagos. Oikos 49:55–65

Grayson FWL, Hassall M (1985) Effects of rabbit grazing on population variables of *Chorthippus brunneus* (Orthoptera). Oikos 44:27–34

Green RE (1984) The feeding ecology and survival of partridge chicks (*Alectoris rufa* and *Perdix perdix*) on arable farmland in East Anglia, UK. J Appl Ecol 21:817–830

Green RG, Evans CA (1940) Studies on a population cycle of snowshoe hares on the Lake Alexander area. J Wildl Manage 4:347–358
Greenblatt JA, Barbosa P (1981) Effects of host diet on 2 pupal parasitoids of the gypsy moth – *Brachymeria intermedia* (Nees) and *Coccygomimus turionellae* (L). J Appl Ecol 18:1–10
Greenblatt JA, Barbosa P, Montgomery ME (1982) Host's diet effects on nitrogen utilization efficiency for two parasitoid species: *Brachymeria intermedia* and *Coccygomimus turionellae*. Physiol Entomol 7:263–267
Greenstone MH (1979) Spider feeding behaviour optimises dietary essential amino acid composition. Nature 282:501–503
Gregory MJ, Lacki MJ, Williams PK (1988) Demographic changes of the eastern chipmunk *Tamias striatus*, with supplemental food. Can Field Nat 102:661–665
Grehan JR (1983) Larval establishment of the borer *Aenetus virescens* (Lepidoptera: Hepialidae) in live trees. N Z Entomol 7:413–417
Griffiths D (1980) The feeding biology of ant-lion larvae: growth and survival in *Morter obscurus*. Oikos 34:364–370
Grime JP, Blythe GM (1969) An investigation of the relationships between snails and vegetation at the Winnats pass. J Ecol 57:45–66
Groeger AW, Schram MD, Marzolf GR (1991) Influence of food quality on growth and reproduction in *Daphnia*. Freshwater Biol 26:11–19
Gross RL, Newberne PM (1980) Role of nutrition in immunologic function. Physiol Rev 60:188–302
Guidi LD (1984) The effect of food composition on ingestion, development, and survival of a harpacticoid copepod, *Tisbe cucumariae* Humes. J Exp Mar Biol Ecol 84:101–110
Gunnarsson T (1987) Selective feeding on a maple leaf by *Oniscus asellus* (Isopoda). Pedobiologia 30:161–165
Gurchinoff S, Robinson WL (1972) Chemical characteristics of jack pine needles selected by feeding spruce grouse. J Wildl Manage 36:80–87
Gurnell J (1983) Squirrel numbers and the abundance of tree seeds. Mammal Rev 13:113–148
Gurr L (1951) Food of the chick of *Notornis hochstetteri*. Notornis 4:114
Guyer C (1988a) Food supplementation in a tropical mainland anole, *Norops humilis*: demographic effects. Ecology 69:350–361
Guyer C (1988b) Food supplementation in a tropical mainland anole, *Norops humilis*: effects on individuals. Ecology 69:362–369
Hagland BM (1980) Proline and valine – cues which stimulate grasshopper herbivory during drought stress? Nature 288:697–698
Hahus SC, Smith KG (1990) Food habits of *Blarina*, *Peromyscus*, and *Microtus* in relation to an emergence of periodical cicadas (*Magicicada*). J Mammal 71:249–251
Hairston NG (1981) Citation classic. ISI Current Contents (AB&ES) 20:20
Hairston NG, Smith FE, Slobodkin LB (1960) Community structure, population control, and competition. Am Nat 94:421–425
Hall-Martin AJ, Basson WD (1975) Seasonal chemical composition of the diet of Transvaal Lowveld giraffe. J South Afr Wildl Manage Assoc 5:19–21
Hamilton J, Coe MJ (1982) Feeding, digestion and assimilation of a population of giant turtles (*Geochelone gigantea* (Schweigger)) on Aldabra atoll. J Arid Environ 5:127–144
Hanley TA (1982) The nutritional basis for food selection by ungulates. J Range Manage 35:146–151
Hanley TA, Cates RG, Van Horne B, McKendrick JD (1987) Forest stand-age-related differences in apparent nutritional quality of forage for deer in southeastern Alaska. Proc 4th Wildland Shrub Symp. USDA For Serv Gen Tech Rep INT-222, USDA, Ogden
Hansen RM, Ueckert DN (1970) Dietary similarity of some primary consumers. Ecology 51:640–648
Hanski I (1987) Pine sawfly population dynamics: patterns, processes, problems. Oikos 50:327–335

Hanski I (1991) Metapopulation dynamics: brief history and conceptual domain. Biol J Linn Soc 42:3–16
Hanski I, Parviainen P (1985) Cocoon predation by small mammals, and pine sawfly population dynamics. Oikos 45:125–136
Hanssen I, Grav HJ, Steen JB, Lysnes H (1979) Vitamin C deficiency in growing willow ptarmigan (*Lagopus lagopus lagopus*). J Nutr 109:2260–2276
Hansson L (1971) Small rodent food, feeding and population dynamics. Oikos 22:183–198
Hansson L (1984) Predation as the factor causing extended low density in microtine cycles. Oikos 43:255–256
Hansson L (1987) An interpretation of rodent dynamics as due to trophic interactions. Oikos 50:308–318
Hardwick DF (1965) The corn earworm complex. Mem Entomol Soc Can 40:3–247
Hardy YJ, Lafond A, Hamel L (1983) The epidemiology of the current spruce budworm (*Choristoneura fumiferana*) outbreak in Quebec (Canada). For Sci 29:715–725
Harris HJ (1970) Evidence of stress response in breeding blue-winged teal. J Wildl Manage 34:747–755
Harris JWE, Alfaro RI, Dawson AF, Brown RG (1985) The western spruce budworm in British Columbia 1909–1983. Pac For Res Centre Info Rep BC-X-257
Harris MP (1969) Food as a factor controlling the breeding of *Puffinus lherminieri*. Ibis 111:139–156
Harris MP (1978) Supplementary feeding of young puffins, *Fratercula arctica*. J Anim Ecol 47:15–23
Harris MP, Wanless S (1990) Breeding success of British kittiwakes *Rissa tridactyla* in 1986–88: evidence for changing conditions in the northern north sea. J Appl Ecol 27:172–187
Hart DD (1985a) Causes and consequences of territoriality in a grazing stream insect. Ecology 66:404–414
Hart DD (1985b) Grazing insects mediate algal interactions in a stream benthic community. Oikos 44:40–46
Hart DD (1986) The adaptive significance of territoriality in filter-feeding larval blackflies (Diptera: Simuliidae) . Oikos 46:88–92
Hart DD (1987a) Experimental studies of exploitative competition in a grazing stream insect. Oecologia 73:41–47
Hart DD (1987b) Feeding territoriality in aquatic insects: cost-benefit models and experimental tests. Am Zool 27:371–386
Hassall M, Dangerfield JM (1990) Density-dependent processes in the population dynamics of *Armadillidium vulgare* (Isopoda: Oniscidae). J Anim Ecol 59:941–958
Hassall M, Jennings JB (1975) Adaptive features of gut structure and digestive physiology in the terrestrial isopod *Philoscia muscorum* (Scopoli) 1763. Biol Bull 149:348–364
Hassall M, Rushton SP (1985) The adaptive significance of coprophagous behaviour in the terrestrial isopod *Porcellio scaber*. Pedobiologia 28:169–175
Hatch MD, Slack CR (1970) Photosynthetic CO_2-fixation pathways. Annu Rev Plant Physiol 21:141–162
Hawkins BA (1988) Do galls protect endophytic herbivores from parasitoids? A comparison of galling and non-galling Diptera. Ecol Entomol 13:473–477
Hawkins BA, Unruh TR (1988) Protein and water levels in *Asphondylia atriplicus* (Diptera: Cecidomyiidae) galls. Southwest Nat 33:114–116
Hays C (1986) Effects of the 1982–83 El Niño on Humboldt penguin colonies in Peru. Biol Conserv 36:169–180
Hebert PDN (1978) The population biology of *Daphnia* (Crustacea, Daphnidae). Biol Rev 53:387–426
Heithaus ER, Fleming TH (1978) Foraging movements of a frugivorous bat, *Carollia perspicillata* (Phyllostomatidae). Ecol Monogr 48:127–143
Hendrichs J, Hendrichs MA (1990) Mediterranean fruit fly (Diptera: Tephritidae) in nature: location and diel pattern of feeding and other activities on fruiting and non-fruiting hosts and non-hosts. Ann Entomol Soc Am 83:632–641

Henry SR, Craig SA (1984) Diet, ranging behaviour and social organization of the yellow-bellied glider (*Petaurus australis* Shaw) in Victoria. In: Smith A, Hume ID (eds) Possums and gliders. Aust Mamm Soc, Sydney, pp 331–341
Henttonen H (1985) Predation causing extended low densities in microtine cycles: further evidence from shrew dynamics. Oikos 45:156–157
Henttonen H, Oksanen T, Jortikka A, Haukisalmi V (1987) How much do weasels shape microtine cycles in the northern Fennoscandian taiga? Oikos 50:353–365
Herbst LH (1986) The role of nitrogen from fruit pulp in the nutrition of the frugivorous bat *Carollia perspicillata*. Biotropica 18:39–44
Hering EM (1951) Biology of the leaf miners. Junk, Dordrecht
Hewson R (1970) Variation in reproduction and shooting bags of mountain hares on two moors in northeast Scotland. J Appl Ecol 7:243–252
Hewson R (1976) A population study of mountain hares (*Lepus timidus*) in northeast Scotland from 1956–1969. J Anim Ecol 45:395–414
Hik DS, Jefferies RL (1990) Increases in the net above-ground primary production of a salt-marsh forage grass: a test of the predictions of the herbivore-optimization model. J Ecol 78:180–195
Hill DA (1985) The feeding ecology and survival of pheasant chicks on arable farmland. J Appl Ecol 22:645–654
Hindell MA, Lee AK (1987) Habitat use and tree preference of koalas in a mixed eucalypt forest. Aust Wildl Res 14:349–360
Hintz HF, Schryver HF, Stevens CE (1978) Digestion and absorption in the hindgut of nonruminant herbivores. J Anim Sci 46:1803–1807
Hirano C (1964) Studies on the nutritional relationships between larvae of *Chilo suppressalis* Walker and the rice plant, with special reference to the role of nitrogen in nutrition of larvae. Bull Nat Inst Agric Sci (Jpn) Ser C 17:103–180
Hirons GJM (1982) The effects of fluctuations in rodent numbers on breeding success in the tawny owl *Strix aluco*. Mammal Rev 12:155–158
Hix DM, Barnes BV, Lynch AM, Witter JA (1987) Relationships between spruce budworm damage and site factors in spruce-fir-dominated ecosystems of western Upper Michigan. For Ecol Manage 21:129–140
Hodkinson ID, Jensen TS, MacLean SF (1979) The distribution, abundance and host plant relationships of *Salix*-feeding psyllids (Homoptera: Psylloidea) in arctic Alaska. Ecol Entomol 4:119–132
Hoffmann RS (1961) The quality of the winter food of blue grouse. J Wildl Manage 25:209–210
Hoogland JL (1985) Infanticide in prairie dogs: lactating females kill offspring of close kin. Science 230:1037–1040
Horn MH, Murray SN, Edwards TW (1982) Dietary selectivity in the field and food preferences in the laboratory for two herbivorous fishes (*Cebidichthys violaceus* and *Xiphister mucosus*) from a temperate intertidal zone. Mar Biol 67:237–246
Horn MH, Neighbors MA, Murray SN (1986) Herbivore responses to a seasonally fluctuating food supply: growth potential of two temperate intertidal fishes based on the protein and energy assimilated from their macroalgal diets. J Exp Mar Biol 103:217–234
Hornfeldt B, Lofgren O, Carlsson BG (1986) Cycles in voles and small game in relation to variations in plant production indices in northern Sweden. Oecologia 68:496–502
Horsfield D (1977) Relationships between feeding of *Philaenus spumarius* (L.) and amino acid concentration in xylem sap. Ecol Entomol 2:259–266
Horsfield D (1978) Evidence for xylem feeding by *Philaenus spumarius* (L.) (Homoptera: Cercopidae). Entomol Exp Appl 24:95–99
Hosking GP, Hutcheson JA (1979) Nutritional basis for feeding zone preference of *Arhopalus ferus* (Coleoptera: Cerambycidae). N Z J For Sci 9:185–192
Hosking GP, Hutcheson JA (1986) Hard beech (*Nothofagus truncata*) decline on the Mamaku plateau, North Island, New Zealand. N Z J Bot 24:263–270
Houde ED (1975) Effects of stocking density and food density on survival, growth and yield of laboratory-reared larvae of sea bream *Archosargus rhomboidalis* (L) (Sparidae). J Fish Biol 7:115–127

Houde ED (1978) Critical food concentrations for larvae of three species of subtropical fishes. Bull Mar Sci 28:395–411

Houde ED, Schekter RC (1978) Simulated food patches and survival of larval bay anchovy, *Anchoa mitchilli*, and sea bream, *Archosargus rhomboidalis*. Fish Bull 76:483–487

Houde SEL, Roman MR (1987) Effects of food quality on the functional ingestion response of the copepod *Acartia tonsa*. Mar Ecol Prog Ser 40:69–77

House HL (1969) Effects of different proportions of nutrients on insects. Entomol Exp Appl 12:651–669

House HL, Barlow JS (1961) Effects of different diets of a host, *Agria affinis* (Fall.) (Diptera: Sarcophagidae), on the development of a parasitoid, *Aphaereta pallipes* (Say) (Hym. Braconidae). Can Entomol 93:1041–1044

Houston D (1977) The effect of hooded crows on hill sheep farming in Argyll, Scotland. Hooded crow damage to hill sheep. J Appl Ecol 14:17–29

Howard DJ, Bush GL (1989) Influence of bacteria on larval survival and development in *Rhagoletis* (Diptera: Tephritidae). Ann Entomol Soc Am 82:633–640

Howell DJ (1974) Bats and pollen: physiological aspects of the syndrome of chiropterophily. Comp Biochem Physiol A 48:263–276

Hume ID (1982) Digestive physiology and nutrition of marsupials. Cambridge University Press, Cambridge

Hume ID (1989) Optimal digestive strategies in mammalian herbivores. Physiol Zool 62:1145–1163

Hume ID, Foley WJ, Chilcott MJ (1984) Physiological mechanisms of foliage digestion in the greater glider and ringtail possum (Marsupiala: Pseudocheiridae). In: Smith A, Hume ID (eds) Possums and gliders. Aust Mamm Soc, Sydney

Hungate RE (1966) The rumen and its microbes. Academic Press, New York

Huntley M, Boyd C (1984) Food-limited growth of marine zooplankton. Am Nat 124:455–478

Hupp JW, Ratti JT, Smith LM (1988) Gray partridge foraging ecology in eastern South Dakota. Great Basin Nat 48:202–205

Hurd LE (1991) Growth efficiency in juvenile mantids: absence of selection for optimization in a food-limited environment (Orthoptera: Mantidae). Proc Entomol Soc Wash 93:748–750

Hurd LE, Eisenberg RM (1984a) Experimental density manipulations of the predator *Tenodera sinensis* (Orthoptera: Mantidae) in an old-field community. I. Mortality, development and dispersal of juvenile mantids. J Anim Ecol 53:269–281

Hurd LE, Eisenberg RM (1984b) Experimental density manipulations of the predator *Tenodera sinensis* (Orthoptera: Mantidae) in an old-field community. II. The influence of mantids on arthropod community structure. J Anim Ecol 53:955–967

Hurd LE, Rathet IH (1986) Functional response and success in juvenile mantids. Ecology 67:163–167

Hurd LE, Eisenberg RM, Washburn JO (1978) Effects of experimentally manipulating density on field populations of the Chinese mantis (*Tenodera ardifolia sinensis* Saussure). Am Midl Nat 99:58–64

Hussain NA, Higuchi M (1980) Larval rearing and development of the brown spotted groper, *Epinephelus tauvina* (Forskal). Aquaculture 19:339–350

Hustler K, Howells WW (1988) The effect of primary production on breeding success and habitat selection in the African hawk-eagle. Condor 90:583–587

Hustler K, Howells WW (1989) Habitat preference, breeding success and the effect of primary productivity on tawny eagles *Aquila rapax* in the Tropics. Ibis 131:33–40

Huxley A (1939) After many a summer. Penguin Reprints, Middlesex, 1961

Iason GR, Duck CD, Clutton-Brock TH (1986) Grazing and reproductive success of red deer: the effect of local enrichment by gull colonies. J Anim Ecol 55:507–515

Ikeda T, Matsumura F, Benjamin DM (1977) Chemical basis for feeding adaptation of pine sawflies *Neodiprion rugifrons* and *Neodiprion swainei*. Science 197:497–498

Ims RA (1987) Responses in spatial organization and behaviour to manipulations of the food resource in the vole *Clethrionomys rufocanus*. J Anim Ecol 56:585–596

Iverson JB (1980) Colic modifications in iguanine lizards. J Morphol 163:79–93

Izhaki I, Safriel UN (1989) Why are there so few exclusively frugivorous birds? Experiments on fruit digestibility. Oikos 54:23–32
Izhaki I, Safriel UN (1990) Weight losses due to exclusive fruit diet – interpretation and evolutionary implications: a reply to Mack and Sedinger. Oikos 57:141–142
Jacques RP (1962) Stress and nuclear polyhedrosis in crowded populations of *Trichoplusia ni* (Hubner). J Insect Pathol 4:1–22
Jaeckle WB, Manahan DT (1989) Feeding by a "nonfeeding" larva: uptake of dissolved amino acids from seawater by lecithrotrophic larvae of the gastropod *Haliotis rufescens*. Mar Biol 103:87–94
Jaksic FM, Jimenez JE, Castro SA, Feinsinger P (1992) Numerical and functional response of predators to a long-term decline in mammalian prey at a semi-arid Neotropical site. Oecologia 89:90–101
Janis C (1976) The evolutionary strategy of the Equidae and the origins of rumen and cecal digestion. Evolution 30:757–774
Jansson RK, Smilowitz Z (1986) Influence of nitrogen on population parameters of potato insects: abundance, population growth, and within-plant distribution of the green peach aphid, *Myzus persicae* (Homoptera: Aphididae). Environ Entomol 15:49–55
Janzen DH (1988) Ecological characterization of a Costa Rican dry forest caterpillar fauna. Biotropica 20:120–135
Jaramillo VJ, Detling JK (1992) Small-scale heterogeneity in a semi-arid North American grassland. II. Cattle grazing of simulated urine patches. J Appl Ecol 29:9–13
Jefferies RL (1980) The role of organic solutes in osmoregulation in halophitic higher plants. In: Rains DW, Valentine RC, Hollaender A (eds) Genetic engineering of osmoregulation. Impact on plant productivity for food, chemicals, and energy. Plenum, New York, pp 135–154
Jenkins D, Watson A, Miller GR (1967) Population fluctuations in the red grouse *Lagopus lagopus scoticus*. J Anim Ecol 36:97–122
Jenkins SH, Bollinger PW (1989) An experimental test of diet selection by the pocket gopher *Thomomys monticola*. J Mammal 70:406–412
Jennings DT, Crawford HS Jr (1989) Predation by red squirrels on the spruce budworm *Choristoneura fumiferana* (Clem) (Lepidoptera: Tortricidae). Can Entomol 121:827–828
Jensen AL, Ball RC (1970) Variations in the availability of food as a cause of fluctuations in predator and prey population densities. Ecology 51:517–520
Jermy T (1985) Is there competition between phytophagous insects? J Zool Syst Evol 23:275–285
Johannes RE, Satomi M (1966) Composition and nutritive value of fecal pellets of a marine crustacean. Limnol Oceanogr 11:191–197
Johnson RG, Boyce MS (1990) Feeding trials with insects in the diet of sage grouse chicks. J Wildl Manage 54:89–91
Joll LM, Phillips BF (1984) Natural diet and growth of juvenile western rock lobsters *Panulirus cygnus*. J Exp Mar Biol Ecol 75:145–170
Jones E, Coman BJ (1982) Ecology of the feral cat, *Felis catus* (L.), in south-eastern Australia. III Home ranges and population ecology in semiarid north-west Victoria. Aust Wildl Res 9:409–420
Jones EN (1990) The effect of forage availability on the home range and population density of *Microtus pennsylvanicus*. J Mammal 71:382–389
Jones GP (1986) Food availability affects growth in a coral reef fish. Oecologia 70:136–139
Jones GP, Norman MD (1986) Feeding selectivity in relation to territory size in a herbivorous reef fish. Oecologia 68:549–556
Jones PJ, Ward P (1976) The level of reserve protein as the proximate factor controlling the timing of breeding and clutch size in the red-billed quelea *Quelea quelea*. Ibis 118:547–574
Jones PJ, Ward P (1979) A physiological basis for colony desertion by red-billed queleas (*Quelea quelea*). J Zool 189:1–20

Jones R, Hall WB (1974) Some observations on the population dynamics of the larval stage in the common gadoids. In: Blaxter JH (ed) The early life history of fish. Springer, Berlin Heidelberg New York. pp 87–102

Jones WT, Waser PM, Elliott LF, Link NE, Bush BB (1988) Philopatry, dispersal, and habitat saturation in the banner-tailed kangaroo rat, *Dipodomys spectabilis*. Ecology 69:1466–1473

Jordan PA, Shelton PC, Allen DL (1967) Numbers, turnover, and social structure of the Isle Royale wolf population. Am Zool 7:233–252

Journet ARP (1980) Intraspecific variation in food plant favourability to phytophagous insects, psyllids on *Eucalyptus blakelyi* M. Ecol Entomol 5:249–262

Joyner K, Gould F (1987) Conspecific tissues and secretions as sources of nutrition. In: Slanski F, Rodriguez JG (eds) Nutritional ecology of insects, mites, spiders, and related invertebrates. Wiley, New York, pp 697–719

Kabayo JP, DeLoach JR, Spates GE, Holman GM, Kapatsa GM (1986) Studies on the biochemical basis of the nutritional quality of tsetse fly diets. Comp Biochem Physiol A 83:133–139

Kahn DM, Cornell HV (1983) Early leaf abscission and folivores: comments and considerations. Am Nat 122:428–432

Kalela O (1962) On the fluctuations in the numbers of arctic and boreal small rodents as a problem of production biology. Ann Acad Sci Fenn Ser A IV 66:1–38

Kalode MB, Pant NC (1967) Studies on the amino acids, nitrogen, sugar and moisture content of maize and sorghum varieties and their relation to *Chilo zonellus* (Swin.) resistance. Indian J Entomol 29:139–144

Kanamitsu K (1966) A life table of a Japanese pine shoot moth, *Evetria cristata* (Walsingham), (Lepidoptera: Tortricidae). Z Angew Entomol 57:190–201

Karasov WH (1982) Energy assimilation, nitrogen requirement, and diet in free-living antelope ground squirrels *Ammospermophilus leucurus*. Physiol Zool 55:378–392

Karasov WH, Levey DJ (1990) Digestive system trade-offs and adaptations of frugivorous passerine birds. Physiol Zool 63:1248–1270

Karban R, Myers JH (1989) Induced plant responses to herbivory. Annu Rev Ecol Syst 20:331–348

Kasuya E, Hibino Y, Ito Y (1980) On "intercolonial" cannibalism in Japanese paper wasps, *Polistes chinensis antennalis* Perez and *P. jadwigae* Dalle Torre (Hymenoptera: Vespidae). Res Popul Ecol 22:255–262

Kavanagh RP, Lambert MJ (1990) Food selection by the greater glider, *Petauroides volans*: is foliar nitrogen a determinant of habitat quality? Aust Wildl Res 17:285–299

Kaya HK, Moon RD (1980) Influence of protein in the diet of face fly on the development of its nematode parasite, *Heterotylenchus autumnalis*. Ann Entomol Soc Am 73:547–552

Kean RI (1956) *Notornis* faeces in evidence on foods as a factor in chick rearing success. Notornis 6:229–240

Kearsley MJC, Whitham TG (1989) Developmental changes in resistance to herbivory: implications for individuals and populations. Ecology 70:422–434

Keeble FW (1910) Plant-animals: a study in symbiosis. Cambridge University Press, Cambridge

Keil RG, Kirchman DL (1991) Contribution of dissolved free amino acids and ammonium to the nitrogen requirements of heterotrophic bacterioplankton. Mar Ecol 73:1–10

Keith JO (1965) The abert squirrel and its dependence on ponderosa pine. Ecology 46:150–163

Keith LB (1983) Role of food in hare population cycles. Oikos 40:385–395

Keith LB, Cary JR (1991) Mustelid, squirrel, and porcupine population trends during a snowshoe hare cycle. J Mammal 72:373–378

Keith LB, Windberg LA (1978) A demographic analysis of the snow-shoe hare cycle. Wildl Monogr 58:1–70

Keller BL, Krebs CJ (1970) *Microtus* population biology. III Reproductive changes in fluctuating populations of *M. ochrogaster* and *M. pennsylvanicus* in southern Indiana, 1965–67. Ecol Monogr 40:263–294

Kelsey JM (1958) Symbiosis and *Anobium punctatum* deGeer. Proc R Entomol Soc Lond 33A:21–24
Kemp GA, Keith LB (1970) Dynamics and regulation of red squirrel (*Tamiasciurus hudsonicus*) populations. Ecology 51:763–779
Kemp WP, Moody UL (1984) Relationship between regional soils and foliage characteristics and western spruce budworm (Lepidoptera: Tortricidae) outbreak frequency. Environ Entomol 13:1291–1297
Kemp WP, Everson DO, Wellington WG (1985) Regional climatic patterns and western spruce budworm (*Choristoneura occidentalis*) outbreaks. US For Serv Tech Bull 0(1693):5–31
Kenagy GJ, Hoyt DF (1980) Reingestion of feces in rodents and its daily rhythmicity. Oecologia 44:403–409
Kenagy GJ, Sharbaugh SM, Nagy KA (1989) Annual cycle of energy and time expenditure in a golden-mantled ground squirrel population. Oecologia 78:269–282
Kennedy JS (1951) Benefits to aphids from feeding on galled and virus-infected leaves. Nature 168:825–826
Kennedy JS (1958) Physiological condition of the host plant and susceptibilty to aphid attack. Entomol Exp Appl 1:50–65
Kennedy JS, Stroyan HLG (1959) Biology of aphids. Annu Rev Entomol 4:139–160
Kennedy PM, Hume ID (1978) Recycling of urea nitrogen to the gut of tammar wallaby (*Macropus eugenii*). Comp Biochem Physiol A 61:117–121
Kenward RE (1982) Goshawk hunting behaviour, and range size as a function of food and habitat availability. J Anim Ecol 51:69–80
Kerr RA (1990) Sunspot-weather link is down but not out. Science 248:684–685
Kidd NAC, Lewis GB, Howell CA (1985) An association between two species of pine aphid, *Schizolachnus pineti* and *Eulachnus agilis*. Ecol Entomol 10:427–432
Kimmerer TW, Potter DA (1987) Nutritional quality of specific leaf tissues and selective feeding by a specialist leafminer. Oecologia 71:548–551
Kimmins JP (1971) Variations in the foliar amino acid composition of flowering and non-flowering balsam fir *[Abies balsamea* (L) Mill.] and white spruce *[Picea glauca* (Moench) Voss] in relation to outbreaks of the spruce budworm *[Choristoneura fumiferana* (Clem)]. Can J Zool 49:1005–1011
King BH (1989) Host-size-dependent sex ratios among parasitoid wasps: does host growth matter? Oecologia 78:420–426
King CM (1981) The reproductive tactics of the stoat, *Mustela erminea*, in New Zealand forests. In: Chapman JA, Pursely D (eds) Proc 1st Worldwide Furbearer Conf, Frostburg, Maryland. The Conference, Frostburg, pp 443–468
King CM (1983) The relationship between beech (*Nothofagus* sp.) seedfall and the populations of mice (*Mus musculus*), and the demographic and dietary responses of the stoat (*Mustela erminea*), in three New Zealand forests. J Anim Ecol 52:141–166
King CM (1985) Interactions between woodland rodents and their predators. Symp Zool Soc Lond 55:219–245
King RD, Bendell JF (1982) Foods selected by blue grouse (*Dendragapus obscurus fuliginosus*). Can J Zool 60:3268–3281
Kiørboe T (1989) Phytoplankton growth rate and nitrogen content: implications for feeding and fecundity in a herbivorous copepod. Mar Ecol Prog Ser 55:229–234
Kirkpatrick TW (1957) Insect life in the tropics. Longmans, London
Klein DR (1970) Food selection by North American deer and their response to over-utilization of preferred plant species. Symp B Ecol Soc 10:25–46
Klimetzek D (1990) Population dynamics of pine-feeding insects: a historical study. In: Watt AD, Leather SR, Hunter MD, Kidd NAC (eds) Population dynamics of forest insects. Intercept, Andover Hampshire, pp 3–10
Kloft W (1957) Further investigations concerning the interrelation between bark condition of *Abies alba* and infestations by *Adelges piceae typica* and *A. nusslini schneideri*. Z Angew Entomol 41:438–442

Kluyver HK, Tinbergen L (1954) Territory and the regulation of density in titmice. Arch Neerl Zool 10:265–289

Kokkinn MJ, Davis AR (1986) Secondary production: shooting a halcyon for its feathers. In: Deckker PDe, Williams WD (eds) Limnology in Australia. CSIRO/ Junk, Melbourne/Dordrecht, pp 252–261

Koller CN, Leonard DE (1981) Comparison of energy budgets of spruce budworm *Choristoneura fumiferana* (Clemens) on balsam fir and white spruce. Oecologia 49:14–20

Komdeur J (1992) Importance of habitat saturation and territory quality for evolution of cooperative breeding in the Seychelles warbler. Nature 358:493–485

Koprowski JL (1991) Response of fox squirrels and gray squirrels to a late-spring early-summer food shortage. J Mammal 72:367–372

Korn H (1989) A feeding experiment with 6-methoxybenzoxazolinone and a wild population of the deer mouse (*Peromyscus maniculatus*). Can J Zool 67:2220–2224

Korn H, Taitt MJ (1987) Initiation of early breeding in a population of *Microtus townsendii* (Rodentia) with the secondary plant compound 6-MBOA. Oecologia 71:593–596

Kotanen P, Jefferies RL (1987) The leaf and shoot demography of grazed and ungrazed plants of *Carex subspathacea*. J Ecol 75:961–975

Krebs CJ, Boutin S, Gilbert BS (1986a) A natural feeding experiment on a declining snowshoe hare population. Oecologia 70:194–197

Krebs CJ, Gilbert BS, Boutin S, Sinclair AER, Smith JNM (1986b) Population biology of snowshoe hares. I. Demography of food-supplemented populations in the southern Yukon, 1976–84. J Anim Ecol 55:963–982

Krohne DT (1989) Demographic characteristics of *Peromyscus leucopus* inhabiting a natural dispersal sink. Can J Zool 67:2321–2325

Kukor JJ, Martin MM (1983) Acquisition of digestive enzymes by siricid woodwasps from their fungal symbionts. Science 220:1161–1163

Kukor JJ, Martin MM (1986) The effect of acquired microbial enzymes on assimilation efficiency in the common woodlouse, *Tracheoniscus rathkei*. Oecologia 69:360–366

Kukor JJ, Cowan DP, Martin MM (1988) The role of ingested fungal enzymes in cellulose digestion in the larvae of cerambycid beetles. Physiol Zool 61:364–371

Kunstyr I, Schiel R, Kaup FJ, Uhr G, Kirchhoff H (1988) Giant gram-negative noncultivable endospore-forming bacteria in rodent intestines. Naturwissenschaften 75:525–527

Kurihara Y, Kikkawa J (1986) Trophic relations of decomposers. In: Kikkawa J, Anderson DJ (eds) Community ecology: patterns and processes. Blackwell, Oxford, pp 127–160

Lack D (1954) The natural regulation of animal numbers. Oxford University Press, Oxford

Laine K, Henttonen H (1983) The role of plant production in microtine cycles in northern Fennoscandia. Oikos 40:407–418

Lake JS (1967) Rearing experiments with five species of Australian freshwater fishes. I. Inducement of spawning. Aust J Mar Freshwater Res 18:137–153

Lambrou PD, Tzanakakis ME (1978) Inhibition of larval growth of *Dacus oleae* (Diptera: Tephritidae) by streptomycin. II. Effects of treating the parents. Entomol Exp Appl 23:163–170

Lampert W, Muck P (1985) Multiple aspects of food limitation in zooplankton communities: the *Daphnia-Endiaptomus* example. Arch Hydrobiol Beih 21:311–322

Lampitt RS, Noji T, von Bodungen B (1990) What happens to zooplankton faecal pellets? Implications for material flux. Mar Biol 104:15–23

Lance AN (1978a) Territories and the food plant of individual red grouse. II. Territory size compared with an index of nutrient supply in heather. J Anim Ecol 47:307–314

Lance AN (1978b) Survival and recruitment success of individual young cock red grouse *Lagopus l. scoticus* tracked by radio-telemetry. Ibis 120:369–378

Lance AN, Mahon G (1975) Foods of a marginal red grouse population in western Ireland. J Wildl Manage 39:183–187

Landsberg J (1987) Feeding preferences of common brushtail possums, *Trichosurus vulpecula*, on seedlings of a woodland eucalypt. Aust Wildl Res 14:361–370

Lane CE (1959) The nutrition of *Teredo*. Ann N Y Acad Sci 77:246–248

Langer P (1988) The mammalian herbivore stomach. Comparative anatomy, function and evolution. Fischer, Stuttgart
Lanyon JM, Sanson GD (1986) Koala (*Phascolarctos cinereus*) dentition and nutrition. II. Implications of tooth wear in nutrition. J Zool Lond (A) 209:169–181
Larson KC, Whitham TG (1991) Manipulation of food resources by a gall-forming aphid: the physiology of sink-source interactions. Oecologia 88:15–21
Larsson S (1985) Seasonal changes in the within-crown distribution of the aphid *Cinara pini* on Scots pine. Oikos 45:217–222
Larsson S, Tenow O (1984) Areal distribution of a *Neodiprion sertifer* (Hym., Diprionidae) outbreak on Scots pine as related to stand condition. Holarct Ecol 7:81–90
Lauckhart JB (1957) Animal cycles and food. J Wildl Manage 21:230–234
Lauckhart JB (1962) Wildlife population fundamentals. North Am Wildl Conf 27:233–242
Laurence GC (1974) Growth and survival of haddock (*Melanogrammus aeglefinus*) larvae in relation to planktonic prey concentration. J Fish Res Board Can 31:1415–1419
Laurence GC (1977) A bioenergetic model for the analysis of feeding and survival potential of winter flounder, *Pseudopleuronectes americanus*, larvae during the period from hatching to metamorphosis. Fish Bull 75:529–546
Laurie WA, Brown D (1990) Population biology of marine iguanas (*Amblyrhynchus cristatus*). III. Factors affecting survival. J Anim Ecol 59:545–568
Law BS (1992a) Physiological factors affecting pollen use by Queensland blossom bats (*Syconycteris australis*). Funct Ecol 6:257–264
Law BS (1992b) The maintenance nitrogen requirements of the Queensland blossom bat (*Syconycteris australis*) on a sugar pollen diet: is nitrogen a limiting resource? Physiol Zool 65:634–648
Lawson DL, Merritt RW, Klug MJ, Martin JS (1982) The utilization of late season foliage by the orange striped oakworm, *Anisota senatoria*. Entomol Exp Appl 32:242–248
Lawson DL, Merritt RW, Martin MM, Martin JS, Kukor JJ (1984) The nutritional ecology of larvae of *Alsophila pometaria* and *Anisota senatoria* feeding on early- and late-season oak foliage. Entomol Exp Appl 35:105–114
Leather SR, Barbour DA (1987) Associations between soil type, lodgepole pine (*Pinus contorta)* provenance, and the abundance of the pine beauty moth, *Panolis flammea*. J Appl Ecol 24:945–952
Lecowicz MJ (1987) Resource allocation by plants under air pollution stress: implications for plant – pest – pathogen interactions. Bot Rev 53:281–300
Leibholz J (1970) The effect of starvation and low nitrogen intakes on the concentration of free amino acids in the blood plasma and on the nitrogen metabolism of sheep. Aust J Agric Res 21:723–734
Lemly AD, Dimmick JF (1982) Growth of young-of-the-year and yearling centrarchids in relation to zooplankton in the littoral zone of lakes. Copeia 1982:305–321
Lenski RE (1984) Food limitation and competition: field experiment with two *Carabus* species. J Anim Ecol 53:203–216
Leopold AS (1953) Intestinal morphology of gallinaceous birds in relation to food habits. J Wildl Manage 17:197–203
Leschine SB, Holwell K, Canale-Parola E (1988) Nitrogen fixation by anaerobic cellulolytic bacteria. Science 242:1157–1159
Leuthold BM, Leuthold W (1978) Ecology of the giraffe in Tsavo East National Park, Kenya. East Afr Wildl J 16:1–20
Levey DJ (1988) Spatial and temporal variation in Costa Rican fruit and fruit-eating bird abundance. Ecol Monogr 58:251–269
Levi MP, Cowling EB (1969) Role of nitrogen in wood deterioration. VII. Physiological adaptation of wood-destroying and other fungi to substrates deficient in nitrogen. Phytopathology 59:460–468
Lewis AC (1984) Plant quality and grasshopper feeding. Effects of sunflower condition on preference and performance in *Melanoplus differentialis*. Ecology 65:836–843
Lewis T (1990) The pelicans' progress. Aust Mag, June 9–10:9–12

Liebhold AM, Elkington JS (1990) Models of the spatial dynamics of epidemic gypsy moth populations. In: Watt AD, Leather SR, Hunter MD, Kidd NAC (eds) Population dynamics of forest insects. Intercept, Andover Hampshire, pp 359–367
Lincoln DE (1985) Host-plant protein and phenolic resin effects on larval growth and survival of a butterfly. J Chem Ecol 11:1459–1468
Lincoln DE, Newton TS, Ehrlich PR, Williams KS (1982) Coevolution of the checkerspot butterfly *Euphydryas chalcedona* and its larval food plant *Diplacus aurantiacus*: larval response to protein and leaf resin. Oecologia 52:216–223
Lindlof B, Lindstrom E, Pehrson A (1974a) On activity, habitat selection and diet of the mountain hare (*Lepus timidus* L) in winter. Viltrevy 9:27–43
Lindlof B, Lindstrom E, Pehrson A (1974b) Nutrient content in relation to food preferred by mountain hare. J Wildl Manage 38:875–879
Lindroth RL, Barman MA, Weisbrod AV (1991) Nutrient deficiencies and the gypsy moth, *Lymantria dispar*: effects on larval performance and detoxification enzyme activity. J Insect Physiol 37:45–52
Lindstrom E (1989) Food limitation and social regulation in a red fox population. Holarct Ecol 12:70–79
Lobel PS (1981) Trophic biology of herbivorous reef fishes; alimentary pH and digestive capabilities. J Fish Biol 19:365–397
Lockwood JA (1989a) Cannibalism in rangeland grasshoppers (Orthoptera: Acrididae): attraction to cadavers. J Kansas Entomol Soc 61:379–387
Lockwood JA (1989b) Ontogeny of cannibalism in rangeland grasshoppers (Orthoptera: Acrididae). J Kans Entomol Soc 62:534–541
Loftin H, Tyson EL (1965) Iguanas as carrion eaters. Copeia 1965:515
Long JL (1984) The diet of three species of parrots in the south of Western Australia. Aust Wildl Res 11:357–371
Lowry JB (1989) Green-leaf fractionation by fruit bats: is this feeding behaviour a unique nutritional strategy for herbivores? Aust Wildl Res 16:203–206
Mabelis AA (1979) Wood ant wars. The relationship between aggression and predation in the red wood ant (*Formica polyctena* Forst.). Neth J Zool 29:451–620
Macedo CA, Langenheim JH (1989) A further investigation of leaf sesquiterpine variation in relation to herbivory in 2 Brazilian populations of *Copaifera langsdorfii*. Biochem Syst Ecol 17:207–216
Mack AL (1990) Is frugivory limited by secondary compounds in fruit? Oikos 57:135–138
Mackie RI (1987) Microbial digestion of forage in herbivores. In: Hacker JB, Ternouth JH (eds) The nutrition of herbivores. Academic Press, Sydney, pp 233–265
Magrath RD (1989) Hatching asynchrony and reproductive success in the blackbird. Nature 339:536–538
Main AR, Shield JW, Waring H (1959) Recent studies on marsupial ecology. Monogr Biol 8:315–331
Manahan DT (1990) Adaptations by invertebrate larvae for nutrient acquisition from seawater. Am Zool 30:147–160
Marcstrom V, Hoglund N, Krebs CJ (1990) Periodic fluctuations in small mammals at Boda, Sweden from 1961 to 1988. J Anim Ecol 59:753–761
Mares MA, Lacher TE Jr (1987) Social spacing in small mammals: patterns of individual variation. Am Zool 27:293–306
Mares MA, Watson MD, Lacher TE Jr (1976) Home range perturbations in *Tamias striatus*. Food supply as a determinant of home range and density. Oecologia 25:1–12
Mares MA, Lacher TE Jr, Willig MR, Bitor NA (1982) An experimental analysis of social spacing in *Tamias striatus*. Ecology 63:267–273
Markin GP (1982) Abundance and life cycles of Lepidoptera associated with an outbreak of the western spruce budworm *Choristoneura occidentalis* (Lepidoptera: Tortricidae) in southeastern Idaho. J Kans Entomol Soc 55:365–372
Martin MM (1983) Minireview: cellulose digestion in insects. Comp Biochem Physiol A 75:313–324

Martin MM, Kukor JJ (1984) Role of mycophagy and bacteriophagy in invertebrate nutrition. In: Klug MJ, Reddy CA (eds) Current perspectives in microbial ecology. American Society of Microbiology, Washington DC, pp 257–263

Martin RW (1985a) Overbrowsing, and decline of a population of the koala, *Phascolarctos cinereus*, in Victoria. I. Food preference and food tree defoliation. Aust Wildl Res 12:355–365

Martin RW (1985b) Overbrowsing, and decline of a population of the koala, *Phascolarctos cinereus*, in Victoria II. Population condition. Aust Wildl Res 12:367–375

Martin RW, Lee A (1984) The koala, *Phascolarctos cinereus*, the largest marsupial folivore. In: Smith A, Hume ID (eds) Possums and gliders. Aust Mamm Soc, Sydney, pp 463–467

Martin TE (1987) Food as a limit on breeding birds: a life-history perspective. Annu Rev Ecol Syst 18:453–488

Mason CF (1970) Food, feeding rates and assimilation in woodland snails. Oecologia 4:358–373

Mason RR (1974) Population change in an outbreak of the Douglas-fir tussock moth, *Orgyia pseudotsuga* (Lepidoptera: Lymantriidae), in central Arizona. Can Entomol 106:1171–1174

Mason RR (1981) Numerical analysis of the cause of population collapse in a severe outbreak of the Douglas-fir tussock moth. Ann Entomol Soc Am 74:51–57

Mason RR, Torgersen TR (1987) Dynamics of non-outbreak populations of the Douglas-fir tussock moth (Lepidoptera: Lymantriidae) in southern Oregon. Environ Entomol 16:1217–1228

Mason RR, Torgersen TR, Wickman BE, Gene PH (1983) Natural regulation of a Douglas-fir tussock moth (Lepidoptera: Lymantriidae) population in the Sierra Nevada. Environ Entomol 12:587–594

Mautz WJ, Nagy KA (1987) Ontogenetic changes in diet, field metabolic rate, and water flux in the herbivorous lizard *Dipsosaurus dorsalis*. Physiol Zool 60:640–658

May RC (1974) Larval mortality in marine fishes and the critical period concept. In: Blaxter JH (ed) The early life history of fish. Springer, Berlin Heidelberg New York, pp 3–19

Maynard-Smith J (1985) Do we need a new evolutionary paradigm? New Sci 105:38

McArthur C, Sanson GD (1991) Effect of tannins on digestion in the common ringtail possum (*Pseudocheirus peregrinus*), a specialized marsupial folivore. J Zool Lond 225:233–251

McCaffery AR (1975) Food quality and quantity in relation to egg production in *Locusta migratoria migratorioides*. J Insect Physiol 21:155–158

McClure MS (1980) Foliar nitrogen: a basis for host suitability for elongate hemlock scale, *Fiorinia externa* (Homoptera: Diaspididae). Ecology 61:72–79

McGugan BM (1954) Needle-mining habits and larval instars of the spruce budworm. Can Entomol 86:439–454

McKillup SC, Butler AJ (1983) The measurement of hunger as a relative estimate of food available to populations of *Nassarius pauperatus*. Oecologia 56:16–22

McNaughton SJ (1979) Grazing as an optimization process: grass-ungulate relationships in the Serengeti. Am Nat 113:691–703

McNaughton SJ (1988) Mineral nutrition and spatial concentrations of African ungulates. Nature 334:343–345

McNaughton SJ, Georgiadis NJ (1986) Ecology of African grazing and browsing mammals. Annu Rev Ecol Syst 17:39–65

McNaughton SJ, Oesterheld M, Frank DA, Williams KJ (1989) Ecosystem-level patterns of primary productivity and herbivory in terrestrial habitats. Nature 341:142–144

McNeill S (1973) The dynamics of the population of *Leptopterna dolobrata* (Heteroptera: Miridae) in relation to its food resources. J Anim Ecol 42:495–507

McQuaid CD (1983) Population dynamics and the effect of food availability on growth of the gastropod *Oxystele variegata* (Anton) on an exposed rocky shore. S Afr J Zool 18:56–61

McTigue TA, Zimmerman RJ (1991) Carnivory vs herbivory in juvenile *Penaeus setiferus* (Linnaeus) and *Penaeus aztecus* (Ives). J Exp Mar Biol Ecol 151:1–16

Merriam HG (1971) Sensitivity of terrestrial isopod populations (*Armadillidium)* to food quality differences. Can J Zool 49:667–674

Merrill W, Cowling EB (1966a) Role of nitrogen in wood deterioration: amounts and distribution of nitrogen in tree stems. Can J Bot 44:1555–1580

Merrill W, Cowling EB (1966b) Role of nitrogen in wood deterioration: amount and distribution of nitrogen in fungi. Phytopathology 56:1083–1090

Merson MH, Kirkpatrick RL (1981) Relative sensitivity of reproductive activity and body-fat level to food restriction in white-footed mice (*Peromyscus leucopus*). Am Midl Nat 106:305–312

Meslow EC, Keith LB (1968) Demographic parameters of a snowshoe hare population. J Wildl Manage 32:812–834

Miller CA (1963) The bionomics of the spruce budworm. Mem Entomol Soc Can 31:12–19

Miller GR (1968) Evidence for selective feeding on fertilized plots by red grouse, hares and rabbits. J Wildl Manage 32:849–853

Miller WE, Epstein ME (1986) Synchronous population fluctuations among moth species (Lepidoptera). Environ Entomol 15:443–447

Millington SJ, Grant PR (1983) Feeding ecology and territoriality of the cactus finch *Geospiza scandens* on Isla Daphne Major, Galapagos. Oecologia 58:76–83

Mills JA, Mark AF (1977) Food preferences of takahe in Fiordland National Park, New Zealand, and the effect of competition from introduced red deer. J Anim Ecol 46:939–958

Mills JA, Lee WG, Mark AF, Lavers RB (1980) Winter use by takahe (*Notornis mantelli*) of the summer green fern (*Hypolepis millefolium*) in relation to its annual cycle of carbohydrates and minerals. N Z J Ecol 3:131–137

Mills JA, Lavers RB, Lee WG, Mara MK (1991) Food selection by takahe *Notornis mantelli* in relation to chemical composition. Ornis Scand 22:111–128

Milne A (1950) The ecology of the sheep tick *Ixodes ricinus* L., spatial distribution. Parasitology 40:35–45

Milton K (1979) Factors influencing leaf choice by howler monkeys: a test of some hypotheses of food selection by generalist herbivores. Am Nat 114:362–378

Milton K (1981) Food choice and digestive strategies of two sympatric primate species. Am Nat 117:496–505

Mittler TE (1967) Effect on aphid feeding of dietary methionine. Nature 214:386

Mols PJM (1987) Hunger in relation to searching behaviour, predation and egg production of the carabid beetle *Pterostichus coerulescens* L.: results of simulation. Acta Phytopathol Entomol Hung 22:107–205

Momot WT (1984) Crayfish production: a reflection of community energetics. J Crustacean Biol 4:35–54

Monaghan P, Uttley JD, Burns MD, Thaine C, Blackwood J (1989) The relationship between food supply, reproductive effort and breeding success in arctic terns *Sterna paradisaea*. J Anim Ecol 58:261–274

Montgomerie RD, Redsell CA (1980) A nesting hummingbird feeding solely on arthropods. Condor 82:463–464

Montgomery ME (1982) Life-cycle nitrogen budget for the gypsy moth, *Lymantria dispar*, reared on artificial diet. J Insect Physiol 28:437–442

Montgomery WL, Gerking SD (1980) Marine macroalgae as foods for fishes: an evaluation of potential food supply quality. Environ Biol Fishes 5:143–153

Montoya R, Ambid L, Agid R (1979) Torpor induced at any season by suppression of food proteins in a hibernator, the garden dormouse (*Eliomys quercinus*). Comp Biochem Physiol A 62:371–376

Mooney HA, Williams KS, Lincoln DE, Ehrlich PR (1981) Temporal and spatial variability in the interaction between the checkerspot butterfly, *Euphydryas chalcedona* and its principal food source, the California shrub, *Diplacus aurantiacus*. Oecologia 50: 195–198

Morgan DL, Frankie GW, Gaylor MG (1983) An evaluation of the gall-forming capacity of live oak in response to the cynipid wasp, *Disholcaspis cinerosa*, in Texas. J Kans Entomol Soc 56:100–108

Moriarty DJW, Barclay MC (1981) Carbon and nitrogen content of food and the assimilation efficiencies of penaeid prawns in the Gulf of Carpenteria. Aust J Mar Freshwater Res 32:245–251
Morita RY (1980) Microbial life in the deep sea. Can J Microbiol 26:1375–1385
Morrison DW (1980) Efficiency of food utilization by fruit bats. Oecologia 45:270–273
Morse DH (1978) Populations of bay-breasted and Cape May warblers during an outbreak of the spruce budworm. Wilson Bull 90:404–413
Morse DH (1979) Prey capture by the crab spider *Misumena calycina* (Araneae: Thomisidae). Oecologia 39:309–319
Morse DH (1980) Behavioural mechanisms in ecology. Harvard University Press, Harvard
Mortensen A, Tindall A (1978) Uric acid metabolism in the caeca of grouse (*Lagopus lagopus*). J Physiol 284:159–160
Mortensen A, Tindall A (1981) Nitrogen recycling in grouse (*Lagopus lagopus).* Parasitology 82:31–33
Morton ES (1973) On the evolutionary advantages and disadvantages of fruit-eating in tropical birds. Am Nat 107:8–22
Morton ES (1978) Avian arboreal folivores: why not? In: Montgomery GC (ed) The ecology of arboreal folivores. Smithsonian Institute, Washington, DC, pp 123–130
Morton ML (1979) Fecal sac ingestion in the mountain white-crowned sparrow. Condor 81:72–77
Moss R (1968) Food selection and nutrition in ptarmigan (*Lagopus mutus*). Symp Zool Soc Lond 21:207–216
Moss R (1977) The digestion of heather by red grouse during the spring. Condor 79:471–477
Moss R, Hanssen I (1980) Grouse nutrition. Nutrit Abstr Rev B 50:555–567
Moss R, Hewson R (1985) Effects on heather of heavy grazing by mountain hares. Holarct Ecol 8:280–284
Moss R, Watson A (1984) Maternal nutrition, egg quality and breeding success of Scottish ptarmigan *Lagopus mutus.* Ibis 126:212–220
Moss R, Weir DN (1987) Demography of capercaillie *Tetrao urogallus* in north-east Scotland. 3. Production and recruitment of young. Ornis Scand 18:141–145
Moss R, Watson A, Parr R (1975) Maternal nutrition and breeding success in red grouse (*Lagopus lagopus scoticus*). J Anim Ecol 44:233–244
Moss R, Welch D, Rothery P (1981) Effects of grazing by mountain hares and red deer on the production and chemical composition of heather. J Appl Ecol 18:487–496
Moss R, Watson A, Parr R (1988) Mate choice by hen red grouse *Lagopus lagopus* with an excess of cocks – role of territory size and food quality. Ibis 130:545–552
Moss R, Trenholm IB, Watson A, Parr R (1990) Plant growth and nitrogen metabolism of red grouse *Lagopus lagopus scoticus* in spring. Ornis Scand 21:115–121
Muck P, Lampert W (1980) Feeding of freshwater filter-feeders at very low concentrations: poor evidence for "threshold feeding" and "optimal feeding" in *Daphnia longispina* and *Endiaptomus gracilis.* J Plankton Res 2:367–379
Murdoch WW (1966) Community structure, population control, and competition – a critique. Am Nat 100:219–226
Murphy DD, Launer AE, Ehrlich PR (1983) The role of adult feeding in egg production and population dynamics of the checkerspot butterfly *Euphydryas editha.* Oecologia 56:257–263
Murray BG (1982) On the meaning of density dependence. Oecologia 53:370–373
Myers JH (1979) The effect of food quality and quantity on emergence time in the cinnabar moth. Can J Zool 57:1150–1156
Myers JH (1980) Is the insect or the plant the driving force in the cinnabar moth-tansy ragwort system? Oecologia 47:16–21
Myers JH (1985) Effect of physiological condition of the host plant on the ovipositional choice of the cabbage white butterfly, *Pieris rapae.* J Anim Ecol 54:193–204
Myers JH (1988) Can a general hypothesis explain population cycles of forest lepidoptera? Adv Ecol Res 18:179–242

Myers JH, Post BJ (1981) Plant nitrogen and fluctuations of insect populations: a test with the cinnabar moth-tansy ragwort system. Oecologia 48:151–156

Myers JH, Monro J, Murray N (1981) Egg clumping, host plant selection and population regulation in *Cactoblastis cactorum* (Lepidoptera). Oecologia 51:7–13

Myllymaki A (1977) Demographic mechanisms in the fluctuating populations of the field vole *Microtis agrestis*. Oikos 29:468–493

Myrberget S (1981) Diet of willow grouse *Lagopus lagopus* chicks on a Norwegian coastal island. Fauna Norv Ser C 4:58–63

Myrberget S (1984) Population cycles of willow grouse *Lagopus lagopus* on an island in northern Norway. Fauna Norv Ser C 7:46–56

Nagy KA (1977) Cellulose digestion and nutrient assimilation in *Sauromalus obesus*, a plant-eating lizard. Copeia 1977:355–362

Nagy KA, Shoemaker VH (1984) Field energetics and food consumption of the Galapagos marine iguana, *Amblyrhynchus cristatus*. Physiol Zool 57:281–290

Nalepa CA (1984) Colony composition, protozoan transfer and some life history characteristics of the woodroach *Cryptocercus punctulatus* Scudder (Dictyoptera: Cryptocercidae). Behav Ecol Sociobiol 14:273–279

Nalepa CA (1990) Early development of nymphs and establishment of hindgut symbiosis in *Cryptocercus punctulatus* (Dictyoptera: Cryptocercidae). Ann Entomol Soc Am 83:786–789

Nash TAM, Jordan AM, Boyle JA (1966) Effect of host pregnancy on pupal production by the tsetse fly. Nature 212:1581–1582

Naylor BJ, Bendell JF (1989) Clutch size and egg size of spruce grouse in relation to spring diet, food supply, and endogenous reserves. Can J Zool 67:969–980

Negus NC, Berger PJ (1977) Experimental triggering of reproduction in a natural population of *Microtus montanus*. Science 196:1230–1231

Nelemans MNE (1987) On the life-history of the carabid beetle *Nebria brevicollis* (F.): egg production and larval growth under experimental conditions. Neth J Zool 37:26–42

Nelemans MNE, Den Boer PJ, Spee A (1989) Recruitment and summer diapause in the dynamics of a population of *Nebria brevicollis* (Coleoptera: Carabidae). Oikos 56:157–169

Neuvonen S (1988) Interaction between geometrid and microtine cycles in northern Fennoscandia. Oikos 51:393–397

Neuvonen S, Saikkonen K, Haukioja E (1990) Simulated acid rain reduces the susceptibility of European pine sawfly (*Neodiprion sertifer*) to its nuclear polyhedrosis virus. Oecologia 83:209–212

Newbery DMcC (1980) Interactions between the coccid, *Icerya seychellarum* (Westw.), and its host tree species on Aldabra atoll. I. *Euphorbia pyrifolia* Lam. Oecologia 46:171–179

Newell R (1965) The role of detritus in the nutrition of two marine deposit feeders, the prosobranch *Hydrobia ulvae* and the bivalve *Macoma balthica*. Proc Zool Soc Lond 144:25–45

Newsome AE (1975) An ecological comparison of the two arid-zone kangaroos of Australia, and their anomolous prosperity since the introduction of ruminant stock to their environment. Q Rev Biol 50:389–424

Newsome AE, Corbett LK (1975) Outbreaks of rodents in semi-arid and arid Australia: causes, preventions, and evolutionary considerations. In Prakash J, Ghosh PK (eds) Rodents in desert environments. Junk, The Hague, pp 117–153

Newsome AE, Parer I, Catling PC (1989) Prolonged prey suppression by carnivores – predator removal experiments. Oecologia 78:458–467

Newton I (1964) Bud-eating by bullfinches in relation to the natural food supply. J Appl Ecol 1:265–279

Newton I (1967) The feeding ecology of the bullfinch (*Pyrrhula pyrrhula* L.) in southern England. J Anim Ecol 36:721–744

Newton I (1980) The role of food in limiting bird numbers. Ardea 68:11–30

Newton I, Marquiss M (1982) Food, predation and breeding season in sparrowhawks (*Accipiter nisus*). J Zool 197:221–240

Newton I, Marquiss M (1986) Population regulation in sparrowhawks. J Anim Ecol 55:463–480
Newton I, Marquiss M, Weir DN, Moss R (1977) Spacing of sparrowhawk nesting territories. J Anim Ecol 46:425–441
Nicotri ME (1977) Grazing effects of four marine intertidal herbivores on the microflora. Ecology 58:1020–1032
Nisbet ICT (1978) Dependence of fledging success on egg-size, parental performance and egg-composition among common and roseate terns, *Sterna hirundo* and *S. dougallii*. Ibis 120:207–215
Nishihara T, Kuroda S (1991) Soil-scratching behaviour by western lowland gorillas. Folia Primatol 57:48–51
Nixon CM (1970) Insects as food for juvenile gray squirrels. Am Midl Nat 84:283
Nixon CM, McClain MW, Donohoe RW (1975) Effects of hunting and mast crops on a squirrel population. J Wildl Manage 39:1–25
Noji TT (1991) The influence of macrozooplankton on vertical particulate flux. Sarsia 76:1–9
Noll UG (1979) Body temperature, oxygen consumption, noradrenalin response and cardiovascular adaptations in the flying fox *Rousettus aegyptiacus*. Comp Biochem Physiol A 63:79–88
Norman MD, Jones GP (1984) Determinants of territory size in the pomacentrid reef fish, *Parma victoriae*. Oecologia 61:60–69
Norris KS (1953) The ecology of the desert iguana *Dipsosaurus dorsalis*. Ecology 34:265–287
North BB, Stephens GC (1971) Uptake and assimilation of amino acids by *Platymonas*. II. Increased uptake in nitrogen-deficient cells. Biol Bull 140:242–254
Oates JF, Whitesides GH, Davies AG, Waterman PG, Green SM, Dasilva GL, Mole S (1990) Determinants of variation in tropical forest primate biomass: new evidence from West Africa. Ecology 71:328–343
Obst BS, Diamond JM (1989) Interspecific variation in sugar and amino acid transport by the avian cecum. J Exp Zool S3:117–126
O'Keeffe JH (1985a) Population biology of the freshwater snail *Bulinus globosus* on the Kenya coast 1. Population fluctuations in relation to climate. J Appl Ecol 22:73–84
O'Keeffe JH (1985b) Population biology of the freshwater snail *Bulinus globosus* on the Kenya coast 2. Feeding and density effects on population parameters. J Appl Ecol 22:85–90
O'Kelly JC, Seifert GW (1969) Relationships between resistance to *Boophilus microplus*, nutritional status, and blood composition in shorthorn×hereford cattle. Aust J Biol Sci 22:1497–1506
Olive CW (1980) Foraging specializations in orb-weaving spiders. Ecology 61:1133–1144
Onstad DW, Reissig WH, Shoemaker CA (1986) Influence of apple cultivar, tree phenology, and leaf quality on the development and mortality of *Choristoneura rosaceana* (Lepidoptera; Tortricidae). Can Entomol 118:123–132
Orskov ER, Kay RNB (1987) Non-microbial digestion of forages by herbivores. In: Hacker JB, Ternouth JH (eds) The nutrition of herbivores. Academic Press, Sydney, pp 267–280
Osawa R (1990) Feeding strategies of the swamp wallaby, *Wallabia bicolor*, on North Stradbroke Island, Queensland. I. Composition of diets. Aust Wildl Res 17:615–621
Osenberg CW (1989) Resource limitation, competition and the influence of life history in a freshwater snail community. Oecologia 79:512–519
Ostfeld RS (1985) Limiting resources and territoriality in microtine rodents. Am Nat 126:1–15
Owen M (1971) The selection of feeding site by white-fronted geese in winter. J Appl Ecol 8:905–917
Owen M (1975) Cutting and fertilizing grassland for winter goose management. J Wildl Manage 39:163–167
Owen-Smith N (1990) Demography of a large herbivore, the greater kudu *Tragelaphus strepsiceros*, in relation to rainfall. J Anim Ecol 59:893–913

Packham JM (1982) *Holcus*, *Holcaphis* and food quality. Proc 5th Symp Insect-Plant Relationships, Wageningen, 1982. Pudoc, Wageningen, pp 429–430

Padhi G, Chatterji SM (1986) Influence of nitrogen in rice varieties on the yellow stem borer (*Scirpophaga incertulas* Wlk) infestation. J Entomol Res 10:171–173

Pahl LI (1987a) Feeding behaviour and diet of the common ringtail possum, *Pseudocheirus peregrinus*, in *Eucalyptus* woodlands and *Leptospermum* thickets in southern Victoria. Aust J Zool 35:487–506

Pahl LI (1987b) Survival, age determination and population age stucture of the common ringtail possum, *Pseudocheirus peregrinus*, in a *Eucalyptus* woodland and a *Leptospermum* thicket in southern Victoria. Aust J Zool 35:625–639

Pant NC, Kapoor S (1964) A note on the nutritional behaviour of the larvae of greater wax moth, *Galleria mellonella* L. (Lepidoptera: Pyralidae). Indian J Entomol 26:125–127

Paris OH (1963) The ecology of *Armadillidium vulgare* (Isopoda: Oniscoidea) in California grassland: food, enemies and weather. Ecol Monogr 33:1–20

Paris OH, Pitelka FA (1962) Population characteristics of the terrestrial isopod *Armadillidium vulgare* in California grassland. Ecoloy 43:229–248

Paris OH, Sikora A (1965) Radiotracer demonstration of isopod herbivory. Ecology 46:729–735

Parra R (1978) Comparison of foregut and hindgut fermentation in herbivores. In: Montgomery GG (ed) The ecology of arboreal folivores. Smithsonian Institute Press, Washington, DC, pp 205–229

Parry WH (1969) A study of the relationship between defoliation of sitka spruce and population levels of *Elatobium abietinum* (Walker). Forestry 42:69–82

Parry WH (1971) Differences in the probing behaviour of *Elatobium abietinum* feeding on sitka and Norway spruces. Ann Appl Biol 69:177–185

Parry WH (1974a) Damage caused by green spruce aphid to Norway and sitka spruce needles. Ann Appl Biol 77:113–120

Parry WH (1974b) The effects of nitrogen levels in sitka spruce needles on *Elatobium abietinum* (Walker) populations in north-eastern Scotland. Oecologia 15:305–320

Parry WH (1976) The effects of needle age on the acceptibility of sitka spruce needles to the aphid *Elatobium abietinum* (Walker). Oecologia 23:297–313

Parry WH (1979) Summer survival of the green spruce aphid *Elatobium abietinum* in north-east Scotland, UK. Oecologia 41:235–244

Parry-Jones K, Augee ML (1991) Food selection by grey-headed flying foxes (*Pteropus poliocephalus*) occupying a summer colony site near Gosford, New South Wales. Wildl Res 18:111–124

Paton DC (1982) The diet of the New Holland honeyeater, *Phylidonyris novaehollandiae*. Aust J Ecol 7:279–298

Pearson DL, Knisley CB (1985) Evidence for food as a limiting resource in the life cycle of tiger beetles (Coleoptera, Cicindelidae). Oikos 45:161–168

Pease JL, Vowles RH, Keith LB (1979) Interaction of snowshoe hares and woody vegetation. J Wildl Manage 43:43–60

Pefaur JE, Yanez JL, Jaksic FM (1979) Biological and environmental aspects of a mouse (*Oryzomys longicaudatus*) outbreak in the semi-arid region of Chile. Mammalia 43:313–322

Pellew RA (1983) The giraffe and its food source in the Serengeti. II. Response of the giraffe population to change in the food supply. Afr J Ecol 21:269–284

Pellew RA (1984a) The feeding ecology of a selective browser, the giraffe (*Giraffa camelopardalis tippelskirchi*). J Zool 202:57–82

Pellew RA (1984b) Food consumption and energy budget of the giraffe. J Appl Ecol 21:141–159

Pendergast BA, Boag DA (1970) Seasonal changes in the diet of spruce grouse in central Alberta. J Wildl Manage 34:605–611

Pennycuik PR, Reisner AH (1989) Food restriction: mechanisms limiting numbers and home site utilization in populations of house mice, *Mus musculus*. Oikos 55:159–164

Perrin RM (1978) The effect of some cowpea varieties on the development and survival of larvae of the seed moth, *Cydia ptychora* (Meyrick) (Lepidoptera: Tortricidae). Bull Entomol Res 68:57–63

Perrins CM (1966) The effect of beech crops on great tit populations and movements. Br Birds 59:419–432

Pfeiffer DG, Burts EC (1983) Effects of tree fertilization on numbers and development of pear psylla (Homoptera: Psyllidae) and on fruit damage. Environ Entomol 12:895–901

Pfeiffer DG, Burts EC (1984) Effects of tree fertilization on protein and free amino acid content and feeding rate of pear psylla (Homoptera: Psyllidae). Environ Entomol 13:1487–1490

Pimm SL, Lawton JH, Cohen JE (1991) Food web patterns and their consequences. Nature 350:669–674

Pittock AB, Nix HA (1986) The effect of changing climate on Australian biomass production – a preliminary study. Clim Change 8:243–256

Plant AR (1984) The cause of green islands induced by the Nepticulidae. Proc Trans Br Entomol Nat Hist Soc 17:82–83

Polis GA (1980) The effect of cannibalism on the demography and activity of a natural population of desert scorpions. Behav Ecol Sociobiol 7:25–35

Polis GA (1981) The evolution and dynamics of intraspecific predation. Annu Rev Ecol Syst 12:225–251

Polis GA (1988) Foraging and evolutionary responses of desert scorpions to harsh environmental periods of food stress. J Arid Environ 14:123–134

Polis GA, Farley RD (1979) Characteristics and environmental determinants of natality, growth and maturity in a natural population of the desert scorpion, *Paruroctonus mesaensis* Stahnke. J Zool (Lond) 187:517–542

Polis GA, Farley RD (1980) Population biology of a desert scorpion: survivorship, microhabitat, and the evolution of life history strategy. Ecology 61:620–629

Polis GA, McCormick SJ (1987) Intraguild predation and competition among desert scorpions. Ecology 68:332–343

Pollard E (1975) Aspects of the ecology of *Helix pomatia* L. J Anim Ecol 44:305–329

Porter KG (1988) Phagotrophic phytoflagellates in microbial food webs. Hydrobiologia 159:89–97

Potter DA, Kimmerer TW (1986) Seasonal allocation of defense investment in *Ilex opaca* Aiton and constraints on a specialist leaf miner. Oecologia 69:217–224

Potts GR (1980) The effects of modern agriculture, nest predation, and game management on population ecology of partridges (*Perdix perdix* and *Alectoris rufa*). Adv Ecol Res 11:1–79

Potts WH, Jackson CHN (1953) The Shinyanga game destruction experiment. Bull Entomol Res 43:365–374

Pough FH (1973) Lizard energetics and diet. Ecology 54:837–844

Poulet SA, Williams R, Conway DVP, Videau C (1991) Co-occurrence of copepods and dissolved free amino acids in shelf sea waters. Mar Biol 108:373–386

Presler RW, Price PW (1988) Host quality and sawfly populations: a new approach to life table analysis. Ecology 69:2012–2020

Prestidge RA, McNeill S (1983) The role of nitrogen in the ecology of grassland Auchenorrhyncha. Symp Br Ecol Soc 22:257–281

Prestwich GD, Bentley BL, Carpenter EJ (1980) Nitrogen sources for neotropical nasute termites. Fixation and selective foraging. Oecologia 46:397–401

Price PW (1989) Clonal development of coyote willow, *Salix exigua* (Salicaceae), and attack by the shoot-galling sawfly, *Euura exiguae* (Hymenoptera: Tenthredinidae). Environ Entomol 18:61–68

Price PW (1991a) Darwinian methodology and the theory of insect herbivore population dynamics. Ann Entomol Soc Am 84:465–473

Price PW (1991b) Insect herbivore population dynamics: is a new paradigm available? Symp Biol Hung 39:177–190

Price PW (1991c) The plant vigor hypothesis and herbivore attack. Oikos 62:244–250

Price PW, Clancy KM (1986) Multiple effects of precipitation on *Salix lasiolepis* and populations of the stem-galling sawfly, *Euura lasiolepis*. Ecol Res 1:1–14

Price PW, Fernandes GW, Waring GL (1987a) Adaptive nature of insect galls. Environ Entomol 16:15–24

Price PW, Roininen H, Tahvanainen J (1987b) Plant age and attack by the bud galler, *Euura micronata*. Oecologia 73:334–337

Price T (1985) Reproductive responses to varying food supply in a population of Darwin's finches: clutch size, growth rates and hatching synchrony. Oecologia 66:411–416

Prins HHT, Ydenberg RC (1985) Vegetation growth and a seasonal habitat shift of the barnacle goose (*Branta leucopsis*). Oecologia 66:122–125

Prins HHT, Ydenberg RC, Drent RH (1980) The interaction of brent geese *Branta bernicla* and sea plantain *Plantago maritima* during spring staging: field observations and experiments. Acta Bot Neerl 29:585–596

Prop J, Vulink T (1992) Digestion by barnacle geese in the annual cycle: the interplay between retention time and food quality. Funct Ecol 6:180–189

Pulliainen E (1970) Winter nutrition of the rock ptarmigan, *Lagopus mutus* (Montin), in northern Finland. Ann Zool Fenn 7:295–302

Pulliainen E (1972) Nutrition of arctic hare (*Lepus timidus*) in northeastern Lapland. Ann Zool Fenn 9:17–22

Pulliainen E (1976) Small intestine and caecal lengths in the willow grouse (*Lagopus lagopus*) in Finnish Lapland. Ann Zool Fenn 13:195–199

Pulliainen E (1979) Animal food of the capercaillie, *Tetrao urogallus*, in the northern Finnish taiga in autumn. Aquilo Ser Zool 19:29–32

Pulliainen E (1982) Habitat selection and fluctuation in numbers in a population of the arctic hare (*Lepus timidus*) on a subarctic fell in Finnish forest Lapland. Z Säugetierkd 47:168–174

Pulliainen E, Iivanainen J (1981) Winter nutrition of the willow grouse (*Lagopus lagopus*) in the extreme north of Finland. Ann Zool Fenn 18:263–269

Pulliainen E, Peiponen V (1981) On the breeding of the redpoll *Carduelis flammea* in NE Finland. Ornis Fenn 58:109–116

Pulliainen E, Salo LJ (1973) Food selection by the willow grouse (*Lagopus lagopus*) in laboratory conditions. Ann Zool Fenn 10:445–448

Pulliainen E, Tunkkari PS (1987) Winter diet, habitat selection and fluctuation of a mountain hare *Lepus timidus* population in Finnish forest Lapland. Holarct Ecol 10:261–267

Pullin AS (1986) Influence of the food plant, *Urtica dioica*, on larval development, feeding efficiencies, and voltinism of a specialist insect *Inachis io*. Holarct Ecol 9:72–78

Pullin AS (1987) Changes in leaf quality following clipping and regrowth of *Urtica dioica*, and consequences for a specialist insect herbivore, *Aglais urticae*. Oikos 49:39–45

Pyke GH (1980) The foraging behaviour of Australian honeyeaters: a review and some comparisons with hummingbirds. Aust J Ecol 5:343–369

Quinn GP (1988a) Ecology of the intertidal pulmonate limpet *Siphonaria diemenensis* Quoy et Gaimard. I. Population dynamics and availability of food. J Exp Mar Biol Ecol 117:115–136

Quinn GP (1988b) Ecology of the intertidal pulmonate limpet *Siphonaria diemenensis* Quoy et Gaimard. II. Reproductive patterns and energetics. J Exp Mar Biol Ecol 117:137–156

Radmer R, Kok B (1977) Photosynthesis: limited yields, unlimited dreams. BioScience 27:599–605

Ramsell J, Paul ND (1990) Preferential grazing by molluscs of plants infected by rust fungi. Oikos 58:145–150

Randall JE (1961) A contribution to the biology of the convict surgeonfish of the Hawaiian islands, *Acanthurus triostegus sandvicensis*. Pac Sci 15:215–272

Rands MRW (1985) Pesticide use on cereals and the survival of grey partridge chicks: a field experiment. J Appl Ecol 22:49–54

Raupp MJ, Werren LH, Sadof CS (1988) Effects of short-term phenological changes in leaf suitability on the survivorship, growth, and development of gypsy moth (Lepidoptera: Lymantriidae) larvae. Environ Entomol 17:316–319

Rausher MD (1980) Host abundance, juvenile survival, and oviposition preference in *Battus philenor*. Evolution 34:342–355

Rausher MD (1981) Host plant selection by *Battus philenor* butterflies: the role of predation, nutrition, and plant chemistry. Ecol Monogr 51:1–20

Raven JA (1983) Phytophages of xylem and phloem: a comparison of animal and plant sap-feeders. Adv Ecol Res 13:135–234

Reader R (1979) Impact of three leaf-feeding insects on three bog ericads. Can J Bot 57:2107–2112

Redford KH, Bouchardet da Fonseca GA, Lacher TE (1984) The relationship between frugivory and insectivory in primates. Primates 25:433–440

Reichman OJ, Van de Graaff KM (1975) Association between ingestion of green vegetation and desert rodent reproduction. J Mammal 56:503–506

Reid DG, Hu Jinchu (1991) Giant panda selection between *Bashania fangiana* bamboo habitats in Wolong Reserve, Sichuan, China. J Appl Ecol 28:228–243

Reid N (1986) Pollination and seed dispersal of mistletoes (Loranthaceae) by birds in southern Australia. In: Ford HA, Paton DC (eds) The dynamic partnership: birds and plants in southern Australia. Government Printer, South Australia, pp 64–84

Remsen JV Jr, Stiles FG, Scott PE (1986) Frequency of arthropods in stomachs of tropical hummingbirds. Auk 103:436–441

Reynolds SE (1990) Feeding in caterpillars: maximizing or optimizing food acquisition. In: Mellinger J (ed) Animal nutrition and transport processes. 1. Nutrition in wild and domestic animals. Karger, Basel, pp 106–118

Reynoldson TB (1966) The distribution and abundance of lake-dwelling triclads – towards a hypothesis. Adv Ecol Res 3:1–71

Reynoldson TB (1977) The population dynamics of *Dugesia polychroa* (Schmidt) (Turbellaria: Tricladidae) in a recently constructed Anglesey pond. J Anim Ecol 46:63–77

Reynoldson TB (1983) The population biology of Turbellaria with special reference to the freshwater triclads of the British Isles. Adv Ecol Res 13:236–326

Richardson BJ, Wood DH (1982) Experimental ecological studies on a subalpine rabbit population. I. Mortality factors acting on emergent kittens. Aust Wildl Res 9:443–450

Richardson KC, Wooller RD (1988) The alimentary tract of a specialist frugivore, the mistletoebird, *Dicaeum hirundinaceum*, in relation to its diet. Aust J Zool 36:373–382

Richardson KC, Wooller RD (1990) Adaptations of the alimentary tract of some Australian lorikeets to a diet of pollen and nectar. Aust J Zool 38:581–586

Richardson KC, Wooller RD, Collins BG (1986) Adaptations to a diet of nectar and pollen in the marsupial *Tarsipes rostratus* (Marsupialia: Tarsipedidae). J Zool(Lond)(A) 208:285–297

Richman S, Dodson SI (1983) The effect of food quality on feeding and respiration by *Daphnia* and *Diaptomus*. Limnol Oceanogr 28:948–956

Riechert SE (1981) The consequences of being territorial: spiders (*Agelenopsis aperta*): a case study. Am Nat 117:871–892

Riley RT (1980) The effects of prolonged starvation on the relative free amino acid composition of the extra-cellular body fluids and protein-bound amino acids in the oyster, *Crassostrea gigas*. Comp Biochem Physiol A 67:279–282

Rimmer DW (1986) Changes in diet and the development of microbial digestion in juvenile buffalo bream, *Kyphosus cornelii*. Mar Biol 92:443–448

Rimmer DW, Wiebe WJ (1987) Fermentative microbial digestion in herbivorous fishes. J Fish Biol 31:229–236

Rivest BR (1983) Development and the influence of nurse egg allotment on hatching size in *Searlesia dira* (Reeve, 1846) (Prosobranchia, Buccinidae). J Exp Mar Biol 69:217–241

Robbins CT (1981) Estimation of the relative protein cost of reproduction in birds. Condor 83:177–178

Roberts CM (1991) Larval mortality and the composition of coral reef fish communities. TREE 6:83–87

Robertson DR (1982) Fish faeces as food on a Pacific coral reef. Mar Ecol Prog Ser 7:253–265
Robinette WL, Baer HC, Pillmore RE, Knittle CE (1973) Effects of nutritional change on captive mule deer. J Wildl Manage 37:312–326
Rogers LL (1987) Effects of food supply and kinship on social behaviour, movements, and population growth of black bears in northeastern Minnesota. Wildl Monogr 97:1–77
Rogers ME, Maisels F, Williamson E, Fernandez M, Tutin CEG (1990) Gorilla diet in the Lope Reserve, Gabon: a nutritional analysis. Oecologia 84:326–339
Rohfritsch O (1987) Different food supply strategies in midge-induced plant galls. In: Labeyrie V, Fabres C, Lachaise D (eds) Insect-plants. Proc 6th Int Symp Insect-Plant Relationships. Junk, Dordrecht, pp 195–200
Room PM, Julien MH, Forno IW (1989) Vigorous plants suffer most from herbivores: latitude, nitrogen and biological control of the weed *Salvinia molesta*. Oikos 54:92–100
Rosenzweig ML (1971) Paradox of enrichment: destabilization of exploitation ecosystems in ecological time. Science 171:385–387
Rosswall T (1983) The nitrogen cycle. In: Bolin B, Cook RB (eds) The major biogeochemical cycles and their interactions. Scope 21. Wiley, New York, pp 46–50
Rousi M, Haggman J, Bryant JP (1987) The effect of bark phenols upon mountain hare barking of winter-dormant Scots pine. Holarct Ecol 10:60–64
Rowe-Rowe DT, Scotcher JS (1986) Ecological carrying-capacity of the Natal Drakensberg for wild ungulates. S Afr J Wildl Res 16:12–16
Rowland SJ (1983) The hormone-induced ovulation and spawning of the Australian freshwater fish golden perch, *Macquaria ambigua* (Richardson) (Percichthyidae). Aquaculture 35:221–238
Rowley I (1970) Lamb predation in Australia: incidence, predisposing conditions, and the identification of wounds. CSIRO Wildl Res 15:79–123
Rowley I (1990) Behavioural ecology of the galah *Eolophus roseicapillus* in the wheatbelt of Western Australia. Surrey Beatty, Chipping Norton
Rubsamen K, Hume ID, Engelhardt WV (1982) Physiology of the rock hyrax. Comp Biochem Physiol A 72:271–277
Ruess RW, Hik DS, Jefferies RL (1989) The role of lesser snow geese as nitrogen processors in a sub-arctic salt marsh. Oecologia 79:23–29
Rusch DA, Reeder WG (1978) Population ecology of Alberta red squirrels. Ecology 59:400–420
Rushton SP, Hassall M (1983a) The effect of food quality on the life history parameters of the terrestrial isopod *[Armadillidium vulgare* (Latreille)]. Oecologia 57:257–261
Rushton SP, Hassall M (1983b) Food and feeding rates of the terrestrial isopod *Armadillidium vulgare* (Latreille). Oecologia 57:415–419
Rushton SP, Hassall M (1987) Effects of food quality on isopod population dynamics. Funct Ecol 1:359–367
Sadleir RMFS (1969) The role of nutrition in the reproduction of wild mammals. J Reprod Fertil Suppl 6:39–48
Sadof CS, Raupp MJ (1991) Effect of variegation in *Euonymus japonica* var *aureus* on two phloem feeding insects, *Unaspis euonymi* (Homoptera: Diaspididae) and *Aphis fabae* (Homoptera: Aphididae). Environ Entomol 20:83–89
Sale PF, Ferrell DJ (1988) Early survivorship of juvenile coral reef fishes. Coral Reefs 7:117–124
Sale PF, Doherty PJ, Eckert GJ, Douglas WA, Ferrell DJ (1984) Large scale spatial and temporal variation in recruitment to fish populations on coral reefs. Oecologia 64:191–198
Sanders EH, Gardner PD, Berger PJ, Negus NC (1981) 6-Methoxybenzoxazolinone: a plant derivative that stimulates reproduction in *Microtus montanus*. Science 214:67–69
Sanders RW, Porter KG (1988) Phagotrophic phytoflagellates. Adv Microb Ecol 10:167–192
Sauer JJC, Skinner JD, Neitz AWH (1982) Seasonal utilization of leaves by giraffes *Giraffa camelopardalis* and the relationship of the seasonal utilization to the chemical composition of the leaves. S Afr J Zool 17:210–219
Savory CJ (1977) The food of red grouse chicks *Lagopus l. scoticus*. Ibis 119:1–9

Savory CJ (1983) Selection of heather age and chemical composition by red grouse (*Lagopus lagopus scoticus*) in relation to physiological state, season and time of day. Ornis Scand 14:135–143

Schaefers GA, Montgomery ME (1973) Influence of Cytokinin (N^6benzyladenine) on development and alary polymorphism in strawberry aphid, *Chaetosiphon fragaefolii.* Ann Entomol Soc Am 66:1115–1119

Schaller GB, Hu Jinchu, Pan Wenshi, Zhu Jing (1985) The giant panda of Wolong. University of Chicago Press, Chicago

Schluter D (1982) Distribution of Galapagos ground finches along an altitudinal gradient: the importance of food supply. Ecology 63:1504–1517

Schoener TW (1982) The controversy over interspecific competition. Is the competitionist view a failed paradigm? Am Sci 70:586–595

Schreiber RW, Schreiber EA (1984) Central Pacific seabirds and the El Niño Southern Oscillation: 1982 to 1983 perspectives. Science 225:713–716

Schultz JC (1988) Plant responses induced by herbivores. TREE 3:45–49

Schweitzer DF (1979) Effects of foliage age on body weight and survival in larvae of the tribe Lithophanini (Lepidoptera: Noctuidae). Oikos 32:403–408

Seager JR (1982) Population dynamics of the antarctic opisthobranch *Philine gibba.* J Exp Mar Biol Ecol 60:163–180

Sedinger JS (1984) Protein and amino acid composition of tundra vegetation in relation to nutritional requirements of geese. J Wildl Manage 48:1128–1136

Sedinger JS (1986) Biases in comparison of proventricular and oesophageal food samples from cackling Canada geese. J Wildl Manage 50:221–222

Sedinger JS (1990) Are plant secondary compounds responsible for negative apparent metabolizability of fruits by passerine birds? A comment on Izhaki and Safriel. Oikos 57:138–140

Sedinger JS, Raveling DG (1984) Dietary selectivity in relation to availability and quality of food for goslings of cackling geese. Auk 101:295–306

Sedinger JS, Raveling DG (1986) Timing of nesting by Canada geese in relation to the phenology and availability of their food plants. J Anim Ecol 55:1083–1102

Seegmiller RF, Ohmart RO (1981) Ecological relationships of ferral burros and desert bighorn sheep. Wildl Monogr 78:1–58

Seiskari P (1962) On the winter ecology of the capercaillie, *Tetrao urogallus*, and the black grouse, *Lyrurus tetrix*, in Finland. Pap Game Res 22:1–119

Semlitsch RD, West CA (1988) Size-dependent cannibalism in noctuid caterpillars. Oecologia 77:286–288

Senrayan R, Annadurai RS (1991) Influence of host's food plant and habitat on *Anastatus ramakrishnae* (Mani) (Hym., Eupelmidae), an egg parasitoid of *Coridius obscurus* (Fab.) (Het., Pentatomidae). J Appl Entomol 112:237–243

Servello FA, Kirkpatrick RL (1988) Nutrition and condition of ruffed grouse during the breeding season in southwestern Virginia. Condor 90:836–842

Shearer JW (1976) Effect of aggregation of aphids (*Periphyllus* spp.) on their size. Entomol Exp Appl 20:179–182

Shellman-Reeve JS (1990) Dynamics of biparental care in the dampwood termite, *Zootermopsis nevadensis* (Hagen): response to nitrogen availability. Behav Ecol Sociobiol 26:389–397

Shepherd RF (1992) Relationships between attack rates and survival of western spruce budworm, *Choristoneura occidentalis* Freeman (Lepidoptera, Tortricidae), and bud development of Douglas-fir, *Pseudotsuga menziesii* (Mirb.) Franco. Can Entomol 124:347–358

Shepherd RF, Bennett DD, Dale JW, Tunnock S, Dolph RE, Thier RW (1988) Evidence of synchronized cycles in outbreak patterns of Douglas fir tussock moth, *Orgyia pseudotsuga* (McDunnough) (Lepidoptera: Lymantriidae). Mem Entomol Soc Can 146:107–121

Sherr EB, Sherr BF (1991) Planktonic microbes: tiny cells at the base of the ocean's food webs. TREE 6:50–54

Shetty S, Kshenoy B, Jacob RT, Hegde SN (1990) Mineral compositiion of pigeon milk. Experientia 46:449–451

Shipp E, Keith K, Hughes RL, Myers K (1963) Reproduction in a free-living population of domestic rabbits, *Oryctolagus cuniculus* (L.), on a sub-arctic island. Nature 200: 858–860
Shore TL, Alfaro RI (1986) The spruce budworm, *Choristoneura fumiferana* (Lepidoptera: Tortricidae), in British Columbia. J Entomol Soc B C 83:31–38
Short HL (1976) Composition and squirrel use of acorns of black and white oak groups. J Wildl Manage 40:479–483
Shorthouse JD (1986) Significance of nutritive cells in insect galls. Proc Entomol Soc Wash 88:368–375
Simon CA (1975) The influence of food abundance on territory size in the iguanid lizard *Sceloporus jarrovi*. Ecology 56:993–998
Simpson SJ, Abisgold JD (1985) Compensation by locusts for changes in dietary nutrients: behavioral mechanisms. Physiol Entomol 10:443–452
Sinclair ARE (1975) The resource limitation of trophic levels in tropical grassland ecosystems. J Anim Ecol 44:497–520
Sinclair ARE, Krebs CJ, Smith JNM (1982) Diet quality and food limitation in herbivores: the case of the snowshoe hare. Can J Zool 60:889–897
Sinclair ARE, Krebs CJ, Smith JNM, Boutin S (1988) Population biology of snowshoe hares. III. Nutrition, plant secondary compounds, and food limitation. J Anim Ecol 57:787–806
Sinclair ARE, Olsen PD, Redhead TD (1990) Can predators regulate small mammal populations? Evidence from house mouse outbreaks in Australia. Oikos 59:382–392
Sinclair ARE, Gosline JM, Holdsworth G, Krebs CJ, Boutin S, Smith JNM, Boonstra R, Dale M (1993) Can the solar cycle and climate synchronize the snowshoe hare cycle in Canada. Evidence from tree rings and ice cores. Am Nat 141:173–198
Singer MC, Ehrlich PR (1979) Population dynamics of the checkerspot butterfly *Euphydryas editha*. Fortschr Zool 25:53–60
Sinha RC, Kanungo MS (1967) Effects of starvation on the scorpion *Palamnaeus bengalensis*. Physiol Zool 40:386–390
Sitch TA, Grewcock DA, Gilbert FS (1988) Factors affecting components of fitness in a gall-making wasp (*Cynips divisa* Hartig). Oecologia 76:371–375
Slansky F, Feeny P (1977) Stabilization of the rate of nitrogen accumulation by larvae of the cabbage butterfly on wild and cultivated food plants. Ecol Monogr 47:209–228
Slater AFG (1988) The influence of dietary protein on the experimental epidemiology of *Heligmosomoides polygyrus* (Nematoda) in the laboratory mouse. Proc R Soc Lond B 234:239–254
Slobodchikoff CN (ed) (1988) Ecology of social behavior. Academic Press, San Diego
Slobodchikoff CN, Schulz WC (1988) Cooperation, aggression, and the evolution of social behavior. In: Slobodchikoff CN (ed) Ecology of social behavior. Academic Press, San Diego, pp 13–32
Slobodchikoff CN, Shields WM (1988) Ecological trade-offs and social behavior. In: Slobodchikoff CN (ed) Ecology of social behavior. Academic Press, San Diego, pp 3–10
Smal CM, Fairley JS (1982) The dynamics and regulation of small rodent populations in the woodland ecosystem of Killarney, Ireland. J Zool 196:1–30
Smirnoff WA (1962) Trans-ovum transmission of virus of *Neodiprion swainei* Middleton (Hymenoptera, Tenthredinidae). J Insect Pathol 4:192–200
Smith AP (1982) Diet and feeding strategies of the marsupial sugar glider in temperate Australia. J Anim Ecol 51:149–166
Smith CC (1968) The adaptive nature of social organization in the genus of tree squirrels *Tamiasciurus*. Ecol Monogr 38:31–64
Smith CC (1981) The indivisible niche of *Tamiasciurus*: an example of nonpartitioning of resources. Ecol Monogr 51:343–363
Smith DC, Douglas AE (1987) The biology of symbiosis. Arnold, London
Smith GT (1966) Observations of the life history of the scorpion *Urodacus abruptus* Pocock (Scorpionidae), and an analysis of its home sites. Aust J Zool 14:383–398

Smith GT, Moore LA (1991) Foods of corellas *Cacatua pastinator* in Western Australia. Emu 91:87–92
Smith JNM, Krebs CJ, Sinclair ARE, Boonstra R (1988) Population biology of snowshoe hares. II. Interactions with winter food plants. J Anim Ecol 57:269–286
Smith M (1979) Behaviour of the koala, *Phascolarctos cinereus* Goldfuss, in captivity. II. Parental and infantile behaviour. Aust Wildl Res 6:131–140
Smith RB, Mommsen TP (1984) Pollen feeding in an orb-weaving spider. Science 226:1330–1332
Snow BK, Snow DW (1971) The feeding ecology of tanagers and honeycreepers in Trinidad. Auk 88:291–322
Soldaat LL, Vrieling K (1992) The influence of nutritional and genetic factors on larval performance of the cinnabar moth *Tyria jacobaeae*. Entomol Exp Appl 62:29–36
Sone K (1987) Population dynamics of the pine needle gall midge, *Thecodiplosis japonensis* Uchida et Inouye (Diptera, Cecidomyiidae). J Appl Entomol 103:386–402
Southern HN (1970) The natural control of a population of tawny owls (*Strix aluco*). J Zool Lond 162:197–285
Southwick CH (1955) The population dynamics of confined house mice supplied with unlimited food. Ecology 36:212–225
Sperber I, Bjornhag G, Riddersrale Y (1983) Function of proximal colon in lemming and rat. Swed J Agric Res 13:243–256
Spidso TK (1980) Food selection by willow grouse *Lagopus lagopus* chicks in northern Norway. Ornis Scand 11:99–105
Spidso TK, Stuen OH (1988) Food selection by capercaillie chicks in southern Norway. Can J Zool 66:279–309
Spight TM (1975) On a snail's chance of becoming a year old. Oikos 26:9–14
Spight TM (1982) Population sizes of two marine snails with a changing food supply. J Exp Mar Biol Ecol 57:195–218
Spradbery JP (1991) Oviposition and the defence of brood in social wasps. In: Bailey WJ, Ridsdill-Smith J (eds) Reproductive behaviour of insects. Chapman and Hall, London, pp 227–263
Sprent JI (1987) The ecology of the nitrogen cycle. Cambridge University Press, Cambridge
Srivastava A (1991) Insectivory and its significance to langur diets. Primates 32:237–242
Stander PE (1992) Foraging dynamics of lions in a semi-arid environment. Can J Zool 70:8–21
Starmer WT, Barker JSF, Phaff HJ, Fogleman JC (1986) Adaptations of *Drosophila* and yeasts: their interactions with the volatile 2-propanol in the cactus-microorganism-*Drosophila* model system. Aust J Biol Sci 39:69–77
Steen JB, Andersen O, Saebo A, Pedersen HC, Erikstad KE (1988) Viability of newly hatched chicks of willow ptarmigan *Lagopus l. lagopus*. Ornis Scand 19:93–96
Steffen JM, Regler GL, Moore AK, Riedesel ML (1980) Urea recycling in active golden-mantled ground squirrels (*Spermophilus lateralis*). Am J Physiol 239:R168-R173
Steller DC (1986) The dietary energy and nitrogen requirements of the grey-headed flying fox, *Pteropus poliocephalus* (Temmink) (Megachiroptera). Aust J Zool 34:339–350
Stephens GC, Schinske RA (1961) Uptake of amino acids by marine invertebrates. Limnol Oceanogr 6:175–181
Stewart WDP, Preston T, Rai AN, Rowell P (1983) Nitrogen cycling. In: Lee JA, McNeill S, Rorison IH (eds) Nitrogen as an ecological factor. Blackwell, Oxford, pp 1–27
Stimson J (1970) Territorial behaviour of the owl limpet, *Lottia gigantea*. Ecology 51:113–118
Stimson J (1973) The role of the territory in the ecology of the intertidal limpet *Lottia gigantea* (Gray). Ecology 54:1020–1030
Storr GM (1964) Studies on marsupial nutrition IV. Diet of the quokka, *Setonix brachyurus* (Quoy & Gairnard), on Rottnest Island, Western Australia. Aust J Biol Sci 17:469–481
Stoszek KJ, Mika PG, Moore JA, Osborne HL (1981) Relationships of Douglas-fir tussock moth defoliation to site and stand characteristics in northern Idaho. For Sci 27:431–442

Straw NA (1989) The timing of oviposition and larval growth by two tephritid fly species in relation to host plant development. Ecol Entomol 14:443–454
Sullivan TP, Sullivan DS (1982) Population dynamics and regulation of the Douglas squirrel (*Tamiasciurus douglasii*) with supplemental food. Oecologia 53:264–270
Sullivan TP, Sullivan DS, Krebs CJ (1983) Demographic responses of a chipmunk (*Eutamias townsendii*) population with supplemental food. J Anim Ecol 52:743–755
Summers RW (1986) Breeding production of dark-bellied brent geese *Branta bernicla bernicla* in relation to lemming cycles. Bird Study 33:105–108
Sutherland WJ (1988) Predation may link the cycles of lemmings and birds. TREE 3:29–30
Sutton RD (1984) The effect of host plant flowering on the distribution and growth of hawthorn psyllids (Homoptera: Psylloidea). J Anim Ecol 53:37–50
Svardson G (1957) The invasion type of bird migration. Br Birds 50:314–343
Swanson GA, Bartonek JC (1970) Bias associated with food analysis in gizzards of blue-winged teal. J Wildl Manage 34:739–746
Swift MJ, Boddy L (1984) Animal-microbial interactions in wood decomposition. In: Anderson JM, Rayner ADM, Walton DWH (eds) Invertebrate-microbial interactions. Cambridge University Press, Cambridge, pp 89–131
Swingland IR, Coe MJ (1979) The natural regulation of giant tortoise populations on Aldabra atoll: recruitment. Philos Trans R Soc Lond B 286:177–188
Tabashnik BE (1982) Responses of pest and non-pest *Colias* butterfly larvae to intraspecific variation in leaf nitrogen and water content. Oecologia 55:389–394
Taitt MJ, Krebs CJ (1981) The effect of extra food on small rodent populations. 2. Voles (*Microtus townsendii*). J Anim Ecol 50:125–138
Tanada Y (1965) Factors affecting the susceptibility of insects to viruses. Entomophaga 10:139–150
Tann CR, Singleton GR, Coman BJ (1991) Diet of the house mouse, *Mus domesticus*, in the Mallee wheatlands of north-western Victoria. Wildl Res 18:1–12
Taylor MFJ (1984) The dependence of development and fecundity of *Samea multiplicalis* on early larval nitrogen intake. J Insect Physiol 30:779–785
Taylor MFJ (1988) Field measurement of the dependence of life history on plant nitrogen and temperature for a herbivorous moth. J Anim Ecol 57:873–891
Taylor MFJ, Forno IW (1987) Oviposition preferences of the salvinia moth *Samea multiplicalis* Guenee (Lep. Pyralidae) in relation to hostplant quality and damage. J Appl Entomol 104:73–78
Taylor RJ (1984) Foraging in the eastern grey kangaroo and the wallaroo. J Anim Ecol 53:65–74
Temby ID, Emison WB (1986) Food of the long-billed corella. Aust Wildl Res 13:57–63
Temrin H, Arak A (1989) Polyterritoriality and deception in passerine birds. TREE 4:106–109
Tenore KR (1977) Growth of *Capitella capitata* cultured on various levels of detritus derived from different sources. Limnol Oceanogr 22:936–941
Tenow O (1972) The outbreaks of *Oporinia autumnata* Bkh. and *Operophthera* spp. (Lep., Geometridae) in the Scandinavian mountain chain and northern Finland 1862–1968. Zool Bidr Upps Suppl 2:1–107
Tenow O, Holmgren B (1987) Low winter temperature and an outbreak of *Epirrita autumnata* along a valley of Finnmarksvidda, the "cold pole" of northern Fennoscandia. UNGT Rep 65:203–216
Thomas DW (1984) Fruit intake and energy budgets of frugivorous bats. Physiol Zool 57:457–467
Thomas G, Croxall JP, Prince PA (1983) Breeding biology of the light-mantled sooty albatross (*Phoebetria palpebrata*) at South Georgia. J Zool 199:123–135
Thomas JD (1990) Mutualistic interactions in freshwater modular systems with molluscan components. Adv Ecol Res 20:125–178
Thomas VJ, Prevett JP (1982) The role of horsetails (Equisetaceae) in the nutrition of northern-breeding geese. Oecologia 53:359–363
Thompson DC (1978) The social system of the gray squirrel. Behaviour 64:305–328

Thompson JN (1981) Reversed animal – plant interactions: the evolution of insectivorous and ant-fed plants. Biol J Linn Soc 16:147–155

Thompson SN (1976) The amino acid requirements for larval development of the hymenopterous parasitoid *Exeristes roborator* Fabricius (Hymenoptera: Ichneumonidae). Comp Biochem Physiol A 53:211–213

Thomson AJ, Shepherd RF, Harris JWE, Silversides RH (1984) Relating weather to outbreaks of western spruce budworm, *Choristoneura occidentalis* (Lepidoptera: Tortricidae), in British Columbia. Can Entomol 116:375–381

Thum M (1988) The significance of carnivory for the fitness of *Drosera* in its natural habitat. I. The reaction of *Drosera intermedia* and *D. rotundifolia* to supplemental feeding. Oecologia 75:472–480

Titze JF (1970) Some aspects of the metabolism of *Amylostereum areolatum*, the fungal symbiont of *Sirex noctilio*, in cultures on wood of *Pinus radiata*. Aust For Res 4:19–22

Titze JF, Turnbull CRA (1970) The effect of club gland secretions of *Sirex noctilio* on the growth of the symbiotic fungus *Amylostereum areoletum*. Aust For Res 4:27–29

Todd AW, Keith LB (1983) Coyote demography during a snowshoe hare (*Lepus americanus*) decline in Alberta (Canada). J Wildl Manage 47:394–404

Tranvik LJ, Porter KG, Sieburth JMcN (1989) Occurrence of bacterivory in *Cryptomonas*, a common freshwater phytoplankter. Oecologia 78:473–476

Troyer K (1982) Transfer of fermentative microbes between generations in a herbivorous lizard. Science 216:540–542

Troyer K (1983) Posthatching yolk energy in a lizard: utilization pattern and interclutch variation. Oecologia 58:340–344

Troyer K (1984a) Diet selection and digestion in *Iguana iguana*: the importance of age and nutrient requirements. Oecologia 61:201–207

Troyer K (1984b) Structure and function of the digestive tract of a herbivorous lizard, *Iguana iguana*. Physiol Zool 57:1–8

Troyer K (1984c) Behavioural acquisition of the hindgut fermentation system by hatchling *Iguana iguana*. Behav Ecol 14:189–193

Trust IJ, Sparrow RAH (1974) The bacterial flora in the alimentary tract of freshwater salmonid fishes. Can J Microbiol 20:1219–1228

Trytek RE, Allen WV (1980) Synthesis of essential amino acids by bacterial symbionts in the gills of the shipworm *Bankia setacea* (Tryon). Comp Biochem Physiol A 67:419–427

Turner JT, Tester PA (1989) Zooplankton feeding ecology: nonselective grazing by the copepods *Acartia tonsa* Dana, *Centropages velificatus* DeOliveira, and *Eucalarus pileatus* Giesbrecht in the plume of the Mississippi river. J Exp Mar Biol Ecol 126:21–43

Turner V (1984a) *Banksia* pollen as a source of protein in the diet of two Australian marsupials *Cercartetus nanus* and *Tarsipes rostratus*. Oikos 43:53–61

Turner V (1984b) *Eucalyptus* pollen in the diet of the feathertail glider, *Acrobates pygmaeus* (Marsupialia: Burramyidae). Aust Wildl Res 11:77–81

Tutin CEG, Fernandez M (1983) Gorillas feeding on termites in Gabon, West Africa. J Mammal 64:530–531

Ueckert DN, Hansen RM (1970) Seasonal dryweight composition in diets of mormon crickets. J Econ Entomol 63:96–98

Ueckert DN, Hansen RM (1971) Dietary overlap of grasshoppers on Sandhill Range in northern Colorado. Oecologia 8:276–285

Ugalde TD, Jenner CF (1990) Substrate gradients and regional patterns of dry matter deposition within developing wheat endosperm. II. Amino acids and protein. Aust J Plant Physiol 17:395–406

Ullrey DE, Robinson PT, Whetter PA (1981a) *Eucalyptus* digestibility and digestible energy requirements of adult male koalas, *Phascolarctos cinereus* (Marsupialia). Aust J Zool 29:847–852

Ullrey DE, Robinson PT, Whetter PA (1981b) Composition of preferred and rejected *Eucalyptus* browse offered to captive koalas, *Phascolarctos cinereus* (Marsupialia). Aust J Zool 29:839–846

Underwood AJ (1978) An experimental evaluation of competition between three species of intertidal prosobranch gastropods. Oecologia 33:185–202

Valiela I, Rietsma CS (1984) Nitrogen, phenolic acids, and other feeding cues for salt marsh detritivores. Oecologia 63:350–356

Van Dijk ThS (1983) The influence of food and temperature on the amount of reproduction in carabid beetles. How to translate the results of laboratory experiments into the reality of the field? Rep 4th Symp Carab, PUDUC, Wageningen, pp 105–123

Van Dijk ThS (1986) How to estimate the level of food availability in field populations of carabid beetles. In: Den Boer PJ, Luff ML, Mossakowski D, Weber F (eds) Carabid beetles, their adaptations and dynamics. Fischer, Stuttgart, pp 371–384

Van Dijk ThS, Den Boer PJ (1992) The life histories and population dynamics of two carabid species on a Dutch heathland. I. Fecundity and the mortality of immature stages. Oecologia 90:340–352

Van Hook RI, Neilsen MG, Shugart HH (1980) Energy and nitrogen relations for a *Macrosiphum liriodendri* (Homoptera: Aphididae) population in an east Tennessee *Liriodendron tulipifera* stand. Ecology 61:960–975

Van Horne B, Hanley TA, Cates RG, McKendrick JD, Horner JD (1988) Influence of seral stage and season on leaf chemistry of southeastern Alaska deer forage. Can J For Res 18:90–99

Van Loon H, Labitzke K (1988) When the wind blows. New Scientist 119:58–60

Van der Meijden E (1979) Herbivore exploitation of a fugitive plant species: local survival and extinction of the cinnabar moth and ragwort in a heterogeneous environment. Oecologia 42:307–323

Vanni MJ (1987) Effects of food availability and fish predation on a zooplankton community. Ecol Monogr 57:61–88

Van Orsdol KG, Hanby JP, Bygott JD (1985) Ecological correlates of lion social organization (*Panthera leo*). J Zool Lond (A) 206:97–112

Varley GC, Gradwell GR (1970) Recent advances in insect population dynamics. Annu Rev Entomol 15:1–24

Vaughan FA (1978) Food habits of the sea bream, *Archosargus rhomboidalis* (Linnaeus), and comparative growth on plant and animal food. Bull Mar Sci 28:527–536

Vaughan MR, Keith LB (1981) Demographic response of experimental snowshoe hare (*Lepus americanus*) populations to overwinter food shortage. J Wildl Manage 45:354–380

Vaughan TA, Czaplewski NJ (1985) Reproduction in Stephens' woodrat: the wages of folivory. J Mammal 66:429–443

Victor BC (1983) Recruitment and population dynamics of a coral reef fish. Science 219:419–420

Victor BC (1986) Larval settlement and juvenile mortality in a recruitment-limited coral reef fish population. Ecol Monogr 56:145–160

Village A (1981) The diet and breeding of long-eared owls in relation to vole numbers. Bird Study 28:215–224

Village A (1982) The home range and density of kestrels in relation to vole abundance. J Anim Ecol 51:413–428

Vines RG (1977) Possible relationships between rainfall, crop yields and the sunspot cycle. J Aust Inst Agric Sci 43:3–13

Vines RG, Tomlinson AI (1985) The Southern Oscillation and rainfall patterns in the Southern Hemisphere. S Afr J Sci 81:151–155

Volney WJA, Cerezke HF (1992) The phenology of white spruce and the spruce budworm in northern Alberta. Can J For Res 22:198–205

Wagner MR, Clancy KM, Kirkbride DM (1987) Predicting number of oocytes in adult western spruce budworm, *Choristoneura occidentalis* (Lepidoptera: Tortricidae). Environ Entomol 16:551–555

Wainhouse D, Gate IM (1988) The beech scale. In: Berryman AA(ed) Dynamics of forest insect populations. Patterns, causes, implications. Plenum, New York, pp 67–85

Wainhouse D, Gate IM, Lonsdale D (1988) Beech resistance to the beech scale: a variety of defenses. In: Mattson WJ, Levieux J, Bernard-Dagan C (eds) Mechanisms of woody plant defenses against insects. Search for pattern. Springer, Berlin Heidelberg New York, pp 277–293

Waldschmidt S (1983) The effect of supplemental feeding on home range size and activity patterns in the lizard *Uta stansburiana*. Oecologia 57:1–5

Wallace AR (1866) Letter to Charles Darwin, 2nd July. In: Marchant J (ed) (1916) Alfred Russell Wallace Letters and reminiscences. Harper & Brothers, New York, pp 140–143

Wallace MP, Temple SA (1988) Impact of the 1982–1983 El Niño on population dynamics of Andean condors in Peru. Biotropica 20:144–150

Wallace RJ, Cotta MA (1988) Metabolism of nitrogen-containing compounds. In: Hobson PN (ed) The rumen microbial ecosystem. Elsevier, Amsterdam, pp 217–249

Wallage-Drees JM, Michielsen NC (1989) The influence of food supply on the population dynamics of rabbits, *Oryctolagus cuniculus* (L.) in a Dutch dune area. Z Säugetierkd 54:304–323

Wallner WE (1987) Factors affecting insect population dynamics: differences between outbreak and non-outbreak species. Annu Rev Entomol 32:317–340

Warburg MR, Linsenmair KE, Berkovitz K (1984) The effect of climate on the distribution and abundance of isopods. Symp Zool Soc Lond 53:339–367

Ward GM, Cummins KW (1979) Effects of food quality on growth of a stream detritivore, *Paratendipes albimanus* (Meigen) (Diptera: Chironomidae). Ecology 60:57–64

Wargo PM (1988) Amino nitrogen and phenolic constituents of bark of American beech, *Fagus grandifolia*, and infestation by beech scale, *Cryptococcus fagisuga*. Eur J For Pathol 18:279–290

Waring GL, Price PW (1988) Consequences of host plant chemical and physical variability to an associated herbivore. Ecol Res 3:205–216

Warren MS, Pollard E, Bibby TJ (1986) Annual and long-term changes in a population of the wood white butterfly *Leptidea sinapis*. J Anim Ecol 55:707–719

Waser PM (1988) Resources, philopatry, and social interactions among mammals. In: Slobodchikoff CN (ed) Ecology of social behaviour. Academic Press, San Diego, pp 109–130

Washburn JO, Frankie GW (1985) Biological studies of iceplant scales, *Pulvinariella mesembryanthemi* and *Pulvinaria delottoi* (Homoptera: Coccidae), in California. Hilgardia 53:1–27

Washburn JO, Grace JK, Frankie GW (1987) Population responses of *Pulvinariella mesembryanthemi* and *Pulvinaria delottoi* (Homoptera: Coccidae) to nitrogen and water conditions of their host plant. Environ Entomol 16:286–295

Waterbury JB, Calloway CB, Turner RD (1983) A cellulolytic nitrogen-fixing bacterium cultured from the gland of Deshayes in shipworms (Bivalvia: Teredinidae). Science 221:1401–1403

Waterhouse DF (1957) Digestion in insects. Annu Rev Entomol 2:1–18

Waterman PG, Ross JAM, Bennett EL, Davies AG (1988) A comparison of the floristics and leaf chemistry of the tree flora in two Malaysian rainforests and the influence of leaf chemistry on populations of colobine monkeys in the Old World. Biol J Linn Soc 34:1–32

Watson A (1965) A population study of ptarmigan (*Lagopus mutus*) in Scotland. J Anim Ecol 34:135–172

Watson A, Moss R (1987) The mechanics of annual changes in ptarmigan numbers: a reply to Bergerud, Mossop, and Myberget. Can J Zool 65:1043–1047

Watson DC, Norton TA (1985) Dietary preferences of the common periwinkle, *Littorina littorea* (L.). J Exp Mar Biol Ecol 88:193–211

Watt AD (1986) The ecology of the pine beauty moth in commercial woods in Scotland. In: Jenkins D (ed) Trees and wildlife in the Scottish Uplands. Institute of Terrestrial Ecology, Huntingdon, pp 79–87

Watt AD, Leather SR, Stoakley JT (1989) Site susceptibility, population development and dispersal of the pine beauty moth in a lodgepole pine forest in northern Scotland. J Appl Ecol 26:147–157

Watts DP (1989) Ant eating behaviour of mountain gorillas. Primates 30:121–125

Webb JW, Moran VC (1978) Influence of host plant on population dynamics of *Acizzia russellae* (Homoptera: Psyllidae). Ecol Entomol 3:313–321

Wegge P (1980) Distorted sex ratio among small broods in a declining capercaillie (*Tetrao urogallus*) population. Ornis Scand 11:106–109

Wei F, Hu J, Xu G, Jiang M, Deng Q, Zhong Z (1989) A study of the life table of wild giant pandas. Acta Theriol Sin 9:81–86

Weins JA (1977) On competition and variable environments. Am Sci 65:590–597

Weis AE, Kapelinski A (1984) Manipulation of host plant development by the gall-midge *Rhabdophaga strobiloides*. Ecol Entomol 9:457–465

Weis AE, Walton R, Grego CL (1988) Reactive plant tissue sites and the population biology of gall makers. Annu Rev Entomol 33:467–486

Welden CW, Slauson WL (1986) The intensity of competition versus its importance: an overlooked distinction and some implications. Q Rev Biol 61:23–44

Wellington GM, Victor BC (1985) El Niño mass coral mortality: a test of resource limitation in a coral reef damselfish population. Oecologia 68:15–19

Wellington WG (1952) Air-mass climatology of Ontario north of Lake Huron and Lake Superior before outbreaks of the spruce budworm, *C. fumiferana* (Clem.) and the forest tent caterpillar *M. disstria* Hbn (Lepidoptera: Tortricidae: Lasiocampidae). Can J Zool 30:114–127

Wellington WG (1962) Population quality and the maintenance of nuclear polyhedrosis between outbreaks of *Malacosoma pluviale* (Dyar). J Insect Pathol 4:285–305

Werner MR, Dindal DL (1987) Nutritional ecology of soil arthropods. In: Slansky F, Rodriguez JG (eds) Nutritional ecology of insects, mites, spiders, and related invertebrates. Wiley, New York, pp 815–836

Werner RA (1979) Influence of host foliage on development, survival, fecundity, and oviposition of the spear-marked black moth, *Rheumaptera hastata* (Lepidoptera, Geometridae). Can Entomol 111:317–322

West C (1985) Factors underlying the late seasonal appearance of the lepidopterous leaf-mining guild on oak. Ecol Entomol 10:111–120

Westigard PH, Lombard PB, Allen RB, Strang JG (1980) Pear psylla: population suppression through host plant modification using Daminozide. Environ Entomol 9:275–277

White J, Strehl CE (1978) Xylem feeding by periodical cicada nymphs on tree roots. Ecol Entomol 3:323–327

White TCR (1969) An index to measure weather-induced stress of trees associated with outbreaks of psyllids in Australia. Ecology 50:905–909

White TCR (1970a) Some aspects of the life history, host selection, dispersal, and oviposition of adult *Cardiaspina densitexta* (Homoptera: Psyllidae). Aust J Zool 18:105–117

White TCR (1970b) The nymphal stage of *Cardiaspina densitexta* (Homoptera: Psyllidae) on leaves of *Eucalyptus fasciculosa*. Aust J Zool 18:273–293

White TCR (1971) Lerp insects (Homoptera: Psyllidae) on red gum (*E. camaldulensis*) in South Australia. S Aust Nat 46:20–23

White TCR (1972) The production of amylose in the faeces of psyllid larvae with special reference to the lerps of *Cardiaspina densitexta*. J Insect Physiol 18:2359–2367

White TCR (1973) Aerial dispersal of adult *Cardiaspina densitexta* (Homoptera: Psyllidae) in South Australia. Trans R Soc S Aust 97:29–31

White TCR (1974) A hypothesis to explain outbreaks of looper caterpillars, with special reference to populations of *Selidosema suavis* in a plantation of *Pinus radiata* in New Zealand. Oecologia 16:279–301

White TCR (1976) Weather, food and plagues of locusts. Oecologia 22:119–134

White TCR (1978) The importance of a relative shortage of food in animal ecology. Oecologia 33:71–86

White TCR (1984) The abundance of invertebrate herbivores in relation to the availability of nitrogen in stressed food plants. Oecologia 63:90–105

White TCR (1985) When is a herbivore not a herbivore? Oecologia 67:596–597

White TCR (1986) Weather, *Eucalyptus* dieback in New England, and a general hypothesis of the cause of dieback. Pac Sci 40:58–78

Whitham TG (1978) Habitat selection by *Pemphigus* aphids in response to resource limitation and competition. Ecology 59:1164–1176

Whitham TG (1979) Territorial behaviour of *Pemphigus* gall aphids. Nature 279:324–325
Whitham TG (1980) The theory of habitat selection: examined and extended using *Pemphigus* aphids. Am Nat 115:449–466
Whitham TG (1986) Costs and benefits of territoriality: behavioural and reproductive release by competing aphids. Ecology 67:139–147
Whitham TG (1987) Evolution of territoriality by herbivores in response to host plant defenses. Am Zool 27:359–370
Whitmore RW, Pruess KP, Gold RE (1986) Insect food selection by 2-week-old ring-necked pheasant chicks. J Wildl Manage 50:223–227
Whysall KDB, Cooper NS, Bigg GR (1987) Long-term changes in the tropical Pacific surface wind field. Nature 327:216–219
Widdowson EM, Whitehead RG (1966) Plasma amino acid ratios and urinary hydroxyproline excretions in rats deficient in protein and calories. Nature 212:683–686
Wiegart RG (1964) The ingestion of xylem sap by meadow spittlebugs, *Philaenus spumarius* (L.). Am Midl Nat 71:422–428
Williams KS, Lincoln DE, Ehrlich PR (1983a) The coevolution of *Euphydryas chalcedona* butterflies and their larval host plants. I. Larval feeding behaviour and host plant chemistry. Oecologia 56:323–329
Williams KS, Lincoln DE, Ehrlich PR (1983b) The coevolution of *Euphydryas chalcedona.* II. Maternal and host plant effects on larval growth, development, and food-use efficiency. Oecologia 56:330–335
Williams PA, Cooper P, Nes P, O'Connor KF (1976) Chemical composition of tall tussocks in relation to the diet of the takahe (*Notornis mantelli* Owen), on the Murchison mountains, Fiordland, New Zealand. N Z J Bot 14:55–61
Williamson EA, Tutin CEG, Rogers ME, Fernandez M (1990) Composition of the diet of lowland gorillas at Lope in Gabon. Am J Primatol 21:265–278
Williamson P, Cameron RAD (1976) Natural diet of the landsnail *Cepaea nemoralis.* Oikos 27:493–500
Williamson P, Cameron RAD, Carter MA (1977) Population dynamics of the landsnail *Cepaea nemoralis* L.: a six year study. J Anim Ecol 46:181–194
Wilson MF, Blackford JH, Blunden CA (1983) Changes in the nutritional composition of two varieties of orchard pear buds during the period of maximum bird damage. J Sci Food Agric 34:794–802
Windberg LA, Keith LB (1976) Snowshoe hare population response to artificial high densities. J Mammal 57:523–553
Winemiller KO, Hong YY (1989) Obligate mucus-feeding in a South American trichomycterid catfish (Pisces: Ostariophysi). Copeia 1989:511–514
Wise DH (1975) Food limitation of the spider *Linyphia marginata*: experimental field studies. Ecology 56:637–646
Wise DH (1983a) Competitive mechanisms in a food-limited species: relative importance of interference and exploitative interactions among labyrinth spiders *(Metapeira labyrinthea)* (Araneae: Araneidae). Oecologia 58:1–9
Wise DH (1983b) The role of competition in spider communities: insights from field experiments with a model organism. In: Strong DR, Simberloff D, Abele LG, Thistle AB (eds) Ecological communities: conceptual issues and the evidence. Princeton University Press, Princeton, pp 42–53
Witter JA, Mattson WJ, Kulman HK (1975) Numerical analysis of a forest tent caterpillar (Lepidoptera: Lasiocampidae) outbreak in northern Minnesota. Can Entomol 107:837–854
Woffinden ND, Murphy JR (1978) Population dynamics of the ferruginous hawk during a prey decline. Great Basin Nat 37:411–425
Wolcott DL, Wolcott TG (1984) Food quality and cannibalism in the red land crab, *Gecarcinus lateralis.* Physiol Zool 57:318–324
Wolcott DL, Wolcott TG (1987) Nitrogen limitation in the herbivorous land crab *Cardisoma guanhumi.* Physiol Zool 60:262–268

Wolda H (1991) The usefulness of the equilibrium concept in population dynamics. A reply. Oecologia 86:144–145

Wolda H, Kreulen DA (1973) Ecology of some experimental populations of the land snail *Cepaea nemoralis* (L.). II. Production and survival of eggs and juveniles. Neth J Zool 23:168–188

Wolff JO (1978) Food habits of snowshoe hares in interior Alaska. J Wildl Manage 42:148–153

Wolff JO (1980) The role of habitat patchiness in the population dynamics of snowshoe hares (*Lepus americanus*). Ecol Monogr 50:111–130

Wood JR, Wood FE (1981) Growth and digestibility for the green turtle (*Chelonia mydas*) fed diets containing varying protein levels. Aquaculture 25:269–274

Wood TG, Thomas RJ (1989) The mutualistic association between Macrotermitinae and Termitimyces. Symp R En Soc Lond 14:69–92

Wooller RD, Russell EM, Renfree MB (1984) Honey possums and their food plants. In: Smith A, Hume ID (eds) Possums and gliders. Aust Mamm Soc, Sydney, pp 439–443

Wooller RD, Richardson KC, Pagendham CM (1988) The digestion of pollen by some Australian birds. Aust J Zool 36:357–362

Wratten SD (1974) Aggregation in the birch aphid *Euceraphis punctipennis* (Zett) in relation to food quality. J Anim Ecol 43:191–198

Wrazen JA, Svendsen GE (1978) Feeding ecology of a population of eastern chipmunks (*Tamias striatus*) in southeast Ohio. Am Midl Nat 100:190–201

Wynne-Edwards VC (1962) Animal dispersion in relation to social behaviour. Oliver and Boyd, Edinburgh

Yamagiwa J, Mwanza N, Yumoto T, Maruhashi T (1991) Ant eating by eastern lowland gorillas. Primates 32:247–254

Ydenberg RC, Prins HHT (1981) Spring grazing and the manipulation of food quality by barnacle geese. J Appl Ecol 18:443–454

Ydenberg RC, Prins HHT, van Dijck J (1984) A lunar rhythm in the nocturnal foraging activities of wintering barnacle geese. Wildfowl 35:93–96

Young AM (1978) "Disappearances" of eggs and larvae of *Helconius* butterflies (Nymphalidae: Heliconiinae) in northeastern Costa Rica. Entomol News 89:81–87

Young BL, Stout IJ (1986) Effects of extra food on small rodents in a south temperate zone habitat: demographic responses. Can J Zool 64:1211–1217

Zimmerman LC, Tracy CR (1989) Interactions between the environment and ectothermy and herbivory in reptiles. Physiol Zool 62:374–409

Ziswiler V (1990) Specialization in extremely unbalanced food: possibilities and limits of its investigation exclusively by functional morphology. Neth J Zool 40:299–311

Zucoloto FS (1987) Feeding habits of *Ceratitis capitata* (Diptera: Tephritidae): can larvae recognise a nutritionally effective diet? J Insect Physiol 33:349–353

Zwickel FC (1980) Surplus yearlings and the regulation of breeding density in blue grouse. Can J Zool 58:896–905

Subject Index

Abies alba 53
Abies balsamea 53, 71, 359
Abies concolor 71
Abies spp 207
absolute shortage of food *see* food, relative v absolute shortage
Acacia articulata 217
Acacia karroo 90
Acacia spp 139
Acanthina spirata 281
Acanthus nigrofuscus 249
Acartia tonsa 113, 115
Accipiter nisus 264
Acer platanoides 120
Acer pseudoplatanus 41, 86
Acizzia russellae 90
Acmaea spp 280
Acrididae 80
Acrobates pygmaeus 187
Acrocephalus sechellensis 284
Adelges piceae 53
Aenetus virescens 54, 107
Agave palmeri 180
age of plant, response of herbivore to *see* herbivore, response to
Agelenopsis aperta 267
Aglais urticae 69
Agria housei 324
Agropyron repens 83
Ailuropoda melanoleuca 142
albatross, mantled sooty 312
Albizia spp 179
Alces alces 306
Alectoris rufa 274
alfalfa (lucerne) 42
algae 1, 109–12, 124, 126–8, 130, 237, 244–7, 277–80, 345
allelochemicals *see* plant defences
Alopex lagopus 306
Alouatta palliata 174
Alsophila pometaria 44
Amaranthus retroflexus 75
Amblyrhynchus cristatus 237
amino acid, absorbed from seawater 109
amino acid quality index 95
Ammospermophilus leucurus 148
Amylostereum areolatum 106
Anabrus simplex 84
Anastatus ramakrishnae 325
anchovy 243, 309
animal food for herbivores *see* food, animal
Anisota senatoria 44
Anobium punctatum 104
ant
  European wood 297
  weaver 177
ant-lion 318
antelope 139
Anthocharis cardamines 70
aphid
  alfalfa 42
  apple 41, 54
  birch 88
  carrot 25
  gall 18, 46, 54, 57, 275
  green spruce 39
  grey pine 38
  redcurrant 56
  spotted pine 38
  strawberry 88
  sycamore 41, 86
Aphis fabae 41, 54, 93
Apodemus spp 162
Apodemus sylvaticus 157, 270
apparent/unapparent plant 23, 25
apple 54, 72, 179
Aquila audax 144
Araneus diadematus 315
Archosargus rhomboidalis 243
Arctiidae 79
Arctium minus 19
Arhopalus ferus 105
Arianta arbustorum 132
Aristolochia reticulata 69
Aristolochia serpentaria 69
Armadillidium vulgare 117
arthropods as food for herbivore *see* food, animal
Artibeus jamaicensis 179

ash 223
Asolcus basilis 266, 275
Ateles geoffroyi 175
Atriplex spp 56
Avena sativa 230

bacteria 341, 350
  as food for fruitfly 97
  herbivores generally *see* food, microorganisms
  molluscs 130
  zooplankton 109, 116
bacterial pathogens 329
Bactrocera jarvisi 98
Bactrocera oleae 98
Bactrocera tryoni 97
balance
  nitrogen *see* nitrogen balance
  of nature *see* population, balance
bamboo 142
Banksia spp 187
barley 151, 230
Bashania fangiana 142
basidiomycete 99, 104
Battus philenor 69
bear, black 271
bedbug 325
beech 53, 225, 272
  American 53, 74
  New Zealand 50, 307
beetle
  bark 33, 107
  carabid (ground) 10, 317
  deathwatch 104
  furniture 104
  leaf 18, 46
  shot-hole borer 106
  tiger 318
  twig-girdling 105
Beta vulgaris 75
Betula papyrifera 45, 315
Betula pendula 120
Betula pubescens 88
Betula pubescens tortuosa 360
Betula spp 203, 351
bilharzia schistosomes 125
biological control 32, 77–9, 365
biotic potential 5, 27, 348
birch 45, 120, 165, 167, 203–4, 315, 351, 360
birth control 20
blueberry 141, 165, 167, 203, 205
Bombus terricola 316
Bonasa umbellus 208, 209
Boophilus annulatus 325
Boophilus microplus 326
Brachychiton populneum 333
bramble 43
Branta bernicla 196
Branta canadensis 196
Branta canadensis minima 198
Branta leucopsis 194, 195
Brassica oleracea var *acephala* 63
bream
  buffalo 247
  sea 243
Bulinus globosus 125
bumble bee 316
bunchberry 141
burdock, lesser 19
burweed 96
butterfly
  checkerspot 65
  colias 70
  helconius 70
  nettle 69
  orange tip 70
  swallowtail 69, 70
  white 17, 63, 70
  wood white 70

C:N ratio 12, 99, 100, 110, 122, 134
C_3 and C_4 plants as food for herbivores 83
Cacatua pastinator 232
cacti 9, 78, 366
Cactoblastis cactorum 78, 337, 366
caecal fermentation *see* microorganisms; gut fermenters
Calluna vulgaris 165, 200, 205, 273
Calypte anna 271
Canis familiaris dingo 136, 343
Canis latrans 306
Canis lupus 306
cannibalism 76, 84, 101, 116, 117, 131–3, 243, 291–300, 316, 320
  and warfare 297
  grazer strategy 293
  human, and warfare 300
  lifeboat strategy 293
capacity of increase, absolute v effective 343, 352, 354
capercaillie 206, 208
Carabidae 317
carbohydrate, excess, in herbivore diet 85, 89, 222, 255 *and see* energy not limiting
Cardiaspina albitextura 36
Cardiaspina densitexta 7, 10, 15, 34, 90, 334
Cardiaspina retator 334
Cardisoma guanhumi 116

Carduelis carduelis 224
Carduelis flammea 225
Carnegeia gigantea 180
carnivory
  by herbivores *see* food, animal
    phytoplankton 109
    plants 12
  cryptic 130, 257
Carollia perspicillata 178
carp 243, 250
Carpobrotus spp 37
carrying capacity 20, 135, 139, 172, 207, 228, 268, 285, 306, 314, 321 *and see* habitat, saturated
Carya illinoensis 58
Carya texana 105
cat, domestic 305, 342, 343
Cavariella aegopodii 25
Cebidichthys violaceus 246
Cecidomyiidae 56, 59
Celmesia petriei 215
Centrocercus urophasianus 209
Centropages hamatus 115
Cepaea nemoralis 130
Cerajocera tussilaginis 19
Cerambycidae 104
Ceratitis capitata 97
Cercartetus nanus 187
  Cercopithecus ascanius 174
Cercopithecus mitis 174
Cervus elaphus 137
Chaetosiphon fragaefolii 88
Chelonia mydas 236
Chen caerulescens atlantica 196
Chen caerulescens caerulescens 196
Chenopodium album 75
Chilo suppressalis 76, 330
Chilo zonellus 76
Chionochloa spp 215
chipmunk 148
  eastern 269
  western 270
Chiropotes satanas 178
Choristoneura fumiferana 71, 324, 330, 359
Choristoneura occidentalis 71, 359
Choristoneura pinus 71
Choristoneura rosaceana 72
Chorthippus brunneus 82
Chortoicetes terminifera 80
Chrysomela confluens 46
chyme 223 *and see* regurgitate
cicada, periodical 96, 157
Cicadellidae 96
Cichlidae 243, 244, 250
Cicindela spp 318
Cinara pini 88
Citheroniidae 44
citrus 92
cladocera 108–12
Cladosporium herbarum 315
Clethrionomys glareolus 157, 270
Clethrionomys rufocanus 270
Clethrionomys spp 162
climate, weather, and outbreaks *see* outbreak
Coccyzus melacorhyphus 228
coexisting species 16 *and see* competitive exclusion
Colias spp 70
Colobus spp 172
community 14
competitive exclusion 16–17, 154, 175
competitive interaction 316
condor, Andean 264
Conolophus pallidus 238
Conolophus subcristatus 238
contingency in evolution 7
Convoluta paradoxa 1
Convoluta roscoffensis 1
Copaifera langsdorfii 44
copepod 108–12, 293, 294
coprophagy 100, 101, 115, 119, 129, 130, 156, 158–9, 165, 167, 177, 182, 185, 224, 238–9, 241, 247, 257
Cordia spp 239
corella 232, 259
Coridius obscurus 325
Corvus cornix 144
Corvus spp 144
cottonwood 46, 275
coyote 306
crab, land 116, 294
cranberry 167
Crataegus monogyna 90
Crataegus spp 223
crayfish 116
"creaming-off" tactic 63, 130, 142, 179, 194, 196, 218, 220, 241, 255
creosote bush 239
cricket, mormon 84
crop milk 221 *and see* regurgitate
crow, Australian and hooded 144
crowberry 205
Cryptocercus punctulatus 101
Cryptococcus fagisuga 53
Cryptomyzus ribis 56
Ctenopharyngodon idella 243
Cubitermes spp 177
cuckoo, Galapagos 228
cycle *see* outbreak
Cyclops spp 293

Cydia pomonella 73, 275
Cydia ptychora 73
Cygnus atratus 200
Cygnus olor 200
Cynips divisa 60
cytokinin 89

Dacus cacuminatus 220
Dacus spp 266
damaged plant, response by herbivore to *see* stress of plant
daminozide 91
Daphnia galeata 111
Daphnia hyalina 111
Daphnia pulex 111
Daphnia rosea 112
Daphnia spp 110, 112
deer
  black-tailed sitka 141
  North American 141
  red 137, 346
  rocky mountain mule 141
defences, plant *see* plant defences
Dendragapus canadensis 207
Dendragapus canadensis franklinii 213
Dendragapus obscurus 208
Dendragapus obscurus fuliginosus 284
density dependence 6, 131, 293, 321, 344, 348, 354
density, optimum *see* population
Dentaria diphylla 63
detritivore 33, 116, 117, 129, 130, 244, 258
Diaptomus kenai 112
Diaptomus spp 112
Dicaeum hirundinaceum 219
Dicrostonyx kilangmiutak 149
dingo 136, 343
Diplacus aurantiacus 67
Dipodomys spectabilis 263
Dipodomys spp 163
Dipsosaurus dorsalis 239
diseases responding to stress of host *see* stress of animals
Disholcaspis cinerosa 60
dispersal 25, 73, 147, 156, 281–6, 321, 361
domestic animals 139, 144, 165, 289, 326, 327, 366
donkey, feral 136, 340
doomed surplus 282 *and see* young, surplus
dormouse, European 149
"double dipping" 57 *and see* senescence v flush feeding
dove, mourning 222
Drepanosiphum platanoidis 41, 86
Drosera spp 12
Drosophila spp 98
drought *see* stress of plant; outbreak and weather
Dugesia polychroa 314
Dugesia spp 314
Dysaphis devecta 41, 54

eagle, Australian wedgetail 144
Ecklonia radiata 246
ecosystem 14
effective strategy for survival *see* survival
effective use of food/resource *see* food; resource; foraging
egg, nurse *see* trophic egg
El Nino Southern Oscillation 150, 190, 226, 238, 243, 263, 264, 309, 312, 363–4
eland 139
Elatobium abietinum 39
Eliomys quercinus 149
Empetrum hermaphroditum 205
endosymbionts 38, 94, 104, 134 *and see* microorganisms
energy not limiting 11, 12, 14, 64, 85, 109, 110, 111, 135, 142, 149, 153, 157, 158–9, 177, 179, 180, 189, 193, 197, 201, 204–5, 209, 218, 219, 222, 224, 239, 246, 255, 300, 303 *and see* carbohydrate, excess
Engraulis mordax 309
Engraulis ringens 243
Enterobacter cloacae 98
Enteromorpha spp 128
environmental resistance 343, *and see* carrying capacity
Eolophus roseicapillus 230
Epiphyas postvittana 73
Epirrita autumnata 360
Epomops buettikoferi 179
equilibrium *see* population
Equisetum fluviatile 197
Equisetum palustre 198
Equisetum spp 167, 196
Equus asinus 136
Erica spp 201
Eriococcus coriaceus 333
Erobium spp 230–32
Erwinia aroideae 330
Eucalyptus andrewsii 184, 185
Eucalyptus blakelyi 89
Eucalyptus camaldulensis 36, 324
Eucalyptus dives 188
Eucalyptus elata 188
Eucalyptus fasciculosa 34, 334

Eucalyptus ovata 327
Eucalyptus radiata 185, 188
Eucalyptus spp 7, 25, 44, 49, 181, 218, 333, 356
Eucalyptus viminalis 188
Euceraphis punctipennis 88
Eulachnus agilis 38
Euonymus fortunei 43
Euonymus japonica var *aureus* 93
Euphorbia pyrifolia 42
Euphydryas chalcedona 67
Euphydryas editha 66
Eupomacentrus rectifraenum 246
euro, hills 190
Eutamias spp 148, 162
Eutamius townsendii 270
Euura lasiolepis 61, 274
Evetria cristata 73
excess carbohydrate in diet *see* carbohydrate
external rumen *see* rumen

faeces as food for herbivore *and see* coprophagy
  caecal 158, 165, 182, 185, 211
  proctodaeal 101–3
Fagus grandifolia 53, 74
Fagus sylvatica 53, 225
Fargesia robusta 142
fast passage through gut *see* "creaming off"
fecundity, potential/actual/realized 336–7, 339
feedback mechanisms *see* negative feedback
feeding deterrents *see* plant defences
Felis catus 305, 342
Fenusa pusilla 45
fern 77, 215
fertilizing of plants, response by herbivore to *see* herbivore
Festuca ovina 82
Festuca rubra 195
Ficus spp 178, 179
finch
  cactus 9, 226, 263
  Darwin's Galapagos 226
  European bull 223
  European gold 224
  ground 18
  medium ground 226
Fiorinia externa 92
fir 207
  balsam 53, 71, 359
  Douglas 71
  silver 53
  white 71
fish
  damsel cortez 246
  damsel giant blue 246
  surgeon 249
flea 313, 325
"floaters" *see* young, surplus
flowering of plant, response of herbivore to *see* herbivore
flush v senescence feeder *see* senescence v flush feeder
fly
  black 267
  caddis 277
  fruit *see* fruitfly
  gall 55
  leaf-mining 50
  parasitic 324
  ragwort seed 18
  syrphid 316
  tephritid 19
  tsetse 10, 326, 328
    vinegar 98
Fomes applanatus 99
food
  animal, for herbivore 76, 84, 116–7, 128–9, 131, 143, 146, 148–9, 156, 157, 161, 173, 174, 177, 178, 187, 190–1, 199, 202–3, 206, 207–8, 209, 216, 218, 222, 225, 226, 229, 234, 237, 238–9, 243–4, 246, 248, 258–60, 272–4, *and see* cannibalism
  effective use of 9, 25, 214, 261, 266, 268, 288, 291, 293, 294, 300, 318–19
  faeces as, for herbivore *see* faeces; coprophagy
  fast passage of *see* "creaming off"
  flowers as, for herbivore 18–19, 70, 75, 79, 81, 146, 150, 158, 203–4, 234, 239 *and see* flowering, response of herbivore to
  indirect cues to 129, 163–4, 253
  identifying in gut/faecal samples 131, 146, 161, 177, 200, 206, 232, 246, 249, 259
  legumes as, for herbivore 31, 118, 139, 154, 158, 173, 179, 207, 217, 240
  microorganisms as, for herbivore 98, 99, 109–12, 119, 124, 130, 134, 245 *and see* microorganisms; fungus
  pollen as, for herbivore 146, 156, 180, 187, 218, 229
  relative v absolute shortage of 10, 11, 23, 31, 108, 110, 119, 124, 202, 242, 243, 255, 263, 269, 283, 290, 302, 304, 309, 310, 314, 315, 316, 317–9, 321, 325–6, 344, 346, 347, 365

food (cont.)
  seed, developing, as, for herbivore 19, 70, 75, 94–5, 145, 146, 148, 150–4, 161, 173, 177, 224, 226, 231, 234, 275, 353
  supplemental feeding of 119, 123–4, 124–5, 141, 147, 152, 155, 162, 168–9, 170, 269, 270, 272, 285, 294, 306–7, 311
  winter staple not the key to success of population 147–8, 150–2, 154, 157, 161–4, 168–9, 171, 173, 193, 225, 229
food web 14
foraging, optimal v effective 246, 303, 316, 323, 343
forest tent caterpillar 329
Formica polyctena 297
fox
  arctic 306
  red 307, 342, 343
Fragaria vesca 88
Fratercula arctica 311
Fraxinus excelsior 223
fruitfly 220
  Mediterranean 97
  olive 98
  Queensland 97
fungus as food for
  grasshoppers 80
  herbivores generally *see* food, microorganisms
  ispopods 120
  spiderlings 315
  squirrels 149
  woodborers 99–106, 134

galah, Australian 230
Galleria mellonella 76
galls 41, 52–62, 254, 274, 275–7
Gecarcinus lateralis 116, 294
Gelachiidae 44
gene pool 6, 15, 20
genes 5, 6, 15, 283, 287, 288, 345
Geochelone gigantea 233
Geometridae 43, 44, 74
Geospiza conirostris 9, 226
Geospiza fortis 226
Geospiza scandens 226
Giraffa camelopardalis tippelskirchi 139
glider
  feathertail 187
  greater 185, 187, 188
  sugar 187
  yellow-bellied 185, 187
Glossina palpalis palpalis 327
Glossina spp 326
Glossopsitta porphyrocephala 218
Glycaspis brimblecombei 36
goatsbeard 224
goldenrod 55
goose 346
  barnacle 194, 195, 198
  black brant 199
  brent 196, 198
  cackling 198
  Canada 196, 198–9
  domestic 199
  greater snow 196
  lesser snow 196
gopher, pocket 149
Gorilla gorilla gorilla 175
Grantiella picta 219
  grass
  arrow 198
  corkscrew 230
  tussock 215
  weed 151–3
grazer strategy *see* cannibalism
grazing, repeated 254
  caddis fly 278
  fish 246
  gall insect 54
  goose 195, 196
  giraffe 139
  hare 165
  limpet 280
  tortoise 234
  turtle 236
grouse
  black 206, 208
  blue 208, 284
  red 200, 211, 212, 213, 272, 366
  ruffed 208, 209
  sage 209
  spruce 207, 213
  willow 203, 211, 212, 272
guano improving plant as nitrogenous food 126, 138, 144, 195, 345 *and see* hot spots
gull, sea 138, 195
gut fermenter 257 *and see* food, microorganims; microorganisms; coprophagy
  fish 247
  galliform 211–12
  glider 185
  goose 194
  green iguanid 240
  hoatzin 217
  isopod 122
  koala 181

lagomorph 165
macropod 189
microtine 158
monkey 172
possum 184
squirrel 145
termite 100
turtle 236
habitat, patchy *see* patchy environment
habitat, saturated 9, 135, 137, 145, 147, 155, 156, 168, 213, 242, 282–3, 285, 303, *and see* carrying capacity
Haematopinus eurysternus 326
Haematopus moquini 127
hare
European mountain 165
North American snowshoe 167, 304, 306
hastened senescence *see* senescence
hatching, asynchronous 264
hawthorn 90, 223
hazard *see* malentity
heather 165, 200, 205, 273
hedgesparrow 265
Helconius spp 70
Helianthus annuus 81
Helicoverpa zea 75, 324
Helix pomatia 131
helpers 281–86, 288 *and see* philopatry
hemlock, eastern 92
herbivore response to plant
age 33, 46, 60, 89, 107, 120, 208,
damage 208 *and see* stress of plant
defences *see* plant defences
fertilization 32, 59, 63, 68, 70, 71, 76–9, 81, 88, 92, 93, 115, 124, 129, 144, 165, 195, 202
flowering 33, 79, 90, 91, 95, 147–8, 161, 223, 225
stress *see* stress of plants
hibernation 149
hickory 105
hoatzin 217
Holcaphis holci 54
Holcus mollis 54, 94
holly 50
Homalodisca coagulata 96
Homo sapiens 20
honeyeater, Australian 219
Hordeum vulgare 230
horse 33
horsetails 196
hot spots 73, 347, 357 *and see* refuges; outbreak centres; guano
hummingbird 217, 271
Hydrobia ulvae 130
Hypolepis millefolium 215
Hyposoter exiguae 324

ice plant 37
Icerya seychellarum 42, 93
Iguana iguana 240
iguanid
Californian desert 239
Galapagos, marine 237
Galapagos, terrestrial 238
green 240
Ilex opaca 50
Ilyanassa obsoleta 129
Inachis io 69
increase, capacity of *see* capacity of increase
infanticide 292
insectivorous plants 12
insects as food for herbivores *see* food, animal
Isoptera 100
Ixodes ricinus 325
Ixodes tasmani 327

kangaroo 351
eastern grey 189
red 190
kelp 246
kittiwake 312
koala 181, 256, 327
kudu, greater 140
kurrajong tree 333, 356
Kyphosus cornelii 247
Kyphosus sydneyanus 247

lagomorphs 164
Lagopus lagopus lagopus 203, 272
Lagopus lagopus scoticus 200, 272
Lagopus mutus 204
Lantana spp 239
Larrea spp 239
Larus argentatus 138
Latrodectus katipo 320
leafhopper 43, 93
legumes as food for herbivore *see* food, legumes
lemming 156, 353
collared 149
lemur, Madagascan 173
Lepidoptera 49, 229, 259, 351, 362
Leptidea sinapis 70
Leptonycteris sanborni 180
Leptopterna dolobrata 94
Lepus americanus 167, 304
Lepus spp 162
Lepus timidus 165

lerp 36, 89, 256
Leucotrichia pictipes 277
lice 313, 326
life boat strategy *see* cannibalism
Limnoria spp 134
limpet 125, 126, 280, 345
Lindbergina aurovittata 43
lion 10, 304
Liriodendron tulipifera 85
Lithocolletis quercus 48
Litoprosopus futilis 297
Littorina littorea 128
lobster, rock 116
locust
  African desert 82
  Australian plague 80
Locusta migratoria 82
Locusta migratoria migratorioides 83
  Lonchocarpus cruciarubieres 217
Lonchocarpus pentaphyllus 240
Loranthaceae 219
lorikeet, Australian 218
Lottia gigantea 280
Lymantria dispar 65, 74, 328
Lymnaea elodes 123
Lynx canadensis 304
Lyrodus pedicellatus 134

Macoma balthica 130
Macquaria ambigua 242
Macropodidae 189
Macropus eugeneii 190
Macropus giganteus 189
Macropus robustus 190
Macropus robustus robustus 189
Macropus rufa 195
Macropygia phasianella 220
Macrosiphum liriodendri 85
Magicicada cassini 96
Magicicada spp 157
maize 76
Malacosoma pluviale 329
malentity 19
Mallophaga 313
managed wildlife *see* widlife, managed
maple 120
Marmota flaviventris 161
Marmota spp 285
mast, seed *see* seed mast
matriline 137, 286
mean *see* population
Medicago sativa 42
Melampus bidentatus 129
Melanoplus differentialis 80
Messa nana 45
Metabacterium spp 249
metapopulation *see* population
6-methoxybenzoxazolinone
  (6-MBOA) 153, 163
microorganisms 33, 108, 129
  as food for herbivore *see* food, microorganism
  gut 76, 94, 100, 117, 122, 145, 158, 172, 177, 182, 194, 211, 236, 247, 256
  nitrogen fixing 11, 101–3, 104, 109, 134, 258
  recycling metabolic nitrogen 13, 101–3, 149, 184, 194, 211, 258
Micropteropus pusillus 179
Microspathodon dorsalis 246
microsporidian 330
Microtus agrestis 157, 159
Microtus californicus 158
Microtus montanus 163
Microtus ochrogaster 157
Microtus pennsylvanicus 159, 270
Microtus pinetorum 159
Microtus spp 162, 270
  Microtus townsendii 270
midge, gall 55
milk-ripe seed 151 *and see* food, seed
millipede 122
mirid 94
mistletoe 219
mistletoebird, Australian 219
Misumena calycina 316
mockingbird, Galapagos 228
monkey 17
  banded leaf 172
  blue 174
  fruit saki 178
  howler 174
  red leaf 172
  spider 175
moose 306
Morter obscurus 319
mosquito 7, 10, 325
moth
  autumnal 360
  cabbage looper 328
  cactus 78, 337, 366
  cinnabar 18, 79
  codling 73, 275
  cossid 192
  cowpea 73
  European corn borer 76
  forest tent 329
  gypsy 65, 74, 328
  hepialid 54, 107, 192
  Japanese pine shoot 73
  light brown apple 73
  noctuid 75, 296, 297, 324

rice stem borer 76, 330
tortricid 73
tussock 73
wax 76
winter 43
mould 315
mountain daisy 215
mouse
Australian smokey 17, 154
deer 270
house 150–4, 307, 343, 351
North American white footed 155
wood 157, 270
Mus musculus 150–4, 307, 343
Mus spp 162
Mustela erminea 307
mustelid 353
mutton bird 305
mycetocytes 94
mycotetes 100
Myzus persicae 89

N^6benzyladenine (N^6B) 89
Nassarius pauperatus 128
Nasutitermes corniger 101
natural selection 6
nature, balance of *see* population, balance
negative feedback mechanisms 6, 344
nematode 328
Neomycta pulicaris 49
Neosomimus parvulus 228
Neotoma spp 156
Nephotettix cincticeps 93
Neuroptera 318
Nezara viridula 266
nitrogen balance 27, 134, 148, 179, 180–1, 185, 327
Noctuidae 75
Norops humilis 265
Nosema fumiferanae 330
Nothofagus spp 215, 307
Nothofagus truncata 50
Notornis mantelli 215
nuclear polyhedrosis 329 *and see* virus
nuthatch 272
nurse eggs *see* trophic eggs
nutrient sink 42, 91 *and see* leaf miners; galls

oak 43, 49, 59, 60, 104, 272
evergreen 48
white 74
oats 230
Odax cyanomelas 246
Odocoileus hemionus 141
Odocoileus hemionus sitkensis 141
Odocoileus virginianus 141
Oecophoridae 44
Oecophylla longinoda 177
oilbird 219
Onchorhyncus kisutch 289
Oncideres singulata 105
Oniscus asellus 120
Operophthera brumata 43
Opisthocomus hoazin 217
optimal foraging *see* foraging
optimum density *see* population
optimum survival *see* survival
Opuntia spp 78, 239, 366
Orconectes virilis 116
Orgyia pseudotsuga 73
Orthoptera 80, 259
Oryctolagus cuniculus 165, 166, 342
outbreak 33, 67, 71, 73–4, 107, 150–2, 166, 190, 212–15, 228, 289, 332–66
and weather 226–30, 356–61, 363–5
centres 357–59 *and see* hotspots; refuges
cyclic 156–7, 159–61, 164, 167–71, 204, 350–5
mini 79, 91
Oxystele variegata 128
oystercatcher, African black 127, 345

panda, giant 142
Panthera leo 304
Panulirus cygnus 116
Papilio polyxenes 70
Paracalanus parvus 108
paradox of enrichment 338
parasite response to stress of host *see* stress
parasitism 1 *and see* predation
parasitoid 323 *and see* predation
Parma victoriae 278
parrot, Australian seed-eating 231
partridge 210–11, 274
Paruroctonus mesaensis 316
Parus major 225
Parus spp 262
passerine 262, 289
patchy environment 108, 111, 158, 253, 334, 344, 361
Patella granularis 126
pecan 58
Pegohylemyia seneciella 18
pelican 337, 362
Australian 310
brown 309
great white African 310
Pelicanus conspicillatus 310
Pelicanus occidentalis californicus 309

Pelicanus onocrotalus roseus 310
Pemphigus betae 46, 58, 275
Pemphigus populivenae 57
Penaeus aztecus 117
Penaeus setiferus 117
perch, golden 242
Perdix perdix 211, 274
Periphyllus acericola 41, 88
periwinkle 128
Perognathus spp 163
Peromyscus leucopus 155
Peromyscus maniculatus 163
Peromyscus spp 162, 270
pests 32, 62, 75, 76, 80, 85, 91, 166, 260, 365, 366
Petauroides volans 185, 187, 188
Petaurus australis 185, 187
Petaurus breviceps 187
Phalacrocorax aristotelis 312
Phascolarctos cinereus 181, 327
Phasianidae 210, 274
Phasianus colchicus 210, 274
pheasant 210–11, 274
Philaenus spumarius 96
Philine gibba 128
philopatry 281–86
Philoscia muscorum 122
Phoebetria palpebrata 312
Phylacteophaga froggatti 49
Phyllostomidae 178
Phylloxera notabilis 58
Phytomyza ilicicola 50
phytoplankton 108–12, 242–3
Picea abies 39
Picea sitchensis 39
Picea spp 207
Pieris oleracea 17
Pieris rapae 17, 63, 70
pigeon, Australian brown 220
pillbugs 117
Pinus banksiana 71
Pinus ponderosa 145, 149
Pinus radiata 102, 105, 106
Pinus spp 56, 59, 165, 207, 283
Pinus sylvestris 38, 88, 206
Piper amalago 178
Planaria spp 314
plant age, damage, fertilizing, flowering, stress, response of herbivore to *see* herbivore
plant defences, lack of response to by herbivore 23, 25, 32, 44, 48, 52, 53, 64, 67, 69, 129, 156, 159, 169–70, 172, 173, 174, 188–9, 220, 229
plant growth regulator 91
Plantago maritima 196
platyhelminth worms 1
Platymonas spp 109
Platypodidae 106
Polistes spp 299
pollen as food for
  herbivore *see* food, pollen
  spiderling 315
Polycelis spp 314
Polyporus versicola 99
poplar 165
population
  balance 5, 262, 341, 344, 349
  equilibrium 5, 291, 341, 349
  increase *see* capacity of increase
  mean 6, 214, 344, 348, 349
  meta 361
  "optimum" density 6
  regulation 5, 321, 341, 344, 349, 354
Populus angustifolia 46, 58, 275
Populus fremonti 57
Porcellio scaber 120, 122
possum
  brushtail 187
  honey 187
  pygmy 187
  ringtail 184, 187
Potamogeton richardsonii 198
praying mantid 319
predation
  human 135, 147, 173, 233–4, 243, 300, 346, 366
  inefficient at limiting prey 53, 55, 59, 60, 61, 67, 68, 70, 74, 78, 80, 90, 91, 92, 94, 111, 112, 119, 123, 124, 126–7, 131, 135, 136, 144, 147, 162, 167, 171, 173, 206, 229, 235, 242, 316, 321, 337, 342, 344, 346, 348, 354
Presbytis entellus 173
Presbytis melalophos 172
Presbytis rubicunda 172
prickly pear 78
primaeval soup 11
Proteaceae 187
proteinase inhibitors 153
protozoa 341, 350
Protyora sterculiae 333
provirus *see* virus
Prunella modularis 265
Prunus spp 96
Pseudocheirus peregrinus 184, 187
Pseudomys fumeus 17, 154
Pseudotsuga menziesii 71, 149
Psiloteredo megotara 134
Psylla palmeni 91
Psylla perigrina 90
Psylla phlebophyllae 91

Psylla pyricola 91
Psylla subferruginea 90
psyllid 7, 25, 34, 333, 334, 346, 356
  citrus 92
  karroo 90
  pear 91
  willow 91
ptarmigan 204, 212
Pteropodidae 178
Pteropus alecto 179
Pteropus poliocephalus 179
Puccinellia maritima 196
Puccinia helianthii 81
puffin, North sea 311
Puffinus lherminieri 312
Pulvinaria delottoi 37
Pulvinariella mesembryanthemi 37
Pyralidae 76–78
Pyrausta nubilalis 76
Pyrrhula pyrrhula 223
Pyrus communis 91

quail 211, 274
Quasi Biennial Oscillation 364
Quelea quelea 222
Quercus alba 74
Quercus calliprinos 48
Quercus robur 43, 49, 60, 104
Quercus spp 43, 60
quokka 190, 256

"r" and "K" strategists 315, 338–9
rabbit 9, 165, 166, 263, 342, 351, 354, 362
ragwort 18, 79
raptor 264
rat
  black or ship 150
  kangaroo 263
  wood 156
Rattus rattus 150
recycling nitrogen *see* microorganisms
red currant 56
red poll, European 225
refection 165 *and see* coprophagy
refuges 151, 156, 166, 168, 169, 357 *and see* hotspots; outbreak centres
regulation *see* population
regurgitate as food for young birds 217, 219, 221, 223, 224, 225, 230, 312
relative shortage *see* food, relative v absolute shortage
resource
  availability of 8–10
  essential v non essential 8–9, 23, 147, 262
  limited 5, 8, 11, 14, 25, 136, 162, 213, 287
responses of herbivore to plant condition *see* herbivore
Rhabdophaga strobiloides 55
Rhodomonas spp 111
Rhynchites pauxillus 50
Rhytisma acerinum 120
rice 76, 93, 151
rice stem borer 76, 330
Rissa tridactyla 312
Rubus spp 43
rumen, external 120, 122
ruminant *see* gut fermenter
Rumex spp 203

Salix chordata 55
Salix lasiolepis 61, 274
Salix spp 91, 203
Salmo gairdneri 250
salmon 289
Salvinia molesta 77
Samea multiplicalis 77
Sargassum spp 238
saturated habitat *see* habitat
sawfly 45, 49, 61
gall-forming 274
scale insect 37, 42–3, 333, 346, 356
beech 53
hemlock 92
Scenedesmus obliquus 111
Schistocerca gregaria 82
Schizaphis graminum 41
Schizolachnus pineti 38
Scirpophaga incertulas 76
Scirpus spp 124
Sciuridae 144
Sciurus aberti 145
Sciurus carolinensis 146, 283
Sciurus niger 146
Scolytidae 107
scorpion 17, 293, 316
Scrophularia californica 68
seagrass 236, 243
Searlesia dira 296
secondary plant chemicals *see* plant defences
seed mast 147, 161, 225, 272, 284, 308, 352, 363
seed, developing, as food *see* food, seed
Selasphorus rufus 271
Selenastrum minutum 112
self-regulating mechanisms 20, 262, 291, 293
Senecio jacobaea 18, 79

senescence
  hastened 33–48, 57, 58, 80, 87, 88–9, 90, 208, 356, 359
  inhibited 89
  v flush feeder 32–52, 66–8, 73, 80, 86, 93, 107, 156 *and see* double dipping
Setonix brachyurus 190
Seychelles warbler 284
shag, North Sea 312
shearwater, Audubon's 312
shipworm, marine 133
shrew 351
shrimp 117
Simulium piperi 267
Siphonaptera 313
Siphonaria diemenensis 125
Sirex noctilio 106
Sitta europaea 272
slater 117
slug 122
snail 122
  carnivorous 128, 281, 296
  freshwater 123–5, 314
  land 130–3
  marine 128
  mud 129
  Roman or edible 131–3, 296
social behaviour *see* territorial and social behaviours
Solanum mauritianum 220
Solidago altissima 55
Sorex spp 351
sorghum 76, 151
sparrowhawk 264
Spartina alterniflora 129
Spermophilus beldingi 148
Spermophilus columbianus 162
Spermophilus parryii 149
Spermophilus richardsonii 149
Spermophilus saturatus 149
spider 17, 267, 315
  crab 316
  katipo 302
  orb-web 16, 315
spittlebug 96
Spodoptera exigua 75, 296, 324
Sporobolus spp 234
spruce 207, 225, 283
  Norway 39
  sitka 39
spruce budworm 70, 289, 324, 330, 351, 359
squirrel
  arctic ground 149
  Beldings ground 148
  Columbian ground 162
  Douglas 270, 283
  fox 146, 147
  golden-mantled ground 149
  gray 146, 147, 283
  ground 148, 285
  red 147, 283
staple food *see* food, winter staple
Stathmorrhopa aphotista 43
Stephanodiscus hantzschii 111
Sterna paradisaea 312
stinging nettle 56, 69
stoat 307
Streatornis caripensis 219
stress
  of animals, response of parasites to 311, 326–30, 363
  of plants, response of herbivore to 24, 33, 60, 61, 73, 80–2, 87–8, 106, 107, 165, 196, 199, 208–9, 223, 356–61, 363
struggle for existence 5, 7, 16
sugar beet 75
sunflower 81
sunspot cycle 364
supplemental feeding *see* food, supplemental
surplus young *see* young, surplus
survival
  curve, type III 27
  differential 5, 11, 15, 16
  of the fittest 6
  optimum v effective strategy for 7, 142
  stuggle for *see* struggle for existance
swan
  Australian black 200
  mute white 200
sweetcorn 76
sycamore 41, 86
Syconycteris australis 181
symbiosis 1, 257

takahe 215
Tamias spp 148, 162
Tamiasciurus douglasii 270, 283
Tamiasciurus hudsonicus 147, 283
Tamiasciurus spp 162
Tamius striatus 269
Tarsipes rostratus 187
Taurotragus oryx 139
Temora longicornis 294
Tenodera sinensis 319
Tenthredinidae 45
Tephritis bardanae 19
Teredinidae 133
Teredo navalis 134

Teredora malleolus 134
termite 100–3, 177
  Californian dampwood 102
tern, arctic 312
territorial and social behaviours 21, 147, 166, 182, 212, 213–15, 261–90
territorial behaviour, female v male 262
Tetrao tetrix 206, 208
Tetrao urogallus 206
Thais emarginata 281
Thais spp 128
Thalassia testudinum 236, 243
Thalassiosira weissflogii 113, 115
Thalassoma bifasciatum 245
Therioaphis maculata 42
Thomomys monticola 149
tick 10
  cattle 325
  koala 182, 327
  sheep 325
Tilapia spp 244, 250, 310
Tisbe cucumariae 110
tit, European great 225
titmice 262
tortoise, giant Aldabra 233
tortoise turf 233–4
Tortricidae 73
Toxomerus marginatus 316
Tragelaphus strepsiceros 140
Tragopogon pratensis 224
Trichodesmium spp 109
Trichoglossus haematodus 218
Trichoplusia ni 328
Trichoptera 17
Trichosurus vulpecula 187
triclad 17, 314
Triglochin palustris 198
Trioza erytreae 92
Triticum aestivum 41, 230
tritrophic flow of nutrients 55, 324
trophic chain 243
trophic eggs 133, 295–6 *and see* cannibalism
trophic level 14, 126
trout, rainbow 250
Tsuga canadensis 92
tulip tree 85
Turbellaria 314
turkey 211
turtle, green 236
Tyria jacobaeae 18, 79

Ulva lobata 237
Ulva spp 128
unapparent plants 23, 25
Unaspis euonymi 43, 93
Ursus americanus 271
Urtica dioica 69

Vaccinium myrtillus 203, 205
Vaccinium spp 202
virus 328, 353, 354, 362
  latent 329, 363
  provirus 329
vole 156, 270, 307, 352, 353, 354
  bank 270
  European field 157, 159
  prairie 157
Vulpes vulpes 307, 342
Vultur gryphus 264

Wallabia bicolor 190
wallaby
  Kangaroo Island tammar 190, 259
  swamp 190
wallaroo 189
warfare and cannibalism 297–300
wasp
  colonial 299
  cynipid 60
  ichneumonid 302, 324
  spider-hunting 302
wasting disease of koala 183
weather, food, and abundance 80, 138, 141, 166–7, 190, 214, 234 *and see* outbreak
weevil, leaf-mining 49
wheat 41, 151, 230
wildlife, managed 135, 165, 200, 366
willow 61, 91, 165, 167, 203, 274
winkle 128
winter food *see* food, winter staple
wolf 306
woodlice 117
woodroach 101
wrasse, bluehead 245

Xanthium strumarium 96
Xestobium rufovillosum 104
Xiphister mucosus 246

yeast 104 *and see* fungus; microorganisms
young, surplus 8, 9, 25, 27, 137, 146, 147, 155, 156, 166, 168, 182, 206, 212, 214, 266, 268, 277, 281–6, 287, 291, 307, 339,

zebra 33
Zenaida macroura 222
zooplankton 108–12, 242–3
Zootermopsis nevadensis 102

MIX
Papier aus verantwortungsvollen Quellen
Paper from responsible sources
FSC® C105338

If you have any concerns about our products,
you can contact us on
ProductSafety@springernature.com

In case Publisher is established outside the EU,
the EU authorized representative is:
Springer Nature Customer Service Center GmbH
Europaplatz 3, 69115 Heidelberg, Germany

Printed by Libri Plureos GmbH
in Hamburg, Germany